ADVANCED GAS CHROMATOGRAPHY – PROGRESS IN AGRICULTURAL, BIOMEDICAL AND INDUSTRIAL APPLICATIONS

Edited by **Mustafa Ali Mohd**

Advanced Gas Chromatography – Progress in Agricultural, Biomedical and Industrial Applications
Edited by Mustafa Ali Mohd

Published by InTech

Janeza Trdine 9, 51000 Rijeka, Croatia

Publishing Process Manager Daria Nahtigal

Technical Editor Teodora Smiljanic

Cover Designer InTech Design Team

Additional hard copies can be obtained from orders@intechopen.com

Advanced Gas Chromatography – Progress in Agricultural, Biomedical and Industrial Applications, Edited by Mustafa Ali Mohd
p. cm.
ISBN 978-953-51-0298-4

For EU product safety concerns:
IN TECH d.o.o., Prolaz Marije Krucifikse Kozulić 3, 51000 Rijeka, Croatia,
info@intechopen.com or visit our website at intechopen.com.

Meet the editor

Mustafa Ali Mohd (PhD) is a professor in the Department of Pharmacology, Faculty of Medicine, University of Malaya, Malaysia. He has vast research and teaching experience in various fields, particularly in chromatographic and mass spectrometric techniques. Heis a member of editorial board of several scientific journals and he holds a position as advisory committee for several high level decision making bodies in food safety and quality, including as expert panel member in WHO expert meetings. Dr. Mustava Ali Mohd coordinates several international projects on analytical techniques such as GCMS and LCMS, and acts as the National project coordinator for the United Nations University Environmental Monitoring programme in East Asia region. His area of expertise includes endocrine disruptors research, clinical pharmacology, environmental monitoring, Natural products research, food safety and toxicology. He holds academic position as the Deputy Dean of the Faculty of Medicine, University of Malaya and currently the Deputy Director of the University of Malaya Medical Centre. He is also the Head and Senior consultant to the Shimadzu-UMMC centre for Xenobiotic Studies (SUCXeS).

Contents

Preface

Gas chromatography has been and it still is, one of the key tools in analytical techniques in many of the advanced research carried out over the globe. This technique has contributed tremendously and was once the main technique in the analysis of specific compounds like volatile compounds, certain pesticides, pharmaceuticals and petroleum products. The advance of this technique has resulted in several tandem instruments with application of other techniques to enhance the results obtained by gas chromatography. Several good and versatile detectors has been developed and plays a pivotal role in many frontline research and industrial applications.

This book is the outcome of contributions by many experts in the field from different disciplines, various backgrounds and expertise. This is a true reflection of the vast area of research that is applicable to this technique and each area has its own strength and focus towards giving highly sensitive and specific identification and quantitation of compounds from different sources and origins. The chapters cover a significant amount of basic concepts and its development till its current status of development. This is followed by chapters on the biomedical applications of this technique which is beneficial as an example of its application in biological and life sciences based research. It also includes some aspects of indoor air assessment, exhaled air analysis and analysis of exudates from natural products.

The chapters include some industrial applications in petroleum, organometallic and pyrolysis applications. These chapters provide some useful information on specific analysis and may be applicable to related industries. The last chapter deals with new dimension and the frontline research and development in gas chromatography. This includes multidimensional and time of flight gas chromatography. I hope that this book will contribute significantly to the basic and advanced users of gas chromatography.

Professor Dr. Mustafa Ali Mohd
Deputy Director of Development
University of Malaya Medical Centre,
University of Malaya
Kuala Lumpur,
Malaysia

Part 1

New Development in Basic Chromatographic and Extraction Techniques

1

Hyphenated Techniques in Gas Chromatography

Xinghua Guo* and Ernst Lankmayr
Institute of Analytical Chemistry and Food Chemistry, Graz University of Technology, Austria

1. Introduction

Hyphenated gas chromatography (GC) (Chaturvedi & Nanda, 2010; Coleman III & Gordon, 2006) mainly refers to the coupling of the high-performance separation technique of gas chromatography with 1) information-rich and sophisticated GC detectors which otherwise can be mostly operated as a stand-alone instrument for chemical analysis, and 2) automated online sample preparation systems. The term "hyphenation" was first adapted by Hirschfeld in 1980 to describe a possible combination of two or more instrumental analytical methods in a single run (Hirschfeld, 1980). It is of course not the case that you can couple GC to any detection systems, although many GC hyphenations have been investigated and / or implemented as to be discussed in this chapter. The aim of this coupling is obviously to obtain an information-rich detection for both identification and quantification compared to that with a simple detector (such as thermal-conductivity detection (TCD), flame-ionization detection (FID) and electron-capture detector (ECD), etc.) for a GC system.

According to the detection mechanism, information-rich detectors can be mainly classified as 1) detection based on molecular mass spectrometry, 2) detection based on molecular spectroscopy such as Fourier-Transform infrared (FTIR) and nuclear magnetic resonance (NMR) spectroscopy, and 3) detection based on atomic spectroscopy (elemental analysis) by coupling with such as inductively-coupled plasma (ICP)-MS, atomic absorption spectroscopy (AAS) and atomic emission spectroscopy (AES), respectively. In addition to these hyphenations mentioned above and which are mounted after a gas chromatograph, it can also include automated online sample preparation systems before a GC system such as static headspace (HS), dynamic headspace, large volume injection (LVI) and solid-phase microextraction (SPME). One of the recent developments is the hyphenation of GC with human beings - so called GC-Olfactometry (GC-O) or GC-Sniffer (Friedrich & Acree, 1998). Hyphenated gas chromatography also include coupling of gas chromatographs orthogonally - multidimensional gas chromatography (MDGC or GC×GC).

2. Hyphenated techniques in gas chromatography

This chapter provides a general overview of the current hyphenated GC techniques with focus on commonly applied GC-MS, GC-FTIR (GC-NMR) for detection of molecular

---

* Corresponding Author

analytes as well as GC-AAS and GC-AES coupling for elemental analysis. Emphasis will be given to cover various GC-MS techniques including ionization methods, MS analysers, tandem MS detection and data interpretation. For more comprehensive overview on their applications, readers are directed to other chapters in this book or other dedicated volumes (Grob & Barry, 2004; Message, 1984; Jaeger, 1987; Niessen, 2001).

2.1 Gas Chromatography-Mass Spectrometry (GC-MS)

In 1957, Holmes and Morrell (Holmes & Morrell, 1957) demonstrated the first coupling of gas chromatography with mass spectrometry shortly after the development of gas-liquid chromatography (James & Martin, 1952) and organic mass spectrometry (Gohlke & McLafferty, 1993). Years later, improved GC-MS instruments were commercialized with the development of computer-controlled quadrupole mass spectrometer for fast acquisition to accommodate the separation in gas chromatograph. Since then, its applications in various areas of sciences has made it a routine method of choice for (bio)organic analysis (Kuhara, 2005).

As its name suggested, a GC-MS instrument is composed of at least the following two major building blocks: a gas chromatograph and a mass spectrometer. GC-MS separates chemical mixtures into individual components (using a gas chromatograph) and identifies / quantifies the components at a molecular level (using a MS detector). It is one of the most accurate and efficient tools for analyzing volatile organic samples.

The separation occurs in the gas chromatographic column (such as capillary) when vaporized analytes are carried through by the inert heated mobile phase (so-called carrier gas such as helium). The driven force for the separation is the distinguishable interactions of analytes with the stationary phase (liquid thin layer coating on the inner wall of the column or solid sorbent packed in the column) and the mobile phase respectively. For gas-liquid chromatography, it depends on the column's dimensions (length, diameter, film thickness), type of carrier gas, column temperature (gradient) as well as the properties of the stationary phase (e.g. alkylpolysiloxane). The differences in the boiling points and other chemical properties between different molecules in a mixture will separate the components while the sample travels through the length of the column. The analytes spend different time (called retention time) to come out of (elute from) the GC column due to their different adsorption on the stationary phase (of a packed column in gas-solid chromatography) or different partition between the mobile phase (carrier gas) and the stationary phase (of a capillary column in gas-liquid chromatography) respectively.

As the separated substances emerge from the column opening, they flow further into the MS through an interface. This is followed by ionization, mass-analysis and detection of mass-to-charge ratios of ions generated from each analyte by the mass spectrometer. Dependent on the ionization modes, the ionization interface for GC-MS can not only ionize the analytes but also break them into ionized fragments, and also detect these fragments together with the molecular ions such as, in positive mode, radical cations using electron impact ionization (EI) or (de-)protonated molecules using chemical ionization (CI). All ions from an analyte together (molecular ions or fragment ions) form a fingerprint mass spectrum, which is unique for this analyte. A "library" of known mass spectra acquired under a standard condition (for instance, 70-eV EI), covering several hundreds of thousand compounds, is stored on a computer. Mostly a database search can identify an unknown component rapidly in GC-MS. Mass spectrometry is a powerful tool in instrumental analytical chemistry because it provides more information about the composition and structure of a substance from less sample than any other analytical technique, and is considered as the only definitive analytical detector among all available GC detection techniques.

The combination of the two essential components, gas chromatograph and mass spectrometer in a GC-MS, allows a much accurate chemical identification than either technique used separately. It is not easy to make an accurate identification of a particular molecule by gas chromatography or mass spectrometry alone, since they may require a very pure sample or standard. Sometimes two different analytes can also have a similar pattern of ionized fragments in their standard mass spectra. Combining the two processes reduces the possibility of identification error. It is possible that the two different analytes will be separated from each other by their characteristic retention times in GC and two co-eluting compounds then have different molecular or fragment masses in MS. In both cases, a combined GC-MS can detect them separately.

2.1.1 Ionization techniques and interfaces

The carrier gas that comes out of a GC column is primarily a pressurized gas with a flow about mL/min for capillary columns and up to 150 mL/min for packed columns. In contrast, ionization, ion transmission, separation and detection in mass spectrometer are all carried out under high vacuum system at approximately 10^{-4} Pa (10^{-6} Torr). For this reason, sufficient pumping power is always required at the interface region of a GC-MS in order to provide a compatible condition for the coupling. When the analytes travel through the length of the column, pass through the transfer line and enter the mass spectrometer they can be ionized by several methods. Fragmentation can occur alongside ionization too. After ion separation by mass analyser, they will be detected, usually by an electron multiplier diode, which essentially turns the ionized analytes/fragments into an electrical signal.

The two well-accepted standard types of the ionization techniques in GC-MS are the prevalent electron impact ionization (EI) and the alternative chemical ionization (CI) in either positive or negative modes.

Electron impact (EI) ionization:

The EI source is an approximately one cubic centimetre device, which is located in the ion source housing as shown in Figure 1. The ion source is open to allow for the maximum conductance of gas from the ion source into the source housing and then into the high-vacuum pumping system. The EI source is fitted with a pair of permanent magnets that cause the electron beam to move in a three-dimensional helical path, which increases the probability of the interaction between an electron and an analyte molecule. Even under this condition, only about 0.01~0.001% of the analyte molecules are actually ionized (Watson, 1997). During the EI ionization, the vaporized molecules enter into the MS ion source where they are bombarded with free electrons emitted from a heated filament (such as rhenium). The kinetically activated electrons (70-eV) collide with the molecules, causing the molecule to be ionized and fragmented in a characteristic and reproducible way.

Due to the very light weight of the electrons (1/1837 that of the mass of a proton or neutron), one collision between an electron and a molecule would bring insufficient internal energy to the molecule to become ionized. The molecule's internal energy is increased by the interaction of the electron cloud after a collision cascade. The energized molecule, wanting to descend to a lower energy state, will then expel one of its electrons. The result is an odd-electron species called a radical cation, which is the molecular ion and has the same integer mass as the analyte molecule. Some of the molecular ions will remain intact and pass through the *m/z* analyser and be detected. The molecular weight of the analyte is represented by the molecular ion peak in the mass spectrum if there are any. The single-charge molecular ion has the same nominal mass of the molecule, but the accurate mass is differed by that of an electron. Most molecular ions will undergo unimolecular decomposition to produce fragment ions.

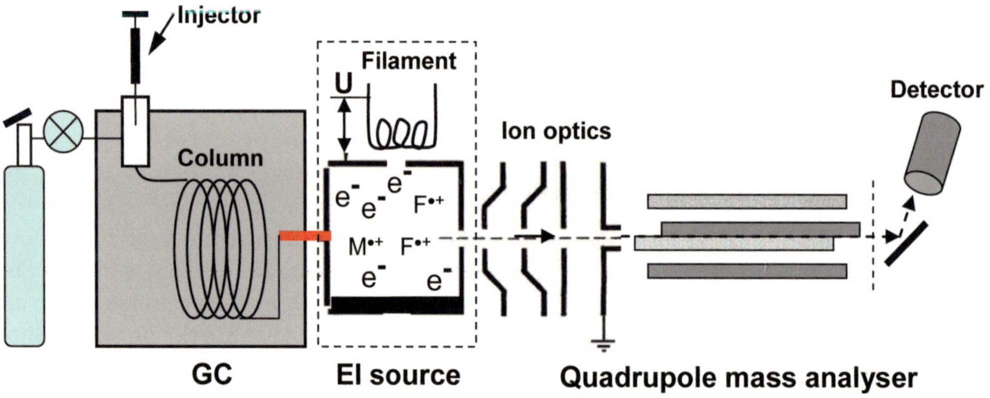

Fig. 1. Schematic drawing of a GC-MS with EI ionization and quadrupole mass analyser.

The reason to apply the 70 eV as the standard ionization energy of electrons is to build a spectrum library with the standard mass spectra and subsequently to perform the MS database searching for unknown identification. The molecular fragmentation pattern is dependant upon the electron energy applied to the system. The excessive energy of electrons will induce fragmentation of molecular ions and results in an informative fingerprint mass spectrum of an analyte (Watson, 1997). The use of the standard ionization energy facilitates comparison of generated spectra with library spectra using manufacturer-supplied software or software developed by the National Institute of Standards (NIST-USA). The two main commercial mass spectra libraries for general purpose are the NIST Library and the Wiley Library. In addition, several small libraries containing a specific class of compounds have also been developed by individual manufacturers or research institutions. This includes libraries for pesticides, drugs, flavour and fragrance, metabolites, forensic toxicological compounds and volatile organic compounds (VOC), just to name a few. Spectral library searches employ matching algorithms such as probability based matching (McLafferty et al., 1974) and dot-product (Stein & Scott, 1994) matching that are used with methods of analysis written by many method standardization agencies. EI is such an energetic process, in some cases, there is often no molecular ion peak left in the resulting mass spectrum. Since the molecular ion can be a very useful structure information for identification, this is sometimes a drawback compared to other soft ionization such as chemical ionization, field ionization (FI), or field desorption ionization (FDI), where the desired (pseudo) molecular ion will be generated and detected. After ionization, the stable ions (which remain without dissociation during its flight from the ion source to the detector) will be pushed out of the EI source by an electrode plate with the same charge polarity, called the repeller. As the ions exit the source, they are accelerated / transmitted into the mass analyser of the mass spectrometer.

Chemical ionization (CI):

CI is a less energetic process than EI. In the latter case, the ionizing electrons accelerated to have some kinetic energy collide directly with gas-phase analyte molecules to result in ionization accompanied by simultaneous fragmentation. However, CI is a low energetic ionization technique generating pseudo molecular ions such as $[M+H]^+$ rather than the conventional $M^{\bullet +}$, and inducing less fragmentation. This low energy usually leads to a simpler mass spectrum showing easily identifiable molecular weight information. In CI a reagent gas such as methane, ammonia or isobutane is introduced into the relatively closed ion source of a mass spectrometer at a pressure of about 1 Torr. Inside the ion source, the reagent gases (and the subsequent reagent ions) are present in large excess compared to the analyte. Depending on the technique (positive CI or negative CI) chosen, the pre-existing reagent gas in the ion source will interact with the electrons preferentially first to ionize the reagent gas to produce some reagent ions such as CH_5^+ and NH_4^+ in the positive mode and OH^- in the negative mode through self-CI of the reagent gas. The resultant collisions and ion/molecule reactions with other reagent gas molecules will create an ionization plasma. When analytes are eluted (from GC) or evaporated (direct inlet), positive and negative ions of the analyte are formed by chemical reactions such as proton-transfer, proton-subtraction, adduct formation and charge transfer reactions (de Hoffmann & Stroobant, 2003), where the proton transfer to produce $[M+H]^+$ ions is predominate during the ionization. The energetics of the proton transfer is controlled by using different reagent gases. Methane is the strongest proton donor commonly used with a proton affinity (PA) of 5.7 eV, followed by isobutane (PA 8.5 eV) and ammonia (PA 9.0 eV) for 'softer' ionization. As mentioned above, in this soft ionization, the main benefit is that the (pseudo) molecular ions closely corresponding to the molecular weight of the analyte of interest are produced. This not only makes the follow-up identification of the molecular ions easier but also allows ionization of some thermal labile analytes.

The sensitivity and selectivity of a mass spectrometer in a hyphenated GC largely depend on the interfacing technique: the ion source and the ionization mode. The sensitivity is related to the ionization efficiency and the selectivity is primarily to the ionization mode. Both sensitivity and selectivity can differ for different classes of compounds. A proper choice of an ionization technique is a key step for a successful GC-MS method development. In most case of EI ionization, the positive mode is much preferred due to its capability efficiently to ionize most analytes and to induce sufficient fragmentation to build up a database with positive mass spectra. The sensitivity in the negative EI mode will be much lower. However, the fragmentation processes in EI can be so extensive that the molecular ion is absent in the mass spectrum of some compounds. As a result, the useful molecular weight information is lost, which is indeed a disadvantage of this ionization technique (Karasek & Clement, 1988; Ong & Hites, 1994). In this aspect, chemical ionization is a reasonable alternative and supplemental to EI as a soft ionization method. Due to the moderate energy transmission to the analytes, the pseudo molecular ions are formed in CI and these even-electron ions are more stable than the odd-electron ions produced with EI

(Harrison, 1992). Unlike EI, both positive chemical ionization (PCI) and negative chemical ionization (NCI) are equally used according to their specific features. The positive mode is best suitable for hydroxyl group-containing alcohols and sugars as well as basic amino compounds. The sensitivity of PCI-MS is comparable with low-resolution EI-MS for most compounds. On the other hand, the negative mode is widely used in environmental analysis because it is highly selective and sensitive to, for instance, organochlorine (halogen-containing) and acidic group-containing compounds (Karasek & Clement, 1988). For this, the sensitivity of the NCI is significantly better compared with that of the PCI. As a result, the NCI has also been equally developed in the sense of the choices of the reagent gases and their combination (Chernetsova et al., 2002). Furthermore, the application of an alternating or simultaneous EI and CI in GC-EI/CI-MS has also been reported (Arsenault et al., 1971 & Hunt et al., 1976). Although it has not been popularly implemented in the commercial GC-MS systems, the advantages of obtaining informative mass spectra in trace analysis of samples with limited quantities are very promising. In addition to the above-mentioned EI and CI ionization interfaces, where the ionization occurs in the vacuum of a mass spectrometer, a recent development has also demonstrated an interfacing technique at the atmosphere. That is the coupling of a GC with a mass spectrometer with the so-called Direct Analysis in Real Time (DART) (Cody et al., 2005 & Cody 2008), where the absence of vacuum interface, electron filament and CI reagent gases offers a robust interfacing technique for rapid analysis.

2.1.2 Molecular weight, molecular ion, exact mass and isotope distribution

For a singly charged analyte, the mass-to-charge ratio of its molecular ion indicates the molecular weight (*a.m.u.*) of the analyte. The molecular ion can be a radical cation $[M]^{\bullet+}$ (positive mode) or a radical anion $[M]^{\bullet}$- (negative mode), derived from the neutral molecule by kicking off / attaching one electron. According to the IUPAC definition (McNaught & Wilkinson, 1997), the molecular mass (*Mr*) is the ratio of the mass of one molecule of that analyte to the unified atomic mass unit *u* (equal to 1/12 the mass of one atom of ^{12}C). It is also called molecular weight (*MW*) or relative molar mass (*Mr*). Knowing the fact that each isotope ion (rather than their average) of an analyte is detected separately in MS mass analyser, one should pay attention to the different definitions of molecular masses in mass spectrometry and GC-MS.

The nominal mass of an ion is calculated using the integer mass by ignoring the mass defect of the most abundant isotope of each element (Yergey et al., 1983). This is equivalent as summing the numbers of protons and neutrons in all constituent atoms. For example, for atoms H = 1, C = 12, O = 16, etc. Nominal mass is also called mass number. To assign elemental composition of an ion in low resolution MS, the nominal mass is often used, which enable to count the numbers of each constituent elements. However, if the sum of the masses of the atoms in a molecule using the most abundant isotope (instead of the isotopic average mass) for each element is calculated (Goraczko, 2005), the monoisotopic mass is obtained (McNaught & Wilkinson, 1997). Monoisotopic mass is also expressed in unified atomic mass units (a.m.u.) or Daltons (Da). For typical small molecule organic compounds with elements C, H, O, N etc., this is the mass that one can measure with a mass spectrometer and it often refers to the lightest isotope in the isotopic distribution. However, for a molecule containing special elements such as B, Fe and Ar, etc., the most abundant isotope is not the lightest one any more. Another important and useful term is the exact mass, which is obtained by summing the exact masses of the individual isotopes of the molecule (Sparkman, 2006). According to this definition, several exact masses can be calculated for one chemical formula depending on the constituent isotopes. However, in practical mass spectrometry, exact mass refers to that corresponding to the monoisotopic mass, which is the sum of the exact masses of the most abundant isotopes for all atoms of the molecule. This is the peak one can measure with high resolution mass spectrometry. This results in measured accurate mass (Sparkman, 2006), which is an experimentally determined mass that allows the elemental composition to be determined either using the accurate mass mode of a high resolution mass spectrometer (Grange et al., 2005). For molecules with mass below 200 u, a 5 ppm accuracy is sufficient to uniquely determine the elemental composition. And at *m/z* 750, an error of 0.018 ppm would be required to eliminate all extraneous possibilities (Gross, 1994). When the exact mass for all constituent isotopes (atoms or molecules) are known, the average mass of a molecule can be calculated. That is the sum of the average atomic masses of the constituent elements. However, one can never measure the average mass of a compound using mass spectrometry directly, but that of its individual isotopes.

Some molecular ions can be obtained using EI ionization. If not, the use of soft-ionization techniques (CI, FI) would help to facilitate identification of the molecular ion. Accurate masses can be determined using high resolution instruments such as magnet/sectors, reflectron TOFs (see Section 2.1.4), Fourier-Transform ion cyclotron resonance (FT-ICR) (Marshall et al., 1998) and Orbitrap (Makarov, 2000) mass spectrometers. Because of isotopes of some elements, molecular ions in mass spectrometry are often shown as a distribution of isotopes. The isotopic distribution can be calculated easily with some programs freely available and allows you to predict or confirm the masses and abundances of the isotopes for a given chemical formula. Isotopic patterns observed are helpful for predicting the appropriate number of special elements (e.g., Cl, Br and S) or even C numbers. When a Library mass spectrum is not available, this will be one of the important means for unknown identification. The elucidation process normally requires not only the combined use of hard and soft ionization techniques but also tandem MS experiments to obtain information of fragments for data interpretation. In some cases, an elemental composition might be proposed for the unknown based on isotope patterns and accurate mass measurements of the molecular and fragment ions (Hernández et al, 2011) (see Section 2.1.4).

2.1.3 Typical mass analyzers and MS detectors in GC-MS

MS instruments with different mass analysers, e.g. magnet/electric sectors, quadrupoles (linear), ion traps (Paul traps & quadrupole linear ion trap), FT-ICR and time-of-flight (TOF) mass analysers have all been implemented for the coupling of GC and MS. Since its conception, linear quadrupoles have been dominating the GC-MS applications.

Quadrupole mass analyser:

A single-stage linear quadrupole mass analyser can be considered as a mass filter and it consists of four hyperbolic metal rods placed parallel in a radial array. A pair of the opposite rods have a potential of $(U+V\cos(\omega t))$ and the other pair have a potential of $-(U+V\cos(\omega t))$. An appropriate combination of direct current (DC) and radio frequency (RF, ~ 1 MHz) electric field applied to the four rods induces an oscillatory motion of ions guided axially into the assembly by means of a low accelerating potential. The oscillating trajectories are mass dependent and ions with one particular mass-to-charge (*m/z*) ratio can be transmitted toward the detector when a stable trajectory through the rods is obtained. At static DC and RF values, this device is a mass filter to allow the ions at this *m/z* pass through and all others will be deflected. Ions of different *m/z* can be consecutively transmitted by the quadrupole mass filter toward the detector when the DC and RF potentials are swept at a constant ratio and oscillation frequencies (de Hoffmann & Stroobant, 2003; Santos & Galceran, 2003).

A single quadrupole mass analyser can be operated in either the full-spectrum scan mode or the selected ion monitoring (SIM) mode. In the scan mode, the ions of a certain *m/z* range will pass through the quadrupole sequentially while scanning the DC and RF potentials to the detector. The advantage is that this allows to record a full mass spectrum containing the information from molecular ion to fragments, which further enables a library searching to compare with the standard mass spectra for unknown identification. It is common to require a scanning of the analyser from about 50 a.m.u. to exclude the background ions from residue air, carrier gases, and CI reagent gases. However, especially in the earlier years, the slow scan speed (250-500 a.m.u. per second) for a quadrupole analyser was an important limiting factor and it is in contrary with the high resolution (narrow peaks about a few seconds wide) of a capillary GC separation. Apparently, a longer dwell time is essentially required to obtain a more sensitive detection. Therefore, the SIM mode is preferred for a selective and sensitive detection in quantification of known analytes, where the DC and RF are set to pre-defined values according to the *m/z* of the analyte (molecular ion or specific fragments) for up to hundred milliseconds. Linear quadrupoles are the most widely used mass analysers in GC–MS, mainly because they make it possible to obtain high sensitivity, good qualitative information and adequate quantitative results with relatively low maintenance. Generally, these instruments are characterised by a bench-top configuration with unit mass resolution and both electron impact and chemical ionisation techniques. The relatively low cost, compactness, moderate vacuum requirements and the simplicity of operation make quadrupole mass spectrometers the most popular mass analyser for GC-MS. The continuous and significant improvements in scanning speed (to allow tandem MS detection), sensitivity and detection limits has also been observed in the past decade. New developments have been implemented in GC–MS instrumentations based on quadrupole technology with regard to the stability of mass calibration, the fast scan-speed with a higher sensitivity. It is now also possible to work simultaneously with full-scan and selected ion monitoring (SIM) modes in a single GC-MS run.

Ion trap (IT) mass analyser:

An ion trap mass spectrometer uses a combination of electric or magnetic fields to capture and store ions in a vacuum chamber. According to its principles of operation, it can be specified as quadrupole ion traps (Paul trap) (Paul & Steinwedel, 1953), quadrupole linear ion traps (Schwartz et al., 2002), FTICR-MS (Penning trap) (Marshall et al., 1998) and Orbitraps (Kington traps) (Kington, 1923; Makarov, 2000), respectively. However, the coupling to GC has been dominated by quadrupole mass analysers and quadrupole ion traps. Due to the applicability and other limitations, only very recently, there have been reports on implementing FTICR-MS (Szulejko & Solouki, 2002) and Orbitrap-MS (Peterson et al., 2010) as detectors in GC-MS.

The three-dimensional Paul ion trap (IT) is composed of a central hyperbolic ring electrode located between two symmetrical hyperbolic end-cap electrodes. The geometry of the device is described by the relationship $r_0^2 = 2\times z_0^2$, where r_0 is defined as the radius of the ring electrode and $2z_0$ is the distance between the end-cap electrodes. The value r_0 is used to specify a trap, which ranges about 1-25 mm. During the ion trapping, an auxiliary oscillating potential of low amplitude is applied across the end-cap electrodes while a RF potential of ~ 1 MHz is applied to the ring electrode. As the amplitude of the RF potential is increased, the ions become more kinetically energetic and they develop unstable trajectories (excited for dissociation or ejected to detector). One of the significant advantages of ion traps compared to the above-mentioned linear quadrupole mass analysers is their high sensitivity in full scan mode. Based on this, many qualitative and quantitative applications have been reported (Allchin et al., 1999 & Sarrion et al., 2000). Beside the conventional full scan mode of operation, an important advancement by application of ion trap in GC-MS has been the capability of an ion trap mass analyser to perform tandem mass spectrometric investigations of ions of interest in collision-induced dissociation (CID) for ion structure elucidation (Plomley et al., 2000) (see the text below for tandem mass spectrometers).

For both linear quadrupole and 3-D mass spectrometers, after mass-resolved ions passing the mass analyser, they continue to travel to the ion detector. It is essential in GC-MS to have fast responding detector with a broad range of magnification. The most popular detector employed is the electron multiplier composed of a series of dynodes. The ions collide with the first one to generate primary electrons. Due to the increasing high voltage (kV) between the dynodes, the electrons further collide with the next dynode and generate more electrons, and so on. Therefore, a small ion current can be finally amplified into a huge signal. Another commonly used detector is the photomultiplier. The ions collide with a phosphor-coated target and are converted into photons that are subsequently magnified and detected. Typically, these detectors are operated at lower voltages (400-700 V) and, therefore, have a longer lifetime than the electron multipliers (using high kV voltages) (Grob, 1995).

Time-of-flight (TOF) mass analyser:

Time-of-flight mass spectrometry (TOF-MS) is based on a simple mass separation principle in which the *m/z* of an ion is determined by a measurement of its flight time over a known distance (Cotter, 1994; Stephens, 1946). Pulsed ions are initially accelerated by means of a constant homogeneous electrostatic field of known strength to have the same kinetic energy (given they have the same charge). Therefore, the square of the velocity of an ion is reversely proportional to its *m/z*, $E_k = (1/2)mv^2$ and the time of arrival *t* at a detector directly indicates its mass (Equation 1).

$$t = \mathrm{d}\times(m/\mathrm{q})^{1/2}\times(2\mathrm{U})^{-1/2} \qquad (1)$$

where d is the length of the flight tube, *m* is the mass of the ion, q is the charge, U is the electric potential difference used for the acceleration. The arrival time *t* can be measured using a transient digitizer or time to digital converter, and is about milliseconds. A TOF mass analyser has theoretically no mass limit for detection and a high sensitivity. However, the spread of kinetic energies of the accelerated ions can lead to different arrival time for the ions with the same *m/z* and result in low mass resolution. This can be overcomed by applying a reflectron flight path (Mamyrin et al., 1973) rather than a linear one to refocus the ions with the same *m/z* value to arrive at the detector simultaneously.

In TOF-MS, ions must be sampled in pulses. This fits well with a laser ionization source. However, when EI or CI are used for GC-MS coupling, a continuous ion beam is produced. Therefore, in most GC-TOF-MS instrumentations, a pulse of an appropriate voltage is applied to deflect and accelerate a bunch of the ions in the orthogonal direction to their initial flight path (so-called oa-TOF-MS). One of the important advantages of TOF-MS as a GC detector is its capability of producing mass spectra within a very short time (a few milliseconds), with high sensitivity. Furthermore, its high mass accuracy (errors in low ppm) has made it an alternative to accurate mass GC-MS using magnet / sector instruments (see Section 2.1.4). In both GC-MS and LC-MS, hybridized TOF-MS instruments with quadrupoles (Q-TOF-MS) (Chernushevich, 2001), ion trap (IT-TOF-MS), and even another TOF (such as TOF/TOF-MS) have all been developed and found more applications than TOF-MS alone (Vestal & Campbell, 2005).

Tandem mass spectrometer as a mass analyser:

Tandem mass spectrometer refers to an arrangement of mass analysers in which ions are subjected to two or more sequential stages of analysis (which may be separated spatially or temporally). The study of ions involving two stages of mass analysis has been termed mass spectrometry/mass spectrometry (MS/MS) (Todd, 1991). For GC applications, it includes mainly triple quadrupole, 3D-ion trap and linear quadrupole ion trap tandem mass spectrometers and only very recently also FTICR (Szulejko & Solouki, 2002 & Solouki et al., 2004) and Orbitrap-MS (Peterson et al., 2010) as detectors in GC-MS.

Triple quadrupole MS/MS involves two quadrupole analysers mounted in a series but operating simultaneously (either as a mass filter or scanner) and with a collision cell between two mass analysers. Since the analysers can be scanned or set to static individually, types of operations include product ion scan, precursor ion scan and neutral loss scan. Product ion scan enables ion structure studies. This is realised by isolation of ions of interest according to their *m/z* using first quadrupole analyser. Then dissociation of the ions after kinetic-energetically activated occurs in the collision cell filled with an inert collision gas. Finally, the product ions will be mass-analysed using the third quadrupole. The fragment information provides further insights into ion structures or functional groups. This can be used to confirm structures of components in question when it is compared with a reference standard or to deduce ion structures for new chemical entities in qualitative identification. Some mass spectra databases include even a library of product ion spectra for some selected compounds to assistant the identification. Furthermore, the precursor ion scan can be utilised to study multiple precursor ions of a certain fragment ion. This is very useful to study a class of compounds producing a common charged fragment during collision-induced dissociation (CID) in the collision cell. For instance, the protonated 1,2-benzenedicarboxylic acid anhydride at *m/z* 149 derived from almost all corresponding phthalate esters can be used as such an ion and, during the precursor scan, all related phthalates will be found. On the other hand, neutral loss scan can be used to study a class of compounds showing a common neutral molecule loss during CID. For instance, the loss of a CO_2 from most deprotonated carboxylic acids or H_2O from protonated alcohols can be applied for this purpose. For dissociation reactions found in either product scan or precursor ion scans, a pair of ions can be selected to detected a specific analyte, which is the so-called selected reaction monitoring (SRM) or multiple reaction monitoring (MRM) when more pairs are chosen in one GC-MS run. SRM and MRM of MS/MS are highly specific and virtually eliminate matrix background due to the two stages of mass selections. It can be applied to quantify trace levels of target compounds in the presence of sample matrices (Santos & Galceran, 2003). This has further secured its important role in modern discovery researches even in the era of fast liquid chromatography tandem mass spectrometry (LC-MS/MS) (Krone et al., 2010).

In a 3-D ion trap MS, the above-mentioned steps such as ion selection, ion dissociation and scanning of product ions occur in a timed sequence in a single trap in contrast to a triple quadrupole MS, where they are at different spatial locations of the instrument while ions travelling through. For a GC-MS/MS coupling, ion trap instruments offer three significant advantages over triple quadrupoles including low cost, easy to operate and high sensitivity for scanning MS/MS (Larsson & Saraf, 1997). Although in the earlier versions of ion trap instruments, a drawback called low-mass cut-off (which affects or discriminates small fragment ions in lower 1/3 mass range during MS/MS scan) has been observed, this has been solved in the modern instruments nowadays.

Although triple quadrupoles and ion traps have significant superiority in tandem mass spectrometry, their resolving powers are limited. Quadrupoles reach about unit resolution and ion trap a bit higher. With them, accurate mass measurements are generally not possible. This is just opposite to high-resolution TOF mass analysers. Another limitation of a triple quadrupole is that can be used only for one step of MS/MS, while an ion trap can perform multiple steps tandem MS experiments, which is sometimes very useful for structure elucidation.

2.1.4 GC with high resolution mass spectrometry (GC-TOF-MS)

Nowadays, the most investigated and applied coupling of GC with high resolution mass spectrometry is GC-TOF-MS, rather than that with sector instruments (Jeol, 2002). Recent advancements in instrumental optics design, the use of fast recording electronics and improvements in signal processing have led to a booming of the TOF-MS for investigation of organic compounds in complex matrices (Čajka & Hajšlová, 2007). GC-TOF-MS with high resolution of about 7000 is capable of achieving a mass accuracy as good as 5 ppm for small molecules. This allows not only isobaric ions to be easily mass-resolved but also the measurements of accurate masses for elemental composition assignment or mass confirmation, which adds one more powerful means for identification in GC-MS besides mass spectrum database searching and tandem mass spectrometry (Hernández et al, 2011). The moderate acquisition speed (maximum rate 10/s) at high resolution and the high acquisition speed (maximum rate of 500/s) at unit resolution can be achieved readily with a linear range of three to four orders of magnitude. High acquisition speed of GC-TOF-MS at unit resolution is very compatible with very narrow chromatographic peaks eluted from a fast and ultra-fast GC or GC×GC (Čajka & Hajšlová, 2007; Hernández et al, 2011; Pasikanti et al., 2008).

High resolution detection in GC-TOF-MS offers not only the high mass accuracy of molecular and fragment ions but also the accurate isotopic distribution with regard to isotope intensities and isotope-resolved information for element assignments. It is extremely helpful for unknown compounds for which no Library spectrum is available for database searching as reviewed recently (Hernández et al, 2011). With the help of software tools, a carbon number prediction filter can be applied to reduce the number of possible elemental compositions based on the relative abundance of the isotopic peak corresponding to ^{13}C (relative to the ^{12}C peak, each ^{13}C isotope contributes 1.1% to the ^{13}C peak). The nitrogen rule can also be used to determine whether the ion is an "even-electron ion" (for instance, protonated or deprotonated molecule) or an "odd-electron ion" (for instance, radical cation or anion) (MaLafferty, 1993). With these considerations, possible elemental compositions can be obtained when it is searched in available databases (e.g., Index Merck, Sigma Aldrich, Chem-Spider, Pubchem, Reaxys) and a chemical structure can be proposed (Hernández et al, 2011; Portolés et al., 2011). Both accurate masses and isotopes of fragment ions should be in agreement with the chemical structure assigned. However, in order to secure this identification, a reference standard will be required in a final step to check the GC retention time and to confirm the presence of fragment ions experimentally by GC-TOF-MS analysis (Hernández et al, 2011).

After the first applications of high resolution GC-TOF-MS such as on extracts of well water (Grange, Genicola & Sovocool, 2002), this approach has been well accepted and has triggered the rapid instrumentation and software developments. GC-TOF-MS has been proven very useful for target screening of organic pollutants in water, pesticide residues in food, anabolic steroids in human urine and xenoestrogens in human-breast tissues as well as non-target screening (Hernández et al, 2011). It has also been successfully applied in metabolomics. As an example, GC-TOF-MS coupled to an APCI source was applied to human cerebrospinal fluid for metabolic profiling (Carrasco-Pancorbo et al., 2009). Moreover, the use of GC×GC coupled to TOF-MS for the metabolic profiling of biological fluids has been discussed in a recent review (Pasikanti et al., 2008). Although it is a very helpful tool, studies have also indicated that the elucidation of unknowns cannot be achieved by following a standardized procedure, as both expertise and creativity are essential in the process (Portolés et al., 2011).

Despite its excellent mass accuracy and sensitivity for qualitative studies, GC-TOF-MS is not as robust as other MS detectors such as triple quadrupoles for quantitation due to its limited dynamic range.

2.1.5 GC coupled with tandem mass spectrometry

The use of capillary gas chromatography coupled with tandem mass spectrometry (GC-MS/MS) can in principle result in a superlative technique in terms of both sensitivity and specificity, as necessary for ultra trace analysis.

For qualitative identification with MS/MS, product ion scan, precursor ion scan and neutral loss with a triple quadrupole or product scan with an ion trap can be used. However, the high-resolution performance of a capillary GC also requires a MS/MS with high scanning speed. Its applications had been limited for a period until the recent decade when the scan speed of a quadrupole has been dramatically improved with an enhanced sensitivity instead of loss. This has greatly extended its applicability for trace unknown identification. For quantification with SRM or MRM, a short dwell time in order to reach sufficient data points for a narrow peak is also required. Improvements on this aspect has been from hundreds milliseconds reduced to a few millisecond for a triple quadrupole instrument. Typical applications are mainly found in contaminations in environment and foods such pesticides and PCBs in foods and biological samples (Krumwiede & Huebschmann, 2008). A large number of applications of tandem mass spectrometry in GC-ITMS for trace analysis can be found in the literature, for example, in environmental analysis (Santos & Galceran, 2003), microorganism characterisation (Larsson & Saraf, 1997) and forensic analysis (Chiarotti & Costamagna, 2000).

2.1.6 Two-dimensional GC coupled with mass spectrometry (GC×GC-MS)

Nowadays a one-dimensional GC offers a peak capacity in the range of 500-1000 (Mondello et al., 2008). Mass spectrometry already adds another dimension (resolving *m/z* of ions) to enhance GC-MS selectivity. However, it is not surprising that samples (e.g., extracts of natural products or metabolites) containing several hundreds or even thousands of volatile constituents are of common occurrence. With the increasing demands and improvements on sensitivity, many trace components have also to be investigated. The fact is that, in a one-dimensional GC separation, analytes are generally not equally distributed along the whole retention time scale but frequently co-elute. As a result, it is required that the system should have much higher peak capacity than the number of sample constituents in order to accommodate all sample components (Davis & Giddings, 1993; Mondello et al., 2008). It is obvious that comprehensive two-dimensional gas chromatography (GC×GC) can significantly enhance the resolving power of a separation (Liu & Phillips, 1991). Its developments and applications have been reviewed recently (Mondello et al., 2008). With thousands of compounds eluting at any time in both dimensions, MS detector has a superior power over any others with regard to sensitivity and specificity. Since the first application of GC×GC-MS (Frysinger & Gaines, 1999), its potential has been gradually exploited for analysis of petroleum, PCBs and food samples and complex extracts of natural products or metabolites (Dallüge et al., 2002; Pasikanti et al., 2008). The advantages of comprehensive two-dimensional gas chromatography in combination with mass spectrometry include unprecedented selectivity (three separation dimensions with regard to volatility, polarity, and mass), high sensitivity (through band compression), enhanced separation power, and increased speed (comparable to ultra-fast GC experiments, if the number of peaks resolved per unit of time is considered) (Mondello et al., 2008). With the development of interfacing techniques, fast MS scanning and data interpretation/presentation methods, it will become an important tool to uncover the wealth of undiscovered information with respect to the complex composition of samples.

2.2 Gas Chromatography-Fourier Transform Infrared Spectroscopy (GC-FTIR)

For novel structures or new chemical entities, it is possible that no matched reference spectra can be found in MS databases. Manual interpretation of mass spectra requires sound knowledge on organic mass spectrometry and dedicated experience and often not possible to suggest any candidate structures. With the help of the molecular spectroscopy FT-IR, information on functional groups or structure moieties with specific infrared absorptions is complementary to MS and can be very valuable for structure elucidation, as already being used as stand-alone.

After the inspiration of the first reported IR spectrum of a trapped GC peak in 1967 (Low & Freeman, 1967) and the first commercial instrument introduced in 1971 by Digilab, there has been a rapid development of hyphenated GC-FTIR techniques. Commonly for on-line GC-

FTIR coupling, the effluent from the GC flows through a heated transfer-line into the light-pipe. A schematic drawing of a typical GC-FTIR is shown in Figure 2. The interferograms are scanned continuously to record either 'on-the-fly' gas-phase vapour IR spectra or trapped component spectra (Jackson et al., 1993). This makes it possible to reconstruct a chromatogram in real time by a vector technique called the Gram-Schmidt method (Malissa, 1983). After the acquisition is finished, the spectra of each GC peak can be normalized and searched by comparison with an IR spectra library (Hanna et al., 1979; Seelemann, 1982). In the earlier days of this technique, there were many discussions and studies about its sensitivity followed by the possible overloading of the columns when a capillary GC is used. Nowadays a GC-FTIR system having *ng* sensitivity of absolute substance amount has been introduced (Bruker Optics, 2009). GC-FTIR has been applied for analysis of polychlrorinated dibenzo p-dioxins, dibenzofurans (Sommer et al., 1997), aromatic polymers (Oguchi et al., 1991), petroleum (Scharma et al., 2009), etc., respectively. The combination of a gas chromatograph with both FT-IR and MS detectors on one instrument allows the simultaneous measurement of one peak by two supplementary detections. In fact, at each retention time, two different chromatograms were obtained. The sample passes the IR detector without destruction and is registered by the subsequent MS detector (Sommer et al., 1997).

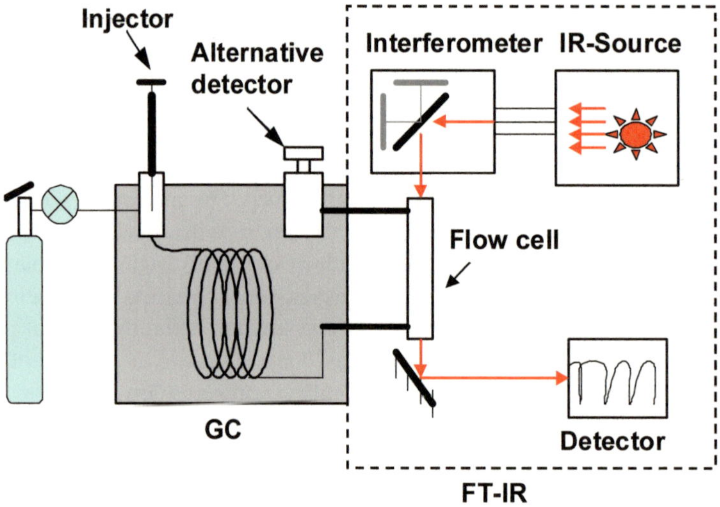

Fig. 2. Schematic drawing of a typical GC-FTIR

2.3 Gas Chromatography-Nuclear Magnetic Resonance spectroscopy (GC-NMR)

Nuclear Magnetic Resonance (NMR) (Rabi, 1938) spectroscopy is considered one of the most powerful analytical techniques for structural elucidation and identification of unknown compounds. It utilizes a physical phenomenon in which magnetic nuclei in a magnetic field absorb and re-emit electromagnetic radiation. Although the most commonly studied nuclei have specific chemical shifts, the abundances of the useful naturally occurring isotopes are generally low. That has significantly limited the sensitivity of NMR. Often mg of pure substance is required in order to obtain a meaningful spectrum. However, organic compounds of synthetic or natural products are often not in the pure form but are found as mixtures. Normally the analysis of such mixtures is performed in two independent steps: separation of the mixture into individual components and followed by identification. An on-line combination with GC separation can apparently accelerate this study.

The investigation on the coupling of GC and NMR has been very promising development because of the valuable information provided by NMR on molecular structure for each separated component (Buddrus & Herzog, 1981; Herzog & Buddrus, 1984). Unlike liquid or solid samples commonly analysed by NMR, the carrier gas in GC causes experimental difficulties in handling and results in low signal-to-noise ratio of the NMR signals obtained at atmospheric pressure. With the applications of Fourier-transform and averaging techniques, and gases can now be studied at fairly low pressures. However, the recording of the NMR spectra of flowing gases, especially those leaving a gas chromatograph containing only trace amount of analytes, imposes some problems. At short dwell times, the line broadening makes the recording of the spectra nearly impossible. Considerable progress towards overcoming its relative lack of sensitivity has been achieved using microcells for the optimum use of available samples and computers to improve signal-to-noise ratio (Milazzo, Petrakis & Brown, 1968). With the aid of stronger magnets and the newly developed microprobes, the first online GC-NMR spectra was recorded recently (Grynbaum et al., 2007; Kühnle et al., 2008). These experiments revealed the high potential of this technique, but it was also shown that the peaks in the GC separation elute up to 10 min, so that the advantage of the high separation performance of GC was lost in the experiment. The required amount for a GC-NMR run was between 1 and 2 mg for each analyte (Grynbaum et al., 2007). Problems arise in the form of partial condensation in the capillary connections (the transfer capillary and the probe head) for analytes with boiling points above 65 °C. It is suggested that the use of a transfer capillary being heated by a bifilar coil constructed from zero-susceptibility wire in combination with a strong magnetic field may solve overcome this problem (Grynbaum et al., 2007). Using stopped-flow measurements, very low sample amounts (~ 100-300 µg at 400 MHz), the potential applications of the hyphenation of high performance capillary GC to microprobe 1H NMR detection with the help of a spectra database has been demonstrated and an identification of stereoisomers in a complex mixture has been achieved (Kühnle et al., 2008). However, nanograms of volatile small compounds such as cockroach sex pheromones, mosquito attractants as well as a model compound geranyl acetate, which were prepared with an off-line GC, have already been investigated successfully using ^{1}H and COSY NMR (Nojima et al., 2011). Since the hyphenation of enantioselective capillary gas chromatography and mass spectrometry is not always sufficient to distinguish between structural isomers, thus requiring peak identification by NMR spectroscopy. The first online coupling of enantioselective capillary gas chromatography with proton nuclear magnetic resonance spectroscopy has been reported (Künhle et al. 2010). NMR allows constitutional and configurational isomers (diastereomers and enantiomers) to be distinguished. Enantiomers display identical spectra at different retention times, which enable an indirect identification of those unfunctionalized alkanes. Further developments in this field are still highly desirable.

2.4 Hyphenated gas chromatographic techniques for elemental analysis

Hyphenated techniques involving ICP-MS, AAS and AES are among the fastest growing research and application areas in atomic spectroscopy. The preferred chromatographic separation techniques include GC, HPLC, capillary electrophoresis (CE) and ion chromatography (IC) as well as field flow fractionation (FFF). General procedure for speciation analysis using GC-hyphenation include separation of the analytes from the sample matrix, formation of volatile derivatives, pre-concentration / cleanup and determination using different elemental analysis techniques.

2.4.1 Gas Chromatography - Inductively Coupled Plasma - Mass Spectrometry (GC-ICP-MS)

Stand-alone ICP-MS does not provide information on the chemical structure of the analyte at molecular level since all forms of the analytes are converted to positively charged atomic ions in the plasma. However, as an excellent elemental analyser (ICP) with resolution on masses (MS), ICP-MS can also be used as gas chromatographic detector. Resulting from this hyphenation, target analytes are separated into their constituent chemical forms or oxidation states before elemental analysis. In GC-ICP-MS, where the sample is gaseous, the transfer line should be inactivated and heated to eliminate sample degradation and condensation and will guide the sample directly into the ICP torch. In this way, the sample is maintained at constant high temperature from the end of the chromatographic column in the GC oven to the tip of the ICP injector (Agilent, 2007). It is almost a universal detector (only H, He, Ag, F, Ne cannot be directly measured), fits perfectly with a wide range of GC carrier gases and flows, and is capable of quantification with isotope dilution. A *pg* level of sensitivity can be achieved. The combination of the high performance chromatographic separation and the ICP atomizer also eliminates some interference of co-eluting compounds. For example in organic analysis, it has been used to study trace halogenated solvent residues in edible oils (Gomez-Ariza & Garcia-Barrera, 2006). GC-ICP-MS is very useful technique for speciation analysis such as sulphur speciation (Gomez-Ariza & Garcia-Barrera, 2006) or organometallic speciation (Bouyssiere, 2001; Kresimon et al., 2001).

2.4.2 Gas Chromatography-Atomic Absorption Spectroscopy (GC-AAS) and -Atomic Emission Spectroscopy (GC-AES)

Atomic absorption spectrometry (AAS) is a well-established detection technique in elemental analysis. In AAS, when the UV light with the right wavelength irradiates on a ground-state free atom, the atom may absorb the light to enter an excited state in a process known as atomic absorption. The wavelength of the light is somehow specific for an element. By measuring the amount of light absorbed, a quantitative determination of the amount of analyte element present can be made. A careful selection of wavelengths allows the specific quantitative determination of individual elements in the presence of others. By supplying enough thermal energy (such as in a flame), analyte compounds dissociate into free atoms. The ease and speed, at which precise and accurate determinations can be achieved, have made AAS one of the most popular methods for the analysis of metals.

Atomic emission spectroscopy (AES), also called optical emission spectroscopy (OES), is another elemental analysis technique in atomic spectroscopy. In this process, analytes are first atomized using either ICP or microwave irradiation (high temperatures), where the atoms are raised to electronically excited states. As they return to lower electronic energy levels, photons are emitted at wavelengths that are characteristic of the particular element. Both the wavelength and intensity of the emitted photons will be recorded. This data may be used to identify the elements present and quantify them after a calibration. GC-AES determines the elemental composition for each GC peak.

The tailored operation condition of a gas chromatograph fits again nicely with the AAS atomizer since the analytes are already present in the gas phase. The coupling can be done by introducing the GC effluent via a heated nickel transfer line into the quartz atomization furnance (Dirkx et al., 1992). Although the interface is rather simple, in practice, several conditions need to be optimized in order to obtain good sensitivity and selectivity for specific analytical problems. Efforts include exploiting the quartz furnace, heating with flame or a thermostat, or using the graphite furnace as the atomization device (Dirkx et al., 1995). The widespread use of organometallic compounds in agriculture, plastics industry and subsequently the release into the environment has led to an increasing concern about their persistence and toxicity. For instance, organotins unlike inorganic tin, it may occur at toxic levels in aquatic and sedimentary environments. The distribution of organotins as well as its bioavailability and toxicity depend critically on the chemical form in which the species are actually present. This is why total tin determination is not able to provide reliable information on the hazards of this element to human health and environment. Therefore, speciation analysis of these organometallic compounds are of great importance (Dirkx et al., 1995). Since its introduction, GC-AAS has been extensively applied to speciation of organometallic compounds such as Sn, Pb and As. Typical examples include speciation of organolead in environment (Baxter & Frech, 1995; Dirkx et al., 1992; Harrison & Hewitt, 1985) and in biological samples (Baxter & Frech, 1995); speciation of organotin compounds in agriculture and plastics industry (Dirkx et al., 1995) as well as arsenic speciation. The reason for the existence of various forms of organometallic compounds is that Sn, Pd and As all have very rich organic chemistry. Just because of this, several derivatization methods have been reported in order to convert them to volatile species for GC analysis and a sensitivity of picogram level can be achieved together with a highly selective detection as the result of the coupling (Dirkx et al., 1995; Harrison & Hewitt, 1985).

As for GC-AES, it has been much less investigated than GC-AAS. Nevertheless, the available reports have indicated that GC-AES coupled with microwave-induced plasma (MIP) can be used to study organic polymers (Oguchi et al., 1991) and to perform speciation analysis of organotin compounds in human urine. This provides a method suitable for rapid sensitive screening of human urine samples without dilution of the sample (Zachariadis & Rosenberg, 2009a, 2009b).

3. Conclusion and remarks

Apparently, the motivation of the enormous efforts on investigating a variety of hyphenated systems is the increasing demands on the wealth of undiscovered information of trace components in a sample. Many stand-alone instrumental analytical techniques are already well developed and able to provide specific information about analytes in samples, although each of them has its advantages and disadvantages. The complexity of real-world sample matrices often exceeds the analytical capabilities of any conventional chromatographic separations. The development and employment of more comprehensive coupling techniques to enable a deeper insight into the composition of natural and synthetic matrices has become a necessity.

The most powerful separation technique in chromatography is still capillary gas-liquid chromatography (one-dimensional or its extension - comprehensive two-dimensional gas chromatography). On the other hand, mass spectrometry including various high resolution MS and tandem MS provides analyte-specific selectivity. Their combination GC-MS is still the most flexible and powerful analytical technique available today for volatile organic substances with many unchallengeable advantages such as unprecedented selectivity, sensitivity, enhanced separation power and speed of analysis (Mondello et al., 2008).

The coupling with other detection techniques further extends the GC applicability. However, the rule of thumb is that the hyphenation of two techniques would only make sense when the analytical powers (separation and detection) of both instruments are enhanced rather than compromised in order to couple. One of the recent good examples is implementation of the very soft ionization APCI interface for GC-TOF-MS (Carrasco-Pancorbo et al., 2009) to generate abundant molecular ions in order to utilize the high resolution and MS/MS capabilities for structure elucidation. Both CI and EI may be too hard to preserve valuable molecular ions. This feature makes this technique very attractive for such as wide-scope screening of a large number of analytes at trace levels. In the future developments, efforts should be put to make hyphenated GC techniques also user-friendly and robust with regard to automatic data interpretation and reporting software, standardized interfaces and even consumables.

4. References

Agilent Technologies. (2007). *Handbook of Hyphenated ICP-MS Applications*, 1st ed. Part No. 5989-6160EN, August 2007

Allcin, C.; Law, R. & Morris, S. (1999). Polybrominated Diphenylethers in Sediments and Biota Downstream of Potential Sources in the UK. *Environmental Pollution*, Vol. 105, No. 2 (May 1999), pp. 197-207, ISSN 0269-7491

Arsenault, G.; Dolhun, J. & Biemann, K. (1971). Alternate or Simultaneous Electron Impact-Chemical Ionization Mass Spectrometry of Gas Chromatographic Effluent. *Analytical Chemistry*, Vol. 43, No. 12 (October 1971), pp. 1720-1722, ISSN 0003-2700

Baxter, D. & Frech, W. (1995). Speciation of Lead in Environmental and Biological Samples. *Pure and Applied Chemistry*, Vol. 67, No. 4 (April 1995), pp. 615-648, ISSN 0033-4545

Bouyssiere, B.; Szpunar, J. & Lobinski, R. (2001). Gas Chromatography with Inductively Coupled Plasma Mass Spectrometric Detection in Speciation Analysis. *Spectrochimica Acta Part B*, Vol. 57, No. 5 (May 2002), pp. 805–828, ISSN 0584-8547

Bruker Optics. (2009). GC-FT-IR in the *ng* Range. Application Note # AN-76E. Available from: http://www.brukeroptics.com/fileadmin/be_user/Applications/ApplicationNotes/AN_76E_GC-FT-IR_in_the_ng_range.pdf

Buddrus, J. & Herzog, H. (1981). Coupling of Chromatography and NMR. 3. Study of Flowing Gas Chromatographic Fractions by Proton Magnetic Resonance. *Organic Magnetic Resonance*, Vol. 15, No. 2 (February 1981), pp. 211-213, ISSN 1097-458X

Čajka, T. & Hajšlová, J. (2007). Gas Chromatography–Time-of-Flight Mass Spectrometry in Food Analysis. *LCGC Europe*, Vol. 20, No. 1 (2007), pp. 25–31, ISSN 1471-6577

Carrasco-Pancorbo, A.; Nevedomskaya, E.; Arthen-Engeland, T.; Zey, T.; Zurek, G.; Baessmann, C.; Deelder, A. & Mayboroda, O. (2009). Gas Chromatography

/Atmospheric Pressure Chemical Ionization-Time of Flight Mass Spectrometry: Analytical Validation and Applicability to Metabolic Profiling. *Analytical Chemistry*, Vol. 81, No. 24 (December 2009), pp. 10071-10079, ISSN 0003-2700

Chaturvedi, K. & Nanda, R. (2010). Hyphenated Gas Chromatography. *International Journal of Pharmaceutical Sciences Review and Research*, Vol. 5, No. 3 (November 2010), pp. 18-27, ISSN 0976-044x

Chernetsova, E.; Revelsky, A.; Revelsky, I.; Mikhasenko, I & Sobolevsky, T. (2002). Determination of Polychlorinated Dibenzo-p-Dioxins, Dibenzofurans, and Biphenyls by Gas Chromatography / Mass Spectrometry in the Negative Chemical ionization Mode with Different Reagent Gases. *Mass Spectrometry Review*, Vol. 21, No. 6 (November 2002), pp. 373- 387, ISSN 1098-2787

Chernushevich, I.; Loboda, A. & Thomson, B. (2001). An Introduction to Quadrupole-Time-of-Flight Mass Spectrometry. *Journal of Mass Spectrometry*, Vol. 36, No. 8, (August 2001), pp. 849–865, ISSN 1096-9888

Chiarotti, M. & Costamagna, L. (2000). Analysis of 11-Nor-9-carboxy-Δ^9-tetrahydro-cannabinol in Biological Samples by Gas Chromatography Tandem Mass Spectrometry (GC/MS-MS). *Forensic Science International*, Vol. 114, No. 1 (October 2000), pp. 1-6, ISSN 0379-0738

Cody, R. (2008). GC/MS with DART Ion Source. Application Note, JEOL Inc., 06.09.2011. Available from http://chromatographyonline.findanalytichem.com/lcgc/data/articlestandard//lcgc/122008/504148/article.pdf

Cody, R.; Laramée J. & Durst H. (2005). Versatile New Ion Source for the Analysis of Materials in Open Air under Ambient Conditions. *Analytical Chemistry*, Vol. 77, No. 8 (March 2005), pp. 2297-2302, ISSN 0003-2700

Coleman III, W. & Gordon, B. (2006). Hyphenated Gas Chromatography, In: *Encyclopedia of Analytical Chemistry*, R.A. Meyers, (Ed.), John Wiley & Sons, ISBN: 9780470027318

Cotter, R. (1994). *Time-of-Flight Mass Spectrometry*. American Chemical Society, ISBN 0-8412-3474-4 , Columbus, USA

Dallüge, J. ; van Stee, L. ; Xu, X. ; Williams, J. ; Beens, J.; Vreuls, R. & Brinkman, U. (2002). Unravelling the Composition of Very Complex Samples by Comprehensive Gas Chromatography Coupled to Time-of-Flight Mass Spectrometry: Cigarette Smoke. *Journal of Chromatography A*, Vol. 974, No. 1 (October 2002), pp. 169–184, ISSN 0021-9673

Davis, J. & Giddings, J. (1983). Statistical Theory of Component Overlap in Multicomponent Chromatograms. *Analytical Chemistry*, Vol. 55, No. 3 (March 1983), pp. 418-424, ISSN 0003-2700

de Hoffmann, E. & Stroobant, V. (2003). *Mass Spectrometry: Principles and Applications (Second Ed.)*. John Wiley & Sons, ISBN 0471485667, Toronto, Canada

Dirkx, W.; van Cleuvenbergen, R & Adams, F. (1992). Speciation of Alkyllead Compounds by GC-AAS: A State of Affairs. *Microchimica Acta*, Vol. 109, No. 1-4 (1992), pp. 133-135, ISSN 0026-3672

Dirkx, W.; Lobinski, R. & Adams, F. (1995). Speciation Analysis of Organotin by GC-AAS and GC-AES after Extraction and Derivatization, In: *Quality Assurance for Environmental Analysis*, Ph. Quevauviller, E.A. Maier and B. Griepink, (Ed.), 357-409, Elsevier, ISSN 0-444-89955-3, Amsterdam, The Netherlands

Friedrich, J. & Acree, T. (1998). Gas Chromatography Olfactometry (GC/O) of Dairy Products. *International Dairy Journal*, Vol. 8, No. 3 (March 1998), pp. 235-241, ISSN 0958-6946

Frysinger, G. & Gaines, R. (1999). Comprehensive Two-Dimensional Gas Chromatography with Mass Spectrometric Detection (GC×GC/MS) Applied to the Analysis of Petroleum. *Journal of High Resolution Chromatography*, Vol. 22, No. 5, (May 1999), pp. 251–255, ISSN 0935-6304

Gohlke, R. & McLafferty, F. (1993). Early Gas Chromatography/Mass Spectrometry. *Journal of the American Society for Mass Spectrometry*, Vol. 4, No. 5 (May 1993), pp. 367-371, ISSN 1044-0305

Gomez-Ariza, J. & Garcia-Barrera, T. (2006). Optimization of a Multiple Headspace SPME-GC-ECD-ICP-MS Coupling for Halogenated Solvent Residues in Edible Oils. *Journal of Analytical Atomic Spectrometry*, Vol. 21, No. 9 (2006), pp. 884-890, ISSN 0267-9477

Goraczko, A. (2005). Molecular Mass and Location of the Most Abundant Peak of the Molecular Ion Isotopomeric Cluster. *Journal of Molecular Modeling*, Vol. 11, No. 4 (April 2005), pp. 271–277, ISSN 1610-2940

Grange, A.; Genicola, F. & Sovocool, G. (2002). Utility of Three Types of Mass Spectrometers for Determining Elemental Compositions of Ions Formed from Chromatographically Separated Compounds. *Rapid Communications in Mass Spectrometry*, Vol. 16, No. 24, (December 2002), pp. 2356–2369, ISSN 1097-0231

Grange, A.; Winnik, W.; Ferguson, P. & Sovocool, G. (2005). Using a Triple-Quadrupole Mass Spectrometer in Accurate Mass Mode and an Ion Correlation Program to Identify Compounds. *Rapid Communications in Mass Spectrometry*, Vol. 19, No. 18 (September 2005), pp. 2699–2715, ISSN 1097-0231

Gross, M. (1994). Accurate Masses for Structure Confirmation. *Journal of the American Society for Mass Spectrometry*, Vol. 5, No. 2 (February 1994), pp. 57-57, ISSN 1044-0305

Grynbaum, M.; Kreidler, D.; Rehbein, J.; Purea, A.; Schuler, P.; Schaal, W.; Czesla, H.; Webb, A.; Schurig, V. & Albert, K. (2007). Online Coupling of Gas Chromatography to Nuclear Magnetic Resonance Spectroscopy: Method for the Analysis of Volatile Stereoisomers. *Analytical Chemistry*, Vol. 79, No. 7 (April 2007), pp. 2708-2713, ISSN 0003-2700

Grob, R. & Barry, E. (2004). *Modern Practice of Gas Chromatography*, Wiley-Interscience, ISBN 0-471-22983-0, New York, USA

Hanna, A.; Marshall, J. & Isenhour, T . (1979). A GC/FT-IR Compound Identification System. *Journal of Chromatographic Science*, Vol. 17, No. 8 (August 1979), pp. 434-, ISSN 0021-9665

Harrison, A. (1992). *Chemical Ionization Mass Spectrometry*, CRC Press, ISBN 0849342546, Boca Raton, USA

Harrison, R. & Hewitt, C. (1985). Development of Sensitive GC-AAS Instrumentation for Analysis of Organometallic Species in the Environment. *International Journal of Environmental Analytical Chemistry*, Vol. 21, No. 1-2 (December 1985), pp. 89-104, ISSN 0306-7319

Hernández, F.; Portolés, T.; Pitarch, E. & López, F. (2011). Gas Chromatography Coupled to High-Resolution Time-of-Flight Mass Spectrometry to Analyze Trace-Level

Organic Compounds in the Environment, Food Safety and Toxicology. *Trends in Analytical Chemistry*, Vol. 30, No. 2 (February 2011), pp. 388-400, ISSN 0165-9936

Herzog, H & Buddrus, J. (1984). Coupling of Chromatography and NMR. Part 5: Analysis of High-Boiling Gas-Chromatographic Fractions by On-line Nuclear Magnetic Resonance. *Chromatographia*, Vol. 18, No. 1, pp. 31-33, ISSN 0009-5893

Hirschfeld, T. (1980). The Hy-phen-ated Methods. *Analytical Chemistry*, Vol. 52, No. 2 (February 1980), pp. 297A-312A, ISSN 0003-2700

Holmes, J. & Morrell F. (1957). Oscillographic Mass Spectrometric Monitoring of Gas Chromatography. *Applied Spectroscopy*, Vol. 11, No. 2 (May 1957), pp. 61-91, ISSN 0003-7028

Hunt, D.; Stafford, Jr. G.; Crow, F. & Russel, J. (1976). Pulsed Positive Negative Ion Chemical Ionization Mass Spectrometry. *Analytical Chemistry*, Vol. 40, No. 14 (December 1976), pp. 2098-2104, ISSN 0003-2700

Jackson, P.; Dent, G.; Carter, D.; Schofield, D.; Chalmers, J.; Visser, T. & Vredenbregt, M. (1993). Investigation of High Sensitivity GC-FTIR as an Analytical Tool for Structural Identification. *Journal of High Resolution Chromatography*, Vol. 16, No. (September 1993), pp. 515-521, ISSN 0935-6304

Jaeger, H. (1987). *Capillary Gas Chromatography-Mass Spectrometry in Medicine and Pharmacology*, Dr. Alfred Huethig, ISBN 3-7785-1375-3, Heidelberg, Germany

James, A. & Martin, J. (1952). Gas-Liquid Partition Chromatography; the Separation and Micro-estimation of Volatile Fatty Acids from Formic Acid to Dodecanoic Acid. *Biochemistry Journal*, Vol. 50, No. 5 (March 1952), pp. 679-690, ISSN 0264-6021

Jeol. (2002). Analysis of Scotch Whiskey and Tequila Samples by Solid-Phase Microextraction and High-Resolution GC/MS. Application Note: MS-11262002-A, 15.09.2011. Available from: http://www.jeolusa.com/PRODUCTS/AnalyticalInstruments/MassSpectrometers/GCmateII/tabid/231/Default.aspx

Karasek, F. & Clement, R. (1988). *Basic Gas Chromatography – Mass Spectrometry Principles & Techniques*. Elsevier, ISBN 0-444-42760-0, Amsterdam, the Netherlands

Kingdon, K. (1923). A Method for the Neutralization of Electron Space Charge by Positive Ionization at Very Low Gas Pressures. *Physical Review*, Vol. 21, No. 4 (April 1923), pp. 408-418, ISSN 0031-899X

Krone, N.; Hughes, B.; Lavery, G.; Stewart, P.; Arlt, W. & Shackleton, C. (2010). Gas Chromatography/Mass Spectrometry (GC/MS) Remains a Pre-Eminent Discovery Tool in Clinical Steroid Investigations Even in the Era of Fast Liquid Chromatography Tandem Mass Spectrometry (LC/MS/MS). *Journal of Steroid Biochemistry & Molecular Biology*, Vol. 121, No. 3 (August 2010), pp. 496–504, ISSN 0960-0760

Krumwiede, D. & Huebschmann, H. (2008). Analysis of PCBs in Food and Biological Samples Using GC Triple Quadrupole GC-MS/MS. Application Note 10262, Thermo Scientific. Available from: http://pops.thermo-bremen.com/pdf /AN10262_PCBs_on_TSQ_Quantum_GC.pdf

Kuhara, T. (2005). Gas Chromatographic-Mass Spectrometric Urinary Metabolome Analysis to Study Mutations of Inborn Errors of Metabolism. *Mass Spectrometry Review*, Vol. 24, No. 6 (November 2005), pp. 814-827, ISSN 1098-2787

Kühnle, M.; Kreidler, D.; Holtin, K.; Czesla, H.; Schuler, P.; Schaal, W.; Schurig, V. & Albert, K. (2008). Online Coupling of Gas Chromatography to Nuclear Magnetic Resonance Spectroscopy: Method for the Analysis of Volatile Stereoisomers. *Analytical Chemistry*, Vol. 80, No. 14 (July 2008), pp. 5481-5486, ISSN 0003-2700

Kühnle, M.; Kreidler, D.; Holtin, K., Czesla, H., Schuler, P., Schurig, V. & Albert, K. (2010). Online Coupling of Enantioselective Capillary Gas Chromatography with Proton Nuclear Magnetic Resonance Spectroscopy. *Chirality*, Vol. 22, No. 9 (October 2010), pp. 808-812, ISSN 1520-636X

Liu, Z. & Phillips, J. (1991). Comprehensive Two-Dimensional Gas Chromatography using an On-column Thermal Desorption Modulator Interface, *Journal of Chromatographic Science*, Vol. 29, No. 6, (June 1991), pp. 227–231, ISSN 0021-9665

Larsson, L. & Saraf, A. (1997). Use of Gas Chromatography-Ion Trap Tandem Mass Spectrometry for the Detection and Characterization of Microorganisms in Complex Samples. *Molecular Biotechnology*, Vol. 7, No. 3 (June 1997), pp. 279-287, ISSN 1559-0305

Low, M. & Freeman, S. (1967). Measurement of Infrared Spectra of Gas-Liquid Chromatography Fractions using Multiple-Scan Interference Spectrometry. *Analytical Chemistry*, Vol. 39, No. 2 (February 1967), pp. 194-198, ISSN 0003-2700

Makarov A. (2000). Electrostatic Axially Harmonic Orbital Trapping: A High-Performance Technique of Mass Analysis. *Analytical Chemistry*, Vol. 72, No.6 (February 2000), pp. 1152-1162, ISSN 0003-2700

Malissa, H. (1983). On the Use of Capillary Separation Columns in GC/FTIR-Spectroscopy and on the Quantitative Evaluation of the Gram-Schmidt Reconstructed Chromatogram. *Fresenius' Journal of Analytical Chemistry*, Vol. 316, No. 7 (January 1983), pp. 699-704, ISSN 0937-0633

Mamyrin, B.; Karataev, V.; Shmikk, D. & Zagulin, V. (1973). The Mass-Reflectron, a New Nonmagnetic Time-of-Flight Mass Spectrometer with High Resolution. *Soviet Physics JETP*. Vol. 37, (July 1973), pp. 45, ISSN 0038-5646

Mamyrin, B. (2001). Time-of-Flight Mass Spectrometry (Concepts, Achievements, and Prospects)". *International Journal of Mass Spectrometry*, Vol. 206, No. 3 (March 2001), pp. 251–266, ISSN 1387-3806

Marshall, A.; Hendrickson, C. & Jackson, G. (1998). Fourier Transform Ion Cyclotron Resonance Mass Spectrometry: a Primer. *Mass Spectrometry Review*, Vol. 17, No. 1 (January 1998), pp. 1-35, ISSN 1098-2787

McLafferty, F.; Hertel, R. & Villwock, R. (1974). Probability Based Matching of Mass Spectra. Rapid Identification of Specific Compounds in Mixtures. *Organic Mass Spectrometry*, Vol. 9, No. 7 (July 1974), pp. 690–702, ISSN 1096-9888

McLafferty, F. & Tureek, F. (1993). *Interpretation of Mass Spectra*. 4th ed.. University Science Books, ISBN 0935702253, Sausalito, USA

McNaught, A & Wilkinson, A. (1997). *IUPAC. Compendium of Chemical Terminology*, 2nd ed. (the "Gold Book"). Blackwell Scientific Publications, ISBN 0-9678550-9-8, Oxford, UK.

Message, G. (1984). *Practical Aspects of Gas Chromatography / Mass Spectrometry*, Wiley, ISBN 0-471-06277-4, New York, USA

Milazzo, B.; Petrakis, L. & Brown, P. (1968). Microsampling Techniques for Combined Gas Chromatography and High-Resolution Nuclear Magnetic Resonance Spectroscopy, *Applied Spectroscopy*, Vol. 22, No. 5 (September 1968), pp. 574-575, ISSN 0003-7028

Mondello, L.; Tranchida, P.; Dugo, P & Dugo, G. (2008). Comprehensive Two-Dimentional Gas Chromatography-Mass Spectrometry: A Review. *Mass Spectrometry Review*, Vol. 27, No. 2 (March 2008), pp. 101-124, ISSN 1098-2787

Niessen, W. (2001). *Current Practice of gas chromatography-mass spectrometry*, Marcel Dekker, ISBN 0-8247-0473-8, New York, USA

Nojima, S; Kiemle, D.; Webster, F.; Apperson, C & Schal, C. (2011). Nanogram-Scale Preparation and NMR Analysis for Mass-Limited Small Volatile Compounds. *Public Library of Science (PLoS) One*, Vol 6, No. 3 (March 2011), e18178, eISSN -1932-6203

Oguchi, R.; Shimizu, A.; Yamashita, S.; Yamaguchi, K & Wylie, P. (1991). Polymer Analysis Using Pyrolysis-GC-FTIR-MS and GC-AED. *Journal of High Resolution Chromatography*, Vol. 14, No. 6 (June 1991), pp. 412–416, ISSN 0935-6304

Ong, V. & Hites, R. (1994). Electron Capture Mass Spectrometry of Organic Environmental Contaminants. *Mass Spectrometry Review*, Vol. 13, No. 3 (May 1994), pp. 259-283, ISSN 1098-2787

Pasikanti K.; Ho, P. & Chan, E. (2008). Gas Chromatography/Mass Spectrometry in Metabolic Profiling of Biological Fluids. *Journal of Chromatography B*, Vol. 871, No. 2 (August 2008), pp. 202–211, ISSN 1570-0232

Paul, W. & Steinwedel, H. (1953). Ein Neues Massenspektrometer ohne Magnetfeld. *Zeitschrift für Naturforschung A*, Vol. 8, No. 7, pp. 448-450, ISSN 0932-0784

Peterson, A.; McAlister, G.; Quarmby, S.; Griep-Raming, J. & Coon, J. (2010). Development and Characterization of a GC-Enabled QLT-Orbitrap for High-Resolution and High-Mass Accuracy GC/MS. *Analytical Chemistry*, Vol. 82, No. 20 (October 2010), pp. 8618-8628, ISSN 0003-2700

Plomley, J; Lausevic, M. & March, R. (2000). Determination of Dioxins/Furans and PCBs by Quadrupole Ion-Trap Gas Chromatography–Mass Spectrometry. *Mass Spectrometry Review*, Vol. 19, No. 5 (September 2000), pp. 305-365, ISSN 1098-2787

Portolés, T.; Pitarch, E.; López, F.; Hernández, F. & Niessen, W. (2011). Use of Soft and Hard Ionization Techniques for Elucidation of Unknown Compounds by Gas Chromatography / Time-of-Flight Mass Spectrometry. *Rapid Communications in Mass Spectrometry*, Vol. (25), No. 11 (June 2011), pp. 1589–1599, ISSN 1097-0231

Rabi, I.; Zacharias, J.; Millman, S. & Kusch, P. (1938). A New Method of Measuring Nuclear Magnetic Moment. *Physical Review*, Vol. 53, No.4 (April 1938), pp. 318-318, ISSN 0031-899X

Santos, F. & Galceran, M. (2003). Modern Developments in Gas Chromatography-Mass Spectrometry-based Environmental Analysis. *Journal of Chromatography A*, Vol. 1000, No. 1 (June 2003), pp. 125-151, ISSN 0021-9673

Sarrion, M; Santos, F. & Galceran, M. (2000). In Situ Derivatization/Solid-Phase Microextraction for the Determination of Haloacetic Acids in Water. *Analytical Chemistry*, Vol. 72, No. 20 (October 2000), pp. 4865-4873, ISSN 0003-2700

Schwartz, J., Senko, M. & Syka, J. (2002). A Two-Dimensional Quadrupole Ion Trap Mass Spectrometer. *Journal of the American Society for Mass Spectrometry*, Vol. 13, No.6 (June 2002), pp. 659-669, ISSN 1044-0305

Seelemann, R. (1982). GC/FTIR Coupling - a modern tool in analytical chemistry. *Trends in Analytical Chemistry*, Vol. 1, No. 14 (October 1982), pp. 333-339, ISSN 0003-2700

Sharma, K; Sharmab, S. & Lahiri, S. (2009). Characterization and Identification of Petroleum Hydrocarbons and Biomarkers by GC-FTIR and GC-MS. *Petroleum Science and Technology*, Vol. 27, No. 11 (June 2009), pp. 1209 - 1226, ISSN 1091-6466

Solouki, T.; Szulejko, J.; Bennett, J. & Graham, L. (2004). A Preconcentrator Coupled to a GC/FTMS: Advantages of Self-Chemical Ionization, Mass Measurement Accuracy, and High Mass Resolving Power for GC Applications. *Journal of the American Society for Mass Spectrometry*, Vol. 15, No. 8 (August 2004), pp. 1191-1200, ISSN 1044-0305

Sommer, S.; Kamps, R.; Schumm, S. & Kleinermanns, K. (1997). GC/FT-IR/MS Spectroscopy of Native Polychlorinated Dibenzo-p-dioxins and Dibenzofurans Extracted from Municipal Fly-Ash. *Analytical Chemistry*, Vol. 69, No. 6 (March 1997), pp. 1113-1118, ISSN 0003-2700

Sparkman, D. (2006). *Mass Spec Desk Reference*, 2nd. ed., Global View Publishing, ISBN 0966081390, Pittsburgh, USA.

Stephens, W. (1946). A Pulsed Mass Spectrometer with Time Dispersion. *Physical Review*, Vol. 69, No. 11 (June 1946), pp. 691-691, ISSN 0031-899X

Stein, S. & Scott, D. (1994). Optimization and Testing of Mass Spectral Library Search Algorithms for Compound Identification. *Journal of the American Society for Mass Spectrometry*, Vol. 5, No. 9 (Sptember 1994), pp. 859-866, ISSN 1044-0305

Szulejko, J. & Solouki, T. (2002). Potential Analytical Applications of Interfacing a GC to an FT-ICR MS: Fingerprinting Complex Sample Matrixes. *Analytical Chemistry*, Vol. 74, No.14 (July 2002), pp. 3434-3442, ISSN 0003-2700

Todd, J. (1991). Recommendations for Nomenclature and Symbolism for Mass Spectroscopy (Including an Appendix of Terms Used in Vacuum Technology). *Pure and Applied Chemistry*, Vol. 63, No. 10 (October 1991), pp. 1541-1566, ISSN 0033-4545

Vestal, M. & Campbell, J. (2005). Tandem Time-of-Flight Mass Spectrometry. *Methods in Enzymology*, Vol. 402, pp. 79–108, ISSN 978-0-12-182807-3

Watson, J. (1997). *Introduction to Mass Spectrometry*, Lippincott-Raven, ISBN 0397516886, Philadelphia, USA

Yergey, J.; Heller, D.; Hansen, G.; Cotter, R. & Fenselau, C. (1983). Isotopic Distributions in Mass Spectra of Large Molecules. *Analytical Chemistry*, Vol. 55, No. 2 (February 1983), pp. 353–356, ISSN 0003-2700

Zachariadis, G. & Rosenberg, E. (2009a). Speciation of Organotin Compounds in Urine by GC-MIPAED and GC-MS after Ethylation and Liquid-Liquid Extraction. *Journal of Chromatography B*, Vol. 877, No. 11-12 (April 2009), pp. 1140-1144, ISSN 1570-0232

Zachariadis, G. & Rosenberg, E. (2009b). Determination of Butyl- and Phenyltin Compounds in Human Urine by HS-SPME after Derivatization with Tetraethylborate and Subsequent Determination by Capillary GC with Microwave-Induced Plasma Atomic Emission and Mass Spectrometric Detection. *Talanta*, Vol. 78, No. 2 (April 2009), pp. 570-576, ISSN 0039-9140

2

Stationary Phases

Vasile Matei, Iulian Comănescu and Anca-Florentina Borcea
Petroleum and Gas University of Ploieşti,
Romania

1. Introduction

Gas chromatography is a physical method of separating and identifying mixtures, with application in chemical practice, scientific investigation, petroleum technology, environmental pollution control, the food industry, pharmacology, biology and medicine. Separation of the sample components takes place by adsorption by a solid or by dissolving in a non-volatile liquid, called the stationary phase. Separation takes place based on different partitions of the sample components between a stationary phase and mobile phase. The mobile phase, in gaseous state, must be insoluble in the stationary phase (solid or liquid), the mobile phase continuously circulating over the stationary phase. Separation and measurement of components in the sample are made in a device called a gas chromatograph. The "heart" of the device is the chromatography column. The chromatography column contains the stationary phase, solid or liquid, as packing material for packed columns, or on the walls of capillary columns. This method of analysis has undergone important developments in recent decades and contributed to the progress of important scientific and applied fields.

2. Requirements of stationary phases

The principle of gas chromatography (GC) is based on the capacity of the stationary phase to produce different separation times upon exiting a chromatographic column that contains, under one form or another, stationary phases for the various mixture components of the sample. Through GC analysis, the sample components are separated through the combined effect of the stationary and mobile phases. The mobile phase is generally a gas, such as H_2, He, N_2, AR etc. The stationary phase is fixed and can produce either adsorption or absorption. A thermodynamic equilibrium is established between the two phases; this is expressed through the theoretic plates. When the sample components are added between the two phases at equilibrium, a difference emerges that can be expressed as a function of the quantities of sample components.

Chromatography is a method of separating multicomponent mixtures. This method relies on differences in partitioning behaviour between a flowing mobile phase and a stationary phase in order to separate the components in a mixture and on different velocities of compounds in the mobile phase. The mobile phase is usually a permanent gas, such as hydrogen, helium, nitrogen, argon etc., as a constant flow with a certain pressure. The stationary phase may be a solid or a liquid that is immobilized or adsorbed in a solid. The stationary phase may consist of particles (porous or solid), the walls of a tube (i.e. capillary) or a fibrous material.

Techniques by physical state of mobile and stationary phase:

Chromatography (C); Liquid chromatography (LC); Gas chromatography (GC); Paper chromatography (PC); Thin-layer chromatography (TLC) Column chromatography (CC) Gas-liquid chromatography (GLC); Gas-solid chromatography (GSC); Liquid-solid chromatography (LSC); Size-exclusion chromatography (SEC); Liquid-liquid chromatography (LLC); Ion exchange chromatography (IEC).

Components separated from the sample are placed in a chromatographic column, a tube in stainless steel, copper, aluminium or glass for packed chromatography, or quartz for capillary chromatography. The column contains the stationary phase as a granular porous solid. The columns are in the oven of the chromatograph, isotherm or temperature programmed. The mobile phase passes through the column at a constant flow, the flow values being correlated with the column type (packed, capillary). Thus, a primary classification could be gas-solid chromatography and gas-liquid-solid chromatography.

Capillary columns are used mostly, but not exclusively. They are the latest and best method, but they cannot replace totally the packed columns. The trends in column chromatography are:

1. Packed columns are still used in 20% of chromatographic analysis.
2. Packed columns are primarily for preparative applications, permanent gas analysis and sample preparation.
3. Packed columns will be used in the future because some applications demand packed columns not capillary columns.

Gas chromatography has several advantages as a physical method of separating gas mixtures, namely:

1. Resolution. The technique is applicable to systems containing components with very close boiling points. By choosing a suitable stationary phase or adsorbent, molecules with similar physical and chemical properties could be separated. Sample components could form in normal distillation, azeotropic mixtures.
2. Sensitivity. The properties of gas chromatographic systems are responsible for their widespread use. The detector based on thermal conductivity of the component can detect picograms of sample. The sensitivity is important considering that a chromatography test is less than 30 minutes. Analysis that typically occurs in about one hour or more can be reduced to the order of minutes due to the high diffusion rate in the gas phase and the phase to fast equilibrium between mobile and stationary phases.
3. Convenience. Operation in gas chromatography is a direct operation. It does not require highly qualified personnel to perform routine separation.
4. Costs. Compared with many other now affordable analytical tools, gas chromatography presents excellent cost value.
5. Versatility. GC is adaptable, from samples containing permanent gas up to liquids having high boiling points and volatile solids.
6. High Separation Power. If you use mobile phases with a low viscosity degree, very long columns will provide a strong separability

7. Assortment of Sensitive Detecting Systems. Detectors used in gas chromatography are relatively simple, with high sensitivity and fast responses given.
8. Ease of Recording Data. Output of the gas chromatograph detector can be conveniently connected to a potentiometer recorder, or integrating systems, computers that can store a large amount of information.
9. Automatization. Gas chromatographs can be used to automatically monitor various chemical processes that allow samples to be taken periodically and injected into the chromatographic columns to be separated and detected.

An ideal stationary phase is selective and has different adsorptivity for each sample component in order to ensure separation, as well as a wide array of operating temperatures. It has to be chemically stable and have a low vapour pressure at high operating temperatures. Some criteria for selecting an adequate stationary phase are:

- Is the stationary phase selective enough in separating the sample components so as to separate them one by one?
- Can there be an irreversible chemical reaction and can the mixture components be separated?
- Does the liquid phase have a vapour pressure low enough at operating temperature?
- Is it thermally stable?

For instance, we quote a separability criterion of sample components known as "component sympathy" for the fixed phase.

Thus, a light n-paraffin mixture with "sympathy" for non-polar stationary phase will separate on grounds of different boiling points; olefins, being polarisable, will show "sympathy" towards a polar stationary phase. The cis- and trans- components of an olefin mixture will separate on a stationary phase consisting of a complex of transitional metals, dissolved in an appropriate solvent, such as polyethylene glycol.

Identifying mixture components can be accomplished using chromatographic etalons - pure substances or known mixtures of components. In these cases, universal detectors may be used (FID, WCD). For unknown mixtures, mass spectrometry is recommended. For the universal detector WCD, only non-fuel components may be used, and for the FID, only fuel components that do not interact chemically with the stationary phase may be used.

The quantitative analysis of a given component is based upon evaluating the chromatographic peak, which is triangle-shaped when columns with filling are used; its surface is measured and divided by the total surface, in different percentages for different types of detectors.

For capillary columns with good resolution, the signal takes the shape of straight lines and calculating the composition of the mixture is done in the order of succession, dividing each individual line by the total number of lines and using an adequate calibration curve, drawn upon determinations of known compounds.

The results can be expressed as a percentage, g/L component in a known mixture, depending on the type of calibration or the calibration units.

Thermal stability is of great importance, since operating under severe circumstances can lead to short analysis intervals and incomplete use of the chromatographic column; for instance, many stationary phases undergo degradation or decomposition at operating temperatures above 250^0 C, resulting in a lower operating life. On the other hand, certain components of unknown mixtures may reversibly or irreversibly poison the stationary phase components, deactivating it over time. In such cases, entry within the stationary phase can be accomplished using guarded columns that retain the poisonous components. It is vital that these components are known so as to prevent poisoning precious columns.

Before usage, the stationary phases placed into the GC columns must be conditioned and activated through streaming the mobile phase - the permanent gas.

Overall, stationary phases constitute a solvent with different selectivity for each component of the mixture in the analyzed sample.

2. Adsorption

Solid adsorbents as powdery material perform two functions in chromatography: the adsorbent itself (GS), in which case the separation takes place by adsorption, and support for stationary phase (non-volatile liquid), liquid stationary phase (GLS) coated on a granular material, in which case the separation takes place by absorption. Commonly, the size of the solid material used in chromatography is in the range of mesh (mesh - the number of openings per linear inch of a screen). Conversion between the Anglo Saxon and European expression (in mm) is presented in Table 1.

U.s. Mesh	Inches	Microns	Millimetres	U.s. Mesh	Inches	Microns	Millimetres
3	0.2650	6730	6.730	40	0.0165	400	0.400
4	0.1870	4760	4.760	45	0.0138	354	0.354
5	0.1570	4000	4.000	50	0.0117	297	0.297
6	0.1320	3360	3.360	60	0.0098	250	0.250
7	0.1110	2830	2.830	70	0.0083	210	0.210
8	0.0937	2380	2.380	80	0.0070	177	0.177
10	0.0787	2000	2.000	100	0.0059	149	0.149
12	0.0661	1680	1.680	120	0.0049	125	0.125
14	0.0555	1410	1.410	140	0.0041	105	0.105
16	0.0469	1190	1.190	170	0.0035	88	0.088
18	0.0394	1000	1.000	200	0.0029	74	0.074
20	0.0331	841	0.841	230	0.0024	63	0.063
25	0.0280	707	0.707	270	0.0021	53	0.053
30	0.0232	595	0.595	325	0.0017	44	0.044
35	0.0197	500	0.500	400	0.0015	37	0.037

Table 1. Mesh to millimetres conversion chart

Hereinafter we present the advantages of adsorption chromatography, which takes place at gas solid interaction, compared to GLS.

The advantages of gas chromatography gas-solid type from gas-liquid-solid can be summarized as follows: 1) stability of stationary phase on a wide range of temperatures; 2) a low detector noise limit; 3) lower values of HETP (height equivalent to a theoretical plate) than for gas-liquid chromatography (the adsorption-desorption process can be faster than the corresponding diffusion process in the liquid phase); 4) increased capacity of structural selectivity in separating geometrical isomers (e.g. using molecular sieves or graphitized carbon black); 5) high chemical selectivity when complexing agents are used as adsorbents; 6) high adsorption capacity allows the separation of gaseous or vapour compounds at room temperature; 7) increased chemical stability of adsorbents provides analysis of aggressive compounds; 8) important techniques have been developed in solids photochemistry and therefore there is better knowledge of adsorption; 9) numerous studies of heterogeneous catalysis were developed.

The limitations of this technique can be summarized as follows: 1) risk of asymmetric area due to the nonlinear adsorption isotherm of some analyzed compounds; 2) low reproducibility of chromatographic characteristics because adsorbents' properties are not as easily standardized as liquid substances; 3) more loss of the analyzed compounds as a result of irreversible adsorption or catalytic conversion of the separation process; 4) limited access to different commercial adsorbents for gas chromatography; 5) strong dependence of retention time of the sample size is common due to nonlinear adsorption, which combines the main advantage of gas-solid chromatography to the efficiency of capillary chromatography.

Adsorbents' selectivity is independent of the type of column used (packed or capillary column). Gas chromatography of adsorption has applications not only in the separation of gas mixtures with low boiling points, but also in the separation of hydrocarbon mixtures, as well as some organic compounds and some aromatic separation.

In the recent past there was a distinct tendency towards solid-gas chromatography. This can be attributed to two factors: first, new achievements in the field of adsorbents and second, using the improved logistics of gas chromatography (better trace analysis, more reproducible techniques for temperatures programming and high coupling techniques). Both solid and liquid stationary phases must have extraordinary properties, among which solid stationary phases are designed to solve many critical issues involved (Rotzsche, 1991). Adsorbents can be classified according to following criteria: chemical structure and geometric structure.

According to the chemical nature classification different interaction types of molecules' samples can take place. Kiselev & Yashin (1985) proposed grouping adsorbates into four groups (A-D), and in three groups (I-III) of adsorbents.

Adsorbents of type I have no functional groups or ions on the surface and thus they are not able to interact specifically with adsorbates. Interaction with all types of molecules A-D occurs non-specifically. Adsorbents are saturated hydrocarbons, graphite or rare gas crystals. The most important representatives of this type is graphitized carbon black (thermal graphitized carbon black) whose properties are close to ideal non-specific adsorbents. Similar to graphite are some inorganic compounds, such as graphite-like boron nitride (BN) or sulphides of metals (i.e. MoS) (Avgul et al., 1975).

Adsorbents of type II develop positive partial charge on the surface adsorbent. Besides these dispersion forces specific interactions develop leading to the orientation and localization of adsorbate molecules on centres having high charges. That concerns the salts having cations with positive charge with small ray, while negative charges are distributed in a relatively large volume (i.e. $BaSO_4$). Thus, the most significant representatives of this type are adsorbents with functional groups of protonated acids, such as hydroxylated silicagel, Lewis acids aprotic centres located on the surface. Molecules A type (saturated, rare gases) are non-specifically adsorbed and they are dispersion forces only. Molecules B, C and D can be adsorbed non-specifically. Type B includes molecules with electron density localized on some bonds or atoms: type II bonds of unsaturated or aromatic compounds, functional groups, atoms having electrons couple (ethers, ketones, tertiary amines, pyridines and nitriles), molecules with high quadrupole moments (N_2 molecules). Interactions between an adsorbent type II and an adsorbed type B develop between centres with electronic high-density (molecules in the sample) and positive charges of adsorbent (i.e. acidic proton of hydroxylated silica gel or a cation of Li, Na, Mg, Ca in acid zeolite or aprotic acids Lewis type on the surface). Type C molecules have a positive charge located on a metal atom and the excess of electron density is distributed on adjacent bonds (organometallic compounds). Type D are molecules containing peripheral functional groups (OH, NH, etc.) whose electronic density increase the density of some atoms (O, N) and reduce the density for other atoms (H). This group includes water, alcohols and primary and secondary amines. Specific interactions of type D adsorbates with Type II adsorbents involves the forces between the adsorbent centres with positive charges and the only couple of electrons of the atoms of O or N of the sample molecules.

Adsorbents type III. Adsorbents types III are specific having centres with high density centres on the surface. In this group are polymers as polyacrylonitrile, copolymers of vinylpyridine and divinylbenzene and polymers group with (C = O) and (–O–) on the surface. This group could include porous polymers based on styrene ethylvinylbenzene cross linked with divinylbenzene, varying by using different polymerization promoters and even non-specific dispersion forces. Adsorbents Type III includes crystalline surfaces formed by anions and chemically modified non-specific adsorbents covered with a monomolecular layer of adequate substance, creating negatively charged surface centres. Adsorbents type III interacts non-specific with adsorbates type A and specific with type B, C or D by forces from the negative charges of the adsorbent surface and from the positive charges of metallic atoms (C) or from functional groups (OH, NH) proton type (D) and dipole or induced dipole type (B).

Classification according to geometric structure

This classification concerns the possibilities to increase the surface. Increasing the surface meets a series of reserves, such as an increase in surface leads to increased dispersion with an increase in heterogeneity. Therefore, increasing contact points between particles will reduce the pore diameter with Knudsen diffusion disadvantage. Kiselev & Yashin (1985) have overcome these difficulties in the development of GCS. The adsorbents are geometrically classified as follows (Rotzsche, 1991).

Type 1 Non-porous adsorbents

Crystalline products with a smooth surface (sodium chloride, graphitized thermal carbon black, BN, MoS,). S_A in the range of 0.1 - 12 m^2/g.

Type 2 Uniformly porous adsorbents with wide pores. Silica gel with pore diameters between 10 and 200 nm (Porasil, Spherasil) and some other forms of silica gel chemical bonded (Durapak a.o.) and other polymers styrene divinylbenzene type with large pores (20 - 400 nm).

Type 3 includes adsorbents with a uniform pores system, but small pores. In this category should be mentioned molecular sieves (zeolites), carbon molecular sieves, porous glass and porous polymers. Pores diameter is around 10 nm.

Type 4 presents adsorbents with non-uniform pores. Among materials of this type there are activated carbon and alumina. Because of geometrical and chemical heterogeneity, having pores in the range 2 20 nm (mezopores) up to 200 nm, these adsorbents are not suitable for GCS. Classification is based on pores size. The difference between porous adsorbents and non-porous adsorbents consists in the free spaces form developed by the porous system. This system is quantitatively characterized by the following parameters (Rotzsche 1991):

- Specific surface area S_A (geometric size of the pore wall area/gram of adsorbent).
- Specific pore volume *V*, (total pore volume/gram of adsorbent).
- Mean pore diameter d_{So} (average diameter of 50% of the pores).
- Pore size distribution.

These parameters could be measured by gas chromatography and other methods, such are mercury porosimetry and reversed size exclusion chromatography. The ratio of pore diameters for porous adsorbents to the molecules diameters of the adsorbate is a significant parameter because the higher ratio is (the molecule diameters are smaller compared with pore diameters), the faster the adsorption equilibrium is reached. For a similar size of pore diameters and molecules diameters the adsorption rate depends equally on the pore shape and the adsorbate molecules' size. For narrow pores the adsorbed molecule on surface atoms of one pore could interact with other pore surface atoms and in this case the adsorbed molecules are trapped, and the transfer of molecules between adsorbent and mobile phase is stopped.

The most widely used chromatography supports are Kieselguhr (diatomaceous earth = light-coloured porous rock composed of the shells of diatoms) based, containing polysilicic acid in the hydrated amorphous silica form having a porous structure and containing varying amounts of metal oxides of Fe, Al, Mg, Ca, Na, K. These supports are known under different commercial terms, the most commonly used type being *Chromosorb*. They are found in several varieties, of which the most important are:

- *Chromosorb P* (pink) is reddish pink and has a surface area situated between 4 -6 m^2 / g. It has the advantage that it can be loaded with a large amount of liquid stationary phase (up to 25 30%) and has high mechanical strength. It is not sufficiently chemically inert and for this reason it is used mainly to separate non-polar substances.
- *Chromosorb W* (white) is white and has a smaller surface area than Chromosorb P, 1-2 m^2 / g. Therefore it can be loaded with a smaller amount of liquid stationary phase, up to 15%. It is chemically inert, but is friable, having a low mechanical strength. It has general usability, so it can be used to separate all classes of compounds, both polar and non-polar.
- *Chromosorb G* surface area is even smaller than Chromosorb W, of only 0.5 m^2 / g and can upload up to 5% stationary phase. But it has the advantage of combining the chemical inaction of type W with the mechanical strength of type P.

There are also supports of other types, but these are much less used. Such supports are based on synthetic polymers (i.e. Teflon), glass and others.

Next will be presented several examples of a few illustrative applications of some adsorbents (Poole, 2003) Table 2:

Stationary Phase		**Maximum Temperature (°C)**	**Usual Applications**
Alumina		200	Alkanes, alkenes, alkines, aromatic hydrocarbons - from C_1 to C_{10}
Silica gel		250	Hydrocarbons (C_1 - C_4), inorganic gases, volatile ethers, esther and ketone
Carbon		350	Inorganic gases, hydrocarbons (C_1-C_5)
Carbon molecular sieves		150	Oxygenated compounds (C_1- C_6)
Molecular sieves (5X and 13X)		350	Hydrogen, oxygen, nitrogen, methane, noble gases. Separation He/Ar and Ar/O_2. Hydrocarbons C_1 - C_3 on 3X, on 13X till C_{12} but no isomers. Cyclodextrine, halocarbons, permanent gases, hydrofluorocarbons, hydrocarbons C_1-C_{10}.
	Q	310	Hydrocarbons C_1-C_{10}, halocarbons C_1- C_2.
	S	250	Volatile organic solvents C_1 - C_6
	U	190	Nitrocompounds, nitrils, water, inorganic gases.

Table 2. Illustrative examples of some adsorbents and temperature values at which they are active

Table 3 briefly presents some gas chromatography main adsorbents' characteristics (Grob & Barry, 2004).

Adsorbent	**Polymeric material**	**Tmax, °C**	**Applications**
HayeSep A	DVB-EGDMA	165	Permanent gases, including: hydrogen, nitrogen, oxygen, argon, CO, and NO at ambient temperature; can separate C2 hydrocarbons, hydrogen sulphide and water at elevated temperatures
HayeSep B	DVB-PEI	190	C1 and C2 amines; trace amounts of ammonia and water

Adsorbent	Polymeric material	Tmax, °C	Applications
HayeSep C	ACN-DVB	250	Analysis of polar gases (HCN, ammonia, hydrogen sulphide) and water
HayeSep D	High purity DVB	290	Separation of CO and carbon dioxide from room air at ambient temperature; elutes acetylene before other C2 hydrocarbons; analyses of water and hydrogen sulphide
Porapak N	DVB-EVB-	190	Separation of ammonia, carbon EGDMA dioxide, water, and separation of 165 acetylene from other C2 hydrocarbons
Porapak P	Styrene-DVB	250	Separation of a wide variety of alcohols, glycols, and carbonyl analytes
Porapak Q	EVB-DVB copolymer	250	Most widely used; separation of hydrocarbons, organic analytes in water and oxides of nitrogen
Porapak R	Vinyl pyrollidone (PM)	250	Separation of normal and branched alcohols
Porapak T	EGDMA (PM)	190	Highest-polarity Porapak; offers greatest water retention; determination of formaldehyde in water
Chromosorb 101	Styrene-DVB	275	Separation of fatty acids, alcohols, glycols, esters, ketones, aldehydes, ethers and hydrocarbons.
Chromosorb 102	Styrene-DVB	250	Separation of volatile organics and permanent gases; no peak tailing for water and alcohols
Chromosorb 103	Cross-linked PS	275	Separation of basic compounds, such as amines and ammonia; useful for separation of amides, hydrazines, alcohols, aldehydes and ketones
Chromosorb 104	ACN-DVB	250	Nitriles, nitroparaffins, hydrogen sulphide, ammonia, sulphur dioxide, carbon dioxide, chloride, vinyl chloride, trace water content in solvents
Chromosorb 105	Crosslinked polyaromatic	250	Separation of aqueous solutions of formaldehyde, separation of acetylene from lower hydrocarbons and various classes of organics with boiling points up to 200° C

Key: DVB – divinylbenzene; EGDMA – ethylene glycol dimethacrylate; PEI – polyethyleneimine; ACN – acrylonitrile; EVB – ethylvinylbenzene.

Table 3. Polymer type adsorbents

3. Types of stationary phase for adsorption

The conditions that liquid stationary phases have to meet are:

- To be a good solvent for the components of the sample, but the solubility of these components has to be differentiated;
- To be practically non-volatile at the temperature of the column (vapour pressure to be less than 0.1 mm Hg);
- Be chemically inert;
- Have a higher thermal stability.

The categories of the stationary liquid phase are:

- *Non-polar stationary phases,* which are compounds, such as hydrocarbons (paraffin) or silicone oils (polysiloxanes) without grafted polar groups. Examples of such stationary phases are: squalane ($C_{30}H_{62}$ hydrocarbon), silicone oils metylsilicon type (names OV 1, SE-30). These phases separate the analyzed compounds in order of increasing boiling points.
- *Polar stationary phases,* which contain a high proportion of polar groups, i.e. the average molecular weight polyethylene glycol (Carbowax 20M), silicone oils with cianopropil groups (OV 225 cianopropyl-methyl-phenyl silicone) diethylene glycol succinate (DEGS), nitrotereftalic ester of polyethylene glycol (FFAP) etc. They differentiate between the non-polar and polar compounds, retaining only those which are polar, and are used especially to separate polar compounds.
- *Intermediate polarity stationary phase,* containing polar groups in the lower concentration or polarizable groups grafted onto a non-polar support. Examples of such stationary phases are phenyl methyl silicone phase (OV 17), dinonyl phthalate, polyethylene glycols having high molecular weight. They are universal stationary phases, which can be used to analyze both polar and non-polar compounds.
- *Specific stationary phases,* which are used in certain cases. They contain compounds that interact only with certain components of the mixture to be analyzed, for example $AgNO_3$ dissolved in polyethylene glycol which forms adduct with olefins.
- *Chiral stationary phases,* containing chiral compounds interacting with only one optical isomer of a pair of enantiomers. Such phases are based on cyclodextrin or certain amino acids.
- Given the polarity of the stationary phase and possible interactions, organic compounds can be grouped in terms of chromatographic separation in the following five classes:
- *Class I* very polar compounds, able to give hydrogen bonds: water, glycerin, glycols, hydroxyacids, and amino acids. These compounds are difficult to separate by gas chromatography, due to high polarity. Excepting water, they derivatized before separation.
- *Class II* polar compounds, which have active hydrogen atoms: alcohols, carboxylic acids, phenols, primary and secondary amines, nitro and nitriles with a hydrogen atom in α position. These compounds are separated in polar stationary phases.
- *Class III* intermediate polarity compounds, without active hydrogen: ethers, esters, aldehydes, ketones, nitro and nitriles without hydrogen atom in α position. These compounds are separated on stationary phases of intermediate polarity.
- *Class IV* compounds with low polarity, but which have active hydrogen: aromatic hydrocarbons, alkenes, chloroform, methylene chloride, dichloroethane, trichloroethane

etc. These compounds are separated on stationary phases of intermediate polarity or non-polar.

- *Class V* non-polar compounds: alkanes, cycloalkanes. These compounds separate in non-polar stationary phases.

This classification is empirical and is only meant to facilitate choosing the most suitable stationary phase for chromatographic analysis. Many polymers contain mixtures of the above functional groups, as indicated in Table 4.

Name	Type	Structure	Density g/ml	Viscosity cP	Average Molecular Weight
OV-1	Dimethysiloxane gum	CH_3	0,975		> 10^4
OV-101	Dimethylsiloxane fluid	CH_3		1500	30000
OV-7	Phenylmethyl dimethyl Siloxane	80 %CH_3 20%C_6H_5	1.021	500	30000
OV-210	Trifluoropropyl methylsiloxane	50%$CH_2CH_2CF_3$ 50%CH_3	1.284	10.000	200000
OV-275	Dicyanoalkyl siloxane			20.000	5000

Table 4. Properties of some commercially available pol-siloxane phases

4. Kovats retention index

A universal approach (Kovats) solved the problems concerning the use, comparison and characterization of gas chromatography retention data. This reporting of retention data as retention time t_R is absolutely meaningless, because all chromatographic parameters and any experimental fluctuation affect the proper measurement of a retention time. Using relative retention data ($\alpha = t'_{R2}/t'_{R1}$) provided improvements, but the default of a universal standard usable for a wide range of temperatures for stationary phases of different polarity has discouraged its use. In the Kovats approach, to retention index I of an alkane is assigned a value equal to 100 times its number of carbon atoms. Therefore, for example, values I of n-octane, n-decane and n-dodecanese, are equal to 800, 1000, 1200, respectively, by definition, and are applicable to any columns, packed or capillary and to any mobile phase, independent of chromatographic condition, including column temperature. However, for any component, the chromatographic conditions, such as stationary phase, its concentration, support and column temperature for packed columns, must be specified. Since retention index is the preferred method of reporting retention data for capillary columns, stationary phase, film thickness and column temperature must also be specified for any component out of n-alkanes, otherwise I values are meaningless. A value I of a component can be determined by mixing a mixture of alkanes with the desired component and chromatography under specified conditions. A plot of the logarithmic adjusted retention time versus retention index is generated and a retention index of the solute considered is determined by extrapolation, as shown in the figure below for isoamyl acetate.

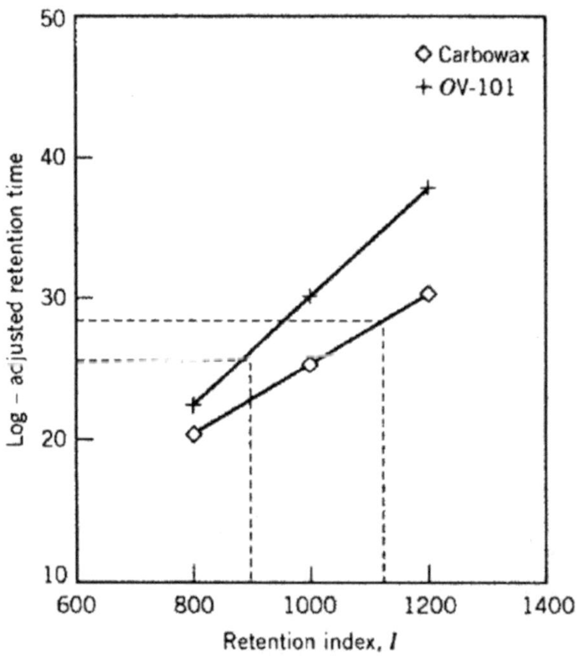

Fig. 1. Plot of logarithm-adjusted retention time versus Kovats retention index: isoamyl acetate at 120◦C (Grob & Barry, 2004).

Selectivity of stationary phases can be determined by comparing the values I of a solute on a non-polar phase, such as squalane OV 101 (I = 872) with corresponding I values of 1128 associated with more polar columns containing Carbowax 20M, for example. This difference of 256 units shows the superior retention of the Carbowax 20M column. Specifically, isoamyl acetate elutes between n-C11 and n-C12 on a Carbowax 20 M columns, but much faster on OV 101, where it elutes after n-octane. Retention indices in GC normalize instrumental variables, allowing retention data obtained by various systems to be compared. For example, isoamyl acetate with a retention index of 1128 will elute between C11 and C23 in the same chromatographic conditions. Retention indices are also very useful for comparing the relative elution order of the series of analytes on a specific column to a given temperature, and comparing the selective behaviour of two or more columns.

5. McReynolds classification of stationary phases

As shown, the most important criteria upon which stationary phases choice is made for a certain analysis is the polarity of the phase and of analyzed compounds. Numerical expression of this polarity, which would facilitate the choice, is not possible because there is no physical size that can be associated directly with the polarity (to some extent the dipole moment could be this size). For this reason, to express the stationary phase polarity they are using some reference compounds, which are arbitrarily chosen. Such a system was developed by Rohrschneider and later developed by McReynolds, based on the Kovats retention indices and using squalane as reference stationary phase, assigning it null polarity. A total of five reference compounds were selected: benzene, 1-butanol, 2-pentanona, nitropropane and pyridine. Each of these compounds can be considered standard for certain classes of substances, which are similar in terms of chromatographic behaviour, as follows:

- *benzene* for unsaturated hydrocarbons and aromatic hydrocarbons;
- *1-butanol* for alcohols, phenols, carboxylic acids;
- *2-pentanona* for aldehydes, ketones, ethers, esters;
- *nitropropane* for nitroderivates and nitriles;
- *pyridine* for aromatic bases and heterocycles with nitrogen.

The steps to determinate the constants McReynolds for a particular stationary phase A, for packed columns, are as follows:

- each reference compound is analyzed on a column with 20% squalane as stationary phase, isothermal at 100 ° C and the corresponding retention indices are determinate;
- reference compounds are analyzed on a column containing 20% stationary phase A, under the same conditions and in such cases Kovats retention indices are established;
- McReynolds constants of the stationary phase A are calculated as follows:
- for benzene (denoted by x '): the difference between the retention index of benzene on stationary phase A and on squalane:

$$x' = I_R(phase\ A) - I_R(squalane) \qquad (1)$$

- for a butanol (denoted by y '): the difference between the retention index on phase butanol on stationary phase A and on squalane;
- The same applies for 2-pentanona (z'), nitropropane (u') and pyridine (s').

These McReynolds constants values were determined for a great number of common stationary phases and they are tabulated. Some values are also given in Table 5.

Name of the stationary phase	**X′**	**Y′**	**Z′**	**U′**	**S′**
Squalane $C_{30}H_{62}$	0	0	0	0	0
Methylsilicone OV-1	16	55	44	64	42
Methylsilicone SE-30	15	53	44	64	41
Methyl-phenyl-silicone (20% phenyl) OV-7	69	113	111	171	128
Methyl-phenyl-silicone (50%phenyl) OV-17	119	158	162	243	202
Cyanopropyl-methyl-phenyl-silicone OV-225	228	369	338	492	386
Carbowax 20M (Polyethylenglycol)	322	536	368	572	510
Nitroterephtalic ester of PEG (FFAP)	340	580	397	602	627
Diethyleneglycol succinate (DEGS)	496	746	590	837	835

Table 5. McReynolds constants values for some usual stationary phases

Currently there are hundreds of stationary phases used in gas chromatography and McReynolds constants allow selecting the one that promises the best separation of the components analyzed. Because obviously there are a number of stationary phases that have similar McReynolds constant values, they can be substituted in between without affecting separation. Thus, if there is a stationary phase recommended in the literature for separation, but is unavailable in a laboratory, it will be replaced with an equivalent.

There was an impulse of consolidating the number of stationary phases used in the mid-1970s. Leary et al. (1973) reported the application of the statistical technique *the nearest neighbour* for the 226 of the stationary phases of McReynolds study and suggested that a total of only 12 phases could replace the 226. Later it has been found that four phases, OV-101, OV-17, OV-225 and Carbowax 20M could provide GC analysis, satisfying 80% of a wide variety of organic compounds, and a list of six favourite stationary phases on which almost all gas-liquid chromatographic analysis could be performed: (1) dimethylpolysiloxane (OV-101, SE-30, SP-210); (2) 50% phenyl-polysiloxane (OV-17, SP-2250); (3) poly-ethylene-glycol of molecular weight> 4000 (Carbowax); (4) diethyleneglycol succinate (DEGS); (5) 3- ciano-propyl-polysiloxane (Silar-10 C, SP-2340) and (6) tri-fluor-propyl-polysiloxane (OV-210, SP-2401).

Another quality of McReynolds constants is guiding the selection of columns to separate compounds with different functional groups, such as ketones by alcohols, ethers by olefins and esters by nitriles. If the analyzer wants a column to elute an ester after an alcohol, the stationary phase should have a value Z' greater than the value Y'. Also, a stationary phase should have a value of Y' greater than Z' in order to make ether elute before the alcohol.

6. Column type

Different chromatography types of column are presented by their diameters. Thus, there are three main types of column: packed columns, capillary columns and micro-packed columns.

Packed columns

The most used column materials are stainless steel, glass, quartz, nickel and polytetrafluorethylene (PTFE). Because of their good thermal conductivity and easy manipulation, stainless steel columns are widely applied. Glass columns are more inert. PTFE columns are used to separate mixtures of halogens and derivates. Classical packed columns' inner diameters range from **1.5** mm to **6** mm. Very important for packed columns is the load; the load is directly proportional to the column cross-section and hence to the square of the inner diameter (Leibnitz & Struppe, 1984).

Capillary columns (open tubular columns)

The terminology used for capillary column includes names such as WCOT (well coated open tubular) which are capillary columns where the stationary phase is deposited directly onto the inner surface of the wall not including substances that could be considered support. SCOT (support coated open tubular) includes those capillaries that have deposited on the inner surface a finely divided support (NaCl, BaCO3, SiO2, etc.) in which stationary phase is submitted. PLOT (porous layer open tubular) columns contain capillaries with stationary phase deposited on a porous support consisting of a finely divided support for packed gas chromatography columns (Celite, Chromosorb, etc.). Porous material is deposited in the capillary tubes from the very first moment of column conditioning. Capillary columns BP, DP, SPB (bond phase) include capillary tubes that have stationary phase chemically immobilized on the surface of the capillary inner wall. When choosing a certain type of capillary column it is necessary to take into account the nature of the sample being analyzed, the equipment available and the effective separation. Capillary columns WCOT have better efficiency in separation than SCOT because the inner column diameter is smaller. SCOT columns can be considered as subgroups of PLOT columns, since all SCOT are PLOT columns, but not vice versa (Ciucanu, 1990). For capillary columns the inner diameter ranges from 0.10 to 0.20 mm.

Wall-coated open-tubular (WCOT) columns contain the stationary phase as a film deposited on the internal surface of the tube wall. The formation of a thin and uniform film along the total length of the column represents the key to capillary columns' efficiency.

Porous-Layer Open Tubular (PLOT)

To increase the sample capacity of capillary columns and decrease the film thickness, porous layers for the inside walls of the column tubing were used. The stationary phase amount is directly related to the efficiency of capillary columns and generates efficiency increasing due to porous layers.

The method presented here used capillary columns of glass and quartz, the most used today. Glass capillaries are drawn from a device consisting of an electric furnace in which glass melts, two traction rollers which pull the column, a curved metallic tube for the spiral of the tube electrically heated at glass melting temperature. Capillary dimensions are set from the ratio of the two rolls' traction speed. Silica (quartz) tubes require temperatures of 1800-2000 ° C. After pulling, the capillary tube is covered with a polymer film in order to increase the tensile strength. Polyamide or polyimide polymers are used, which are stable up to temperatures of 350-400 ° C.

For the glass preparation, silica mixed with some metal oxides (Na_2O, CaO, MgO, B_2O_3, Al_2O_3 etc.) are used that break covalent bonds. In general, glass is chemically inert; however, because of impurities and the superficial structure, capillaries present catalytic activity and adsorption of the sample component. Metal oxides will act as Lewis acid. Molecules of analyzed compounds containing π electrons or non-participating electrons will react with Lewis acids. Adsorption and catalytic properties of glass and silica are due to silanol groups and siloxane bridges on the surface of the capillary tube. Also, water in the atmosphere adsorbs to the surface by hydrogen bonds. Heating the tube, water can be removed in order to form stable siloxane bridges. For a capillary tube to become a chromatographic column, the inner surface of the tube must be covered with a uniform and homogenous film of stationary phase.

By silanization of silanol groups, the critical surface tension (one that is established between the solid surface and stationary phase) is significantly reduced, but has the same value for glass and quartz. Silanizated surfaces have a surface tension so small that only non-polar stationary phases (OV-101) can "wet" them. Polar stationary phases (Carbowax 20M) do not give uniform films. The basic elements of glass capillary tubes' chemistry are alkaline metal ions, silanol groups and siloxane bridges. These are active centres that catalyze at high temperatures the decomposition reactions of sample components as well as of the stationary phase. For these reasons these active centres have to be deactivated.

Capillary tube surface deactivation involves removing the metal ions and blocking out the silanol groups with no catalytic activity groups. Removal of metal ions of silica and glass capillary tubes is achieved by acid washing in a static or dynamic regime. An acidic wash causes not only metal removing, but also increases the number of silanol groups, by breaking the siloxane bonds on the surface of the tube. In the past, tubes' surface deactivation was with surfactants, but they have low thermal stability and are no longer used. Current methods use disilazane, cyclic siloxanes and polysiloxanes. In all cases siloxane groups turn into silicate esters. Methods to deactivate with silazane realize the blocking of silanol groups by reaction with hexamethyldisilazane, di-fenil-mehyl-disalazane, dibutiltetramethyldisilazane, dihexyltetramethyldisilazane etc.

Reaction for deactivation with disilazane is:

$$Si - OH + NH(SiR_3)_2 \rightarrow Si - O - SiR_3 + NH_2 - SiR_3$$

$$Si - OH + NH_2 - SiR_3 \rightarrow Si - O - SiR_3 + NH_3$$

Uniform films of stationary phase can be obtained when the adhesion forces between stationary phase and capillary column surface are greater than the stationary phase molecules. Cohesion forces are stronger with increased similarity between organosiloxanic groups and functional groups of stationary phase. Methyl groups have high thermal stability, are chemically inert and are used to deactivate capillary columns with non-polar stationary phase. Phenyl groups have lower thermal stability than methyl groups, but allow the deposition of medium polarity stationary phase. Cyanopropyl and trifluoropropyl groups are strong polar and it is difficult to bond on the glass by reaction with disilazane, because disilazane have a reduced thermal stability. Another method of blocking the activity of silanol groups is based on reaction with cyclosiloxanes. Deactivation with cyclosiloxanes allows bonding some functional groups similar to those of the stationary phase.

Another method for modifying the surface chemistry of glass and silica capillary tubes is based on the stationary phase, by blocking the surface active centres, as well as getting a very stable film of stationary phase. Carbowax 20M was the first stationary phase used to deactivate glass capillary tubes. This step is followed by a heating treatment.

Another method of decomposition of the stationary phase is to use radicalic initiators. The initiator used for immobilization of polyethylene glycol is dicumilperoxide. Another possibility of immobilization of Carbowax 20 M is based on the reaction between OH groups of polyethylene glycol with diisocyanate in the presence of dibutyltindilauril. A chemically inert and thermal stability capillary column improves also by immobilizing the stationary phase polysiloxanes type with terminal hydroxyl groups OH, such as OV-17-OH, OV-31-OH, OV-240-OH.

Changing the surface chemistry by immobilizing the stationary phase can be done with a large number of siloxane polymers. A simple method of immobilization is polymerization in a column of silicon monomers consisting in the formation of bonds Si–O–Si between siloxane polymers and the capillary column wall. Immobilization can be done in the presence of radicalic initiators. Immobilization of the stationary phase can be initiated also by gamma radiation at room temperature. Stationary phases without vinyl groups require a high dose of radiation. Immobilization in the presence of gamma radiation eliminates the danger of degradation of the stationary phase and allows very precise control of reaction.

Micro-Packed Columns

Due to their parameters concerning mass capacity, phase ratio, resolution and short analysis time, micro-packed columns are very attractive. They are micro pore tubes having inner diameters ranging from 0.3 - 1 mm and at lengths varying from 1 to 15 m, packed with particles 0.007-0.3 mm in diameter. Table 6 presents geometrical characteristics for packed columns and micro-packed columns.

	Irregular micro-packed column	Classical packed columns	Regular micro-packed columns
Tube i.d.*(d,)* [mm]	0.3-0.5	≥ 2	0.3-1.5
Particle size *(d,)* mm	0.05-0.15	0.12-0.30	0.04-0.3
d_p/d_c	0.04-0.10	0.07-0.25	0.25-0.5
sample capacity µg	5-20	>1000	1-10
h (mm)	0.4-1.2	>0.5	0.15-0.40
L ,m	1.8 (6 ft)	4.5 (15 ft)	4.5 (15 ft)
Preferred applications	Short columns permit high-speed analyses coupled techniques with MS.	Simple separation problems; trace analysis; coupled techniques with spectrometry	Multi-component analysis; high-resolution coupled technique with MS; trace analysis

Table 6. Geometrical characteristics for packed columns and micro-packed columns

Micro-packed columns are suitable for the analysis of multi-component mixtures and trace constituents, especially when coupling with mass spectrometry, thus allowing quickly analyses.

7. Operation column

Several parameters can be used to evaluate the operation of a column and to obtain information about a specific system. An ideal gas chromatographic column is considered to have high resolving power, high speed of operation and high capacity.

Column Efficiency

Two methods are available for expressing the efficiency of a column in terms of HETP (height equivalent to a theoretical plate): measurement of the peak width (1) at the baseline, $N=16(V_R/w_b)^2$ and the peak width at half the peak height (2), $N=5.54(VR/w_k)^2$, where N is the height equivalent to theoretical plate (HETP) and w has the significance from the graphic below (Figure 2).

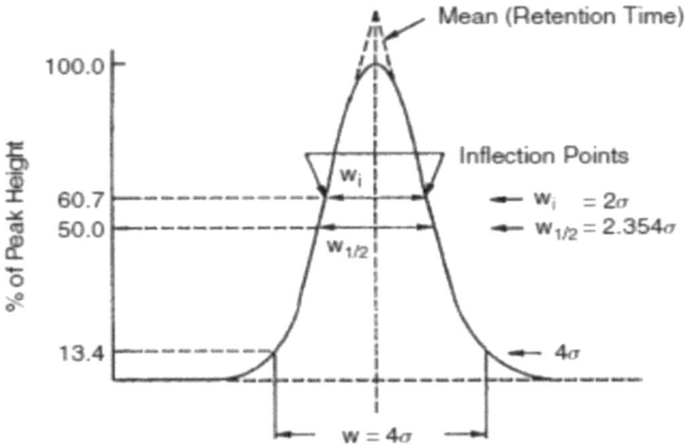

Fig. 2. Methods for expressing the efficiency of a column in terms of HETP

Effective Number of Theoretical Plates

Open-tubular columns generally have a larger number of theoretical plates. The *effective number of theoretical plates*, N_{eff}, characterizes open-tubular columns.

$$N_{eff}=16(V_R'/w_b)^2 \quad (2)$$

where V'_R the adjusted retention volume.

Separation Factor

The *separation factor* (S) describes the efficiency of open-tubular columns:

$$S= 16(V_R'/w_b)^2 =16(t'_R/w_b)^2 \quad (3)$$

where V'_R is the adjusted retention volume and t'_R is the adjusted retention time.

8. Preparation of columns

The most used technique for coating supports with high concentrations (>15%) of viscous phases is solvent evaporation. This technique leads to a uniform phase deposition. The steps to follow are presented below (Grob & Barry, 2004).

1. Prepare a solution of known concentration of liquid phase in the suitable solvent.
2. Add the desired amount of solid support in a known volume of the solution.
3. Transfer the mixture in a Büchner funnel, in order to remove the solvent excess.
4. Measure the volume of filtrate.
5. Dry the "wet" packing to remove residual solvent.
6. Calculate the mass of liquid phase retained on the support.

This technique allows obtaining a uniform coating of a support and minimizes the oxidation of the stationary phase during column.

9. The technology of capillary columns

The widespread use of fused silica for capillary columns is because it is inert compared to other glasses. In 1986 Jennings published a comparison of fused silica with other glasses, such as soda-lime, borosilicates and lead. Fused silica is formed by introducing pure silicon tetrachloride into a high-temperature flame followed by reaction with the water vapour generated in the combustion (Jennings, 1997). On the surface they have distinguished three types of silanol groups: adsorptive, strong, weak or none. The composition of some glass and silica are presented in Table 7.

Glass	SiO_2	Al_2O_3	Na_2O	K_2O	CaO	MgO	B_2O_3	PbO	BaO
Soda-lime	68	3	15	-	6	4	2	-	2
Borosilicate (Pyrex 7740)	81	2	4	-	-	-	13	-	-
Potash soda-lead (Corning 120)	56	2	4	9	-	-	-	29	-
Fused silica	100								

Table 7. Composition of some glass and silica (Jennings, 1986)

10. Column performances

Column performances need to take into account van Deemter expression, carrier gas choice, effect of carrier gas viscosity on linear velocity and phase ratio. Column efficiency, the resolution and sample capacity of a capillary column depend on construction material of the column, the inner diameter and the film thickness of the stationary phase. The distribution coefficient K_D could be interpreted as a function of chromatographic parameters. For a given temperature column of a tandem solute-stationary phase, K_D is constant. In this situation, K_D represents the ratio between the concentration of solute in stationary phase and the concentration of solute in carrier gas:

$$K_D = k\beta = r/2d_f \quad (3)$$

where $\beta = r/2d_f$ is the phase ratio, r is the radius of the column and d_f is the film thickness of stationary phase.

11. Selecting stationary phases

Using packed columns for gas chromatography needs to have a "collection" of available columns packed with different stationary phases and variable lengths for different purposes. The non-volatile liquid phases for packed columns have a good selectivity, but the thermal stability as well as the difference of thickness of the film at elevated temperatures is low and this will affect the inertness of stationary phase and the efficiency of the column.

For the best choice of a stationary phase it is important to remember the motto "like dissolves like". A polar phase could lead to a lower column efficiency compared to a non-polar phase. The maximum temperature limits decrease, as well as the operation lifetime if operating temperature is high (Grob & Barry, 2004). The lower thermal stability of polar phases can be avoided using a thinner film of stationary phase and a shorter column for lower elution temperatures.

12. Installing, conditioning, exhausting and regeneration of chromatographic columns

On column installing it is important to select the proper ferrule, meaning that the inner diameter of the ferrule fits to the outer diameter of the column, in order to avoid loss of carrier gas. Table 8 presents the size of ferrules suitable to column size (inner diameter).

Ferrules inner diameter, mm	Columns inner diameter, mm
0.4	0.25
0.5	0.32
0.8	0.53

Table 8. Ferrules inner diameter compatible with columns inner diameter

In gas chromatograph the chromatography column is placed in the oven, after the injector and before the detector. Between the injector and the column a *retention gap* or *guard column* is often installed. Its length is 0.5-5.0 m and it is actually a deactivated fused-silica tube. In this retention gap the condensed solvent resides after injection, but it is removed by vaporization. The role of a guard column is to collect the non-volatile compounds and particulate matter from samples in order to avoid their penetration in the chromatography column and maintain the lifetime of the column.

Conditioning of chromatography columns is recommended for residual volatiles' removal. For capillary column conditioning there are three steps to follow:

1. Carrier gas flow has to be constant when column temperature is higher than room temperature.
2. The maximum temperature of the stationary phase or of the column must not be transcended because the chromatography column could be damaged.
3. For capillary column, in order to obtain a steady baseline, it is indicated only to purge carrier gas flow for 30 min at ambient temperature and then, with a rate around 4°C/min, elevate the temperature slightly over the maximum temperature limit and maintain for several hours. The first two procedures are also applicable for conditioning of a packed column.

When non-volatiles and particulate matter are accumulating on the inlet of the column or on the injector liner, the chromatography column is contaminated and it is no longer working to its full potential. This is reflected mostly in peak tailing and changes in the retention characteristics of the column. Column reactivation could be realized either by removing the inlet part of the column (1-2 meters) or by turning the column around and then applying step 3 for a longer time. The extreme solution is to remove the solvent.

13. Practical application of GC analysis for mixtures of hydrocarbons

One of the fields where GC is increasingly used is in the oil and gas industry, and specially the refinery sector. In refineries the GC is used not only for identifying the components from a mixture, and then concentration, but also as an integrate tool for loop control.

The gas mixture that you may find on a refinery is usually called "refinery gas", and is a mixture of various gas streams produced in refinery processes. It can be used as a fuel gas, a final product, or a feedstock for further processing. An exact and fast analysis of the components is essential for optimizing refinery processes and controlling product quality. Refinery gas stream composition is very complex, typically containing hydrocarbons, permanent gases and sulphur.

In this practical example of GC use in a refinery, the hydrocarbon mixture was an industrial C5 cut separated from a Fluidized Catalytic Cracking (FCC) plant. Table 9 (Comanescu & Filotti, 2010).

The gas chromatograph was a HP 5890 Series II apparatus, with a configuration and operating parameters similar to those for PONA analysis (ASTM D 6293-98). Commercially available standard (Agilent) for refinery gas analysis was employed for calibration of both hydrocarbons retention times and concentrations.

Component	Normal boiling point, °C	Raw refinery C5 cut (approx. average) % mass
C4 hydrocarbons	- - -	1.88
isopentan	27.85	45.76
1-pentene (α-amylene)	29.97	5.37
n-pentane	36.07	8.36
isoprene	34.07	0.25
2-pentene (β-amylene)	36.07 (trans) ; 36.94 (cis)	12.58
piperylenes	42.0 (trans) ; 44.0 (cis)	0.65
isoamylenes	31.16 (2-methyl-1-butene) ; 38.57 (2-methyl-2-butene) ; 20.1 (3-methyl -1-butene)	7.26 12.06 1.34
cyclopentene	44.24	2.33
C5+ hydrocarbons	- - -	2.16
Specific weight (d_{15}^{15})	- - -	6.652

Table 9. Composition of C5 fractions and boiling points of main C5 hydrocarbons

14. Analysis of natural products

Gas chromatography coupled with MS represents a fast and cheap method for the separation and identification of compounds from natural product mixtures. Alkylation of animal fats and vegetable oils for biofuel fabrication is one of the industrial processes that needs analysis of natural products, both for raw materials and for products (Christie, 1989).

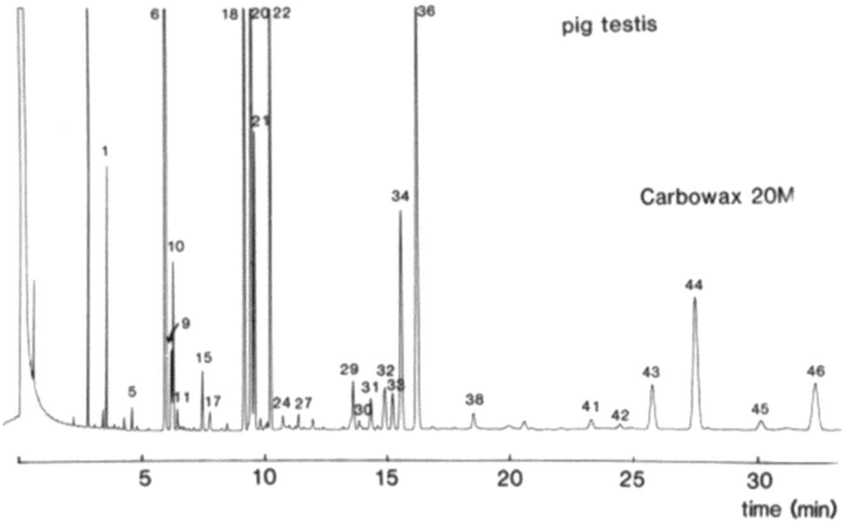

Fig. 3. Separation of the fatty acids (methyl esters) of pig testis on a fused silica column (25 m x 0.22 mm) coated with Carbowax 20M™ (Chrompak UK).

The separation of the methyl esters of the fatty acids of pig testis lipids on the Carbowax 20M™ column is illustrated in Figure 3. A gas chromatograph Carlo Erba Model 4130 split/splitless injection system was used. The carrier gas was hydrogen (1 ml/min), programme temperature was 165ºC for 3 min and then temperature was raised at a rate of 4ºC/min to 195ºC and maintained for 23 min.

The three 16:1 isomers are well separated and distinct from the C_{17} fatty acids. The C_{18} components are separated from a minor 19:1 fatty acid and C_{20} unsaturated constituents. The 20:3(n-3), which co-chromatographs with 20:4(n-6) are just separable on a slightly more polar Silar 5CP™ column. C_{22}, important fatty acids are cleanly separated (Christie, 1989). Peaks can be identified in Table 10 (Christie, 1989).

Equivalent chain-lengths of the methyl ester derivatives of some fatty acids					
No	Fatty acid	Silicone	Carbowax	Silar 5CP	CP-Sil 84
1.	14.00	14.00	14.00	14.00	14.00
2	14-isobr	14.64	14.52	14.52	14.51
3	14-anteiso	14.71	14.68	14.68	14.70
4	14:1(n-5)	13.88	14.37	14.49	14.72
5	15:0	15.00	15.00	15.00	15.0
6	16:0	16.00	16.00	16.00	16.00
7	16-isobr	16.65	16.51	16.51	16.50
8	16-anteiso	16.73	16.68	16.68	16.69
9	16:1(n-9)	15.76	16.18	16.30	16.48
10	16:1(n-7)	15.83	16.25	16.38	16.60
11	16:1(n-5)	15.92	16.37	16.48	16.70
12	16:2(n-4)	15.83	16.78	16.98	17.47
13	16:3(n-3)	15.69	17.09	17.31	18.06
14	16:4(n-3)	15.64	17.62	17.77	18.82
15	17:0	17.00	17.00	17.00	17.00
16	17:1(n-9)	16.76	17.20	17.33	17.50
17	17:1(n-8)	16.75	17.19	17.33	17.51
18	18:0	18.00	18.00	18.00	18.00
19	18:1(n-11)	17.72	18.14	18.24	18.40
20	18:1(n-9)	17.73	18.16	18.30	18.47
21	18:1(n-7)	17.78	18.23	18.36	18.54
22	18:2(n-6)	17.65	18.58	18.80	19.20
23	18:2(n-4)	17.81	18.79	18.98	19.41
24	18:3(n-6)	17.49	18.85	19.30	19.72
25	18:3(n-3)	17.72	19.18	19.41	20.07
26	18:4(n-3)	17.55	19.45	19.68	20.59
27	19:1(n-8)	18.74	19.18	19.32	19.47
28	20:1(n-11)	19.67	20.08	20.22	20.35
29	20:1(n-9)	19.71	20.14	20.27	20.41
30	20:1(n-7)	19.77	20.22	20.36	20.50

Equivalent chain-lengths of the methyl ester derivatives of some fatty acids					
31	20:2(n-9)	19.51	20.38	20.58	20.92
32	20:2(n-6)	19.64	20.56	20.78	21.12
33	20:3(n-9)	19.24	20.66	20.92	21.43
34	20:3(n-6)	19.43	20.78	21.05	21.61
35	20:3(n-3)	19.71	20.95	21.22	21.97
36	20:4(n-6)	19.23	20.96	21.19	21.94
37	20:4(n-3)	19.47	21.37	21.64	22.45
38	20:5(n-3)	19.27	21.55	21.80	22.80
39	22:1(n-11)	21.61	22.04	22.16	22.30
40	22:1(n-9)	21.66	22.11	22.23	22.36
42	22:3(n-6)	21.40	22.71	22.99	23.47
43	22:4(n-6)	21.14	22.90	23.21	23.90
44	22:5(n-6)	20.99	23.15	23.35	24.19
45	22:5(n-3)	21.18	23.50	23.92	24.75
46	22:6(n-3)	21.04	23.74	24.07	25.07

*This document is part of the book "*Gas Chromatography and Lipids*" by William W. Christie and published in 1989 by P.J. Barnes & Associates (The Oily Press Ltd), who retain the copyright

Table 10. Equivalent chain-lengths of the methyl ester derivatives of some natural fatty acids*

15. Conclusions

1. Gas chromatography is an old and still current analytical method used in research and applied fields.
2. Separation of components in the sample is based on the difference of partition (adsorption or absorption) between a mobile phase in continuous motion, a permanent gas and a fixed phase called stationary phase, solid powder (separation by adsorption) or non-volatile liquid for separation by absorption.
3. Stationary phase is deposited in chromatography column, which is the heart of the gas chromatography apparatus.
4. Chromatography columns can be made of different materials: stainless steels, copper alloy, nickel, aluminium, plastics, glass, silica etc.
5. Chromatography columns can be packed, micro-packed, capillary, with differences in diameter, length, thickness of the packed-bed and, ultimately, with different column performance.
6. The materials most used for columns are glasses and silica, especially for capillary columns.
7. Using capillary columns have led to a considerable enlargement of the column length, shortening analysis time, increasing separation performance, the finding of stationary phases with improved performances both in terms of separability as well temperature resistance.
8. Efforts have increased to find the best stationary phase, allowing a good separation of geometric and optical isomers, for example polysiloxanes, and mixtures of different types and proportions with improved thermal resistance.

16. References

Avgul, N. N., Kiselev, A. V. & Poshkus, D. P. (1975). *Adsorbtsiya Gazovi Parovna Adnorodnikh Poverchnostyakh*, Khimiya, Moscow

Ciucanu, I. (1990). *Cromatografia de gaze cu coloane capilare*, Editura Academiei Romane, Bucharest

Christie, W. W. (1989). *Gas Chromatography and Lipids. A Practical Guide*, The Oily Press, Dundee, Scotland. URL: *www.lipidlibrary.aocs.org*

Comanescu, I. & Filotti, L. (2010). Vapour-liquid Equilibrium Data for C5 Saturated-Unsaturated Hydrocarbons Solutions with Solvents Monopropylene Glycol, Dipropylene Glycol or N-Methyl-2-Pyrrolidone, *Revista de Chimie* Vol.61: (10): 986-991

Grob, R. L. & Barry, E. F. (2004). *Modern Practice of Gas Chromatography*, Wiley-Interscience, New-York Jennings, W. G. (1986). *Comparisons of Fused Silica and Other Glasses Columns in Gas Chromatography*, Alfred Heuthig, Publishers, Heidelberg, Germany, 12-21

Jennings, W. G. (1997). *Analytical Gas Chromatography*, 2nd ed., Academic Press, San Diego, CA, 36-45

Kiselev, A. K. & Yashin, Ya. I. (1985). *Gas- and Flussigkeits-Adsorptionchromatographie*,VEB Deutscher Verlag der Wissenschaft, Berlin

Leary, J. J., Justice, J. B., Tsuge, S., Lowry, S. R. & Isenhour, T. L. (1973). Correlating gas chromatographic liquid phases by means of nearest neighbor technique: a proposed set of twelve preferred phases, *Journal of Chromatography Science* Vol.(11): 201-206

Leibnitz, E. & Struppe, H. G. (1984). Ed. *Handbuch der Gas-Chromatographie Vol. 2*, Akademische Verlagesellschaft Geest & Portig, Leipzig

Poole, C. F. (2003). *The Essence of Cromatography*, Elsevier, Amsterdam

Rotzsche, H. (1991). *Stationary Phases in Gas Chromatography*, Elsevier, Amsterdam

Design, Modeling, Microfabrication and Characterization of the Micro Gas Chromatography Columns

J.H. Sun, D.F. Cui, H.Y. Cai,
X. Chen, L.L. Zhang and H. Li
State Key Laboratory of Transducer Technology, Institute of Electronics, Chinese Academy of Sciences, Beijing, China

1. Introduction

Gas chromatography (GC) systems can separate different components of gaseous mixtures, and are important analytical tools for a variety of disciplines, including environmental analysis, methane gas probes, and homeland security and pollution monitoring. However, most modern GC devices suffered the problems of low detection speed, sensitivity, and poor stability. In addition, these conventional GC devices were bulky and fragile, which ruled out the possibility of the in-field use. In some cases, gas samples were collected in the field and analyzed in the laboratory using a conventional GC, which was inconvenient and inefficient.

Combined with micro-detector (such as micro thermal conductivity detector, micro photoionization detector, etc.), Micro GC columns can be developed into a miniaturized chromatographic system because of its very small size. This integrated GC system, with a small size, light weight, rapid analysis, high sensitivity, easy to use, etc., can be widely applied to environmental pollution, home safety, pesticide residues, food safety, pre-diagnosis of cancer and other areas for achieving on-site and on-line rapid testing.

In this chapter, several micro gas chromatography columns were designed for building micro μGC systems, and this chapter will contain the following sections: Structural design, modeling analysis, microfabrication and characterization of the micro gas chromatography columns.

2. Structural consideration

Structure of the μGC columns can be designed in accordance with the requirements of the designer. Several various structural designs are appeared in research papers, but their differences in-depth analysis of these designs are almost absence. In fact, structure is an important factor for affecting the separation performance of the μGC columns, because the shape of the column, especially the mutations sections (such as the corner of the channel),will change the airflow velocity, pressure distribution, the thickness of the stationary phase film and other important factors, moreover, these factors are key factors for affecting the separation performance of μGC columns.

In this work, two frequently-used configurations (the spiral channel and the serpentine channel, as shown in Fig.1) of columns were designed for GC analysis, the shape of cross-section is rectangular, and their sizes are consistent. Then, the effect of airflow rate and pressure distribution in the channel was simulated using ANSYS analysis. Especially, the effect of these factors in the corners was the focus.

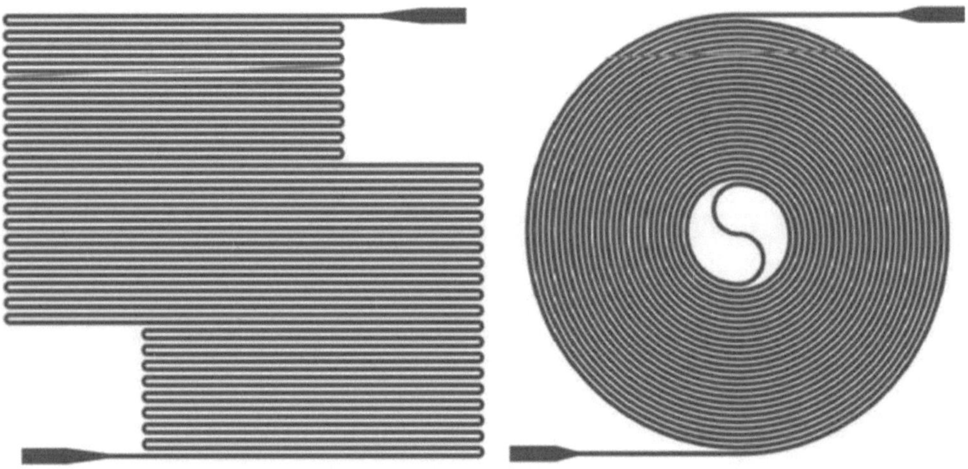

(a) (b)

Fig. 1. The structure design, (a) the serpentine channel, (b) the spiral channel, for μGC column.

3. Theoretical consideration and simulation

3.1 Theoretical consideration

If the volume and the concentration of a sample are small enough (about μmol) and in the linear range of adsorption isotherm, the elution curve equation can be determined by equation (1) based on the plate theory.

$$C = \frac{\sqrt{n}}{\sqrt{2\pi}} \bullet \frac{m}{V_r} \bullet \exp[-\frac{n}{2}(1-\frac{V}{V_r})^2] \tag{1}$$

Where C is the sample concentration at any point of the elution curve, m is the weight of the solute, V is the retention volume at any point of the elution curve, V_r is the retention volume of the solute, n is the number of theoretical plates. When $V = V_r$, C reaches its max value:

$$C_{\max} = \frac{\sqrt{n} \bullet m}{\sqrt{2\pi} \bullet V_r} \tag{2}$$

Derived from the elution curve equation, the number of theoretical plates n can be defined by the following equation:

$$n = 5.54(\frac{t_r}{w_{1/2}})^2 \quad (3)$$

Where t_r is the retention time, and $w_{1/2}$ is the width of the peak at half height.

And the theoretical plate height (H) can be determined by:

$$H = \frac{L}{n} \quad (4)$$

Where L is the length of the column. According to the equation (3) and (4), as the chromatography peak's bottom width (w) decreases, the number of theoretical plates (n) increases and the theoretical plate heigh (H) decreases accordingly, which results in higher column efficiency. Hence, n and H are the index of column efficiency.

Due to some effect factors for H can't be shown in equation (4), the theoretical plate height (H) can also be given by the following formula:

$$H = \frac{2D_g}{\bar{u}} f_1 f_2 + \frac{(1+9k+25.5k^2)}{105(k+1)^2} \frac{w^2 \bar{u}}{D_g} \frac{f_1}{f_2} + \frac{2}{3} \frac{k}{(k+1)^2} \frac{(w+h)^2 d_f^2}{D_s h^2} \bar{u} \quad (5)$$

where D_g and D_s are the binary diffusion coefficients in the mobile and stationary phases, respectively, d_f is the thickness of the stationary phase, w and h are the channel width and height, respectively, k is the retention factor, and f_1 and f_2 are the Giddings-Golay and Martin-James gas compression coefficients, respectively. The average linear airflow velocity is given by

$$\bar{u} = \frac{w^2 p_0 (P^2 - 1)}{24 \eta L} f_2 \quad (6)$$

Where P_0 is the outlet pressure, P is the ratio of the inlet to outlet pressure, L is the column length, and η is the carrier gas viscosity.

From the equation (5) and (6), the airflow rate and pressure are key factors for the separation performance of the μGC columns.

Resolution is called overall separation efficiency, which is defined as the difference of retention time between two adjacent chromatography peaks divided by the half of the sum of these two peak's bottom width:

$$R = \frac{t_{r_2} - t_{r_1}}{\frac{1}{2}(w_1 + w_2)} = \frac{2(t_{r_2} - t_{r_1})}{w_1 + w_2} \quad (7)$$

The definition of resolution (R) in equ (7) does not reflect all the factors which influence resolution, because resolution is actually determined by column efficiency (n), selectivity factor (α) and capacity factor (k), hence the resolution also can be described by equation (8):

$$R = \frac{\sqrt{n}}{4}\left(\frac{\alpha - 1}{\alpha}\right)\left(\frac{k}{1+k}\right) \tag{8}$$

After the stationary phase is chosen, the selectivity factor could be correspondingly fixed, which means that the resolution is only affected by n. For a column with a certain theoretical plate height, square of the resolution is proportional to the length of column:

$$(\frac{R_1}{R_2})^2 = \frac{n_1}{n_2} = \frac{L_1}{L_2} \tag{9}$$

Therefore the major approach to improve resolution is to increase the column length.

3.2 Simulation

Base on chromatography theory, the airflow rate and pressure are key factors for the separation performance of the μGC columns. An obviously different variation of the airflow velocity, pressure distribution and other factors will be appeared in these μGC columns due to their different structural, and these variations will affect the separation performance of μGC columns. So a detailed comparative analysis for the two frequently-used configurations of columns is made in this paper, which provides a theoretical basis for designing ideal μGC columns.

In this chapter, the effects on the airflow rate and pressure were simulated using ANSYS analysis. Because the major effect of airflow rate is come from the mutation section (such as the corner of GC-channel) of these channels, the corner of the chromatographic channel are selected for the simulation. In the simulation, the first step is modelling. The parameters of the modelling are consistent with the size of the actual channel (The cross-section is rectangular, the width of the channel is 150 μm, and the depth of the channel is 100 μm).

After modelling, boundary conditions were set as follows: the airflow rate in the gas inlet was set to 18 cm/s, the pressure of the gas outlet was set to 0, the displacement of the other part was set to 0).

Fig.2 (a) and (b) show the solutions of the simulation, the airflow rate in the corner of serpentine channel is 15.73 cm/s, the difference of airflow rate between in the corners of serpentine channel and in the gas inlet reaches 3.13 cm/s. However, the airflow rate in the mutation section of the spiral channel is 16.93 cm/s, the difference of airflow rate between in the corners of spiral channel and in the gas inlet is only 1.33 cm/s. Moreover, the changes in distribution of airflow rate in the corners of the serpentine channel are obvious and relatively large. But the distribution of airflow rate in the corners of the spiral channel is relatively uniform. Fig.3 (a) and (b) show the solutions of the pressure distribution in the serpentine channel and spiral channel, the simulation results show that there exists a pressure gradient in these two kinds of channel, but the change rate of pressure in serpentine channel is obviously much larger than that of the spiral channel, the maximum pressure value in spiral channel is only 32.80 Pa, however, the maximum pressure value in the serpentine channel is over 150.0 Pa, the value is close to 5 times compared to the former. Seen from the above analysis, the impact on airflow rate and pressure in serpentine channel is very significant, and these effects will lead to deterioration of separation performance. Because the difference of the airflow rate in the channel would change the thickness of the stationary phase film. Consequently, the variation of the thickness of the stationary phase film in the channel would lead bad tailing peaks.

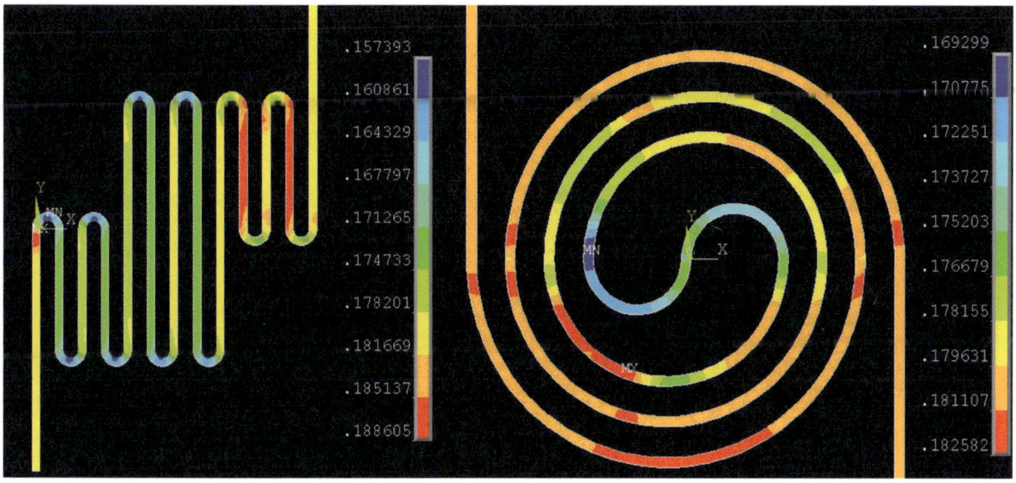

(a) (b)

Fig. 2. The effect of the airflow rate in the corners of the channel (a)the serpentine channel,(b) the spiral channel, were simulated using ANSYS analysis.

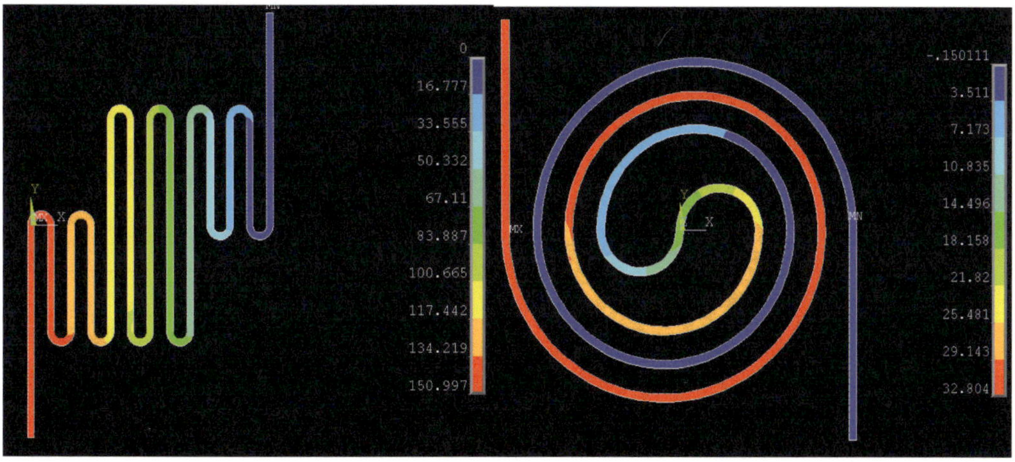

(a) (b)

Fig. 3. The pressure distribution in the GC channel (a)the serpentine channel,(b) the spiral channel, were simulated using ANSYS analysis.

4. Microfabrication

4.1 Column fabrication

Fabrication of the μGC column includes aluminium deposition, photolithography and deep reactive-ion etching (DRIE) (the fabrication process is illustrated in Fig.4). Firstly, a 2-μm-thick electron-beam evaporation aluminium film was deposited on a p-type <100> silicon wafer which served as the etch mask in following steps. Secondly, a thickness of approximately 2 μm AZ1500 photoresist was coated on the wafer and patterned as an etch mask for aluminium. Subsequently, aluminium without the protection of photoresist was etched away by an etchant (H_3PO_4) and the silicon surface was exposed. Then, a DRIE process, instead of the anisotropically KOH chemical etching process, was utilized to form the rectangular micro channels. Finally, all aluminium masks were removed and the silicon wafer and pyrex7740 glass were held together and heated to approximately 400℃. A 1000 V potential was then applied between the glass and the silicon, and the resulting electrostatic force pull the wafers into intimate contact.

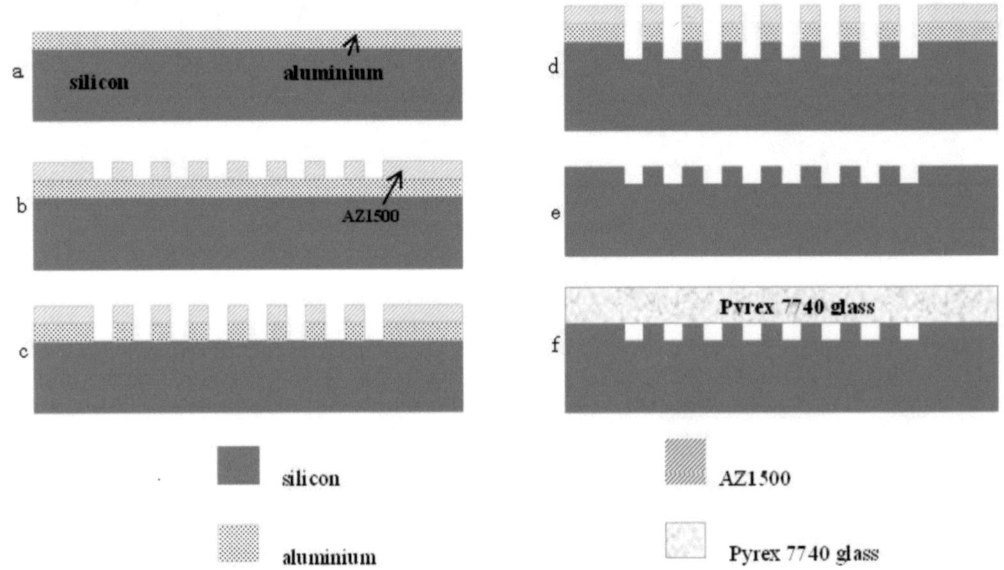

Fig. 4. The GC column process flow.

4.2 The stationary phase coating

The uniformity, stability and thickness of the stationary phase film are important factors that can affect the separation efficiency. In this section, the selection principle of stationary phase, coating methods and coating procedure are introduced.

There is no regularity to follow for choosing the stationary phase. Generally, "Like dissolves like" principle has been identified as the basic theory for selecting the stationary phase. In the application, the selection of stationary phase should be determined according to the actual situation.

i. Separation of non-polar compounds: Generally, non-polar stationary phase is used. Each component flows over the GC column according to the order of boiling point. The component with a low boiling point is the first one out of the GC column, and followed by the component with a high boiling point.
ii. Separation of polar compounds: Generally, polar stationary phase is used. The component with a small polar is the first one out of the GC column, and followed by the component with a relatively large polar.
iii. Separating non-polar and polar compounds: Generally, polar stationary phase is used. The non-polar component is the first one out of the GC column, and followed by the polar component.
iv. Separation of complex and difficult compounds: Two or more mixed stationary phase can be used.

Stationary phase coating methods are generally including static and dynamic coating. Static coating procedure is defined as: The stationary phase is filled with the GC column, one end is sealed, and the other end is connected a vacuum pump, the solvent is slowly evaporated under the pressure of vacuum pump, until all of the solvent is completely evaporated, and a stationary phase film with a thickness of 0.1-0.2 microns is left on the channel.

Dynamic coating procedure is defined as: A stationary phase solvent is injected into the GC column, and the stationary phase solvent flows through the GC column under pressure, the thickness of Stationary phase can be controlled by changing the flow rate and the concentration of the stationary phase solvent. The stationary phase solvent is pushed out from the other side of the GC column, and the nitrogen gas was delivered through the column for several hours to completely evaporate the solvent, and a stationary phase film with a thickness of 0.1-0.2 microns is left on the channel.

In this work, separation of benzene and homologue of benzene was taken as for example. In order to completely separate the mixture, OV-1 or OV-101 was the optimal stationary phase. So OV-1 was selected to coat the micro GC column using a dynamic coating procedure. The process was shown as following: Firstly, about 5 μl OV-1 was dissolved in 2.0 ml mixtures of n-pentane and dichloromethane with a volume ratio of 1:1. The mixture was agitated for 30 minutes to ensure full dissolution. Secondly, the inlet of the column was connected with a capillary, and the outlet was connected with a laboratory made micro-pump, which was used to inject the stationary phase solution into the GC column. After the GC column was full of the solution, the micro-pump was turned off for 30 minutes to make sure that it was long enough for the stationary phase to attach to the channel wall. Then, nitrogen gas was delivered through the column for a few hours to completely evaporate the n-Pentane and dichloromethane. Subsequently, the column was put into an oven under a nitrogen atmosphere in which the temperature of the oven was firstly increased gradually by $5\,^{\circ}C\,/\min$ until $100\,^{\circ}C$ and then the temperature of the oven was kept at $100\,^{\circ}C$ for 4 hours. Fig. 5 shows an SEM view of the stationary phase film coated on the column wall, in which the film is uniform and adheres well to the channel. The composition analysis of the film (the selected area of the film on the channel wall can be seen from Fig. 6(a)), by X- Ray Photoelectron Spectroscopy, is described in the Fig. 6(b) which shows that the composition agrees with the composition of OV-1.

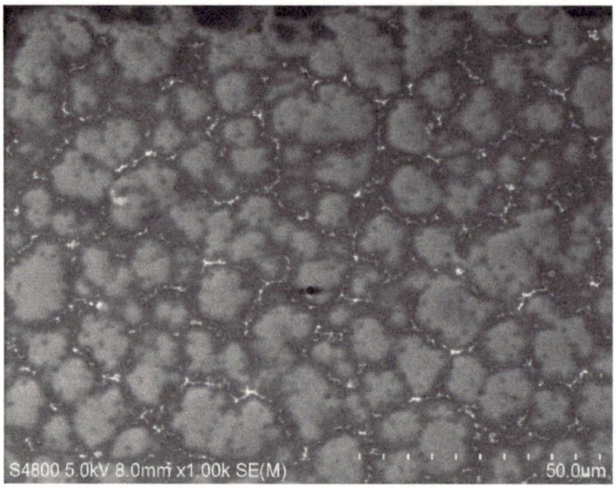

Fig. 5. The SEM view of the stationary phase film.

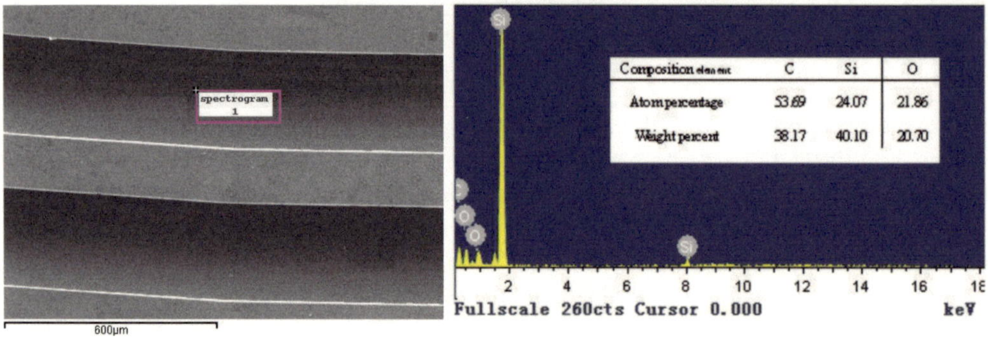

(a) (b)

Fig. 6. The composition analysis of the film which operated by X- Ray Photoelectron Spectroscopy, (a) the selected area of the film on the channel wall, (b) the percent of the composition element in the film.

4.3 Interface technology

The inlet/outlet interface technology of MEMS-based GC columns is a key factor. If not properly tackled, some problems, such as instability of the interface, air leakage from the interface and the airflow jamming would happen, which usually led to the failure of the GC columns. In order to solve these problems, a fixed base (the base dimensions is 8.0 mm long, 4.0 mm wide and 5.0 mm high, See Fig.7) was used to connect the tubes with the inlet /outlet of the column. The major steps are briefly covered: Firstly, a film of heat-resistant adhesive was coated on the bottom surface of the base, and then the base was aligned with the inlet or the outlet carefully and bonded with the column; After 30 minutes, more heat-resistant adhesive was coated around the base with a steel tube (Tubing, steel, $1/32''\times 0.25mm\ ID$) into the base, and then some heat-resistant adhesives were coated around the joint points. Subsequently, a burst pressure test with 0.4 MPa pressure was applied on the column and the interface remained undamaged. Fig. 8 shows a photograph of the completed GC columns including the connected base.

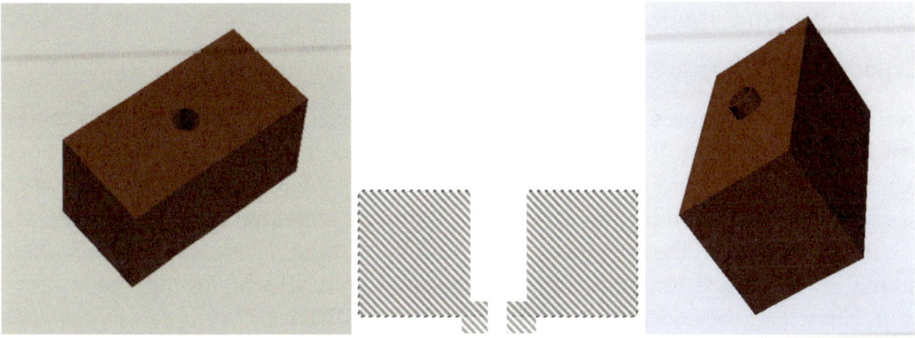

Fig. 7. The picture of the fixed base.

(a) (b)

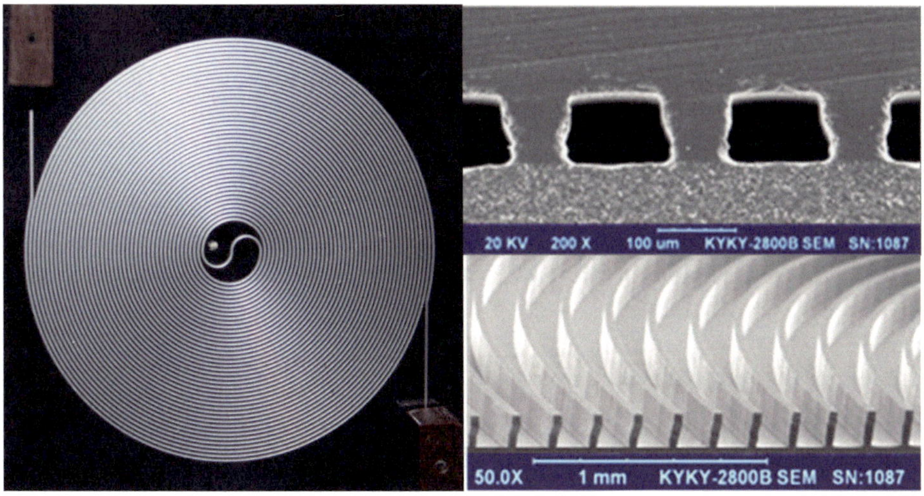

(c) (d)

Fig. 8. The photograph of the completed GC columns, (a) 0.75m serpentine channel column, (b) 0.75m spiral channel column, and the chips dimension are all 2.2cm×2.2cm, (c) 3 m spiral channel column and the chip dimension is 3.5cm × 3.5, (d) SEM micrograph of the section and the channel.

5. Results and discussions

5.1 Experimental setup and test condition

GC performances were measured through the flame ionization detector (FID) method on an Agilent 7890A GC system. In the test, the carrier gas was helium and the original operating temperature of columns was 40℃, in order to reduce the total analysis time of the component mix, temperature programming, which the temperature is increased from 40°C to 80°C at a rate of 5°C /min, was used in the process of separating the sample. These samples were dissolved in 1.0 ml CS_2, and the volume of each component sample, which was injected at time zero, was 1 μl. The top-air method was used with a split ratio of 1:10.

5.2 Experimental results

In this section, the effect of the structure the flow rate and the length for separation performance were analyzed, and then the experiments for separating complex mixtures were studied.

5.2.1 Effect of the structure for separation performance

These columns were operated at the velocity of 18 cm/s, the sample is a 3-component mixture including benzene, toluene, o-xylene, and the chromatogram graphs were shown in Fig.9 (a) and (b). In Fig. 9 (a), the serpentine channel GC columns separated the gaseous mixture with a resolution of 8.9 (calculated by equation (7)) between toluene and o-xylene and yielded about 1900 theoretical plates (calculated by equation (3)). However, the chromatographic peak emerged bad tailing peaks. The inconsistent of the film thickness were the primary cause for the tailing peaks. However, the spiral channel GC columns yielded about 3900 theoretical plates and separated the gaseous mixture with the resolution of 10.97 (as shown in Fig. 9 (b)). Moreover, the chromatographic peaks were greatly improved. So the structure of GC channel, especially the design of the corner, was a key factor for the separation performance. The spiral channel possessed streamlined shape. As a result, the airflow couldn't be baffled, and the separation experiment also verified that the spiral channel GC column showed excellent overall separation performance and separation efficiency for the gas mixture. Thus, the spiral channel column is superior to the serpentine channel column for GC analysis. Therefore, on this basis, we designed a 3 m spiral column for GC application, then, the effect of the flow rate and the length for separation performance was analyzed and its separation performance for complex mixture was studied.

5.2.2 Effect of the flow rates for separation performance

Because the flow rate was a key factor for separation performance, different flow rates were used in the process of separating the sample (the sample is a 3-component mixture including benzene, toluene, o-xylene). Firstly, the 3 m spiral column was operated at the velocity of 42 cm/s, and the chromatogram graph was shown in Fig.10 (a). From experimental data, the number of plates can be calculated as given in equation (3). The 3 m spiral column yielded 3410 plates. The resolution, R, can be calculated by equation (7).The resolution between benzene and toluene was 4.05, and the resolution between toluene and o-xylene was 9.27. While, if the flow rate was appropriately reduced and the column was operated at the velocity of 18 cm/s, the column yielded approximately 6160 plates (the experimental data was shown in Fig.10 (b)), moreover, the resolution between benzene and toluene was improved from 4.05 to 8.02, and the resolution between toluene and o-xylene was improved from 9.27 to 14.3. The separation efficiency of the gas mixture was strongly improved.

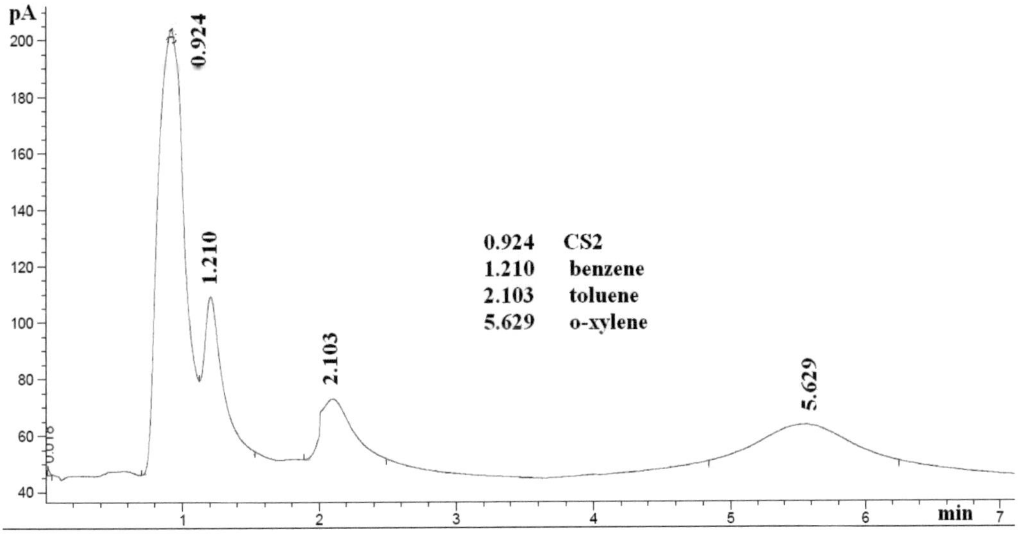

(a)

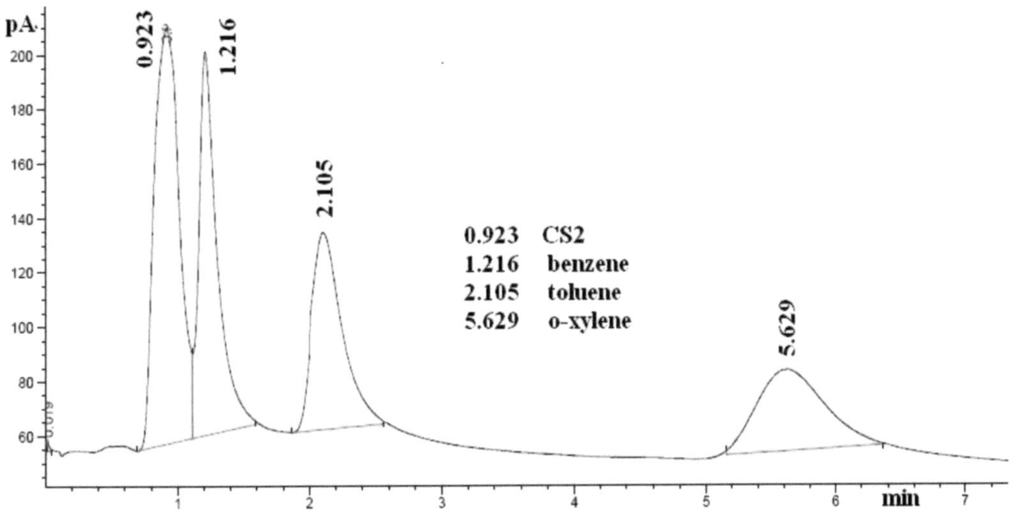

(b)

Fig. 9. Chromatogram of a gas mixture achieved using the (a) 0.75m serpentine channel column, (b) 0.75m spiral channel column.

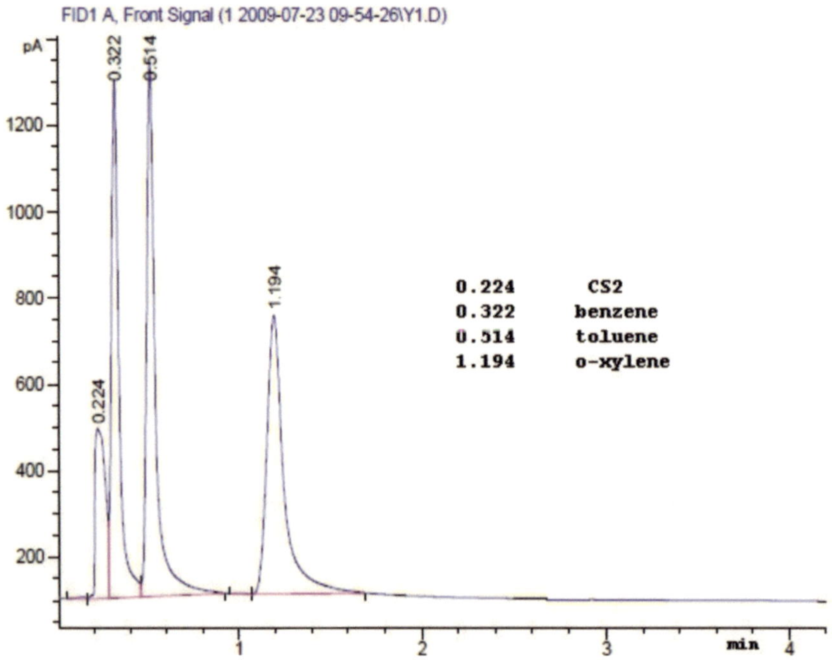

(a)

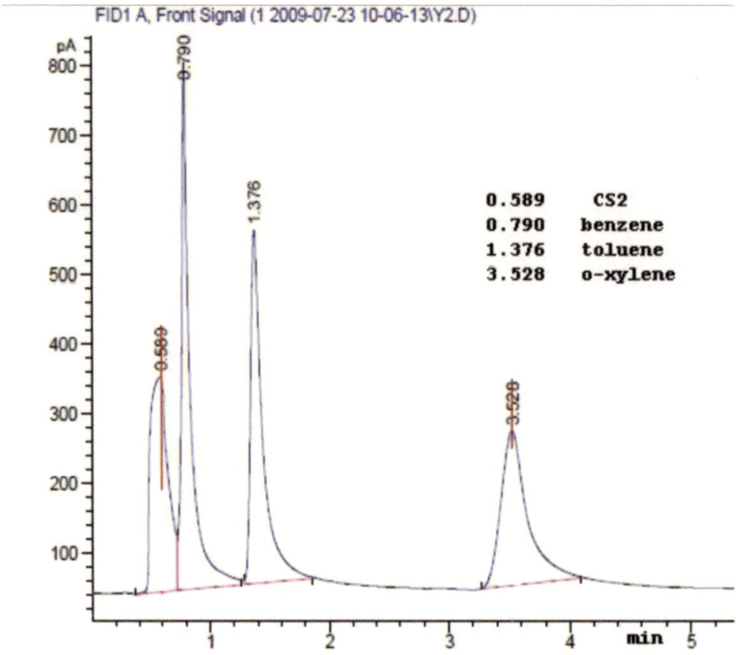

(b)

Fig. 10. Chromatogram of a gas mixture containing benzene, toluene and o-xylene achieved using the 3m column coated with OV-1, (a) operated at the flow rate of 42 cm/s, (b) operated at the optimal flow rate of 18 cm/s.

5.2.3 The effect of the column length for separation performance

Based on the formula (9), the major approach to improve resolution is to increase the column length.

So in this paper, the separation performance of GC columns with different length (such as the length of the column is 0.5 m, 1 m and 3 m, but the section width and depth of the column are the same) were compared with, and the separation experiments were all operated at the velocity of 18 cm/s and used the same sample (the sample including 3-component mixture: benzene, toluene, o-xylene). The columns yielded approximately 2400, 3370 and 6160 theoretical plates (the actual test values), respectively.

The theoretical predictions values could be also roughly computed 2500,5000 and 15000 plates(derived from the equation 1 and equation 4), respectively, The actual test values were about 41% of theoretical prediction values When the column length increased to 3 meters (Assuming the maximum retention factors is less than 10, the diffusion coefficients in the gas and liquid phases are about 0.1~0.2 cm^2/m, and the thickness of the stationary phase film is 0.2 μm), The comparison curve of the number of theoretical plates between the theoretical prediction values and the actual values was shown in Fig.11.

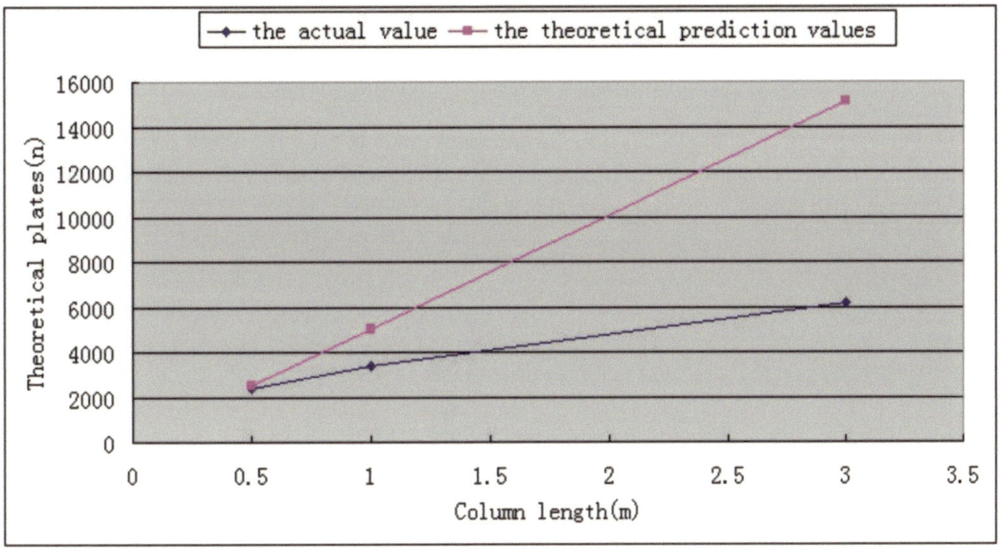

Fig. 11. The comparison curve of the number of theoretical plates between the theoretical prediction values and the actual values.

The discrepancy with the predicted number of the theoretical plates can be attributed to stationary phase film nonuniformity with the increase in column length. Consequently, the theoretical plates could be significantly increased by lengthening the GC column. Moreover, the separation performance of GC column can be greatly improved. The GC column can acquire a good overall separation performance when the length of the GC column is increased to 3 m.

5.2.4 Separation of the complex mixtures using 3 m spiral column

A separation experiment was performed to separate the sample mixture of benzene, toluene, ethylbenzene, p-xylylene and o-xylene using the microfabrication GC column. The column was operated at the velocity of 18 cm/s, and the chromatogram graph was shown in Fig.12. From experimental data, the spiral 3 m column yielded 7100 theoretical plates, and the analysis time was less than 200 sec. The GC column acquired a good overall separation performance.

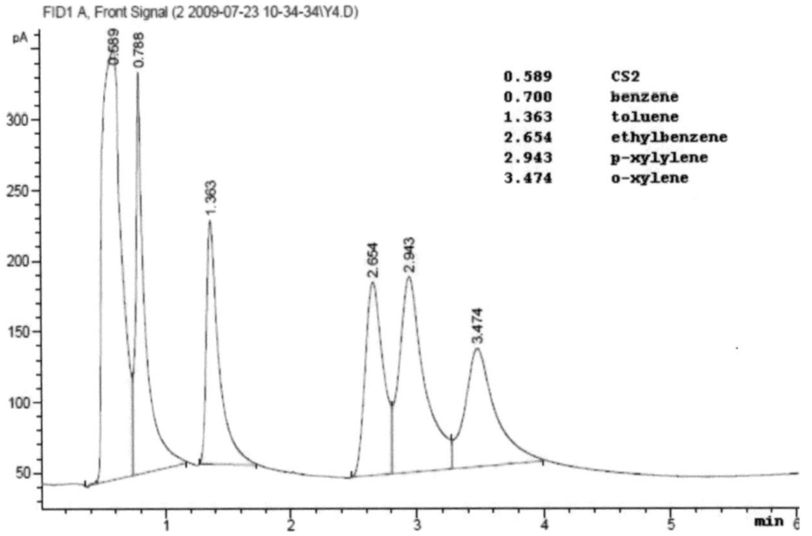

Fig. 12. Chromatogram of a gas mixture containing benzene, toluene, ethylbenzene, p-xylylene and o-xylene, achieved using the 3m column coated with OV-1 which operated at the flow rate of 18 cm/s.

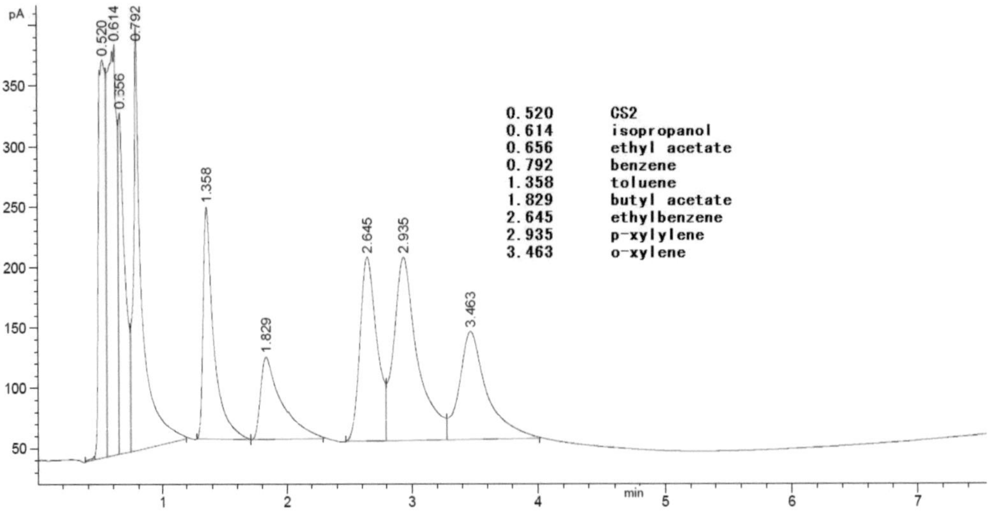

Fig. 13. Chromatogram of a gas mixture containing benzene, toluene, ethylbenzene, p-xylylene, o-xylene, butyl acetate, isopropanol and ethyl acetate achieved using the 3m column coated with OV-1 which operated at the flow rate of 18 cm/s.

Then, three other components were mixture into the sample, of which two kinds of components were polar components (isopropanol and ethyl acetate), and another sample was the butyl acetate. The separation experiment was carried out under the same conditions, the chromatograms was shown in the Fig.13, these polar components and solvents came out almost simultaneously, which resulted in these two polar components couldn't be separated by non-polar stationary phase, but the acid butyl ester was perfectly separated. The result indicated the non-polar column with high selectivity and specificity.

6. Conclusion

In summary, this chapter has presented the simulation, fabrication and experimental results for the microfabricated GC columns. Prototypes of devices have been fabricated successfully and the stationary phase is tackled properly. From the result of the simulation, the changes in distribution of airflow rate in the corners of the serpentine channel are obvious and relatively large, and the change rate of pressure in serpentine channel is obviously larger than that of the spiral channel, the value is close to 5 times compared to the former. Seen from the experimental data, the impact on airflow rate and pressure in serpentine channel is very significant, and this effect leads to deterioration of separation performance. Moreover, the difference of the airflow rate in the channel changes the thickness of the stationary phase film when the stationary phase is deposited. Consequently, the variation of the thickness of the stationary phase film in the channel leads bad tailing peaks. So, the spiral channel column is superior to the serpentine channel column for GC analysis. A series of spiral GC columns are designed according to the above analysis. The 3 m yields approximately 7100 plates and perfectly separates the complex mixture in less than 200 sec. Combined with micro-detector and micro sampler, Micro GC columns can be developed a miniaturized chromatographic system and serve as a platform technology for gas mixture separations.

Currently, the developed micro-columns can not achieve its own heating, but the design and fabrication process of the integrated micro-heater GC columns have been completed, its performance will be reported in subsequent papers. Micro- thermal conductivity detector, micro-optical ionization detector and micro-meter will be developed in our further work. A new generation of micro-chromatographic system will be developed by integrating micro GC column with the micro detectors. The micro GC system will be applied to environmental pollution, home safety, pesticide residues, food safety, pre-diagnosis of cancer and other areas for achieving on-site rapid testing.

7. Acknowledgment

The authors greatly acknowledge the financial support from the National Science Foundation of China under Grant number 61176112, 60976088 and 60701019. This work is sponsored by the Major National Scientific Research Plan (2011CB933202). The authors would like to thank Z.W.Ning and Y.N. Zhang and L.D.Du for their technical supports.

8. References

A. Bhushan, D. Yemane, D.Trudell, et al., (2007). Fabrication of micro-gas chromatograph columns for fast chromatography, Microsyst Technol 2007, Vol. 13, pp.361-368.

A.D. Radadia, R.I. Masel, M.A. Shannon, J.P. Jerrell, and K.R. Cadwallader (2008).Micromachined GC columns for fast separation of organophosphonate and organosulfur compounds,Analytical Chemistry, Vol. 80, No. 11, pp. 4087–4094.

G. R. Lambertus, C. S. Fix, S. M. Reidy, R. A. Miller, D. Wheeler, E. Nazarov, R. Sacks (2005),Silicon Microfabricated Column with Microfabricated Differential Mobility Spectrometerfor GC Analysis of Volatile Organic Compounds, Anal. Chem ,Vol.77,pp.7563-7571.

Golay, M. J. E. (1958), Gas Chromatography1958, D. Desty, Ed. Butterworths Sci. Pub., London, pp. 36–55.

J.H Sun, D.F.Cui, H.Y. Cai, H.Li, X. Chen, L.D. Du and L.L Zhang,Design, Simulation and fabrication of High Performance Gas Chromatography Columns for Analysis of Volatile Organic Compounds, Sensor Letters,Vol.9,No.2,pp.655-658. ISSN 1546-198X.

J.H Sun, D.F.Cui, H.Y. Cai, X. Chen, H. Li and L.D. Du(2010), Simulation and Evaluation of the Silicon-Micromachined Columns for Gas Chromatography, Chinese journal of chemistry,Vol. 28,pp.2315-2317. ISSN 1001-604X.

J.H.Sun, D.F.Cui, Y.T Li, L.L Zhang, J. Chen, H. Li and X.Chen,(2009).A High Resolution MEMS-based Gas Chromatography Column for the Analysis of Benzene and Toluene Gaseous Mixtures, Sensors and Actuators B: Chemical,Vol.141, No. 2, pp.431-435. ISSN 0925-4005.

R. L. Grob.(1977), Introduction,Modem Pracrice of Gas Chromatography. John Wiley & Sons, Inc., ISBN 0-471-59700-7, New York.

Radadia. A, R. I. Masel and M. A. Shannon.(2007). New Column Designs for Micro-GC,Transducers '07, Lyon, France, Jun. pp.10-14.

S.-I. Ohira, K. Toda, (2008).Micro gas analyzers for environmental and medical applications, Analytica Chimica Acta, Vol. 619,No. 2, pp. 143–156

Zampolli, I. Elmi, J. Stürmann, S. Nicoletti, L. Dori, G.C. Cardinali (2005). Selectivity enhancement of metal oxide gas sensors using a micromachined gas chromatographic column, Sensors and Actuators B: Chem, Vol. 105 ,No.2, pp. 400-406, ISSN 0925-4005.

4

Porous Polymer Monolith Microextraction Platform for Online GC-MS Applications

Samuel M. Mugo, Lauren Huybregts, Ting Zhou and Karl Ayton
Grant MacEwan University, Edmonton, Canada

1. Introduction

One of the key steps in chemical analysis is sample preparation. It is a very important step prior to analysis, otherwise the analyte signal could be suppressed by sample matrix interference. In most cases sample preparation entails some form of extraction of the analyte(s) of interest from the interfering species. Conventional extraction techniques include liquid-liquid extraction (LLE) and solid phase extraction (SPE). LLE involves mixing two immiscible solvent together, and based on solubility equilibria the analyte or the interfering species partitions preferentially in the organic solvent. Ideally, the partition coefficient is very high to allow for efficient extraction (Park et al., 2001). In SPE, analytes of interest are extracted from solution by passing the liquid through a solid phase, such as a C-18 SPE cartridge. Based on different physical and chemical properties, the desired solute either is adsorbed onto the solid phase, from whence it can later be eluted, or remains in solution while impurities are retained on the solid phase (Körner et al., 2000). The LLE and SPE extraction methods are laborious and demand the use of copious amounts of samples and solvents, therefore driving up the cost of chemical analysis. For LLE, emulsion formation can be a nuisance. These methods are also limited by factors like cartridge clogging in SPE, single use of the SPE and the need for toxic and polluting solvents (such as halogenated solvents like chloroform and dichloromethane) in LLE.

Other emerging extraction techniques that are based on miniaturization of LLE in effort to minimize solvent use and simplify the sample preparation process include: single-drop microextraction (SDME), hollow fibre liquid-phase microextraction (HF-LPME) and dispersive liquid–liquid microextraction (DLLME). Briefly, SDME, introduced by Cantwell and Jeannot in 1996, entails exposing a microdrop of extractant placed on the end of a Teflon rod to the sample solution or sample headspace. The HF-LPME uses two phase sampling mode, a hydrophobic hollow fibre is used to support the extractant and the fibre is exposed to the sample and the analyte is extracted in the organic extractant. On the other hand, the DLLME is based on rapidly injecting a few microliters of high density solvents mixed with a disperser solvent into the aqueous sample. The turbulence from the rapid injection causes large interfacial area emulsion droplets to form and distribute in the sample solution. Analytes partition into this emulsion droplets; and after centrifugation, they sediment at the bottom of a vial and then are sent for analysis. In general these miniaturized LLE techniques are convenient, efficient (given a high partition coefficient), inexpensive and requires very little solvent. They are also easily coupled to gas chromatography (GC) (Jeannot & Cantwell, 1996). An extensive discussion of these methods is available and covered in the review by Pena-Pereira et al., 2009. An obvious disadvantage with these methods is that only a few organic solvents are usable; and so the methods lack versatility and are only applicable to the extraction of a limited number of analytes.

Solid phase microextraction (SPME), developed by Pawliszyn in the 1990s (Arthur & Pawliszyn, 1990), surmounts some of the limitations of the SPE, LLE and miniaturized LLE. SPME method employs a fused silica fiber coated with a polymeric coating to extract organic compounds from their sample matrix and directly transfer analytes into GC, eliminating offline sample extraction steps (Z. Zhang & Pawliszyn, 1995). Different polymer stationary phases have been used in SPME for the analysis of many different analyte classes. Common commercially-available SPME stationary phases include polyacrylate, polydimethylsiloxane (PDMS), divinylbenzene (DVB)/PDMS, carboxen/PDMS, and DVB/Carboxen/PDMS. However, SPME fibres are generally very fragile and cases of them breaking off in the GC port are very common. Therefore they have to be used with extreme care, and sometimes they have to be replaced fairly often, which makes them expensive. Another disadvantage of SPME is the limited variety of available stationary phases, which limits the range of analyte classes that can be extracted. There have been publications of novel SPME stationary phases being developed to address this.

Second and third generation variants of SPME have been developed, such as the needle-trap devices. The need-trap extraction method involves packing a needle with polymer beads or carbon based sorbents. A gaseous sample is drawn through the needle housing the sorbent plug and the analyte of interest is adsorbed and then directly injected and desorbed into GC (Ueta et al., 2010). The needle-trap extraction method possesses the benefits of microscale extraction (*i.e.* low cost, high efficiency, low-volume solvent use) while overcoming SPME's problem of fragility issues. However, this method is only applicable to headspace extraction of volatile samples and so can only be used to extract a limited range of analytes.

A new class of microextraction technique is polymer monolith microextraction (PMME) introduced by Feng and coworkers in 2006. Feng and coworkers fabricated their device using a regular plastic syringe (1 ml) and a poly (methacrylic acid-co-ethylene glycol dimethacrylate) monolith entrained in a silica capillary (2 cm × 530 µm I.D.) (Zhang et al. 2006). PPME shares the advantages of SPME while solving most of its limitations. In general the most important component of the PPME device is the porous polymer monolith usually formed in the confines of a silica capillary, although other materials such as polyether ether ketone (PEEK) and titanium tubes can used (Wen et al. 2006).

Porous polymer monoliths are relatively new materials synthesized by carrying out free radical (radicals generated by heat or ultra-violet radiation following destabilization of azo initiators) polymerization of cross-linking and monovinyl monomers, in the presence of suitable porogenic solvents. The generated radicals initiate a rapid polymer chain growth at what become nucleation sites, which continue to grow as the reaction proceeds. As polymer molecular weight increases, the solubility decreases and a two-phase system of solid polymer and liquid solvent results. The resulting monolith microstructure consists of an agglomeration or globules, whose size directly impacts resulting pore size distribution.

Globule size is influenced by many factors, including the number of nucleation sites present, monomer concentration, solubility, and degree of cross-linking. Thus, polymer microstructure can be controlled by rate of reaction, monomer/porogen ratio, type of porogenic solvents and fraction of cross-linking monomers. Detailed descriptions of various chemistries, and characterization of resulting monoliths can be found in references (Gibson et al. 2008 & Svec, 2010).

The polymer monoliths used in making the PPME platforms can easily be made from different types of monomers with different polarities depending on the analytes of interests. The large surface area provided by the tailor-able macropores in monolith structure may help to improve the extraction efficiency. The convection flow provided by the flow through channel within the monolith also helps in accelerating mass transfer. Furthermore, PPME device is very easy to prepare, which also makes it a better choice than particle packed extraction cartridges.

There have been numerous literatures reporting on the application of this PMME method for extraction and coupling to GC and LC, with the latter being more common. For example, Wen et.al detected and quantified a series of sexual hormones (testosterone, methyltestosterone and progesterone) in liquid cosmetics extracted by a poly (methacrylic acid-co-ethylene glycol dimethacrylate) monolithic capillary. By simple dilution and filtering after extraction, the samples were directly injected into HPLC system. The limit of detection was 2.3 to 4.6 μg/L. Calibration curves showed good linearity in the concentration range of 10 to 1000 μg/L with R^2 above 0.996 (Wen et al., 2006).

A variation on conventional PPME involves the production of molecularly-imprinted polymers (MIPs) to make up the extraction monolith (Bravo et al., 2005; Sanbe & Haginaka, 2003; Schweitz, 2002; Zhou et al., 2010). The imprints left by template molecules act as "artificial receptor-like binding sites" which are selective to the analyte of interest. Zhou and coworkers synthesized a molecularly imprinted solid-phase microextraction monolith (MIP-PPME) for selective extraction of pirimicarb in tomato and pear. They prepared a pirimicarb MIP monolith in a micropipette tip using methacrylic acid as the functional monomer, ethylene dimethacrylate as the cross-linker and the mixture of toluene-dodecanol as the porogenic solvent. The dynamic linear ranges were from 2.0 to 1400 μg/kg for pirimicarb in tomato and pear (Zhou et al., 2010).

Until now polymer monolithic materials used in PMME have mainly been poly-(methacrylic acid-co-ethylene dimethacrylate) (MAA-co-EDMA) (Liu et al., 2011) and most documented applications have been for LC. Our group has demonstrated the use of many other types of acrylate, epoxy-based, and acrylamide polymer monoliths for PMME and have shown that these simple PPME devices can be fabricated for online-GC applications. A schematic set up on how they are coupled to GC-MS is shown in Figure 1, which is similar to a commercial SPME operation. The fabrication and application of these newly developed PMME materials will be discussed in this chapter. In particular we shall demonstrate the application of PPME to the extraction of caffeine, polyaromatic hydrocarbons (PAHs) and hormones in water. We have also evaluated the use of porous layer open tubular (PLOT) monoliths as PPME platforms. The potential advantage of PLOT (otherwise will be referred in this chapter as open tubular polymer monolith) is the faster mass transfer and consequently faster extraction than PPME. These materials are still being evaluated in our research group.

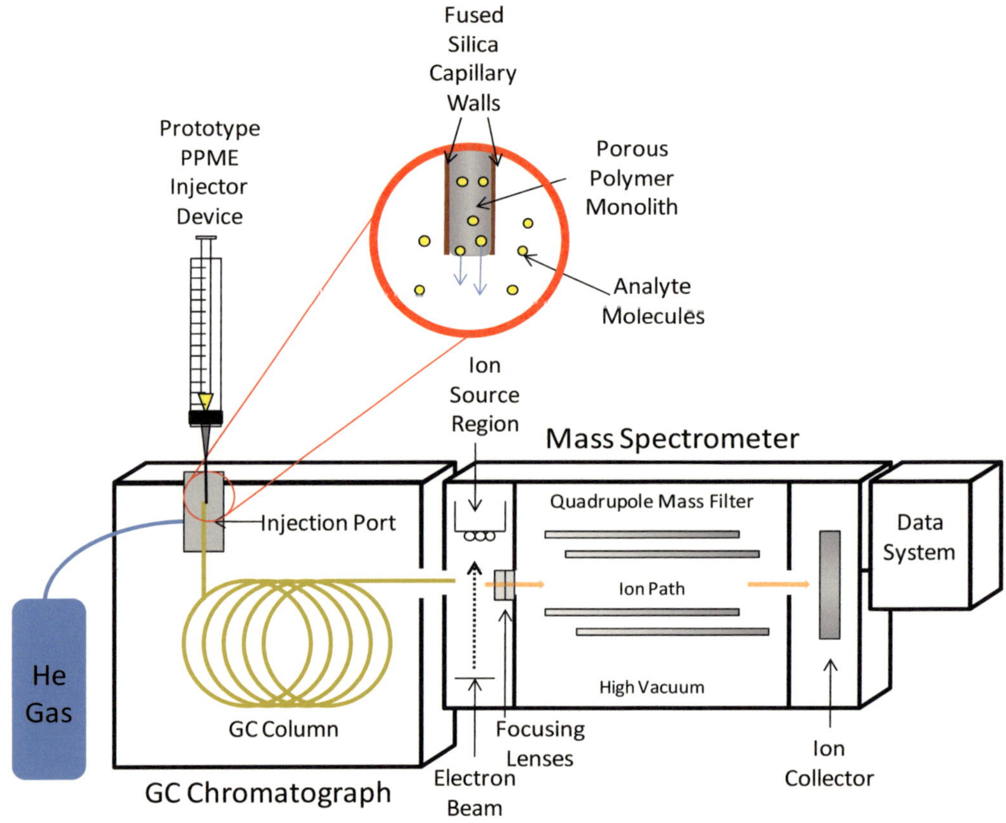

Fig. 1. Schematic of a PPME at the Injection Port of a GC-MS system.

2. Chemicals and materials

Acrylamide (99%, AA), styrene (>99%), 4-vinylpyridine (99%, 4-VP), glycidyl methacrylate (97%, GMA), ethylene glycol methacrylate (98%, EDMA), divinylbenzene (80%, DVB), 4,4′-azobis(4-cyanovaleric acid) (75+%), 1,1′-carbonyldiimidazole, 3-(trimethylsilyl) propyl methacrylate (98%), cyclohexanol (97%), 1-dodecanol (98+%), 3-aminopropyltriethoxylsilane (99%, APTES), ethanol (85%), naphthalene (≥99.7%), 2,6-dimethylnaphthalene (99%), phenanthrene (≥99.5%), HPLC grade acetonitrile, fluorene (≥99.0%), caffeine, isotopically-labelled (trimethyl ^{18}C) caffeine, megestrol acetate, 17β-estradiol, acetic acid (≥99%) and methanol were purchased from Sigma-Aldrich (St. Louis, MO, USA). Sodium acetate (99+%) was purchased from Alfa Aesar (Ward Hill, MA, USA). 99.7% NaOH and citric acid monohydrate were purchased from Fisher Scientific (Fair Lawn, NJ, USA).

The structures of analytes selected for extraction and selected reagents used for making the polymer monoliths are shown in Figure 2. All through the experiment deionized (D.I.) water used was prepared using a Millipore Elix 10 water purification system.

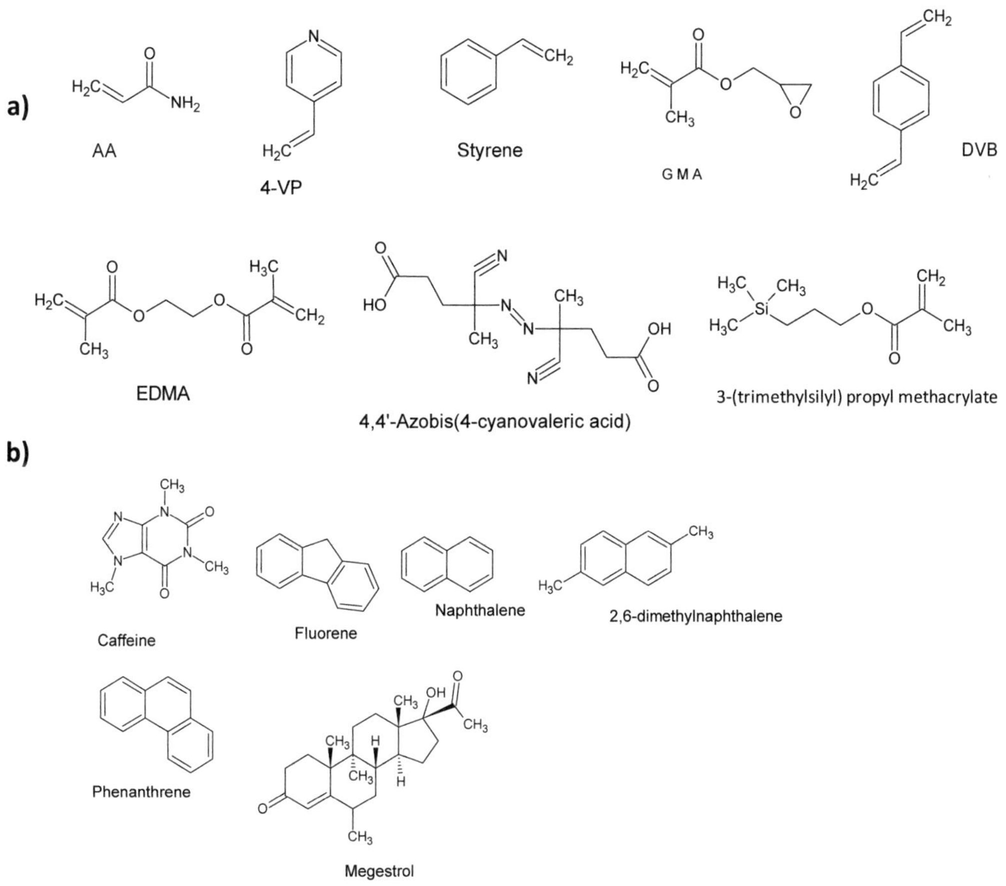

Fig. 2. a) Structures of some reagents used to make the polymer monoliths and, b) structures of some of the analytes that were tested.

3. Methods

3.1 Formation of porous polymer monoliths on fused silica capillary

3.1.1 Poly(Acrylamide-co-EDMA)

The poly(acrylamide-co-EDMA) monolithic capillary was formed inside a polyimide-coated fused silica capillary (10 cm x 250 µm, i.d., Polymicro Technologies, USA) by heat-initiated polymerization. To begin, the capillary was conditioned by flushing (using a Harvard Apparatus PHD Ultra syringe pump) it with 1.0 M NaOH at a flow rate of 10 µL/min for 3 h, 0.1 M HCl for 1 h, washed with deinionized (D.I.) water for 1 h, then dried with a flow of nitrogen. Next, the capillary was flushed at 10 μL/min with 20% (v/v) 3-(trimethoxysilyl)propyl methacrylate in toluene overnight (17 h), then infused (for 5 minutes at 5 μL/min) with the pre-polymer mixture, consisting of 13.0 mg acrylamide, 179 μL EDMA, 7 mg 4,4'-azobis(4-cyanovaleric acid) initiator, and 500 μL 1/1 (v/v) ethanol/D.I. water. Both ends of the capillary were capped and the capillary was placed in a 70°C oven overnight (16 h). Once monolith formation was complete, the capillary was flushed with 1/1 (v/v) methanol/D.I. water to remove any unreacted reagents.

3.1.2 Poly (Styrene-co-Divinylbenzene)

The poly(styrene-co-divinylbenzene) monolithic capillary was formed the same way as the preparation of poly(acrylamide-co-EDMA) monolith except the composition of prepolymer mixture, which consists of 200 μL styrene, 80 μL divinyl benzene, 600 μL ethanol, and 5 mg 4,4'-Azobis(4-cyanovaleric acid) initiator.

3.1.3 Poly (4Vinylpyridine-co-EDMA)

The poly (4Vinylpyridine-co-EDMA) monolith was formed the same as documented above for the others. The pre-polymer mixture, consisted of 179 μL of EDMA, 13 μL of 4-vinylpyridine, 7 mg of 4,4'-Azobis(4-cyanovaleric acid), 250 μL of ethanol, and 250 μL of D.I. water.

3.1.4 Poly (Glycidylmethacrylate-co-EDMA)

The poly (glycidylmethacrylate-co-EDMA) monolith was formed by polymerization of a mixture of 80 μL of EDMA, 240 μL of glycidyl methacrylate, 7 mg of 4,4'-azobis(4-cyanovaleric acid), 400 μL of cyclohexanol, and 180 μL of dodecanol.

3.1.5 Poly (styrene-co-divinylbenzene) PLOT

The poly(styrene-divinylbenzene) PLOT capillary was formed inside a 250 μm i.d., 10 cm long fused silica capillary as outlined schematically in Figure 3. This method was adapted from the literature (Schweitz, L. 2002). Briefly, the capillary was conditioned by flushing with 1.0 M NaOH for 3 h at a flow rate of 10 μL/min. This was followed by a D.I. water wash for 1 h, then dried with a flow of nitrogen. Next, the capillary was filled with 20% (v/v) APTES in toluene overnight (16 h), then rinsed with acetone for 1 h and then dried. The initiator solution, consisting of 8.4 mg of 4,4'-azobis(4-cyanovaleric acid) initiator, 9.7 mg of 1,1'-carbonyldiimidazole, 1.5 mL of methanol, and 1.5 mL of 50 mM sodium acetate-acetic acid buffer (pH ~5.5) was prepared and the pH of the final solution was adjusted to 5.5 using acetic acid. The capillary was infused at 10 μL/min with the initiator solution overnight (16 h), then washed with 1/1 (v/) methanol/D.I. water for 30 minutes and dried with a flow of nitrogen. Next, the prepolymer mixture, consisting of 200 μL styrene, 200 μL DVB, and 600 μL ethanol, was prepared and infused into the capillary for 5 min at 5 μL/min. Both ends of the capillary were capped and placed in a 70°C oven overnight (16 h) for polymerization to ensue. The PLOT column was washed with 1/1 (v/v) methanol/water.

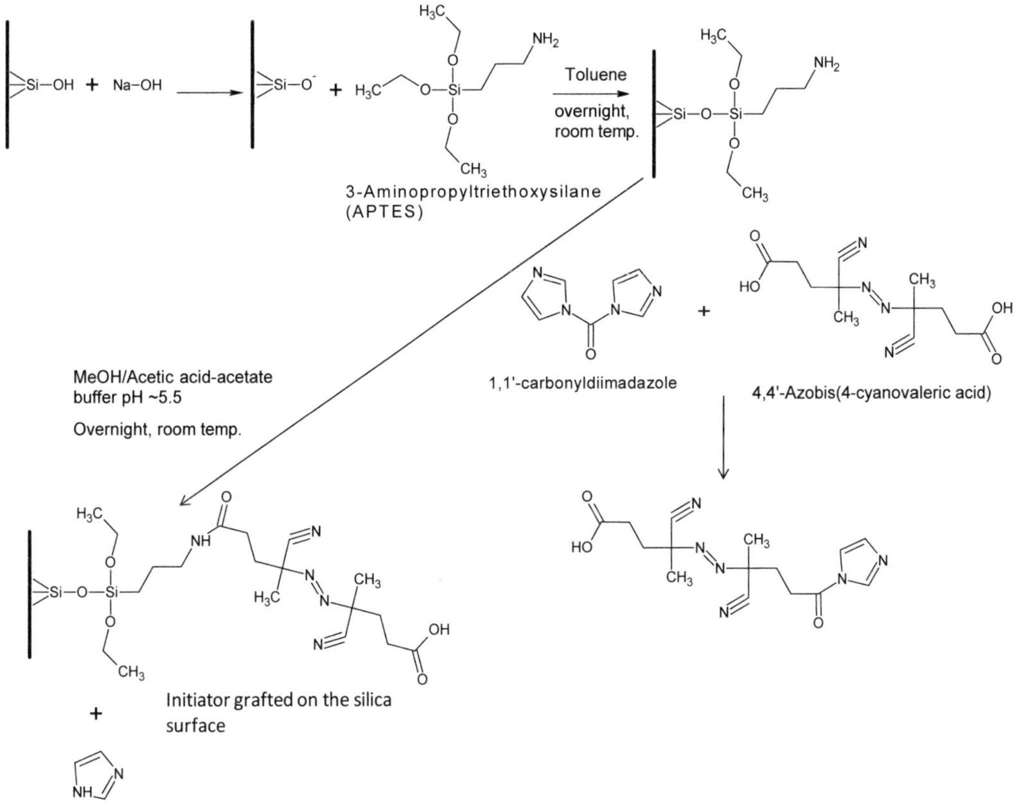

Fig. 3. Reaction schematic for grafting initiator onto the capillary wall for open tubular monolith formation

3.2 PPME device fabrication

A prototype PPME device was prepared using the porous polymer monolith described above, disposable plastic syringes with removable needles, micropipette tips, and fast-drying epoxy. A disposable plastic syringe was dismantled by removing the inner plunger. The rubber stopper was removed from the end of the plunger and the tip of the stopper was trimmed off, creating a sort of rubber washer. A 10-100 µL yellow micropipette tip was trimmed so that its wide end fits snugly over the end of the hard plastic plunger, the stem of which had previously been trimmed to fit inside the pipette tip. The pipette tip was filled with thoroughly-mixed epoxy glue then secured onto the trimmed plunger stem. A capillary containing the desired porous polymer monolith was then inserted into the narrow end of the pipette tip attached to the plunger and the rubber washer was slipped over the capillary and secured to the base of the pipette tip with epoxy. The prototype was left to dry overnight then reinserted into the syringe outer barrel. With the PPME capillary withdrawn into the syringe barrel, a needle was attached to the syringe. This enables the PPME device to puncture a GC septum effectively before the PPME capillary is injected for desorption. A schematic of the in-house fabricated PPME device is shown in Figure 4.

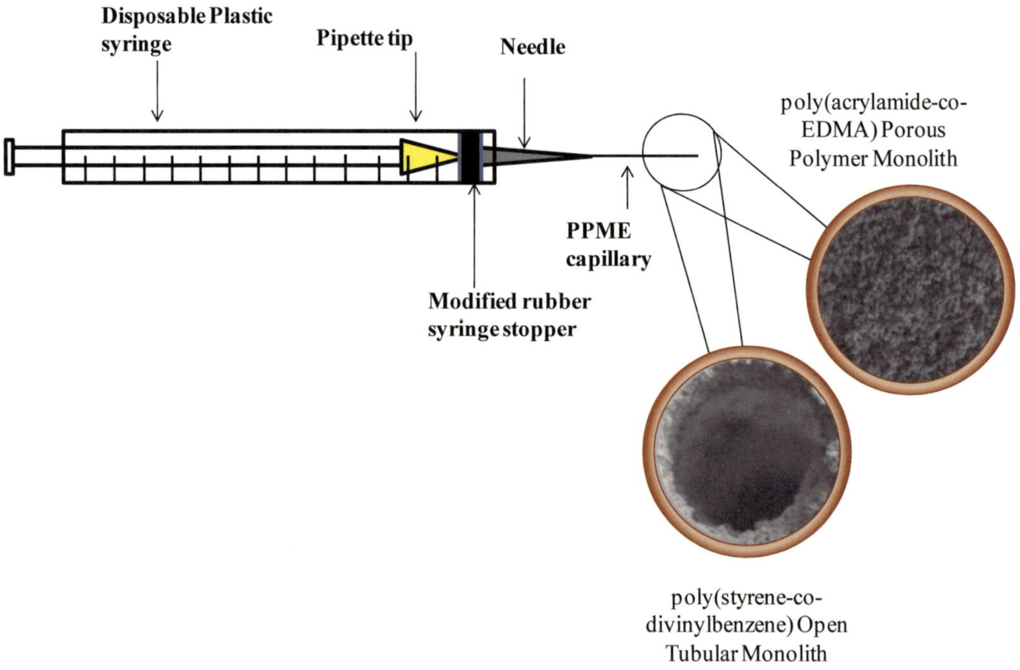

Fig. 4. A schematic of the in-house fabricated PPME device with an inset of SEM images of poly(acrylamide-co-EDMA) and poly (styrene-co-divinylbenzene) PLOT platforms.

3.3 Extraction of caffeine and PAHs using PPME and analysis by GC-MS

3.3.1 Caffeine extraction

3 mL of caffeine standards with the concentration of 100, 200, 300, and 400 µg/mL were prepared in 4 mL vials and 75 µL of the internal standard, isotopically-labelled (trimethyl ^{13}C) caffeine, was added. The poly(acrylamide-co-EDMA) PPME capillary was chosen for this extraction as it posses fairly high polarity stationary material that could preferentially adsorb caffeine. The PPME was conditioned by inserting it in the GC injection port for 30 minutes at 270˚C, then cooled at room temperature. Starting with the 100 µg/mL caffeine standard, the vial was placed on a stir plate and stirred at a controlled rate. The PPME capillary was then immersed in the standard solution for 7 minutes to allow caffeine to adsorb onto the polymer monolith, then the PPME was injected into the GC port and desorbed at the GC injector for 3 minutes at 200˚C. This procedure was repeated for the 200, 300, and 400 µg/mL standards with the PPME capillary being cooled at room temperature between runs. For comparison, the above procedure was also repeated using a commercially available DVB/Carboxen/PDMS StableFlex™ SPME fibre (Supelco).

3.3.2 Polyaromatic hydrocarbon (PAH) extraction

1.5 mL of PAH standards containing naphthalene, 2,6-dimethylnaphthalene, and phenanthrene with the concentration of 20, 50, 100, and 200 µg/mL were prepared in 2 mL vials with 75 µL of the internal standard fluorene (480 µg/mL) added. Based on polarity matching considerations, the poly (styrene-co-divinylbenzene) PPME device was conditioned in the GC injection port for 8 minutes at 280°C then cooled at room temperature. Starting with the 20 µg/mL standard, the vial was placed on a stir plate and stirred at a controlled rate. The PPME capillary was immersed in the standard solution for to allow the PAHs to adsorb onto the polymer monolith. The adsorption times that were evaluated included 3 minutes, 1 minute, and 30 seconds. After extraction the PPME was injected into the GC injection port (200°C) for 3 minutes' desorption. This procedure was repeated for the 50, 100, and 200 µg/mL standards with the PPME capillary being cooled at room temperature for 8 minutes between runs.

3.3.3 Evaluation of capability of the porous monoliths for hormone extraction

Hormones in the environment emerging contaminants of scientific interest. Conventional analysis approach is to extract the hormones by a C-18 SPE cartridge and elute with an appropriate solvent followed by analysis by HPLC-DAD-MS. Analysis of hormones by GC-MS often demands a derivatization process. Different derivatization reagents have been used including silylating reagent usually, N,O-bis(trimethylsilyl)trifluoroacetamide and Heptafluorobutyric acid. Our goal was to explore the use of PPME to extract the hormones and carry out on fiber derivation followed by injection of the PPME into the GC. This has been demonstrated largely on acrylate SPME fibers by several authors (Yang et al. 2006, Pan et al. 2008).

Before the attempt to carry out on-porous polymer derivatization, it was crucial first to determine which porous polymer monolith was best suited for the extraction of hormones. Megestrol standards with the concentration of 0.1, 1, 10, and 20 ppm in D.I. water were prepared. Poly(GMA-co-EDMA) and poly(4VP-co-EDMA) porous polymer monoliths were chosen to evaluate their ability to extract the hormone. Both monoliths were preconditioned by washing with 1:1 (v/v) methanol/D.I. water at 25 µL/min for 40 minutes then drying with nitrogen. Starting with the 0.1 ppm megestrol standard, each porous polymer monolith in a capillary (10 cm long) was infused with the hormone solution at 25 µL/min for 40 minutes and the eluate solution was collected in a 2 mL vial. The monoliths were then dried with nitrogen and eluted with 9:1 (v/v) methanol/D.I. water at a flow rate of 25 µL/min. The eluate was collected in a 2 mL vial. This procedure was repeated for the 1, 10, and 20 ppm hormone standards and the collected solutions were analyzed by HPLC-DAD for both the monoliths.

4. Instrumentation

4.1 GC-MS instrumentation conditions

GC analysis was carried out on an Agilent 6890N gas chromatography system with an Agilent 5975C VL MSD (Agilent Technologies, USA). Analytes were separated on a 5% phenyl methyl siloxane capillary column (30 m x 250 µm I.D., 0.25 µm film thickness, Agilent Technologies, USA). The column oven temperature was initially set at 180°C for 3 minutes and then was increased to 205°C at 5°C/min. This temperature was held till the end of the analysis. The MS detector temperature was set at 200°C. The injection port temperature was set at 280°C with a split-less injection mode. Helium was used as carrier gas set at the flow rate of 24.1 mL/min.

4.2 HPLC -DAD Instrumentation and conditions

Analysis of megestrol eluted from the porous polymer monolith were performed on a Thermo LCQ fleet HPLC-DAD system consisting of a LC pump, an autosampler, a 6-port injection valve with a 20-μL injection loop, and a diode array detector (DAD) with the scan range from 250 to 400 nm. The temperature of column oven was set at 30 °C. The temperature of autosampler tray was set at 6 °C to prevent degradation of the hormones. An Agilent Zorbox Eclipse PAH RPLC column (250 x 3.0 mm I.D., 5 μm) (Agilent, Santa Clara, CA, USA) was used for the separation of hormone. The mobile phase used was acetonitrile (C) and 0.1% formic acid in H_2O (D) at a flow rate of 500.0 μL/min. The eluent gradient setting is shown in Table 1.

%M.P. / Time(min)	Mobile phase C %	Mobile phase D %
0	20	80
5	55	45
6	95	5
10	95	5
12	20	80
15	20	80

Table 1. HPLC gradient setting for the separation of antibiotics/hormones.

5. Results and discussion

5.1 Monolithic structure of PPME and PLOT

The structure of different PPME and PLOT was investigated by scanning electron microscope (SEM), shown in Figure 3. As can be seen in Figure 3 (inset picture consist of SEM image of poly(acrylamide-co-EDMA) and poly (styrene-co-DVB) PLOT), there were many macropores and flow-through channels in the network skeleton , which can provide the high porosity and good mass transfer desired for a microextraction platform. The SEM image of poly (styrene-DVB) PLOT confirms that using the initiator capillary wall grafting procedure outlined in Figure 3, the polymer monolith only formed around the walls of the capillary. Using this procedure many other different types of open tubular polymer monoliths can be fabricated. It is proposed that the PLOT structure could potentially result in a faster mass transfer and efficient analyte extraction compared to the conventional polymer monolith. At present, research is still being conducted to confirm and test this hypothesis.

5.2 Applications of PPME

5.2.1 Caffeine extraction

Different concentrations of caffeine standards from 100 to 400 μg/mL, with isotopically-labelled (trimethyl ^{13}C) caffeine was used as the internal standard, were extracted using a poly (acrylamide-co-EDMA) PPME. The calibration curve obtained is shown in Figure 5 a). The y-axis on the calibration curve is the peak area ratio of caffeine (m/z 194) to the internal standard (m/z 197), obtained from extracted ions chromatogram. The linear regression coefficient (R^2) achieved with the PPME was 0.9771. This satisfactory R^2 indicates that this in-house built poly (acrylamide-co-EDMA) PPME is a suitable analytical tool for extracting caffeine in a wide range of concentration. We compared the extraction efficiency of the in-house fabricated PPME with the commercial SPME (2 cm-50/30µm DVB/Carboxen™/PDMS StableFlex ™ purchased from Supelco (Bellefonte, PA, USA)). The linear regression coefficient for the commercial SPME was 0.999 (as shown in Figure 5 b), which certainly rivals the in-house made PPME. This could partly be due to potential better mass transfer for the commercial SPME due to the thin film thickness (30 µm) of the stationary phase. However, it is expected by fabricating open tubular polymer monolith based PPME the mass transfer effect could improve and the results could compare with the commercial one. However, the results obtained with the PPME herein shows good usability especially with possible benefits over the commercial SPME, such as better sample capacity and less prone to breakage and thus can be reused for a much longer time.

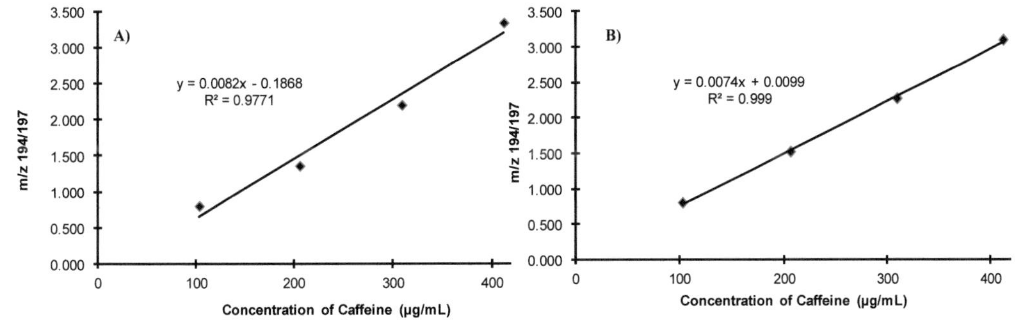

Fig. 5. Calibration curve of caffeine extracted using a) Poly (acrylamide-co-EDMA) PPME and b) commercial SPME.

The successful extraction of caffeine by PPME suggests an obvious extension of this extraction method to analysis of other compounds such as, theobromine, and theophylline and other compounds in the methylxanthine class. Theobromine is found in chocolate, tea leaves, and the cola nut, while theophylline is found in cocoa beans as well as in therapeutic drugs for respiratory disorders. Because of their presence in food and drugs it is clear that an effective and efficient extraction method for methylxanthines would be a relevant application. Using a multi-step extraction, which included the use a ^{18}C extraction tube SPE followed by filtration, Aresta's group reported the determination of caffeine, theobromine, and theophylline, among other compounds, in human milk by HPLC-UV (Aresta & Zambonin, 2005). Since it has been shown that PPME is a viable method for caffeine extraction, it is reasonable to suggest that PPME may also be used to analyze other methylxanthines, and other polar like compounds of interest in pharmaceutical, nutraceutical and environmental fields. Moreover, PPME may be more efficient, reusable, could reduce analysis cost by reducing the steps needed in analysis process and in reducing the solvents required.

5.2.2 Extraction of polyaromatic hydrocarbons

With the successful extraction of caffeine using PPME, the extraction of another class of analytes, polyaromatic hydrocarbons (PAHs) using poly(styrene-co-DVB) PPME (a fairly non polar material whose polarity could potentially march the PAHs) were also attempted. A mixture of four different PAHs at different concentrations were tested. The PAHs tested included naphthalene, 2,6-dimethylnaphthalene, phenanthrene and florene, where florene was also used as an internal standard. While good results were obtained at 3 min, 1 min adsorption time, it was found that even with a 30-second adsorption, satisfactory extraction of the PAHs was still achieved. Figure 6 showed the GC-MS chromatogram of four PAHs extracted by poly(styrene-co-DVB) PPME with 30- second adsorption time. Good baseline separation was achieved as attested in the chromatogram, with enough resolution to allow separation of more PAHs. This potential is important in environmental samples which can have numerous PAHs present. The PAHs peaks on the PPME showed some tailing, possibly attributable to analytes slow mass transfer. We propose this could be alleviated by the use of a PPME fabricated poly(styrene-co-DVB) PLOT platform. This work is in progress.

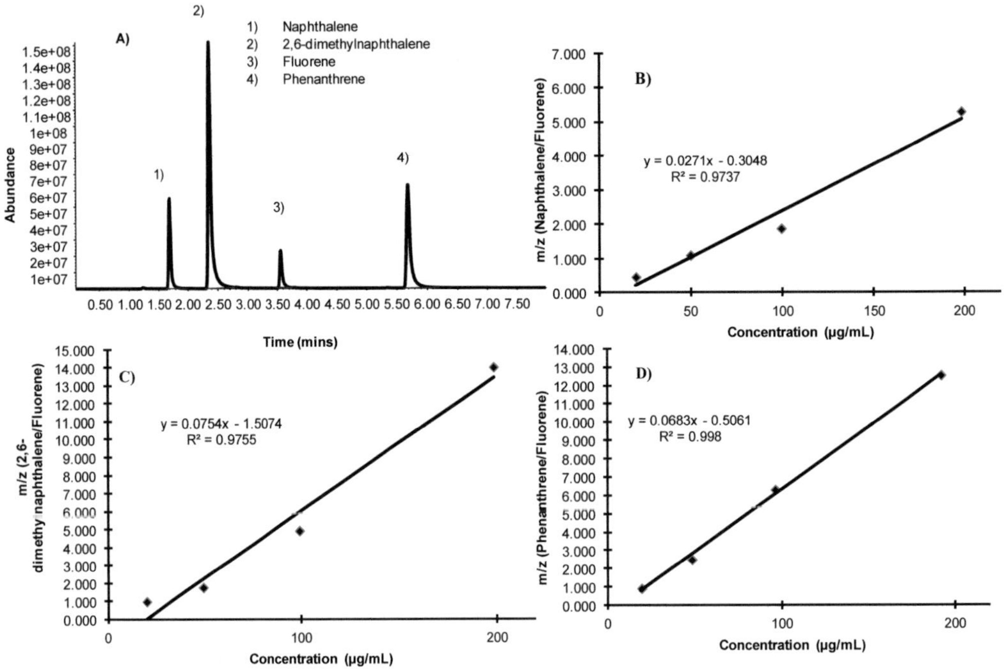

Fig. 6. A) GC-MS chromatogram showing separation of 100 µg/mL of three PAHs including; 1) naphthalene, 2) 2,6-dimethylnaphthalene, 3) fluorene, and 4) phenanthrene extracted by poly(Styrene-co-DVB) PPME; Calibration curves of B)naphthalene; C) 2,6-dimethylnaphthalene; D) phenanthrene extracted by poly(Styrene-co-DVB) PPME with 30 second adsorption and fluorene as an internal standard.

Calibration curves of the three PAHs extracted by poly(Styrene-co-DVB) PPME are shown in Figure 7 b,c,d, with analytes concentration range from 0 to 200 µg/mL. All the three PAHs showed satisfactory linear calibration curves with R^2 higher than 0.997 ($R^2 = 0.9737$ for naphthanlene, 0.9755 for 2,6-dimethylnaphthalene and 0.998 for phenanthrene). Clearly, the results obtained using the in-house fabricated poly (styrene-co-DVB) PPME revealed several things. First, the well-defined peaks in chromatograms obtained from GC-MS analysis showed that PAHs, even in low concentrations, are readily extracted and desorbed from the PPME capillary, making PPME a suitable method for PAH extraction. Second, this method showed feasibility for environmental monitoring of PAHs in bodies of water such as lakes, rivers, and ponds. The PPME, being more robust and rugged than conventional SPME fibers, can be taken to monitoring sites, which could make sampling process much easier. Third, this method can be extended to the analysis of any hydrophobic anyalytes immersed in an aqueous solution, since the analytes would very readily adsorb onto the hydrophobic polymer . With this knowledge, a plethora of polymers can be fabricated, each with a specific targeted analytes class to be extracted from a given sample matrix.

5.2.3 Extraction efficiency of megestrol by the polymer monoliths

There is growing public concern for the presence of hormones in the ecosystem, and the possible pathways by which these substances can enter into human and animal food chains. The concern focuses particularly on the common use of these substances in animal husbandry especially as growth promoting agents. Due to their stability most of the hormones are excreted unchanged or as active metabolites. Most commonly used processes are based on the extraction of hormones from a sample matrix and analysis by liquid chromatography (Mitani et al., 2005; Yi et al., 2007). Analysis by GC-MS could however be faster, but the hormones require derivatization to make them more volatile. This has been actualized typically by extraction of the hormones from sample matrix by C-18 cartridges, followed by pre-concentration, derivatization and then GC-MS analysis. A new method called on-fiber derivatization allows the hormones to be extracted, derivatized on the solid-phase and subsequently analyzed directly by GC-MS. Carpinteiro and coworkers combined SPME and on-fiber derivatization to detect and quantify estrogens in water by GC-MS. Samples were first extracted by SPME, and then were silylated on the headspace of a vial containing the derivatization agent. The MS signal intensity were dramatically increased by the derivatization step. The study of matrix effects demonstrated that the method is applicable for the determination of estrogens in both surface and sewage water (Carpinteiro et al., 2004). This on-fiber derivatization method eliminates the further sample preparation step before injecting into GC-MS. We propose PPME could have even better versatility in carrying out extraction and on-porous polymer derivatization of the hormones.

To test the hypothesis, two types of monoliths, poly(4VP-co-EDMA) and poly (GMA-co-EDMA) were first tested for their extraction efficiency of megestrol. We evaluated the % recovery of megestrol using HPLC-DAD analysis. Both poly(4VP-co-EDMA) and poly (GMA-co-EDMA) were found to have very good adsorptive capacity for megestrol with calculated % recovery of 98% and 107% respectively. An overlapped chromatogram indicating megestrol peak for poly(4VP-co-EDMA) , poly (GMA-co-EDMA) and directly injected 10 ppm megestrol standard is shown in Figure 8. These two polymers were found to be good stationary phases for analysis of hormones.

At the point of publication of this chapter, the demonstration of the on-fiber derivatization of the hormones using PPME fabricated from poly(4VP-co-EDMA) and poly (GMA-co-EDMA) was ongoing.

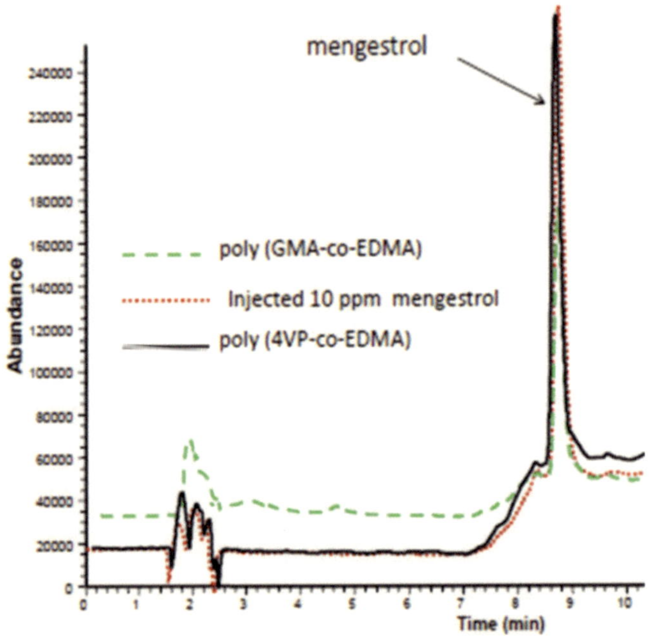

Fig. 8. Overlapped chromatogram indicating megestrol peak for poly(4VP-co-EDMA) and poly (GMA-co-EDMA) and directly injected 10 ppm megestrol standard.

6. Conclusions

Polymer monolith microextraction (PPME) has been demonstrated to be feasible, facile and versatile platform for analysis, preconcentration and extraction of analytes and for online coupling to GC-MS. The monoliths that have been tested including poly (styrene-co-divinylbenzene), poly(acrylamide-co-EDMA), poly (4-vinylpyridine-co-EDMA) and poly (glycidyl methacrylate-co-EDMA, have been found to work well when matched with analytes of corresponding polarity. The polymer monoliths based materials provided for satisfactorily fast mass transfer and thus good extraction efficiency. Good linear calibration curves were achieved in a wide concentration range when using poly(acrylamide-co-EDMA) for analysis of caffeine, poly (styrene-co-divinylbenzene) for analysis of PAHs. Both poly (4-vinylpyridine-co-EDMA) and poly (glycidyl methacrylate-co-EDMA were tested and found to be good sorbents for extraction of hormones such as megestrol with good % recoveries of 98% and 107% respectively, which suggests the potential of applying this PPME method to a wider selection of hormones and to a more complicated sample matrix.

An analogous PLOT polymer monolith was successively prepared and is being tested as a microextraction platform.

7. References

Aresta, A.; Palmisano, F. & Zambonin, C. G. (2005). Simultaneous determination of caffeine, theobromine, *Food Chemistry*, Vol. 93, Issue 1, (November 2005), pp. 177-181, ISSN 0308-8146.

Arthur, C.L. & Pawliszyn, J. (1990). Solid phase microextraction with thermal desorption using fused silica optical fibers. *Analytical Chemistry*, Vol. 62, No. 19, (October 1990), pp. 2145–2148, ISSN 0003-2700.

Bravo, J. C.; Fernández, P. & Dúrand, J. S. (2005). Flow injection fluorimetric determination of β-estradiol using a molecularly imprinted polymer. *Analyst*, Vol. 130, No.10, (August 2005), pp. 1404-1409, ISSN 0003-2654.

Carpinteiro, J.; Quintana, J.B.; Rodriguez, I.; Carro, A.M.; Lorenzo, R.A. & Cela, R. (2004). Applicability of solid-phase microextraction followed by on-fiber silylation for the determination of estrogens in water samples by gas chromatography-tandem mass spectrometry. *Journal of Chromatography A*, Vol. 1056, No.1-2, (November 2004), pp. 179-185, ISSN 0021-9673.

Fan, Y.; Zhang, M. & Feng, Y. Q. (2005). Poly (acrylamide-vinylpyridine-N, N′-methylene bisacrylamide) monolithic capillary for in-tube solid-phase microextraction coupled to high performance liquid chromatography. *Journal of Chromatography A*, Vol. 1099, No. 1-2, (September 2005), pp. 84-91, ISSN 0021-9673.

Gibson, G.T.; Mugo, S.M.& Oleschuk, R.D. (2008). Surface-mediated effects on porous polymer monolith formation within capillaries. *Polymer*, vol. 49, No. 13-14, (June 2008), pp.3084–3090, ISSN 0032-3861

Görög, S.(2011). Advances in the analysis of steroid hormone drugs in pharmaceuticals and environmental samples. *Journal of Pharmaceutical and Biomedical Analysis*, Vol. 55, No. 4, (November 2010), pp.728–743, ISSN 0731-7085.

Jeannot, M. A. & Cantwell, F. F. (1996). Solvent Microextraction into a Single Drop. *Analytical Chemistry*, Vol. 68, No. 13, (July 1996), pp. 2236-2240, ISSN 0003-2700.

Körner, W.; Bolz, U.; Süßmuth, W.; Hiller, G.; Schuller, W.; Hanf, V. & Hagenmaier, H. (2000). Input/output balance of estrogenic active compounds in a major municipal sewage plant in Germany. *Chemosphere*, Vol. 40, No. 9-11, (May 2000), pp. 1131-1142, ISSN 0045-6535.

Liu, L.; Cheng, J.; Matsadiq, G. & Li, J. K. (2011). Novel polymer monolith microextraction using a poly-(methyl methacrylate-co-ethylene dimethacrylate) monolith and its application to the determination of polychlorinated biphenyls in water samples to the determination of polychlorinated biphenyls in water samples. *Chemosphere*, Vol. 83, No. 10, (April 2011), pp. 1307-1312, ISSN 0045-6535.

Mitani, K.; Fujioka, M. & Kataoka, H. (2005). Fully automated analysis of estrogens in environmental waters by in-tube solid-phase microextraction coupled with liquid chromatography-tandem mass spectrometry. *Journal of Chromatography A*, Vol. 1081, No. 2, (July 2005), pp. 218-224, ISSN 0021-9673.

Pan, Y.P. & Tsai, S. W. (2008). Solid phase microextraction procedure for the determination of alkylphenols in water by on-fiber derivatization with N-tert-butyl-dimethylsilyl-N-methyltrifluoroacetimide. *Analytica Chimica Acta*, Vol. 624, No. 2, (August 2008), pp. 247-252, ISSN 0003-2670.

Park, C. H.; Park, S. J.& Shin, H. S. (2001). Sensitive determination of bisphenol A in environmental water by gas chromatography with nitrogen–phosphorus detection after cyano methylation. *Journal of Chromatography A*, Vol.912, No.1, (March 2001), pp. 119-125, ISSN 0021- 9673.

Planeta, J.; Moravcová, D.; Roth, M.; Karásek, P.& Kahle, V. (2010). Silica-based monolithic capillary columns- Effect of preparation temperature on separation efficiency.

Journal of Chromatography A, Vol. 1217, No. 36, (July 2010), pp. 5737-5740, ISSN 0021-9673.

Pena-Pereira, F.; Lavilla, I.& Bendicho, C. (2009). Miniaturized preconcentration methods based on liquid–liquid extraction and their application in inorganic ultratrace analysis and speciation: A review. *Spectrochimica Acta Part B: Atomic Spectroscopy*, Vol. 64, No. 1, (January 2009), pp. 1-15, ISSN 0584-8547.

Sanbe, H.& Haginaka, J. (2003). Uniformly sized molecularly imprinted polymers for bisphenol A and β-estradiol; retention and molecular recognition properties in hydro-organic mobile phases. *Journal of Pharmaceutical and Biomedical Analysis*, Vol. 30, No. 6, (2003), pp. 1835-1844, ISSN 0731-7085.

Schweitz, L. (2002). Molecularly imprinted polymer coatings for open-tubular capillary electrochromatography prepared by surface initiation. *Analytical Chemistry*, Vol. 74, No.5, (January 2002), pp. 1192-1196, ISSN 0003-2700.

Svec, F. (2010) Porous polymer monoliths: amazingly wide variety of techniques enabling their preparation. *Journal of Chromatography A*, Vol. 1217, No. 6, (February 2010), pp. 902-924, ISSN 0021-9673.

Ueta, I.; Saito, Y.; Teraoka. K.; Matsuura, H.; Fujimura, K.& Jinno, K. (2010). Novel Fire Investigation Technique Using Needle Extraction in Gas Chromatography. *Analytical Sciences*, Vol. 26, No. 11, (November 2010), pp. 1127-1132, ISSN 0910-6340.

Wen, Y.; Zhou, B. S.; Xu, Y.; Jin, S. W. & Feng, Y. Q. (2006). Analysis of estrogens in environmental waters using polymer monolith in polyether ether ketone tube solid-phase microextraction combined with high-performance liquid chromatography. *Journal of Chromatography A*, Vol. 1133, No. 1-2, (November 2006), pp. 21-28, ISSN 0021-9673.

Yang, L.; Lan, C.; Liu,H.; Dong, J. & Luan, T. (2006). Full automation of solid-phase microextraction/on-fiber derivatization for simultaneous determination of endocrine-disrupting chemicals and steroid hormones by gas chromatography–mass spectrometry. *Analytical and Bioanalytical Chemistry*, Vol. 386, No.2, (July 2006), pp. 391-397, ISSN 1618-2642.

Yi, W.; Ying, W.; Bing-Sheng, Z.; Xu, Y. & Yu-Qi, F. (2007). Determination of sexual hormones in liquid cosmetics by polymer monolith microextraction coupled with high performance liquid chromatography. *Chinese Journal of Analytical Chemistry*, Vol. 35, No. 5, (May 2007), pp. 681-684, ISSN 1872-2040.

Zhang, M.; Fang, W.; Zhang, Y.F.; Nie, J. & Feng, Y.Q. (2006). Novel polymer monolith microextraction using a poly (methacrylic acid-ethylene glycol dimethacrylate) monolith and its application to simultaneous analysis of several angiotensin II receptor antagonists in human urine by capillary zone electrophoresis. *Journal of Chromatography A*, Vol. 1102, No. 1-2, (January 2006), pp. 294–301, ISSN 0021-9673.

Zhang, Z. & Pawliszyn, J. (1995). Quantitative extraction using an internally cooled solid phase microextraction device. Analytical Chemistry, Vol. 67, No. 1, , pp. 34-43, ISSN 0003-2700.

Zhou, J.; Ma, C.; Zhou, S.; Ma, P.; Chen F.; Qi, Y. & Chen, H. (2010). Preparation, evaluation and application of molecularly imprinted solid-phase microextraction monolith for selective extraction of pirimicarb in tomato and pear. *Journal of Chromatography A*, Vol. 1217, No. 48, (November 2010), pp. 7478-7483, ISSN 0021-9673.

5

Derivatization Reactions and Reagents for Gas Chromatography Analysis

Francis Orata
Masinde Muliro University of Science and Technology, Kenya

1. Introduction

Derivatization reactions are meant to transform an analyte for detectability in Gas Chromatography (GC) or other instrumental analytical methods. Derivatization in GC analysis can be defined as a procedural technique that primarily modifies an analyte's functionality in order to enable chromatographic separations. A modified analyte in this case will be the product, which is known as the derivative. The derivative may have similar or closely related structure, but not the same as the original non-modified chemical compound.

Volatility of sample is a requirement for GC analysis. Derivatization will render highly polar materials to be sufficiently volatile so that they can be eluted at reasonable temperatures without thermal decomposition (Knapp, 1979) or molecular re-arrangement (Kühnel et al., 2007; Blau and King 1979). Understanding the chemistry of the analytes, derivatizing reagents used in sample preparation, and the detailed functionality of Gas Chromatography are important to get reliable results. For GC analysis, compounds containing functional groups with active hydrogens such as -SH, -OH, -NH and -COOH are of primary concern because of the tendency of these functional groups to form intermolecular hydrogen bonds (Zaikin and Halket, 2003). These intermolecular hydrogen bonds affect the inherent volatility of compounds containing them, their tendency to interact with column packing materials and their thermal stability (Sobolevsky et al., 2003). Since GC is used to separate volatile organic compounds, modification of the functional group of a molecule by derivatization enables the analysis of compounds that otherwise can not be readily monitored by GC. Derivatization process either increases or decreases the volatility of the compound of interest. It also reduces analyte adsorption in the GC system and improves detector response, peak separations and peak symmetry.

In addition to particular analytes such as pharmaceuticals, biomolecules such us organic acids, amides, poly-hydroxy compounds, amino acids, pesticides and other persistent organic compounds, new classes of compounds of interest for example fluorinated alkylated substances and polycyclic aromatic hydrocarbons continue to emerge. It is necessary to develop and/or improve on chemical analytical methods and hence the need to familiarize with derivatization methods that are applicable to GG analysis. Generally derivatization is aimed at improving on the following aspects in Gas Chromatography.

i. Suitability

Suitability is the form of compounds that is amenable to the analytical technique. For GC, it is a requirement that the compound to be analyzed should be volatile with respect to gas chromatographic analysis conditions, as compared to liquid chromatography (LC), where the compound of interest should be soluble in the mobile phase. Therefore, derivatization procedure modifies the chemical structure of the compounds so that they can be analyzed by the desired technique.

ii. Efficiency

Efficiency is the ability of the compound of interest to produce good peak resolution and symmetry for easy identification and practicability in GC analysis. Interactions between the compounds themselves or between the compounds and the GC column may reduce the separation efficiency of many compounds s and mixtures (Knapp, 1979). Derivatization of analyte molecules can reduce these interactions that interfere with analysis. Also, compounds that co-elute or have poor resolution from other sample components during separation in GC can frequently be resolved by an appropriate derivative.

iii. Detectability

Detectability is the outcome signal that emanates from the interaction between the analyte and the GC detector. Increasing the amounts of materials will impact the range at which they can be detected in Gas chromatography. This can be achieved either by increasing the bulk of the compound or by introducing onto the analyte compound, atoms or functional groups that interact strongly with the detector and hence improve signal identification. For example the addition of halogen atoms to analyte molecules for electron capture detectors (ECD) and the formation of trimethylsilyl (TMS) ether derivatives to produce readily identifiable fragmentation patterns and mass ions (Knapp, 1979).

1.1 Derivatization reagent

Derivatization reagent is the substance that is used to chemically modify a compound to produce a new compound which has properties that are suitable for analysis in GC or LC. The following criteria must be used as guidelines in choosing a suitable derivatization reagent for GC analysis.

i. The reagent should produce more than 95 % complete derivatives.
ii. It should not cause any rearrangements or structural alterations of compounds during formation of the derivative.
iii. It should not contribute to loss of the sample during the reaction.
iv. It should produce a derivative that will not interact with the GC column.
v. It should produce a derivative that is stable with respect to time.

1.2 Objectives for derivatization

The following outlined objectives among others can be achieved by application of proper derivatization procedures;

i. Improvement of resolution and reduce tailing of polar compounds which may contain -OH, -COOH, =NH, $-NH_2$, -SH, and other functional groups.

ii. Analysis of relatively nonvolatile compounds.
iii. Reduction of volatility of compounds prior to GC analysis.
iv. Improvement of analytical efficiency and hence increase detectability.
v. Stabilization of compounds for GC analysis.

2. Types of derivatization reactions

Derivatization reactions used for gas chromatography (GC) fall into three general reaction types namely; Alkylation of which the general process is esterification, Acylation and Silylation. Through these three processes, highly polar materials such as organic acids, amides, poly-hydroxy compounds, amino acids are rendered suitable for GC analysis by making them sufficiently volatile. These general processes are discussed below.

2.1 Alkylation

Alkylation is mostly used as the first step for further derivatizations or as a method of protection of certain active hydrogens in a sample molecule. It represents the replacement of active hydrogen by an aliphatic or aliphatic-aromatic (e.g., benzyl) group in process referred to as esterification. Equation 1 below shows the general reaction equation representing the esterification process.

$$RCOOH + PhCH_2X \rightarrow RCOOCH_2Ph + HX$$

Equation 1: General reaction for esterification process: X = halogen or alkyl group R, H = another alkyl group R.

The principal chromatographic use of this reaction is the conversion of organic acids into esters, especially methyl esters that produce better chromatograms than the free acids. Alkylation reactions can also be used to prepare ethers, thioethers and thioesters, N-alkylamines, amides and sulphonamides (Danielson, 2000). In general, the products of alkylation are less polar than the starting materials because active hydrogen has been replaced by an alkyl group. The alkyl esters formed offer excellent stability and can be isolated and stored for extended periods if necessary. In esterification an acid reacts with an alcohol to form an ester. In the reaction, a catalyst more often an inorganic acid such as hydrochloric acid or thionyl chloride (Zenkevich, 2009) is recommended for example, in the trans-esterification of fats or oils (Sobolevsky et al., 2003).

2.1.1 Derivatization reagents used in alkylation

Common derivatization reagents for the Alkylation type of reactions are Dialkylacetals, Diazoalkales, Pentafluorobenzyl bromide (PFBBr), Benzylbromide, Boron trifluoride (BF_3) in methanol or butanol and Tetrabutylammonium hydroxide (TBH) among others. Alkylation reagents can be used alone to form esters, ethers and amides or they can be used in conjunction with acylation or silylation reagents. The reaction conditions can vary from strongly acidic to strongly basic with both generating stable derivatives. However, it must be noted that the reagents are more limited to amines and acidic hydroxyls and that the reaction conditions are frequently severe while the reagents are often toxic. Some derivatization reagents and their respective derivatization procedures in alkylation reactions are discussed below.

2.1.1.1 Dialkylacetals

Dimethylformamide (DMF) is an example of dialkylacetals with a general formula $CH_3CH_3NCHOROR$ are used to esterify acids to their methyl esters. Dialkylacetals have a wider applicability for the derivatization of a number of functional groups containing reactive hydrogens. Because the principal reaction product is dialkylacetals (DMF), the isolation of the derivative is not required and the reaction mixture can be injected directly into the gas chromatograph (Regis, 1999). This reagent is an excellent first choice for derivatization of a compound for which there is no published method available. The reaction between N, N-dimethylformamide dimethylacetal and Carboxylic acid is as follows (Equation 2).

$$CH_3CH_3NCHOROR + R`COOH \rightarrow R`COOR + ROH + CH_3CH_3NCHO$$

Equation 2: The reaction between N, N-dimethylformamide dimethylacetal and Carboxylic acid.

Although carboxylic acids, phenols, and thiols react quickly with DMF, to give the corresponding alkyl derivatives, hydroxyl groups are not readily methylated. During derivatization procedure, care should be taken because *N*, *N*-dimethylformamide dimethylacetals are moisture sensitive. The reagents work quickly with derivatization taking place just upon dissolution. The reaction is suitable for flash alkylation, where derivatization takes place in the injection port. Since adjustment of polarity and volatility of the sample is possible, this can allow change of retention time. It is worth noting that the reagents will react with water to give the corresponding alcohol though traces of water will not affect the reaction as long as you have an excess of the acid.

A rapid procedure for the derivatization of both carboxylic and amino acids using DMF-Dialkylacetal reagents is by dissolving the sample in 0.5 mL of a 1:1 solvent of choice/ reagent mixture by heating to 100 °C. For example, it is advisable to use pyridine for fatty acids, acetonitrile for amino acids as solvent of choice (Thenot and Horning, (1972).

Alternative method comprises of combining 50 mg fatty acid and 1 ml DMF in a reaction vial. Cap the vial and heat at 60 °C for 10 - 15 minutes or until dissolution is complete and analyze the cooled sample in gas chromatography (Thenot et al., 1972).

2.1.1.2 Diazoalkales

The main reagent in the diazoalkales group is diazomethane. Diazomethane (N_2CH_2) is the quickest and cleanest method available for the preparation of analytical quantities of methyl esters. The reaction of diazomethane with a carboxylic acid is quantitative and essentially instantaneous in ether solutions. In the presence of a small amount of methanol as catalyst, diazomethane reacts rapidly with fatty acids, forming methyl esters. Elimination of gaseous nitrogen drives the reaction forward. The reaction for the conversion of carboxylic acids to methyl esters (Equation 3) is outlined below:

$$RCOOH + {}^{-}CH_2N^{+}N \rightarrow RCOOCH_3 + N_2$$

Equation 3: The reaction for the conversion of carboxylic acids to methyl esters

The yield is high and side reactions are minimal. Sample sizes of 100 - 500 μl are easily derivatized and the isolation of the methyl esters is simple and quantitative when dealing with acids having chain lengths from C8 to C24. However, care should be taken in handling diazomethane because it is carcinogenic, highly toxic, and potentially explosive.

2.1.1.3 Pentafluorobenzyl bromide (PFBBr) and Pentafluorobenzyl-hydroxylamine hydrochloride (PFBHA)

Pentafluorobenzyl bromide ($C_7H_2F_5Br$) and also Pentafluorobenzyl-hydroxylamine hydrochloride can be used to esterify phenols, thiols, and carboxylic acids. Equation 4 below is an example for PFBBr derivatization process.

$$R`OR + C_7H_2F_5Br \rightarrow R`OC_7H_2F_5 + RBr$$

Equation 4: Reaction of PFBBr with either phenols, thiols or carboxylic acids: R = Hydrogen

Derivatization can be done with O-(2,3,4,5,6-pentafluorobenzyl)-hydroxylamine hydrochloride (PFBHA) for carbonyls and pentafluorobenzyl bromide (PFBBr) for carboxylic acid and phenol groups for gas (GC/MS) in an electron impact mode (EI) and a gas chromatograph/ion trap mass spectrometry (GC/ITMS) in both chemical impact and EI modes. To confirm different isomers, the PFBHA-derivatives of analytes can be rederivatized by silylation using N, O-bis (trimethylsilyl)-trifluoroacetamide (BSTFA) (Jang and Kamens, 2001).

Example for the application of this method is from the work of Allmyr et al., (2006) where the analysis of Triclosan in human plasma and milk was accomplished by the conversion of Triclosan into its pentafluorobenzyl ester by adding 2 ml of H_2O (Milli-Q), 50 µl of 2M KOH, 10 µl pentafluorobenzyl chloride in 10% toluene and 0.3 g NaCl (more sodium chloride was added if emulsion formed upon mixing) to the sample extract and shaking the tube vigorously for 2 minutes. This was followed by extraction of the aqueous phase with 2 ml of n-hexane. Then, 3 ml of 98 % H_2SO_4 was added to the extract, and the tube inverted 60 times and another extraction using 2 ml of n-hexane followed. The final extract was reduced to 2 ml under a gentle stream of nitrogen gas at room temperature. Approximately 0.5 ml of the extract was injected to GC for analysis.

i. Organic Acids derivatisation

The following procedure by Chien et al., (1998) and Galceran et al., (1995) is a variation of methods previously utilized for organic acid and phenol analysis. In the method, sample extracts for organic acid/phenol derivatization are first dried by passing a gentle stream of nitrogen in the sample. Once the solvent has been evaporated, acetone is added to bring each sample to a volume of 500 µL. Each sample is added 20 µL of 10% PFBBr solution and 50 µL of 1,4,7,10,13,16-hexaoxacyclooctadecane $[C_2H_4O]_6$ (18-crown-6 ether solution), both in acetone. Approximately 10 mg of potassium is added to each extract, and the extracts are capped and sonicated for three hours. Upon completion of sonication the acetone is evaporated by passing a gentle stream of nitrogen, and the residue dissolved into hexane for GC analysis.

Equation 5 below shows the chemical reactions for the conversion of organic acids and phenols into their pentafluorobenzyl esters and ether respectively using pentafluorobenzyl bromide (PFBBr) derivatization.

i. $$ROCOH + C_7H_2F_5Br \xrightarrow[18-c-6\ ether]{K_2CO_3} C_7H_2F_5OCOR + HBr$$

$$\text{ii.}\quad PhOH + C_7H_2F_5Br \xrightarrow[\text{18-c-6 ether}]{K_2CO_3} C_7H_2F_5OPh + HBr$$

Equation 5: Reactions for the conversion of organic acids and phenols into their pentafluorobenzyl esters (i) and ether (ii) respectively using pentafluorobenzyl bromide (PFBBr).

In another example of pentafluorobenzylation procedure (Naritisin and Markey, 1996 & Kawahara, 1968), 20 µl of PFBBr is added to the vial containing already extracted and cleaned sample in methyl chloride. The vial is then capped and shaken for 20 - 30 minutes. Gas chromatography analysis can then be accomplished using either of the two methods depending on the available detector. A portion of methylene chloride phase can be injected into the chromatograph for FID analysis or it can be evaporated using a gentle stream of nitrogen gas and re-dissolve in benzene for ECD analysis. Reagents containing fluorinated benzoyl groups are good for ECD analysis.

ii. Carbonyls

The derivatization process performed by Spaulding et al., (2003), Destaillats et al., (2001) & Rao et al., (2001) for the analysis of carbonyl involves first, the reduction in volume of sample extracts to < 50 µL under a gentle stream of nitrogen. Once the extract volume has been reduced, a 9:1 (v/v) mixture of carbonyl-free Acetonitrile: Dichloromethane (DCM) is added to bring the sample to a volume of 500 µL. This is followed by the addition of 50 $mgmL^{-1}$ solution of *O*-(2, 3, 4, 5, 6-pentafluorobenzyl) hydroxylamine hydrochloride (PFBHA) in methanol that will result to a target PFBHA concentration of 5 mM. The sample is left at room temperature for a period of 24 hours then subsequently analyzed using GC. The balanced chemical reaction where *O*-(2, 3, 4, 5, 6-pentafluorobenzyl) hydroxylamine hydrochloride (PFBHA) is used for the conversion of carbonyls into their pentafluorobenzyl oximes is provided below (Equation 6).

$$R'OCR'' + C_6F_5\text{-}ONH_2 \rightarrow C_6F_5\text{-}ONR'CR'' + \text{Isomer(s)} + H_2O$$

Equation 6: Chemical reaction for the conversion of carbonyls into pentafluorobenzyl oximes using *O*-(2, 3, 4, 5, 6-pentafluorobenzyl) hydroxylamine hydrochloride (PFBHA).

The PFBHA derivatization process for most of the carbonyl compounds could be completed in 2 h at room temperature (Bao et al., 1998).

2.1.1.4 Benzylbromide

Benzyl bromide reacts with the acid part of an alkyl acid to form an ester, and therefore increase the volatility of the analyte of interest. Orata et al., (2009) determined long chain perfluorinated acids namely; perfluoro-n-octanoic acid (PFOA), 2H-perfluoro-2-octenoic acid (FHUEA), 2H-perfluoro-2-decenoic acid (FOUEA) and 2H-perfluoro-2-dodecenoic acid (FNUEA) in biota (fish) and abiota (water) by derivatization and subsequent analysis by GC/MS. The method involved derivatization of long chain perfluorinated compounds using benzyl bromide solution and acetone to form benzylperfluorooctanoate (benzyl ester) as presented by the equation 7 below which shows the modification of perfluorooctanoic acid to the respective ester.

$$CF_3(CF_2)_6COOH + C_7H_7Br \rightarrow C_{15}H_7F_{15}O_2 + HBr$$

Equation 7: Convertion of Perfluorooctanoic acid to Benzylperfluorooctanoate using benzyl bromide.

Long Chain Perfluorinated alkyl acids derivatization procedure by Orata et al (2009) is as follows: Following extraction and sample clean up procedure, the sample is dried by a gentle stream of nitrogen. The sample residue is then dissolved in 0.5 mL of acetone and 0.1 mL of benzyl bromide solution is added to it. The tube is heated at 80 °C for 15 min, cooled, and evaporated to dryness with nitrogen. The residue is then dissolved in 1 mL of methylene chloride for GC analysis. Results obtained in the study (Orata et al, 2009) demonstrate that GC/MS can be an alternative to the LC/MS method for quantification of perfluorinated acids in contaminated areas, where expected higher concentration of the analyte is expected. It was in the view of the author (Orata et al, 2009) that the method can be further improved to lower the detection limits.

2.1.1.5 Tetrabutylammonium hydroxide (TBH)

Derivatization of a carboxylic acid with tetrabutylammonium hydroxide (TBH) forms butyl ester, which will allow a longer retention times in a GC column (Lin et al., 2008). The reagent is most commonly used for low molecular weight acids and is especially suitable for low molecular weight amines (Regis, 1999). Equation 8 represents the derivatization reaction for the conversion of carboxylic acid to alkyl esters using TBH.

$$[N\ (CH_3)_4]^+[OH]^- + RCOOH \rightarrow RCOOC_4H_9$$

Equation 8: Conversion of carboxylic acid to alkyl esters using TBH.

The following derivatization procedure can be used for flash alkylation which is suitable for biological fluids and thermally stable fatty acids analysis. Biological fluid or tissue is first extracted using toluene. Then 4 mL of the extract is transferred into a nipple tube and evaporate under a gentle stream of nitrogen gas at 60 °C. 25 µL of TBH solution is added to dissolve the residue. After 30 minutes, 4 µL of the dissolved residue is injected directly onto the chromatograph. The injection port temperature is set at 260 °C or above in GC instrument.

2.1.1.6 Boron trifluoride (BF3) in methanol or butanol

Boron trifluoride (BF_3) in methanol or *n*-Butanol has a general formula is F_3B: HO-C_nH_{2n+1} where (n = 1 or 4). This reagent is convenient and inexpensive method for forming esters. It is most commonly used to form methyl (butyl) ester by reacting it with acids, as shown by the following general equation 9.

$$F_3B{:}\ HO\text{-}C_nH_{2n+1} + RCOOH \rightarrow RCOOC_nH_{2n+1}$$

Equation 9: General equation for formation of alkyl esters using Boron trifluoride (BF_3) as a derivatization reagent.

Boron trifluoride reagent forms fluoroboron compounds by reaction with atmospheric oxygen and methanol and therefore making it prone to instability (Christie 1993). Therefore, it should always be stored at the refrigerator temperature discarded after a few months use.

In a study by Lough (1964) were polyunsaturated fatty acids were analyzed, it was observed that boron trifluoride-methanol will cleave the rings in cyclopropane fatty acids (commonly encountered in microorganisms), and it causes cis-trans isomerization of double bonds in conjugated fatty acids. In addition, it reacts with the antioxidant Butylated hydroxytolune (BHT) to produce spurious peaks in chromatograms.

The following esterification procedure can be used in derivatization using Boron trifluoride. To a 100 mg organic acid in a vial, 3 mL BF_3 in methanol or BF_3 in n-butanol is added and heated to 60 °C for 5 to 10 minutes. Heating temperature and time may vary i.e. Ribeiro et al., (2009) used 90 °C for 10 minutes for the process. This is followed by cooling and transferring the mixture to a separating funnel with 25 mL hexane. Further, the sample is washed 2 times with saturated NaCl solution, and then dried over anhydrous Na_2SO_4. The solvent is evaporated to concentrate the sample and lastly injected onto column for GC analysis.

2.2 Silylation

Silylation is the most prevalent derivatization method as it readily volatizes the sample and therefore very suitable for non-volatile samples for GC analysis. Silylation is the introduction of a silyl group into a molecule, usually in substitution for active hydrogen such as dimethylsilyl [$SiH(CH_3)_2$], t-butyldimethylsilyl [$Si\ (CH_3)_2C(CH_3)_3$] and chloromethyldimethylsilyl [$SiCH_2Cl(CH_3)_2$]. Replacement of active hydrogen by a silyl group reduces the polarity of the compound and reduces hydrogen bonding (Pierce, 1968). Many hydroxyl and amino compounds regarded as nonvolatile or unstable at 200 – 300 °C have been successfully analyzed in GC after silylation (Lin et al., 2008 & Chen et al., 2007). The silylated derivatives are more volatile and more stable and thus yielding narrow and symmetrical peaks (Kataoka, 2005).

2.2.1 Silylation reaction and mechanism

The silylation reaction is driven by a good leaving group, which means a leaving group with a low basicity, ability to stabilize a negative charge in the transitional state, and little or no back bonding between the leaving group and silicon atom (Knapp, 1979). The mechanism involves the replacement of the active hydrogens (in -OH, -COOH, -NH, -NH_2, and -SH groups) with a trimethylsilyl group. Silylation then occurs through nucleophilic attack (SN_2), where the better the leaving group, the better the siliylation. This results to the production of a bimolecular transition state (Kühnel et al., 2007) in the intermediate step of reaction mechanism. The general reaction for the formation of trialkylsilyl derivatives is shown by equation 10. The leaving group in the case of trimethylchlorosilane (TMCS) is the Cl atom.

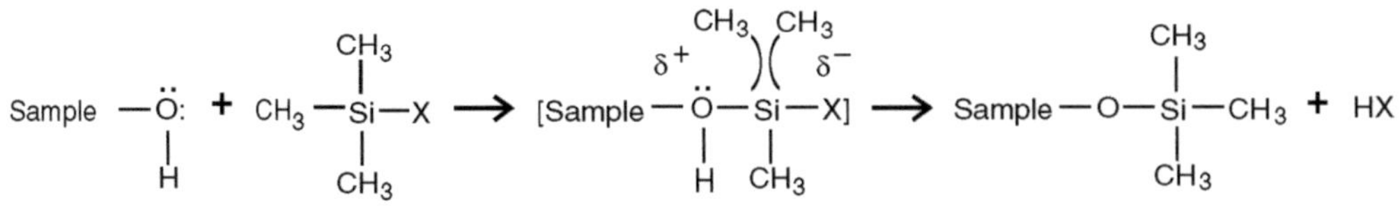

Equation 10: General reaction mechanism for the formation of trialkylsilyl derivatives for trimethylchlorosilane, X = Cl

In silylation derivatisation, care must be taken to ensure that both sample and solvents are dry. Silyl reagents generally are moisture sensitive, and should be stored in tightly sealed containers (Sobolevsky et al., 2003) and therefore the solvents used should be as pure and as little as possible. This will eliminate excessive peaks and prevent a large solvent peak. In silylation, pyridine is the most commonly used solvent. Although pyridine may produce peak tailing, it is an acid scavenger and will drive the reaction forward. In many cases, the need for a solvent is eliminated with silylating reagents. The completion of the derivatization process in silylation is usually observed when a sample readily dissolves in the reagent. According to Regis (1999) the ease of reactivity of the functional group toward silylation follows the order:

Alcohol > Phenol > Carboxyl > Amine > Amide /hydroxyl

For alcohols, the order will be as follows:

Primary > Secondary > Tertiary

Many reagents will require heating that is not in excess of 60 °C for about 10 - 15 minutes, to prevent breakdown of the derivative. Although hindered products may require long term heating.

2.2.2 Derivatization reagents used in Silylation

Reagents used for the silylation derivatization process include Hexamethyldisilzane (HMDS), Trimethylchlorosilane (TMCS), Trimethylsilylimidazole (TMSI), Bistrimethylsilylacetamide (BSA), Bistrimethylsilyltrifluoroacetamide (BSTFA), N-methyl-trimethylsilyltrifluoroacetamide (MSTFA), Trimethylsilyldiethylamine (TMS-DEA), N-methyl-N-t-butyldimethylsilyltrifluoroacetamide (MTBSTFA), and Halo-methylsilyl derivatization reagents. Halo-methylsilyl derivatization reagents which although not discussed in this chapter can produce both silylated and halogenated derivatives for ECD. Silyl reagents will react with both alcohols and acids to form trimethylsilyl ethers and trimethylsilyl esters respectively. These derivatives formed are volatile and for the most part, are easily separated (Scott, 2003). Silyl reagents are compatible with most detection systems but, if they are used in excess, they can cause difficulties with flame ionization detectors (FID) (Sobolevsky et al, 2003). Silyl reagents are influenced by both the solvent system and the addition of a catalyst. A catalyst (e.g., trimethylchlorosilane or pyridine) increases the reactivity of the reagent.

Reagents that introduce a t-butyldimethylsilyl group in place of the trimethylsilyl group were developed to impart greater hydrolytic stability to the derivatives. These t-butyldimethylsilyl derivatives not only have improved stability against hydrolysis, but they also have the added advantage of distinctive fragmentation patterns, which makes them useful in a variety of GC/MS applications. Most trimethylsilyl and t-butyldimethylsilyl derivatives have excellent thermal stability and are amenable to a wide range of injection and column conditions (Pierce 2004). Silylation gives ability to derivatize a wide variety of compounds for GC analysis. In addition a large number of silylating reagents available.

2.2.2.1 Bis(trimethylsilyl)-acetamide (BSA)

Bis(trimethylsilyl)-acetamide (BSA) was the first widely used silylating reagent. The strength of BSA as a strong silylating reagent is enhanced more because acetamide is a good leaving group. BSA reacts under mild conditions and produces relatively stable by-products. However, the by-product which is trimethylsilyl -acetamide sometimes produces peaks that overlap those of other volatile derivatives as shown in the equation 11 below.

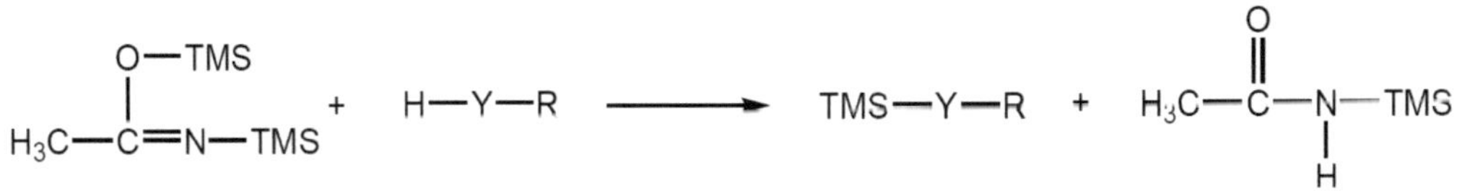

Equation 11: Equation showing the by-product which is trimethylsilyl -acetamide from Bis(trimethylsilyl)-acetamide silylation reagent: TMS = $Si(CH_3)_3$, Y = O, S, NH, NR`, COO, R, R` = Alk, Ar.

BSA forms highly stable trimethylsilyl derivatives with most organic functional groups under mild reaction conditions but their mixtures also oxidize to form silicon dioxide, which can foul FID detectors. Experimentally, BSA has been found to be very efficient and only a small amount of reagent is required. Moreso the derivatization process takes a shorter time at room temperature. Saraji and Mirmahdieh, (2009) studied the influence of different parameters on the in-syringe derivatization process using BSA were three parameters, the amount of the BSA, the reaction time and the reaction temperature were investigated to achieve the highest derivatization reaction yield. The results showed that 0.2 μL volume of BSA derivatization reagent is enough to obtain the maximum derivatization reaction yield for all compounds in 2 min time at room temperature.

2.2.2.2 Bis(trimethylsilyl)trifluoroacetamide (BSTFA)

N, N-bis(trimethyl-silyl)trifluoro-acetamide (BSTFA) just like bis(trimethylsilyl)-acetamide (BSA) are the two most popular reagents for Silylation type of derivatization. They react rapidly with organic acids to give high yields. Gerhke, (1968) developed a method in which a few milligram of an acid is placed in a vial and about 50 μl of BSA or BSTFA is added to it. The reaction can be expected to be complete directly within the solution, but the mixture can also be heated for 5 to 10 min. at 60 °C to ensure that reaction is really complete, before GC analysis. BSTFA reacts faster (as shown using the reaction 12 below) and more completely than BSA, due to presence of trifluoroacetyl group.

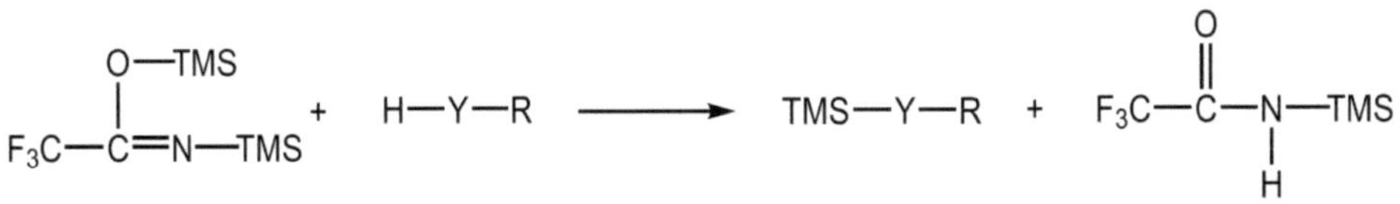

Equation 12: Silylation reaction using N, N-bis(tri methyl-silyl)trifluoro-acetamide: TMS = $Si(CH_3)_3$, Y = O, S, NH, NR`, COO, R, R` = Alk, Ar.

The high volatility of BSTFA and its byproducts results in separation of early eluting peaks. In addition, its stable products result in low detector noise and fouling. Addition of trimethylchlorosilane (TMCS) catalyzes reactions of hindered functional groups in secondary alcohols and amines.

In the method used by Schauer et al., (2002), sample extracts for analysis of hydroxylated polycyclic aromatic hydrocarbons (Hydroxy-PAHs), the sample extracts are first reduced in volume to 200 µL under a gentle stream of nitrogen. To each sample 20 µL of freshly prepared solution of 10% (v/v) Trimethylchlorosilane (TMCS) in N-O-bis (trimethylsilyl)-trifluoroacetamide (BSTFA) is added. The samples are capped, wrapped with Teflon tape, and heated at 45 °C for 24 hours to convert the targeted analytes to their trimethylsilyl derivatives. The balanced reaction (Equation 13) for this conversion to trimethyl silyl (TMS) ethers using N-O-bis (trimethylsilyl)-trifluoroacetamide (BSTFA) is provided in below.

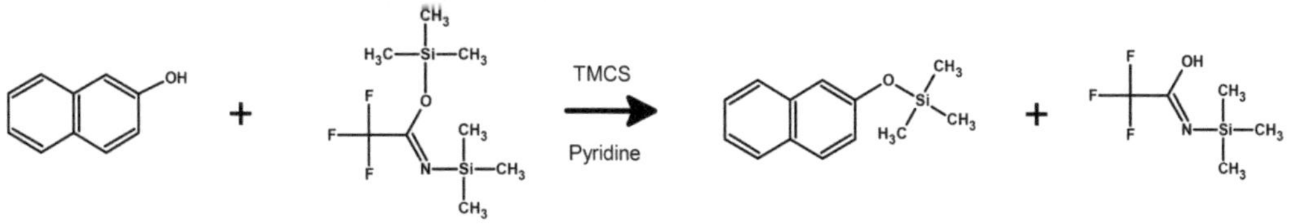

Equation 13: Derivatization of Hydroxylated polycyclic aromatic hydrocarbons (Hydroxy-PAHs) into trimethyl silyl (TMS) ethers using BSTFA.

Caution must be exercised to prevent any water from entering the samples as this will lead to hydrolysis of the BSTFA reagent and prevent any of the targeted analytes from undergoing derivatization.

Szyrwińska et al., (2007) and Kuo and Ding, (2004) used N, O-bis-(trimethylsilyl)trifluoroacetamide (BSTFA) containing 1% trimethylchlorosilane (TMCS), and bromoacetonitrile (BAN) reagents to analyze standard solutions of Bisphenol-A (BPA) and extracts of powdered milk.

The method was by derivatization using Trimethylsilylation with the derivatizing reagent which was BSTFA + 1% TMCS. Bisphenol-A standard solution (200 µL) was placed in a vial (1 mL) and evaporated to dryness under a gentle stream of nitrogen at 60 °C. Silylating agent (BSTFA containing 1 % TMCS; 100 µL) was added to the residue and the vial was vortex mixed and heated at 80 °C for 30 min. After cooling, the derivatized solution was evaporated to dryness and the residue was re-dissolved in 100 µL chloroform. This solution (1 µL) was analyzed by GC–MS.

2.2.2.3 N-methyl-trimethylsilyltrifluoroacetamide (MSTFA)

N-methyl-trimethylsilyltrifluoroacetamide (MSTFA) is the most volatile of the trimethylsilyl acetamides. It is most useful for the analysis of volatile trace materials where the derivatives may be near the reagent or by-product peak. The general equation of the reaction (Equation 14) using MSTFA is shown below;

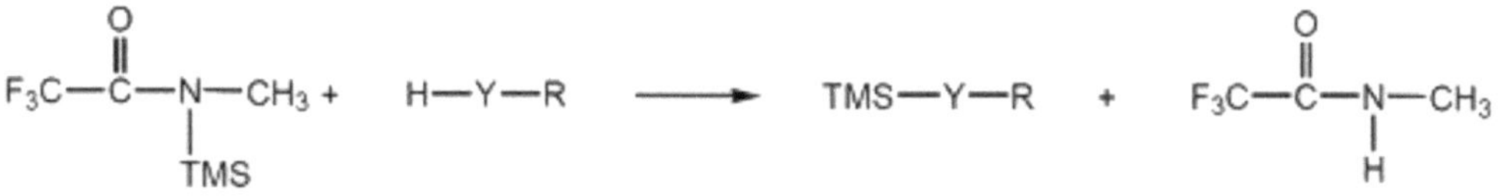

Equation 14: Silylation reaction using N-methyl-trimethylsilyltrifluoroacetamide (MSTFA): TMS = $Si(CH_3)_3$, Y = O, S, NH, NR`, COO, R, R` = Alk, Ar.

Silylation derivatization procedures by Butts, (1972) using BSTFA, BSA, or MSTFA formed the basis for derivatisation application that was used to analyze over 200 organic compounds including carboxylic acids, amines, alcohols, phenols, and nucleic acids. The basis of the procedure is as follows: to a 1 - 5 mg sample, 100 µL each of pyridine as a solvent and silylating reagent are added. The mixture is capped and heated to 60 °C for 20 minutes. For moderately hindered or slowly reacting compounds, BSTFA is recommended for use with 1 or 10 % TMCS catalyst. Under extreme conditions compounds may require heating for up to 16 hours. When Ketones are derivatized using this procedure, they may form 15 - 20 % enol trimethylsilyl esters. The esters can be eliminated by first forming a methoxime. In the case of Amino acids the reaction may be required to be in a sealed tube or vial (Butts, 1972). The samples are heated cautiously, near the boiling point of the mixture until a clear solution is obtained.

2.2.2.4 Hexamethyldisilzane (HMDS)

Hexamethyldisilzane (HMDS) is a weak donor, as it has symmetry. If it is used for derivatization, it will attack only easily silylated hydroxyl groups. Equation 15 shows the reaction using hexamethyldisilzane.

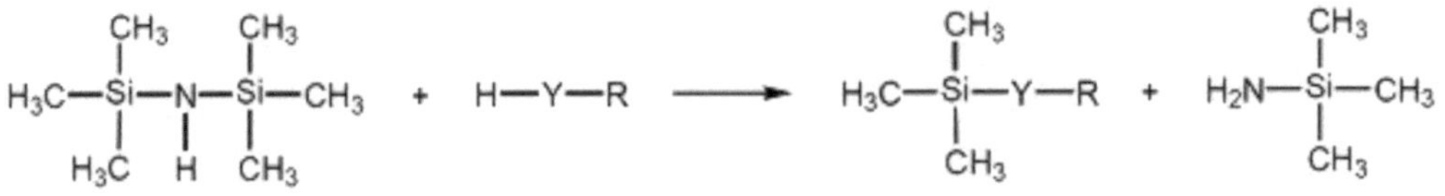

Equation 15: Silylation reaction using Hexamethyldisilzane (HMDS): Y = O, S, NH, NR`, COO, R, R` = Alk, Ar.

Hexamethyldisilzane is a weak TMS donor that can be used for silylation of carbohydrates. It can be used as mixture with pyridine and trifluoroacetic acid. The derivatization procedure is simple as shown in this procedure where 2 ml of hexamethyldisilazane, 2 ml of pyridine and 175 µL of trifluoroacetic acid are added to the samples. Silylation is then performed at 60 °C for 1 hour.

2.2.2.5 Trimethylchlorosilane (TMCS)

Trimethylchlorosilane (TMCS) is also a weak donor. In addition, it produces hydrochloric acid as a by product with is acidic. It is therefore not commonly used. However, it is often found as a catalyst to increase TMS donor potential. An example of derivatization reaction using Trimethylchlorosilane (TMCS) is shown in equation 16.

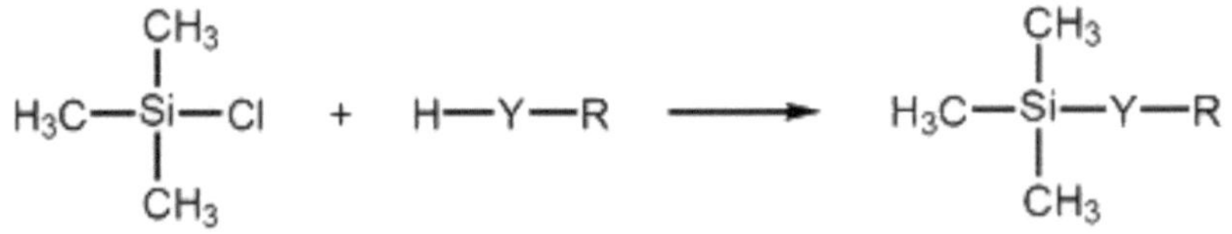

Equation 16: Silylation reaction using Trimethylchlorosilane (TMCS): Y = O, S, NH, NR`, COO, R, R` = Alk, Ar.

2.2.2.6 Trimethylsilylimidazole (TMSI)

Trimethylsilylimidazole (TMSI) is not a weak donor, but it is selective as it reacts with alcohols and phenols but not amines or amides (nitrogen groups). Since it is selective, it will target the hydroxyls in wet sugars and also derivatize the acid sites of amino acids. It will leave the amino group free for fluorinated derivatization. An example of reaction equation using TMSI is shown below (Equation 17).

Equation 17: Silylation reaction using Trimethylsilylimidazole (TMSI): TMS = R, R\` = Alk, Ar.

The derivatives produced are suitable for ECD analysis. Isoherranen and Soback, (2000) used the following derivatization procedure for the determination of Gentamicin. In the procedure, the sample solution was evaporated to dryness under a stream of nitrogen in an auto sampler vial and the dry residue was dissolved in 50 ml of anhydrous pyridine. To the pyridine solution, 100 ml of TMSI were added and the vial was closed and incubated for 15 min at 60 °C. Thereafter, 70 ml of Trifluoroacetic anhydride (TFAA) were added and the vial was closed and incubated for 60 min at 60 °C.

2.2.2.7 Trimethylsilyldiethylamine (TMS-DEA)

Trimethylsilyldiethylamine (TMS-DEA) reagent is used for derivatizing amino acids, antibiotics, urea-formaldehyde condensates, steroids and carboxylic acids and it also targets hindered compounds. Hydrolysis of TMS derivatives and reagents produces hexamethyldisiloxane [$(CH_3)_3SiOSi(CH_3)$]. Hexamethyldisiloxane is quite inert and does not interfere in the reaction or produce byproducts with the sample. The reaction by product diethylamine is very volatile and the reaction can be driven to completion by evaporating the diethylamine produced. Equation 18 shows the derivatisation reaction using TMS-DEA reagent.

$$\text{TMS-N }(C_2H_5)_2\text{+ H-Y-R} \rightarrow \text{TMS-Y-R + H-N }(C_2H_5)_2$$

Equation 18: Silylation reaction using Trimethylsilyldiethylamine (TMS-DEA) reagent: TMS = $Si(CH_3)_3$, Y = O, S, NH, NR\`, COO, R, R\` = Alk, Ar.

Because of its high volatility, it is eluted with the solvent or reagent and usually does not interfere with the chromatogram.

2.2.2.8 N-methyl-N-t-butyldimethylsilyltrifluoroacetamide (MTBSTFA)

Silylation derivatisaion using N-methyl-N-t-butyldimethylsilyltrifluoroacetamide (MTBSTFA) replaces the active hydrogen with tert-Butyldimethylsilyl (t-BDMS) group. The tert-Butyldimethylsilyl derivatives which are more resistant to hydrolysis and can be up to 10,000 times more stable than TMS derivatives. N-methyl-N-t-butyldimethylsilyltrifluoroacetamide will target sulfonic and phosphoric groups if present and it is suitable for GC with Mass spectroscopy detector as it produces easily interpreted mass spectra. A typical (shown as equation 19) reaction involving MTBSTFA as the derivatization reagent is shown below.

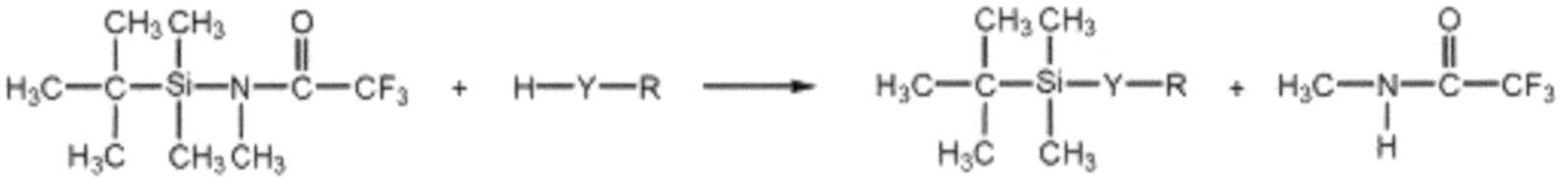

Equation 19: Silylation reaction using N-methyl-N-t-butyldimethylsilyltrifluoroacetamide (MTBSTFA) reagent: Y = O, S, NH, NR`, COO, R, R` = Alk, Ar.

Silylation derivatization procedure using MTBSTFA can be carried out in the laboratory as follows. To a 1 - 5 mg sample, 100 µL each of reagent and solvent is added and shaken on a vortex mixture for five minutes. In case of hindered compounds, they should be heated in a closed vial at 60 °C for 1 hour. For alcohols and amines, MTBSTFA with 1 % tert-butyl dimethylchlorosilane (t-BDMCS) catalyst is used.

In another method where dicarboxylic acids are produced, 150 µL of sample, 100 µL of acetonitrile and 10 uL of MTBSTFA are added. The vial is sealed and the mixture is allowed to stand overnight. Then 100 µL of water is added to hydrolyze any unreacted MTBSTFA, followed with 250 µL of hexane. Vortex mixing and centrifuging follows before the upper hexane layer is decanted and dried to approximately 5 µL under a gentle stream of nitrogen. The sample is finally injected onto GC column for analysis.

2.3 Acylation

Derivatization by acylation is a type of reaction in which an acyl group is introduced to an organic compound. In the case of a carboxylic acid, the reaction involves the introduction of the acyl group and the loss of the hydroxyl group. Compounds that contain active hydrogens (e.g., -OH, -SH and -NH) can be converted into esters, thioesters and amides, respectively, through acylation (Zenkevich, 2009). Acylation is also a popular reaction for the production of volatile derivatives of highly polar and in volatile organic materials (Zaikin and Halket, 2003). Acylation also improves the stability of those compounds that are thermally labile by inserting protecting groups into the molecule. Acylation can render extremely polar materials such as sugars amenable to separation by GC and, consequently, are a useful alternative or complimentary to the silylation. Equation 20 shows an example of an acylation is the reaction between acetic anhydride and an alcohol.

$$CH_3OCOCOCH_3 + HOR \xrightarrow{H^+\text{(Catalyst)}} CH_3OCOR' + HOCOCH_3$$

Equation 20: Reaction between acetic anhydride and an alcohol to produce acetate ester and acetic acid. (Blau and Halket, 1993).

Acylation has the following benefits in GC analysis.

i. It improves analyte stability by protecting unstable groups.
ii. It can provide volatility on substances such as carbohydrates or amino acids, which have many polar groups that they are nonvolatile and normally decompose on heating.

iii. It assists in chromatographic separations which might not be possible with compounds that are not suitable for GC analysis.
iv. Compounds are detectable at very low levels with an electron capture detector (ECD).

In addition, acyl derivatives tend to produce fragmentation patterns of compounds in MS applications which are clear to interpret and provide useful information on the structure of these materials. Acylation can be used as a first step to activate carboxylic acids prior to esterfication. However, acylation derivatives can be difficult to prepare especially because of interference of other reaction products (acid by-products) which need to be removed before GC separation. Acylation reagents are moisture sensitive, hazardous and odorous.

2.3.1 Derivatization reagents used in acylation

Common reagents for the Alkylation process are Fluoracylimidazoles, Fluorinated Anhydrides, N-Methyl-bis(trifluoroacetamide) (MBTFA), Pentafluorobenzoyl Chloride (PFBCI) and Pentafluoropropanol (PFPOH). Acylating reagents readily target highly polar, multi-functional compounds, such as carbohydrates and amino acids. In addition, acylating reagents offer the distinct advantage of introducing electron-capturing groups (Kataoka, 2005), and therefore enhancing detectability during analysis. These reagents are available as acid anhydrides, acyl derivatives, or acyl halides. The acyl halides and acyl derivatives are highly reactive and may be suited for use where issues of steric hindrance may be a factor. Acid anhydrides are available in a number of fluorinated configurations, which can improve detection. These fluorinated anhydride derivatives are used primarily for electron capture detection (ECD), but can also be used for flame ionization detection (FID). Fluorinated anhydrides are often used in derivatizing samples to confirm drugs of abuse. Despite this special use, their acidic nature requires that any excess or byproducts be removed prior to GC analysis to prevent deterioration of the column. Because of the acidic byproducts, the derivatization process has been carried out in pyridine, tetrahydrofuran or another solvent capable of accepting the acid by-product.

2.3.1.1 Fluorinated anhydrides

Fluorinated Anhydrides include the following compounds; Trifluoroacetoic Anhydride (TFAA), Pentafluoropropionic Anhydride (PFPA) and Heptafluorobutyric Anhydride (HFBA) which are suitable for both Flame ionization Detectors (FID) and Electron Capture Detectors (ECD). Equations i, ii, and iii representing acylation reactions using HFBA, PFPA and TFAA derivatization reagents respectively are shown below (Equation 21).

$$\text{i.}\quad C_3F_7OCOCOC_3F_7 + H\text{-}Y\text{-}R \rightarrow C_3F_7OC\text{-}Y\text{-}R + C_3F_7OC\text{-}OH$$

$$\text{ii.}\quad C_2F_5OCOCOC_2F_5 + H\text{-}Y\text{-}R \rightarrow C_2F_5OC\text{-}Y\text{-}R + C_2F_5OC\text{-}OH$$

$$\text{iii.}\quad CF_3OCOCOCF_3 + H\text{-}Y\text{-}R \rightarrow CF_3OC\text{-}Y\text{-}R + CF_3OC\text{-}OH$$

Equation 21: Derivatization reactions using HFBA, PFPA and TFAA respectively: Y = O, NH, NR´, R, R´ = Alk, Ar.

The perfluoroacid anhydrides and acyl halide reagents react to form acidic byproducts which must be removed prior to the GC analysis in order to prevent damage to the chromatography column. The reagents basically react with alcohols, amines, and phenols to produce stable and highly volatile derivatives. Bases, such as triethylamine, can be added as an acid receptor and promote reactivity (Lin et al., 2008). Amine bases also may be used as catalysts/acid acceptors.

The following procedure by Palmer et al., (2000) for derivatization using Fluorinated Anhydrides involves the use of Triethylamine (TMA), as a catalyst. In a 5 mL vial, 50 µg (or 250 µg in case of FID) of sample is dissolve in 0.5 mL benzene. Then, 0.1 mL of 0.05 M triethylamine in benzene is added followed by 10 µL of desired anhydride such as heptafluorobutyric Anhydride (HFBA). The vial is capped and heat to 50 °C for 15 minutes. This is followed by cooling and addition of 1 mL of a 5 % aqueous ammonia solution. The cool mixture is shaken for 5 minutes, to separate the benzene layer, and injected directly onto the GC column.

2.3.1.2 Fluoracylimidazoles

Fluoracylimidazoles include Trifluoroacetylimidazole (TFAI), Pentafluoropropanylimidazole (PFPI) and Heptafluorobutyrylimidazole (HFBI) which reacts under mild conditions. The by-products (imidazole and/or N-methytrifluoroacetamide) are not acidic and therefore do not harm the column. Care must be taken because these reagents react violently with water. Fluoracylimidazoles work best with amines and hydroxy compounds (Kataoka, 2005). For example Heptafluorobutyrylimidazole readily forms derivatives with phenols, alcohols and amines (as shown in equation 22) and the derivatives are suitable for ECD.

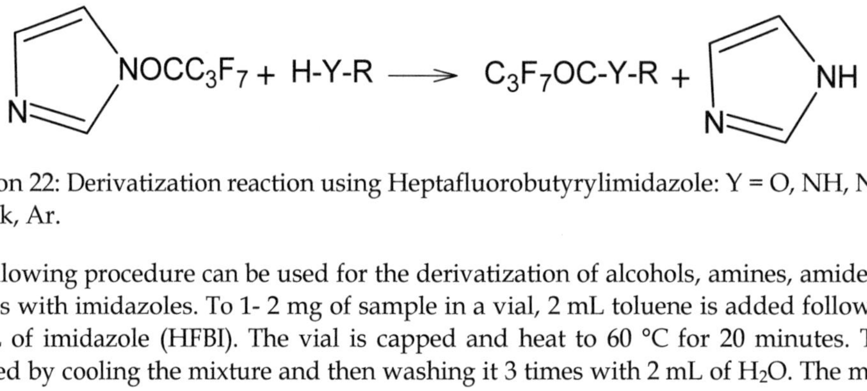

Equation 22: Derivatization reaction using Heptafluorobutyrylimidazole: Y = O, NH, NR´, R, R´ = Alk, Ar.

The following procedure can be used for the derivatization of alcohols, amines, amides, and phenols with imidazoles. To 1- 2 mg of sample in a vial, 2 mL toluene is added followed by 0.2 mL of imidazole (HFBI). The vial is capped and heat to 60 °C for 20 minutes. This is followed by cooling the mixture and then washing it 3 times with 2 mL of H_2O. The mixture is then dried over $MgSO_4$ and inject onto the GC for separation.

2.3.1.3 N-Methyl-bis(trifluoroacetamide) (MBTFA)

N-Methyl-bis(trifluoroacetamide) (MBTFA) reagent reacts rapidly with primary and secondary amines, and also slowly with hydroxyl groups and thiols. Reaction conditions are mild with relatively inert and non acidic by-products and therefore do not damage the GC column (Lelowx et al., 1989). The general reaction is presented in equation 23:

$$F_3COCN\ (CH_3)\ OCCF_3 + H\text{-}Y\text{-}R \rightarrow F_3OC\text{-}Y\text{-}R + CH_3NHOCCF_3$$

Equation 23: Representative reaction of the derivatization of amines, hydroxyl groups and thiols using N-Methyl-bis(trifluoroacetamide) (MBTFA) reagent: Y = O, S, NH, NR´, R, R´ = Alk, Ar.

N-Methyl-N-bis(trifluoroacetamide is recommended for the analysis of sugars (Matsuhisa, 2000; Hannestad, 1997) and as an acylation reaction is often used for amine drugs, such as stimulants, amino acids, and alcohols.

In the method used by Matsuhisa, (2000), 20 µL of D-glucose standard aqueous solution (10 µg/mL) was sampled and freeze-dried. Trifluoroacetylation was performed by adding 10 µL of pyridine and 10 µL of MBTFA (N-methyl-bis-trifluoroacetamide) to the dried sample, and then heating at 60 °C for 60 minutes. The cooled sample was ready for GC analysis.

Donike, (1973) applied the following procedure for the rapid derivatization of amines. To 10 mg sample, 0.2 mL MBTFA and 0.5 mL solvent of choice is added in a reaction vial. The vial is capped and heat to 60 - 100 °C for 15 to 30 minutes. It was also observed that some compounds with steric hindrance require additional heating.

2.3.1.4 Pentafluorobenzoyl Chloride (PFBCI)

Pentafluorobenzoyl chloride (PFBCI) is used in making derivatives of alcohols and secondary amines of which secondary amines are the most highly reactive, forming the most sensitive ECD derivatives of amine and phenol. Phenols are the most receptive site for this reagent (Regis, 1999). Pentafluorobenzoyl chloride (PFBCI) is suitable for functional groups that are sterically hindered. A base such as NaOH is often used to remove the HCl that is produced as byproduct. This derivatization procedure which is presented by the reaction below (equation 24) basically uses a pentafluorobenzoyl chloride (PFBCl) to provide rapid formation of the derivatives of amines and phenols.

$$C_6\,F_5\text{-OCCl} + C_6H_5\text{-OH} \rightarrow C_6\,F_5\text{-OCO-}\,C_6H_5 + HCl$$

Equation 24: Formation of derivatives of amines and phenols using Pentafluorobenzoyl chloride (PFBCI).

The following derivatization procedure can be used for PFBCI reagent. In a 5 mL vial, 25 - 50 mg of sample and 2.5 mL of 2.5 N NaOH are dissolved. Then, 0.1 g of PFBCl, is added and the vial is capped, shaken vigorously for 5 minutes. The derivative formed is extracted into methyl chloride and dried over $MgSO_4$. The sample is then injected directly onto the GC column.

2.3.1.5 Pentafluoropropanol (PFPOH)

Pentafluoropropanol ($CF_3CF_2CH_2OH$) is used in combination with pentafluoropropionic anhydride (PFPA) and is applied commonly with polyfunctional bio-organic compounds (Regis, 1999). For example, 2,2,3,3,3-Pentafluoropropanol is used in combination with PFPA to make derivatives of the most common functional groups, especially polyfunctional bio-organic compounds (Regis, 1999). The formed derivatives are highly suitable for ECD.

Sample analysis and derivatization can be done through the following sequential derivatization procedure, in which 1 mg of sample is added in a reaction vial, followed by the addition of 50 µL PFPOH and 200 µL of PFPA. The sample is then heated to 75 °C for 15 minutes and then evaporated to dryness under N_2. An additional 100 µL PFPA is added and the sample is again heated to 75 °C for 5 minutes. The sample is finally evaporated to dryness and dissolved in ethyl acetate prior to injection.

2.3.1.6 4-Carbethoxyhexafluorobutyryl chloride (4-CB)

4-Carbethoxyhexafluorobutyryl chloride (4-CB) forms stable products with secondary amines, such as methamphetamine, allowing the removal of excess agent by adding protic solvents. 4-Carbethoxyhexafluorobutyryl chloride decreases the net charge of the peptides and increased hydrophobicity (Klette, 2005). The following equation 25 shows the derivatization of an amine using 4-CB reagent.

$$ClOCC_3F_6OCOC_2H_5 + R\text{-}NH_2 \rightarrow RNHOCC_3F_6OCOC_2H_5 + HCL$$

Equation 25: Amine derivatization using 4-Carbethoxyhexafluorobutyryl chloride (4-CB) reagent.

The procedure for derivatization follows the one used by Dasgupta et al., (1997). In the procedure, 50 µL of 4-CB is added to dried sample (dried by passing of a gentle stream of nitrogen gas to remove the organic phase) followed by incubation at 80 °C for 20 minutes. After derivatization the excess 4-CB is evaporated and the derivatized sample is reconstituted in 50 µL ethyl acetate of which 1 - 2 µL is injected to GC for analysis.

3. GC Chiral Derivatization

Chiral derivatization involves reaction of an enatiomeric molecule with an enatiomerically pure chiral derivatizing agent (CDA) to form two diastereomeric derivatives that can be separated in this case using GC. A solution in which both enantiomers of a compound are present in equal amounts is called a racemic mixture. Diastereomers are stereoisomers (they have two or more stereo centers) that are not related as object and mirror image and are therefore not enantiomers. In other word, unlike enatiomers which are mirror images of each other and non-sumperimposable, diastereomers are not mirror images of each other and non-superimposable. Diastereomers can have different physical properties and reactivity.

Any molecule having asymmetric carbon is called as chiral molecule. Chirality of analyte molecules requires special consideration in their analysis and separation techniques. Scientists and other regulatory authorities are in the demand of data on concentrations and toxicity of the chiral pollutants, and therefore chiral derivatization is becoming an essential, urgent and demanding field. However the derivatization procedures are tedious and time consuming due to the different reaction rates of the individual enantiomers (Schurig, 2001).

Generally, there are two ways of separating enantiomers by chromatography:

i. Separation on an optically active stationary phase
ii. Preparation of diastereomeric derivatives that can be separated on a non chiral stationary phase.

The second option (ii above) requires derivatization of the analyte molecule. The presence of a suitable functional group in a pollutant is a condition for a successful derivatization of a chiral molecule. The indirect chromatographic analysis of racemic mixtures can be achieved by derivatization with a chiral derivatizing agent resulting into the formation of diastereoisomeric complex/salt (Hassan et al., 2004). From the resulting chromatograms, calculations are made to determine the enantiomeric concentration of the analyte. The diastereoisomers having different physical and chemical properties can be separated from each other by an achiral chromatographic method. Nowadays, the chromatographic methods are the most popular for enantiomeric analysis of environmental pollutants (Hassan et al 2004).

3.1 Gas chromatography chiral derivatization reagents

Gas Chromatography analysis of enantiomeric compounds on nonracemic or achiral stationary phases requires the use of enantiopure derivatization reagents (Hassan et al, 2004). Enantiopure compounds refer to samples that contain molecules having one chirality within the limits of detection. These reagents generally target one specific functional group to produce diastereomers of each of the enantiomeric analytes in GC to produce chromatograms. Some of the most common Gass Chromatography chiral derivatization reagents are: (-) menthylchloroformate (MCF), (S)-(-)-N-(Trifluoroacetyl)-prolylchlorides (TPC), (-)-α-Methoxy-α-trifluoromethylphenylacetic acid (MTPA). Examples of the use of these derivatizing agents were employed in analysis of major drugs of forensic interest, and for optically active alcohols (Tagliaro et al., (1998).

3.1.1 N-trifluoroacetyl-L-prolyl chloride (TPC)

The reagent is used for optically active amines, most notably amphetamines as represented in the following reaction (equation 26) where N-Trifluoroacetyl-L-prolyl chloride couples with amines to form diastereomers which can be separated on GC columns as it increases the sample volatility.

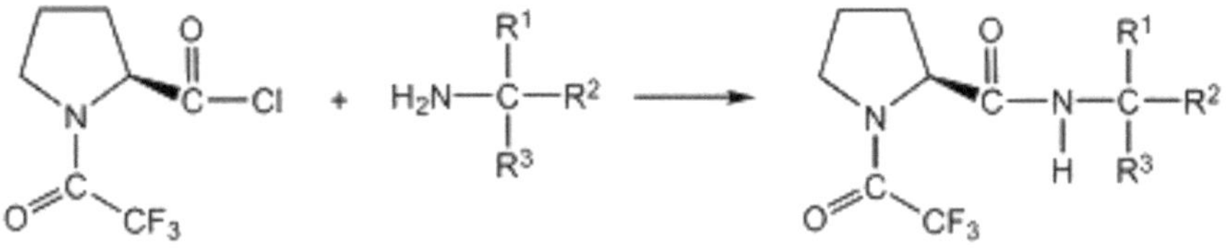

Equation 26: N-Trifluoroacetyl-L-prolyl chloride derivatization of amines

The standard TPC derivatization procedure is as follows; following sample clean up and extraction procedure, 1.0 mL of the TPC reagent is added to the clean sample. The mixture is allowed to stand for 5 min before the addition of 20 μL of triethylamine to take up excess unreacted TPC. After 15 min of intermittent shaking, 1.0 mL of 6 N HC1 is used to remove the ammonium salt. The mixture is finally washed with 1 mL of distilled water and then dried over anhydrous magnesium sulfate before dilution and analyzed by GC.

3.1.2 (S)-(–)-N-(Trifluoroacetyl)-prolylchloride (l-TPC)

(S)- (-)-N-(Trifluoroacetyl)-prolylchloride (l-TPC) is widely used for amine drugs analysis. Qiao Feng Tao & Su Zeng (2002) analyzed the enantiomers of chiral amine drugs by using stereo selective methods to separate enantiomers on an achiral capillary gas chromatography by pre-column chiral derivatization with S-(_)-N-(fluoroacyl)-prolyl chloride. In the study, it was noted that the stereo selectivity and sensitivity can be improved by chiral derivatization. The method has been used to determine S-(+) - methamphetamine in human forensic samples and to analyze enantiomers of amphetamine and fenfluramine in rat liver microsomes. Two analytical procedures used by Qiao Fang Tao and Su Zing (2002) for aqueous and organic phase derivatization methods are described below.

Aqueous phase derivatization method: Urine or liver homogenate was added to a 10 ml Teflon-lined, screw-capped test tube. The pH of the sample was adjusted to 9 by 10 M NaOH, then 0.5 ml of chilled 1 M sodium bicarbonate/sodium carbonate buffer (pH 9.0). After mixing, 40 µL of S-(_)-TFAPC was added, and the tube was vortex mixed for 30 min at room temperature. The resulting mixture was saturated with NaCl and adjusted to pH 9, followed by addition of 0.5 ml of chilled 1 M sodium bicarbonate/sodium carbonate buffer (pH 9.0) and extracted with 4 ml of ethyl acetate by gently shaking for 15 min. After the phases separate, the organic layer was washed with 2 ml of deionized water. The organic layer was evaporated to dryness under a gentle stream of air at 40 °C. The residue was cooled to room temperature and reconstituted with ethyl acetate. An aliquot of 1 µL of analyte was analyzed by GC/MS.

Organic phase derivatization method: A 1.0 ml sample of microsomal mixture spiked with chiral amine was piped into a 15 ml screw capped test tube. The pH of the sample was adjusted to 12 - 13 with 40% NaOH. The mixtures were extracted with 2 ml of chloroform by rotatory shaking for 1 min. After centrifugation at 4000 rpm for 10 min, the aqueous phase was removed by aspiration. The remaining organic phase was dried by anhydrous sodium sulfate, and then transferred to another clean screw-capped test tube. A 10 µL sample of triethylamine as a catalytic agent and 40 µL S-(_)-TFAPC were added into the test tube and mixed. The tube was capped and allowed to react for 15 min at room temperature by gentle shaking. The chloroform layer was washed with 2 ml distilled water and evaporated to dryness at under a water bath and a gentle stream of air 50° C. The residues were allowed to cool to room temperature and reconstituted with 40 µL ethyl acetate prior to injection into the GC/FID system (Qiao Feng Tao & Su Zeng, 2002).

3.1.3 (–)-α-Methoxy-α-trifluoromethylphenylacetic acid (MTPA)

(-)-α-Methoxy-α-trifluoromethylphenylacetic acid is also mostly used in drug analysis. In a study by S-M Wang et al., (2005) pharmaceutical drugs were analyzed using compounds that were derivatized with (-)-α-Methoxy-α-trifluoromethylphenylacetic acid (MTPA) as the derivatizing reagent. These types of derivatization reactions are represented by equation 27. In the reaction the hydroxyl group on the MTPA molecule is lost in the derivative formation.

$$\mathrm{HO{-}\overset{O}{\overset{\|}{C}}{-}\underset{\underset{CF_3}{|}}{\overset{C_6H_5}{\overset{|}{C}}}{-}O{-}CH_3 + R{-}NH_2 \longrightarrow RHN{-}\overset{O}{\overset{\|}{C}}{-}\underset{\underset{CF_3}{|}}{\overset{C_6H_5}{\overset{|}{C}}}{-}O{-}CH_3 + H_2O}$$

Equation 27: Derivatization reaction between (-)-α-Methoxy-α-trifluoromethylphenylacetic acid and an amine group of an analyte molecule.

The derivatization using MTPA can be achieved as follows: after extraction, clean up and drying the sample under nitrogen, the residue is added 50 µL N, N-dicyclohexycarbodiimide and 100 µL MTPA. The reaction mixture is then thoroughly mixed, then incubated at 70 °C for 20 min before GC analysis.

4. Summary

The choice of the derivatization technique for analysis of compounds will depend on the available reagent and reaction types that can produce derivatives that give desirable results in GC. The derivatives must be suitable, detectable and efficient for GC analysis. The evaluation of the functional group of the analyte, the GC detector and even the by products of the reaction among others considerations will guide the choice of derivatization technique.

For example, insertion of perfluoroacyl groups into a molecule enhances its detectability by electron capture. The presence of a carbonyl group adjacent to halogenated carbons in an analyte enhances the electron capture detector (ECD) response. For mass spectroscopic detector, acyl derivatives tend to direct the fragmentation patterns of compounds and therefore, providing useful information on the structure of molecules.

For acid analytes, the first choice for derivatization is esterification. Acids are reactive compounds and also more polar to be separated well by gas chromatography. The underivatized acids will tend to tail because of the adsorption and non-specific interaction with the column. Esterification is used to derivatize carboxylic acids and other acidic functional groups. The reaction involves the condensation of the carboxyl group of an acid and the hydroxyl group of an alcohol, with the elimination of water. Consequently, the polarity of the molecule is reduced.

Nearly all functional groups which present a problem in gas chromatographic separation (hydroxyl, carboxylic acid, amine, thiol, phosphate) can be derivatized by silylation reagents. The derivatives of the silylation reactions are generally less polar, more volatile and more thermally stable. The introduction of a silyl group(s) can also serve to enhance mass spectrometric properties of derivatives, by producing either more favorable diagnostic fragmentation patterns of use in structure investigations, or characteristic ions of use in trace analyses in other related techniques.

In considering the functional group or compound type, dimethylformamide-dialkylacetals (DMF-DEA) or similar type of reagents are recommended for sterically hindered aldehydes, amines, carboxylic acids, and phenols. Shorter chain reagents like DMF- DEA will produce more volatile derivatives than longer chain reagents. The by products of the reaction will also determine the choice of reagent. In this view, perfluoroacylimidazole is a better choice rather than perfluoroacid anhydrides. Perfluoroacid anhydrides produce acidic byproducts which must be removed from the reaction mixture before the derivatives are injected onto the GC column. With perfluoroacylimidazole there are no acid byproducts.

The complexity surrounding chiral derivatization is generally because of different reaction rates of the individual enantiomers. Derivatization with a chiral derivatizing agent resulting into the formation of diastereoisomeric complexes is an option in GC analysis, and has been applied to analyze various analytes (Hassan et al., 2004 & Schurig, 2001). Table 1 gives a summary guide of derivatization technique and reagent selection based on the functional group of the analyte molecule.

Functional group	Reaction type	Derivatization reagent
Alcohols and Phenols	Silylation	Bis(trimethylsilyl)–acetamide, Bistrimethylsilyltrifluoroacetamide, N-methyl-N-t-butyldimethylsilyltrifluoroacetamide
	Acylation	Heptafluorobutyrylimidazole, Pentafluoropropionic Anhydride, Trifluoroacetic anhydride, N-Methyl-bis(trifluoroacetamide)
	Alkylation	Dimethylformamide, Pentafluorobenzyl bromide
Carboxylic acids	Silylation	Bis(trimethylsilyl)–acetamide, Bistrimethylsilyltrifluoroacetamide, Trimethylsilylimidazole, N-methyl-N-t-butyldimethylsilyltrifluoroacetamide
	Acylation	Pentafluoropropanol / pentafluoropropionic anhydride
	Alkylation	Dimethylformamide, Tetrabutylammonium hydroxide
Active hydrogens	Silylation	Bis(trimethylsilyl)–acetamide, Bistrimethylsilyltrifluoroacetamide / Trimethylchlorosilane, Hydrox-Sil, N-methyl-trimethylsilyltrifluoroacetamide,
	Acylation	Pentafluoropropanol / pentafluoropropionic anhydride
	Alkylation	Dimethylformamide, Tetrabutylammonium hydroxide
Carbohydrates and Sugars	Silylation	Hexamethyldisilzane, Trimethylsilylimidazole
Amides	Silylation	Bis(trimethylsilyl)–acetamide, N, O-bis-(trimethylsilyl)-trifluoroacetamide
	Acylation	Heptafluorobutyrylimidazole
	Alkylation	Dimethylformamide
Amines	Silylation	Bistrimethylsilyltrifluoroacetamide, N-methyl-N-t-butyldimethylsilyltrifluoroacetamide
	Acylation	Trifluoroacetic anhydride, Pentafluorobenzoyl chloride, Heptafluorobutyrylimidazole
	Alkylation	Dimethylformamide (Diacetals)
Amino acids	Silylation	Bistrimethylsilyltrifluoroacetamide, Trimethylsilylimidazole
	Acylation	Heptafluorobutyrylimidazole
	Alkylation	Dimethylformamide, Tetrabutylammonium hydroxide
Catecholamines	Silylation	Trimethylsilylimidazole
	Acylation	Pentafluoropropionic Anhydride, Heptafluorobutyrylimidazole,
Inorganic anions	Silylation	Bistrimethylsilyltrifluoroacetamide, N-methyl-N-t-butyldimethylsilyltrifluoroacetamide
Nitrosamine	Silylation	Bistrimethylsilyltrifluoroacetamide
	Acylation	HFBA, Pentafluoropropionic anhydride, Trifluoroacetic anhydride
	Alkylation	Dimethylformamide, Pentafluorobenzyl bromide
Sulfonamides	Acylation	Trifluoroacetoic & Heptafluorobutyric Anhydride, Pentafluorobenzyl bromide
Sulfides	silylation	Trimethylsilylimidazole

Table 1. A summary guide for the derivatization techniques and reagents selection based on sample types.

5. References

Allmyr M, McLachlan MS, Sandborgh-Englund G, Adolfsson-Erici M. (2006). Determination of Triclosan as its pentafluorobenzoyl ester in human plasma and milk using electron capture negative ionization mass spectrometry. Anal Chem. 2006; 78: 6542–6.

Bao Ming-Liang, Pantani Francesco, Griffini Osvaldo, Burrini Daniela, Santianni Daniela, Barbieri Katia (1998). Determination of carbonyl compounds in water by derivatization–solid-phase micro extraction and gas chromatographic analysis. Journal of Chromatography A. 809 (1998) 75–87

Blau, K., King, G. (1997). Handbook of Derivatives for Chromatography; Heyden & Sons Ltd.; London, 1979

Butts W.C. (1972). Two-column gas chromatography of trimethylsilyl derivatives of Biochemically significant compounds. Analytical Biochemistry. Volume 46, Issue 1, March 1972, Pages 187-199

Chen, B.-G., Wang, S.-M. and Liu, R. H. (2007). GC-MS analysis of multiply derivatized opioids in urine. Journal of mass spectrometry JMS (2007). Volume: 42, Issue: 8, Pages: 1012-1023

Chien, C.-J.; Charles, M. J.; Sexton, K. G.; Jeffries, H. E. (1998). Analysis of airborne carboxylic acids and phenols as their pentafluorobenzyl derivatives: gas Chromatography/Ion Trap Mass Spectrometry with a novel chemical ionization reagent, PFBOH, Environmental Science And Technology. 1998, 32, 299-309.

Christie, W.W. (1993): preparation of ester derivatives of fatty acids for chromatographic analysis. Advances in Lipid Methodology - *Two*, pp. 69-111 (1993) (Ed. W.W. Christie, Oily Press, Dundee),

Danielson, Neil D., Gallagher Patricia A., and Bao James J. (2000). Chemical Reagents and Derivatization Procedures in Drug Analysis. Encyclopedia of Analytical Chemistry. John Wiley & Sons Ltd, Chichester, 2000. R.A. Meyers (Ed.). pp. 7042–7076.

Dasgupta, Amitava, Blackwell Walter and Burns Elizabeth (1997). Gas Chromatographic-mass spectrometric identification and quantitization of urinary phenols after derivatization with 4-carbethoxyhexafluorobutyryl chloride, a novel derivative. Journal of Chromatography B: Biomedical Sciences and Applications Volume 689, Issue 2, 21 February 1997, Pages 415-421

Destaillats, H.; Spaulding, R. S.; Charles, M. J. (2002). Ambient air measurement of acrolein and other carbonyls at the Oakland-San Francisco Bay Bridge toll plaza. Environmental Science and Technology 2002, 36, 2227-2235.

Donike, M. (1973). Acylation with bis (acylamides). N-methyl-bus (trifluoroacetamide) and bis (trifluoroacetamide), two new reagents for trifluoroacetylation. Journal of Chromatography A. Volume 78, Issue 2, 25 April 1973, Pages 273-279.

Galceran, M. T.; Moyano, E.; Poza, J. M. (1995). Pentafluorobenzyl derivatives for the gas-chromatographic determination of hydroxy-polycyclic aromatic hydrocarbons in urban aerosols. Journal of Chromatography A 1995, 710, 139-147.

Gehrke, Charles W. (1968). Quantitative gas-liquid chromatography of amino acids in proteins and biological substances: Macro, semimicro, and micro methods. Analytical Biochemistry Laboratories (Columbia, Mo). 1st edition. 101-289-264

Hannestad, U.; Lundblad, A. (1997). Accurate and Precise Isotope Dilution Mass Spectrometry Method for Determining Glucose in Whole Blood. Clinical Chemistry, 1997, 43, 5, 794-800.

Hassan, Y. Aboul-Enein & Imran Ali (2004): Analysis of the chiral pollutants by Chromatography, Toxicological & Environmental Chemistry, 86:1, 1-22

Isoherranen, Nina and Soback Stefan (2000). Determination of gentamicin after Trimethylsilylimidazole and trifluoroacetic anhydride derivatization using gas Chromatography and negative ion chemical ionization ion trap mass spectrometry. Analyst, 2000, 125, 1573–1576

Jang Myoseon and Kamens Richards, M. (2001). Characterization of Secondary Aerosol from the Photooxidation of Toluene in the Presence of NOx and 1-Propene Environ. Sci. Technol. 2001, 35, 3626-3639

Kataoka, Hiroyuki (2005). Gas chromatography of amines as various derivatives. Quantitization of amino acids and amines by chromatography - methods and protocols. Journal of Chromatography Library. Volume 70, 2005, Pages 364-404.

Kawahara, F.K. (1968). Micro determination of pentafluorobenzyl ester derivatives of Organic acids by means of electron capture gas chromatography.

Klette, L Kevin., Jamerson H. Matthew, Morris-Kukoski L. Cynthia, Kettle R. Aaron, and Snyder J. Jacob (2005). Rapid Simultaneous Determination of Amphetamine, Methamphetamine, 3, 4-Methylenedioxyamphetamine, 3, 4-Methylenedioxymethamphetamine, and 3, 4-Methylenedioxyethyla. Journal of Analytical Toxicology. Volume Number: 29: Issue 7: 669–674

Knapp, Daniel R. (1979). Handbook of Analytical Derivatization Reaction; pages 10, 10; Wiley & Sons; New York, 1979.

Kühnel, E., Laffan Dr, D.D.P., Lloyd-Jones, G.C., Martínez del Campo, T., Shepperson, I.R. Slaughter,J.L.,(2007). Mechanism of Methyl Esterification of Carboxylic Acids by Trimethylsilyldiazomethane. Angewandte Chemie International Edition. Volume 46, Issue 37, pages 7075–7078, September 17, 2007.

Kuo, H.W., Ding, W.H. (2004). Trace determination of bisphenol A and phytoestrogens in infant formula powders by gas chromatography-mass spectrometry. J. Chromatogr. A, 1027, 67.

Lelowx, M.S. De jong E.G and Meas RA.A. (1989). Improved screening method for beta-blockers in urine using solid- phase extraction and capillary gas chromatography—mass spectrometry. Journal of Chromatography. 488 (1989). 357-.367.

Lin, D.-L. Wang, S.-M. Wu, C.-H. Chen, B.-G. Liu, R.H. (2008). Chemical Derivatization for the Analysis of Drugs by GC- MS — A Conceptual Review. Journal of Food and Drug Analysis, Vol. 16, No. 1, 2008, Pages 1-1

Matsuhisa, M.; Yamasaki, Y.; Shiba, Y.; Nakahara, I.; Kuroda, A.; Tomita, T. (2000). Important Role of the Hepatic Vagus Nerve in Glucose Uptake and Production by the Liver Metabolism, 2000, 49, No 1 (January), 11-16.

Naritisin, D.B. and Markey, S.P. (1996). Assessment of DNA oxidative damage by Quantification of thymidine glycol residues using gas chromatography/electron Capture negative ionization mass spectrometry. Anal. Biochem. 241:35-41.

Orata, F. Quinete, N. Wilken, R.D. (2009). Long Chain Perfluorinated Alkyl Acids Derivatization and Identification in Biota and Abiota Matrices Using Gas

Chromatography. Bulletin of Environmental Contamination and Toxicology 2009 Vol. 83 No. 5 pp. 630-635

Palmer, R.B, Kim N.H. and Dasgupta A (2000). Simultaneous determination of fenfluramine and phentermine in urine using gas chromatography-mass spectrometry with pentafluoropropionic anhydride derivatisation. Ther. Drug Monit. 22 (2000). 418-422.

Pierce, Alan E. (1968). Silylation of Organic Compounds. Pierce Chemical Company, 1968. Pierce, (2004) .GC Derivatization. .Applications Handbook & Catalog.

Qiao Feng Tao and Su Zeng (2002). Analysis of enantiomers of chiral phenethylamine drugs by capillary gas chromatography/mass spectrometry/flame-ionization detection and pre-column chiral derivatization. J. Biochem. Biophys. Methods 54 (2002) 103–113.

Rao, X.; Kobayashi, R.; White-Morris, R.; Spaulding, R.; Frazey, P.; Charles, M. J. (2001). J GC/ITMS measurement of carbonyls and multifunctional carbonyls in PM2.5 particles emitted from motor vehicles.Journal of AOAC International 2001, 84, 699-705.

Ribeiro Bárbara, Guedes de Pinho Paula, Andrade B. Paula, Baptista Paula, Valentão Patrícia. (2009). Fatty acid composition of wild edible mushrooms species: A comparative study. Micro chemical Journal 93 (2009) 29–35 Regis (1999). Chromatography Catalog, pages 89, 90

Saraji, M. and Mirmahdieh, S. (2009). Single-drop micro extraction Followed by in-syringe derivatization and GC-MS detection for the determination of parabens in water and cosmetic products. J. Sep. Sci. 2009, 32, 988 - 995

Schauer, J. J.; Kleeman, M. J.; Cass, G. R.; Simoneit, B. R. T. (2002). Environ. Science and Technology 2002, 36, 1169-1180.

Schurig V (2001). Chiral separation of amino acids is of great importance in biology, Pharmaceutics and agriculture. This has been performed traditionally with Chromatographic methods such as gas chromatography (GC). J. Chromatogr. A 906 (2001) 275.

Scott, R. P.W. (2003). Gas chromatography. book 2 Chrom-Ed Book Series. http://www.library4science.com/eula.html. 2

S-M Wang, Lewis j. Russell, Canfield Dennis, Li Tien-Lai, Chen Chang-Yu, Liu H. Ray (2005). Enantiomeric determination of ephedrines and norephedrines by chiral derivatization gas chromatography-mass spectrometry approaches. *J. Chromatogr. B 825 (2005) 88–95*

Sobolevsky, T.G. Alexander I.R. Miller, B. Oriedo, V., Chernetsova, E.S., Revelsky, I. A. (2003). Comparison of silylation and esterification/acylation procedures in GC-MS analysis of amino acids. Journal of Separation Science. Volume 26, Issue 17, pages 1474–1478, November 2003

Spaulding, R. S.; Schade, G. W.; Goldstein, A. H.; Charles, M. J. Journal of Geophysical Research, [Atmospheres] 2003, 108, ACH7/1-ACH7/17.

Szyrwińska, K. Kołodziejczak A., Rykowska I., Wasiak W., and Lulek J. (2007). Derivatization and Gas chromatography- low-resolution Mass Spectrometry of Bisphenol A. Acta Chromatographica, No. 18, 2007

Tagliaro Franco, Smith Frederick P, Tedeschi Luciano, Castagna Franca, Dobosz Marina, Boschi Ilaria and Pascali Vincenzo (1998). Toxicological and Forensic Applications Chapter 21. Journal of Chromatography Library Volume 60, 1998, Pages 917-961

Thenot, J.-P., Horning, E.C. (1972). Amino acid *N*-dimethyllaminomethylene alkyl esters. New derivatives for GC and GC-MS studies. Anal. Lett.5 (8): 519-29.

Thenot, J. P, Stafford M. and Horning M. G. (1972). Fatty acid esterification with N, N-Dimethylformamide dialkyl acetals for GC analysis. Anal. Lett. 5(4): 217-23.

Zaikin, V.G., Halket, J.M (2003). Derivatization in mass spectrometry--2. Acylation. Eur J Mass Spectrum. 2003; 9(5):421-34.

Zenkevich, Igor, G. (2009). Acids: Derivatization for GC Analysis Encyclopedia of Chromatography, Third Edition 2009, DOI: 10.1081/E-ECHR3-120045222

6

Parameters Influencing on Sensitivities of Polycyclic Aromatic Hydrocarbons Measured by Shimadzu GCMS-QP2010 Ultra

S. Pongpiachan[1,2,*], P. Hirunyatrakul[3], I. Kittikoon[3] and C. Khumsup[3]
[1]*NIDA Center for Research and Development on Disaster Prevention and Management, School of Social and Environmental Development, National Institute of Development, Administration (NIDA), Sereethai Road, Klong-Chan, Bangkapi, Bangkok,*
[2]*SKLLQG, Institute of Earth Environment, Chinese Academy of Sciences (IEECAS), Xi'an,*
[3]*Bara Scientific Co., Ltd., Bangkok,*
[1,3]*Thailand*
[2]*China*

1. Introduction

Throughout the decades of analytical instrument designs and developments there have been many incidents and scientific improvements, which have assisted to generate major design trend changes. A unique and innovative technology enabled analytical instrument designed to meet the needs of quantitative chemical analysis of polycyclic aromatic hydrocarbons (PAHs) in various environmental compartments. PAHs are a class of very stable organic molecules made up of only carbon and hydrogen and contain two to eight fused aromatic rings. PAHs are formed during incomplete combustion of organic materials such as fossil fuels, coke and wood. These molecules were oriented horizontal to the surface, with each carbon having three neighboring atoms much like graphite. The physicochemical properties and structures of a variety of representative PAHs can be seen in Table 1-2. Epidemiological evidence suggests that human exposures to PAHs, especially Benzo[a]pyrene are high risk factors for carcinogenic and mutagenic effects. There are hundreds of PAH compounds in the environment, but only 16 of them are included in the priority pollutants list of US EPA (EPA, 2003). Many PAHs have also been identified as cancer-inducing chemicals for animals and/or humans (IARC, 1983).

In 1775, the British surgeon, Percival Pott, was the first to consider PAHs as toxic chemicals with the high incidence of scrotal cancer in chimney sweep apprentices (IARC, 1985). Occupational exposure of workers by inhalation of PAHs, both volatile and bound to respirable particulate matter, and by dermal contact with PAH-containing materials, occurs at high levels during coke production, coal gasification, and iron and steel founding. Coke oven workers have a 3- to 7- fold risk increase for developing lung cancer (IARC, 1984 and IARC, 1987).

* Corresponding Author

Congeners	MW (g/mol)	MP (°C)	BP (°C)	P_S	P_L	Log K_{ow}	H	Log K_{OA}
Ace	154.2	96	277.5	0.3	1.52	3.92	12.17	6.23
Ac	150.2	92	265-275	0.9	4.14	4.00	8.40	6.47
Fl	166.2	116	295	0.09	0.72; 0.79[a]	4.18	7.87	6.68
Ph	178.2	101	339	0.02	0.11; 0.06	4.57	3.24	7.45
1-MePh	192.3	123	359			5.14		
An	178.2	216	340	1.00E-03	7.78E-02	4.54	3.96	7.34
Pyr	202.3	156	360	6.00E-04	1.19E-02; 8E-03	5.18	0.92	8.61
Flu	202.3	111	375	1.23E-03	8.72E-03	5.22	1.04	8.60
B[*a*]F	216.3	187	407			5.40		
B[*b*]F	216.3	209	402			5.75		
Chry	228.3	255	448	5.70E-07	1.07E-04	5.86	6.50E-02	10.44
Tri	228.3	199	438	2.30E-06	1.21E-04	5.49	1.20E-02	10.80
p-terp	230.1	213		4.86E-06		6.03		
B[*a*]A	228.3	160	435	2.80E-05	6.06E-04	5.91	0.58	9.54
B[*a*]P	252.3	175	495	7.00E-07	2.13E-05	6.04	4.60E-02	10.77
B[*e*]P	252.3	178		7.40E-07	2.41E-05		0.02	
Per	252.3	277	495	1.40E-08		6.25	3.00E-03	12.17
B[*b*]F	252.3	168	481			5.80		
B[*j*]F	252.3	166	480					
B[*k*]F	252.3	217	481	5.20E-08	4.12E-06	6.00	1.60E-02	11.19
B[*g,h,i*]P	268.4	277			2.25E-05	6.50	7.50E-02	11.02
D[*a,h*]A	278.4	267	524	3.70E-10	9.16E-08	6.75		
Cor	300.4	>350	525	2.00E-10		6.75		

Source: http://www.es.lancs.ac.uk/ecerg/kcjgroup/5.html
MP (°C): Melting Point, K_{ow} : Octanol-water partition coefficient, BP (°C): Boiling Point,
H: Henry 's Law Constant, P_S: Vapour pressure of solid substance
K_{oa}: Octanol-air partition coefficient, P_L: Vapour pressure of subcooled liquid

Table 1. Physiochemical properties of PAHs

For this reason, the monitoring of PAHs in environmental media is a reasonable approach to assess the risk for adverse health effects. Since the fate of PAHs in the natural environment is mainly governed by its physiochemical properties, the study of general properties of the compounds is of great concern. It is well known that aqueous solubility, volatility (e.g. Henry's law constant of air/water partition coefficient, octanol/air partition coefficient), hydrophobicity or lipophilicity (e.g., n-octanol/water partition coefficient) of PAHs vary widely (Mackay and Callcot, 1998), the differences among their distribution in aquatic systems, the atmosphere, and the soil are significant. Molecular weight and chemical structure influence physical and chemical properties between individual PAHs. The vapor pressure and water solubility basically decrease with the increasing molecular weight. The fate of PAHs in the environment is largely determined by its physiochemical properties; as a result, high mobility of low molecular weight (LMW) species can be expected (Wild and Jones, 1995).

Congener	Abbreviation	M.W. [g]	Chemical Structure
Acenaphthylene	Ac	152	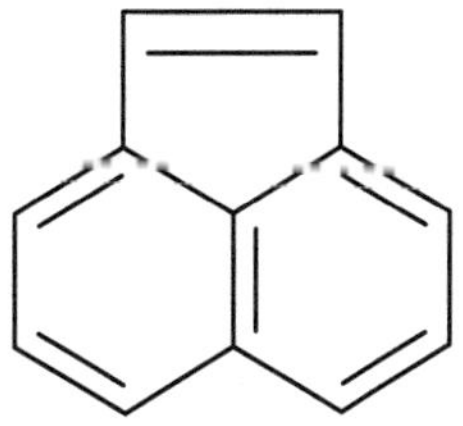
Acenaphthene	Ace	154	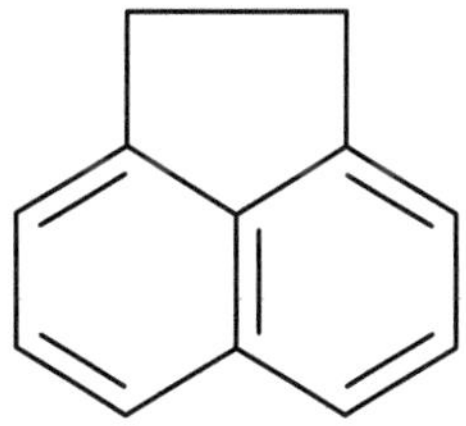
Fluorene	Fl	166	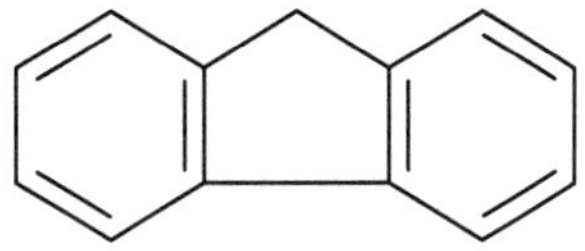
Phenanthrene	Ph	178	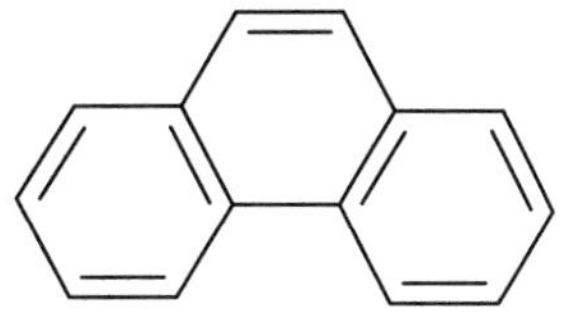
Anthracene	An	178	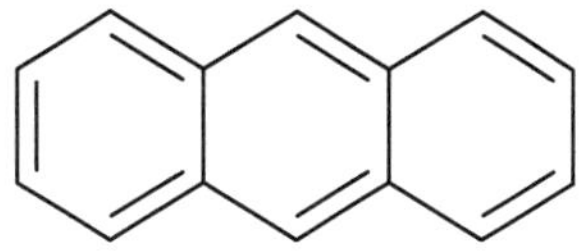
3-Methyl Phenanthrene	3-MePh	192	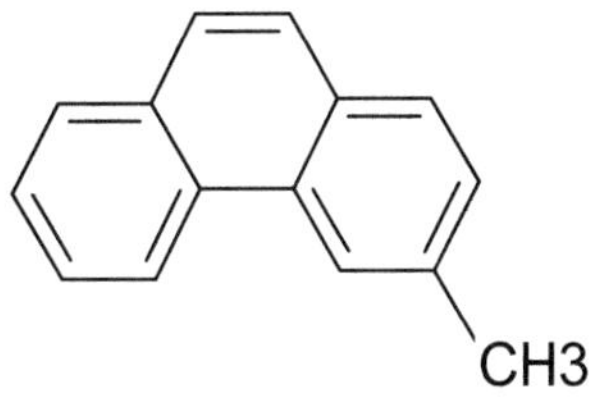

Congener	Abbreviation	M.W. [g]	Chemical Structure
9-Methyl Phenanthrene	9-MePh	192	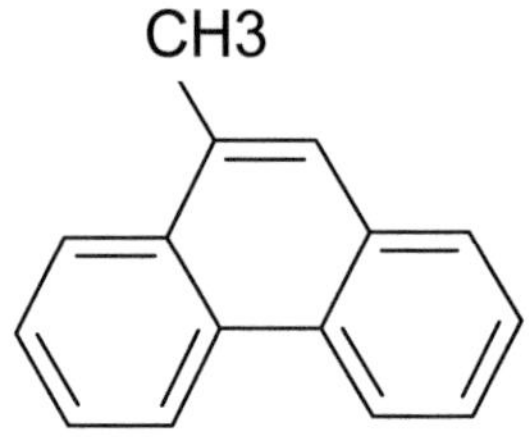
1-Methyl Phenanthrene	1-MePh	192	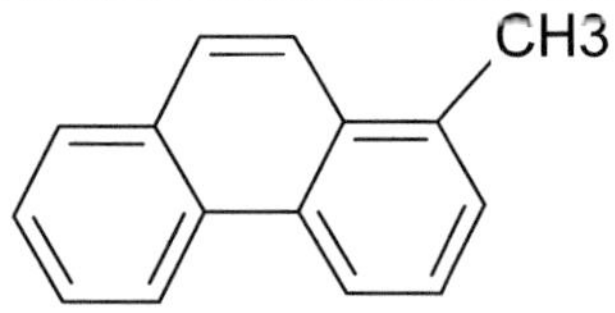
2-Methyl Phenanthrene	2-MePh	192	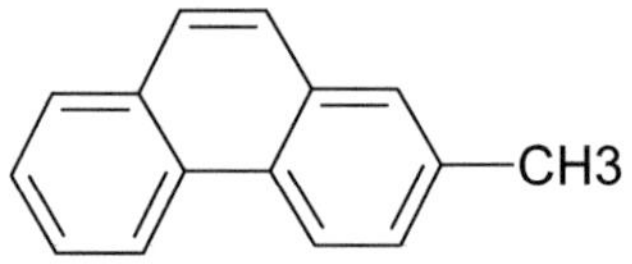
1-methyl-7-isopropyl phenanthrene (Retene)	Ret	234	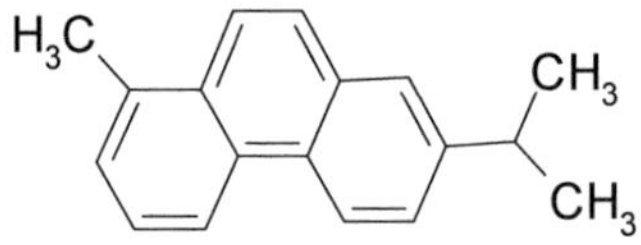
Fluoranthene	Fluo	202	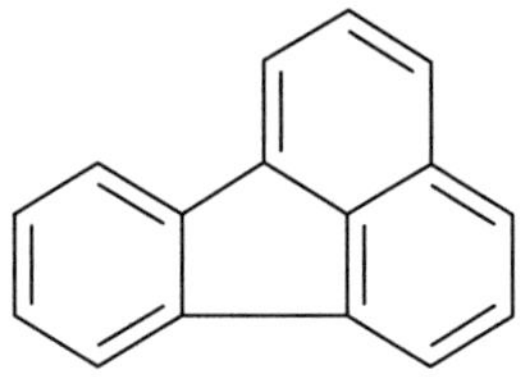
Pyrene	Py	202	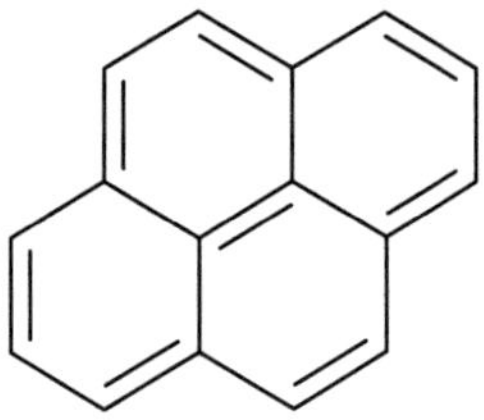
Benz[*a*]anthracene	B[*a*]A	228	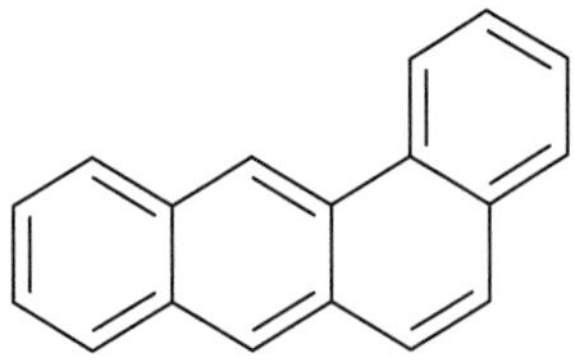

Congener	Abbreviation	M.W. [g]	Chemical Structure
Chrysene	Chry	228	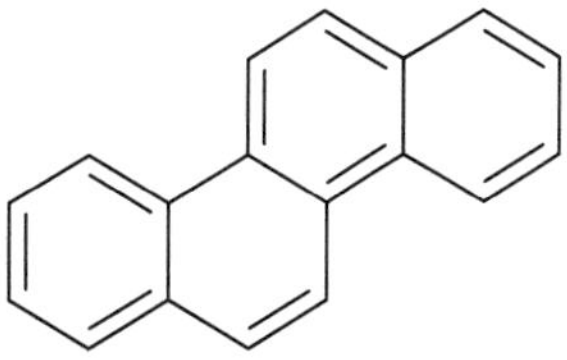
Triphenylene	Tri	228	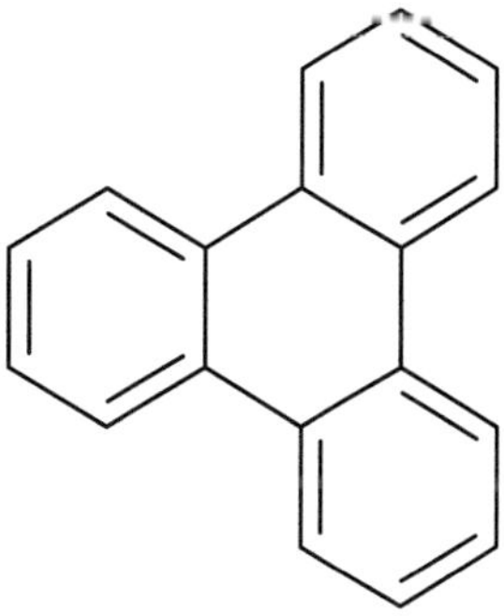
Benzo[*b*]fluoranthene	B[*b*]F	252	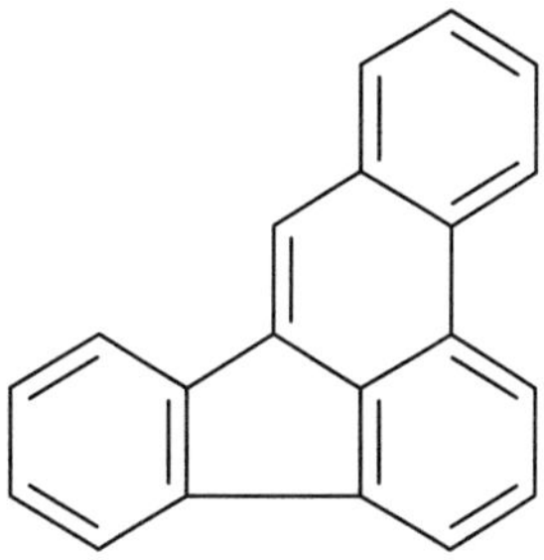
Benzo[*j*]fluoranthene	B[*j*]F	252	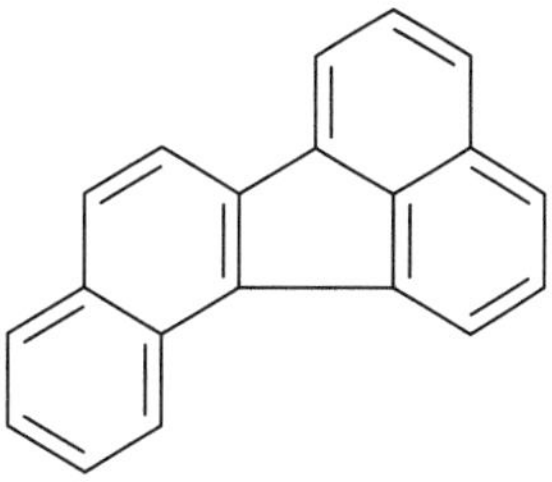
Benzo[*k*]fluoranthene	B[*k*]F	252	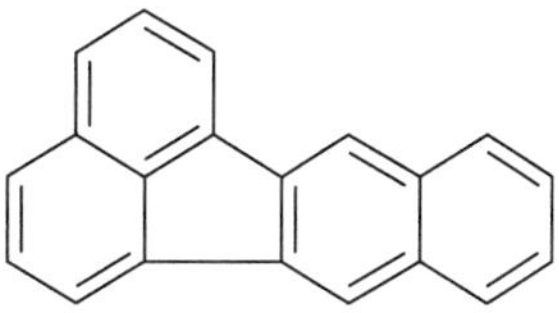

Congener	**Abbreviation**	**M.W. [g]**	**Chemical Structure**
Benzo[*e*]pyrene	B[*e*]P	252	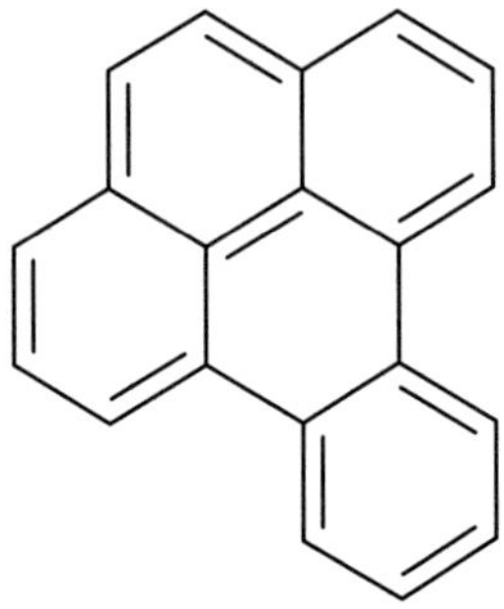
Benzo[*a*]pyrene	B[*a*]P	252	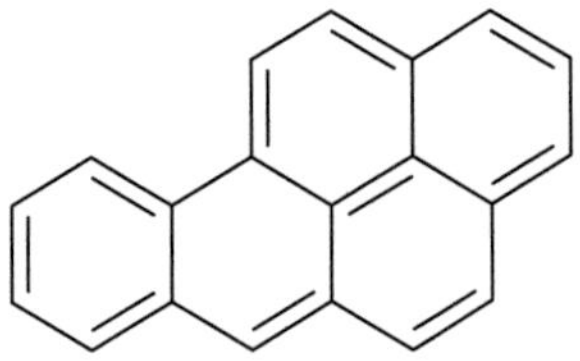
Perylene	Per	252	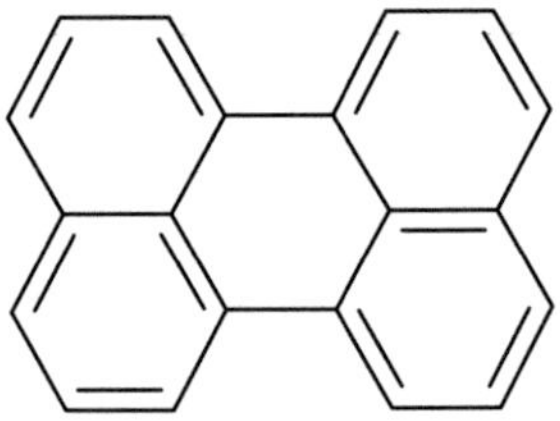
Indeno[*1,2,3-cd*]pyrene	Ind	276	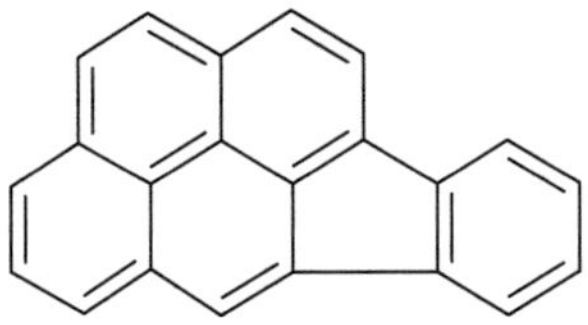
Benzo[*g,h,i*]perylene	B[*g,h,i]*P	276	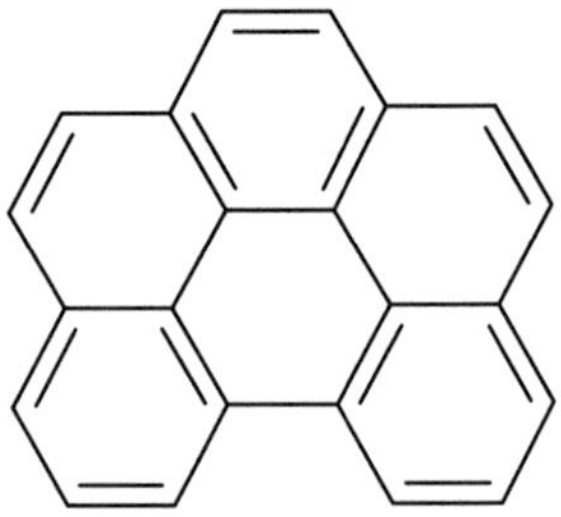
Anthanthrene	Ant	276	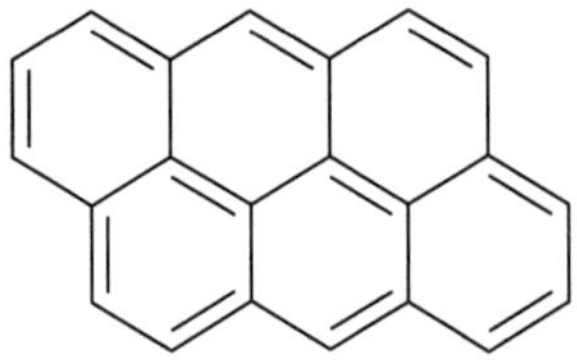

Congener	Abbreviation	M.W. [g]	Chemical Structure
Dibenzo[*a,h*]anthracene	D[*a,h*]A	278	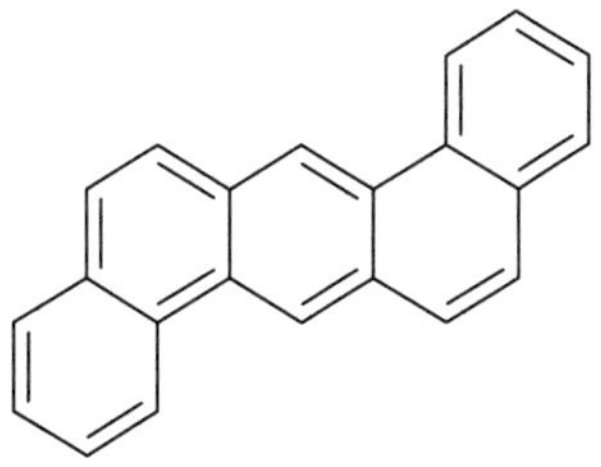
Coronene	Cor	300	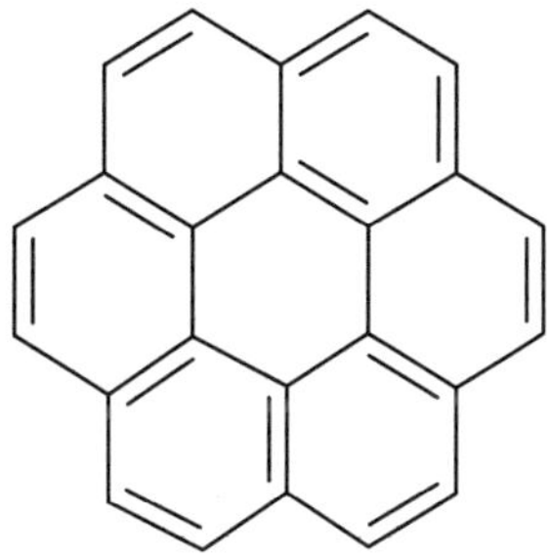

Table 2. Chemical structures of PAHs (Harvey *et al.*, 1991)

On the other hand, PAHs are also quite involatile, and have relatively low vapor pressure and resistance to chemical reactions. As a consequence PAHs are persistent in the environment and demonstrate a tendency to accumulate in biota, soils, sediments, and are also highly dispersed by the atmosphere. Furthermore, PAHs become more resistant to biotic and abiotic degradation as the number of benzene rings increase (Bossert & Bartha, 1986). In addition, PAHs are generally not biomarkers as inferred by various authors (Bernstein et al., 1999; Ehrenfreund, 1999; Blake and Jenniskens, 2001), because they can form from abiological carbonaceous matter by free radical buildup (Simoneit and Fetzer, 1996) or by aromatization of biomarker natural products (Simoneit, 1998). The abiogenic PAHs occur primarily as the parent compounds without alkyl or alicyclic substituents as is the case for the aromatic biomarkers. Thus, phenanthrene is not a biomarker, but 1-methyl-7-phenanthrene (i.e., retene) is, because the latter is generated from numerous natural product precursors by dehydrogenation.

The 1952 Nobel Prize winners in chemistry, Archer John Porter Martin and Richard Laurence Millington Synge, had established the partition coefficient principle, which can be employed in paper chromatography, thin layer chromatography, gas phase and liquid-liquid applications. This technique is widely recognized as Hydrophilic Interaction Chromatography (HILIC) in HPLC or High-Performance Liquid Chromatography (sometimes referred to as High-Pressure Liquid Chromatography). HPLC characteristically exploits various kinds of stationary phases, a pump that drives the mobile phase(s) and analyte through the column, and a detector to afford a distinctive retention time for the analyte. Additional information related to the analyte can be achieved by using specific types of detectors (e.g. UV/Visible/Fluorescence Spectroscopic Detectors). It is also worth to mention that the analyte retention time is a function of its affinity with the stationary phase, the ratio/composition of solvents used, and the flow rate of the mobile phase. Although HPLC has the advantage of analyzing several organic compounds in a single chromatogram (e.g. analysis of PAHs), the sensitivity is not high enough to detect target compounds in part per billion (ppb) levels. To overcome the instrumental detection limit, Rolland Gohlke and Fred McLafferty had developed the application of a mass spectrometer as the detector in gas chromatography, which later on called as Gas Chromatography Mass Spectrometry (GC-MS) (Gohlke, 1959; Gohlke and Fred, 1993). In 1964, Robert E. Finnigan had extended the potential of GC-MS by developing a computer controlled quadrupole mass spectrometer under the collaboration with Electronic Associates, Inc. (EAI), a leading U.S. supplier of analog computers. By the 2000s computerized GC/MS instruments using quadrupole technology had become both essential to chemical research and one of the foremost instruments used for organic analysis. Today computerized GC/MS instruments are widely used in environmental monitoring of water, air, and soil; in the regulation of agriculture and food safety; and in the discovery and production of medicine.

Recently, GC/MS featuring ion trap systems, based on a selected parent ion and a whole mass spectrum of its daughter ions, offering the broadest selection of innovative features to expand the working mass range of typical GC/MS system to 900 amu with higher performances and specificities (Jie and Kai-Xiong, 2007). This technique minimizes the ratio between the numbers of ions from matrix background to those of target compound and thus enhances the analytical sensitivity of GC/MS system. Identifications of PAHs associated with atmospheric particulate matter, coastal and marine sediments are extremely elaborate since the congeners are generally present at trace levels in a complex mixture. Application of GC/MS featuring ion trap systems with low detection limits has allowed significant progress in characterization of organic component of particulate matter distributed in all environmental compartments. However, there are still some difficulties associated with the identification process using GC/MS featuring ion trap systems.

It is well known that the mass number of positively charged ions travelling through the quadrupole filter relies on the voltage applied to the quadrupole filter. As a consequence, the detection of a predetermined mass range can be conducted by altering (i.e. scanning) the applied voltage in a particular scope. Based on this theory, it is evident that the faster the scanning speed is, the shorter the detection time of a scanning cycle is, and the greater the number of scanning cycles within a predetermined time period can be achieved. This implies that a higher time resolution can be acquired by enhancing the scanning speed. However, the relatively fast scanning speed can lead to a major sensitivity drop due to the modification in the applied voltage while the positively charged ion is passing through the quadrupole. Assuming that a positively charged ion takes the time t1 to travel though the length L of the quadrupole filter, the kinetic energy of the ion plays an important role on the time t1 when the ion passes through the quadrupole filter. Since the voltage applied to the quadrupole while the ion passes through is positively correlated with the voltage scanning speed, the extremely high scanning speed deteriorates the instrument sensitivity as the scanning speed is enhanced.

GC-MS-QP 2010 Ultra was developed by Shimadzu engineers to resolve this problem. Advanced Scanning Speed Protocol technology (ASSP technology) was invented by using the electrical field to control the kinetic energy of the positively charged ions generated from the ion source while they are passing through the quadrupole filter. ASSP technology was specially designed for adjusting the kinetic energy of positively charged ions travelling through the quadrupole filter. The shorter the travelling time is, the alteration of voltage applied to rod electrodes in the course of the passage through the quadrupole filters is smaller and thus a large number of positively charged ions can enter to the quadrupole filter. This finally leads to an improvement of instrumental sensitivity. In addition, FASST (Fast Automated Scan/SIM Type) was deliberately designed for collecting Scan data and SIM data in a single measurement. This data acquisition technique can be employed in both gathering qualitative spectral data (Scan) and analyzing quantitative data (SIM). It is also worth to note that ASSP has amended this procedure by permitting the SIM dwell times to be reduced by as much as five times without losing sensitivity, letting the user to observe more SIM channels.

Without any doubt, ASSP coupled with FASST is an innovative data acquiring technique that integrates the qualitative assessment (i.e. Scanning of spectral data) with the quantitative analysis (SIM). Despite its high speed scanning performance, little is known about the factors governing relative response factors (RRFs) detected by GC-MS-QP 2010 Ultra (Shimadzu). Since the data quality is greatly reliant on the values of RRFs, it is therefore crucial for the analyzer to know the factors affecting sensitivity and stability of RRFs. An enormous number of papers have appeared on the topic of GC/MS analysis of PAHs using single quadrupole and quadrupole ion trap, most of which have not advanced knowledge to investigate the factors influence the sensitivity of RRFs determined by ASSP and/or FASST data acquiring techniques. Although the principles of the instrumental operation and data quality control are well established and straightforward, there are some uncertainties persisted and need to be clarified. For instance, there are several factors that can significantly affect the values of RRF detected by GC-MS ion trap such as "manifold temperature drop", "glass injection port linear contamination" and "column degradation" (Pongpiachan et al., 2009). However, it remained unclear to what extent other parameters such as ion source temperature (I.S.T), interface temperature (I.T) and scanning speed (S.S) could affect sensitivity of RRFs detected by using ASSP and FASST data acquiring techniques. Hence, the objectives of this study are

1. To statistically determine the effect of "ion source temperature" on fluctuations of RRFs of PAHs
2. To quantitatively assess the influence of "interface temperature" on alterations of RRFs of PAHs
3. To investigate the "scanning speed" effect on variations of RRFs of PAHs.

2. Materials and methods

2.1 Materials and reagents

All solvents are HPLC grade, purchased from Fisher Scientific. A cocktail of 15 PAHs Norwegian Standard (NS 9815: S-4008-100-T) (phenanthrene (Phe), anthracene (An), fluoranthene (Fluo), pyrene (Pyr), 11h-benzo[a]fluorene (11H-B[a]F), 11h-benzo[b]fluorene (11H-B[b]F), benz[a]anthracene (B[a]A), chrysene (Chry), benzo[b]fluoranthene (B[b]F), benzo[k]fluoranthene (B[k]F), benzo[a]pyrene (B[a]P), benzo[e]pyrene (B[e]P), indeno[1,2,3-cd]pyrene (Ind), dibenz[a,h]anthracene (D[a,h]A), benzo[g,h,i]perylene (B[g,h,i]P); each 100 µg mL-1 in toluene: unit: 1×1 mL) and a mix of recovery Internal Standard PAHs (d12-perylene (d12-Per), d10-fluorene (d10-Fl); each 100 µg mL-1 in xylene: unit: 1×1 mL) were supplied by Chiron AS (Stiklestadveine 1, N-7041 Trondheim, Norway). Standard stock solutions of 4 µg mL-1 of deuterated PAHs (used as internal standard) and 100µg mL-1 of native PAHs were prepared in nonane. Working solutions were obtained by appropriate dilution in n-cyclohexane. All solutions were stored in amber colored vials at -20 °C.

2.2 Analytical apparatus

A mass spectrometer is an instrument that separates ions according to the mass-to-charge ratio (m/z) and measures their relative abundance. The instrument is calibrated against ions of known m/z. All mass spectrometers operate by separating gas phase ions in a low-pressure environment by the interaction of magnetic or electrical fields on the charged particles. In this study, the analyses were performed using a Shimadzu GCMS-QP2010 Ultra system comprising a high-speed performance system with ASSP function (i.e. achieving maximum scan speed of 20,000 u sec-1) and an ultra-fast data acquisition speed for comprehensive two-dimensional gas chromatography (GC × GC). The rod bias voltage is automatically optimized during ultra high-speed data acquisition, thereby minimizing the drop in sensitivity that would otherwise occur above 10,000 u sec-1. The GCMS-QP2010 Ultra achieves a level of sensitivity better than five times that of conventional instruments, and is particularly effective for scan measurement in applications related to fast-GC/MS and comprehensive GC/MS (Patent: US6610979).

The target compounds were separated on a 60 m length × 0.25 mm i.d. capillary column coated with a 0.25 µm film thickness (phase composition: cross-linked/surface bonded 5% phenyl, 95% methylpolysiloxane. Specified in EPA methods 207, 508, 515, 515.2, 524.2, 525, 548.1, 680, 1625, 1653, 8081, 8141, 8270 and 8280) stationary phase (Agilent JW Scientific DB-5 GC columns). Helium (99.999%) was employed as carrier gas at a constant column flow of 1.0 mL min-1 and a pressure pulse of 25 psi with a duration of 0.50 min. All injections (1 µL) were performed through a universal injector in the splitless mode and the standards were introduced using a 10 µL Hamilton syringe. The GC oven temperature was programmed as follows: 1 min at at 40 °C, heated at 8 °C min-1 to 300 °C and held for 45 min. By employing these chromatographic conditions it was possible to qualitatively distinguish between Ph/An, B[a]A/Chry and 11H-B[a]F/11H-B[b]F, three pairs of isomers that are commonly co-eluted by gas chromatographic systems.

2.3 Compound identification and quantification

In order to quantify PAHs in environmental samples, all the detected compounds are normally identified by comparing the retention time and mass spectra of the authentic standards. Appropriately selected quantification ions can be beneficial to distinguish a particular mass spectrum of an individual compound from the co-eluted complex. In general, the most abundant ion serves as the quantification ion, which is the case for PAHs in this study. The molecular markers are identified by comparing first the retention times with authentic standards within a range of ± 0.2 min, secondly the quantification ions.

Quantification of the compounds is based upon the Internal Standard (IS) method. One of the fundamental requirements of using an IS is that it displays similar physiochemical properties or the same type of substitution as the analytes because be similar to each other. A relative response factor (RRF) for each native analyte was first determined. This is used for quantification, as the relative response between the internal standard (IS) and the native analyte should remain constant. It is a convenient method because recovery losses of the compound during extraction and analysis are assumed to match those of the IS. It is calculated employing the appropriate IS using the following equation: The calculation of relative response factor (RRF) is described as follows;

$$F = \frac{A_{nat}}{A_{is}} \times \frac{C_{is}}{C_{nat}} \tag{1}$$

Where A_{nat} = Peak area of the native compound in the standard; C_{nat} = Concentration of the native compound in the standard; A_{is} = Peak area of internal standard; C_{is} = Concentration of the internal standard. The RRF_{STD} used for quantifying samples are the mean of those calculated for the two quantification standards run on the same day. Concentration (C) of analytes in sample extracts is calculated using the following formula:

$$C = \frac{A_{nat}}{A_{is}} \times \frac{1}{RRF_{STD}} \times \frac{W_s}{W_{is}} \tag{2}$$

Where W_{is} = weight of IS added to the sample, W_s = weight or volume of the sample analyzed. A recovery determination standard (RDS) was used for the calculation of both internal standards (IS) and the sampling efficiency standard (SES) of recoveries during sample preparation and extraction/purification. A known amount of RDS was added at the final stage prior to GC/MS analysis and was assumed to suffer zero loss.

$$\left[\left(\frac{A_{is}}{A_{RDS}}\right)_S \times \left(\frac{A_{RDS}}{A_{is}}\right)_{STD}\right] \times \left[\left(\frac{C_{is}}{C_{RDS}}\right)_{STD} \times \left(\frac{C_{RDS}}{C_{is}}\right)_S\right] \times 100\% \tag{3}$$

Where A_{RDS} = Peak area of recovery determination standard; C_{RDS}: = Concentration of the recovery determination standard. Recoveries of IS were used as an indication of the anlyte losses during extraction, pre-concentration, cleanup/fractionation and blow down stages. The calculation of the sampling efficiency by using the sampling efficiency standard (SES) is described as follows:

$$\left[\left(\frac{A_{is}}{A_{SES}}\right)_{STD} \times \left(\frac{A_{SES}}{A_{is}}\right)_S\right] \times \left[\left(\frac{C_{is}}{C_{SES}}\right)_S \times \left(\frac{C_{SES}}{C_{is}}\right)_{STD}\right] \times 100\% \tag{4}$$

Where A_{SES} = Peak area of the sampling efficiency standard, C_{SES} = Concentration of the sampling efficiency standard. Recoveries of SES were used as an indication of analyte losses during sampling as opposed to the analysis. In this study, the IS d10-Fl and d12-Per were employed to calculate the values of RRFs of Group 1 PAHs (i.e. Phe, An, Fluo, Pyr, 11H-B[a]F, 11H-B[b]F, B[a]A, Chry) and Group 2 PAHs (i.e. B[b]F, B[k]F, Benzo[a]pyrene, B[e]P, Ind, D[a,h]A, B[g,h,i]P) respectively. In addition, the factors governing variations of RRFSTD of 15 PAHs Norwegian Standard (NS 9815: S-4008-100-T) as illustrated in Equation 1 had been carefully investigated. The results and details will be discussed in a later part of this chapter.

2.4 Statistical analysis

In this chapter, Pearson correlation analysis, multiple linear regression analysis (MLRA), hierarchical cluster analysis (HCA) and principal component analysis (PCA) were conducted by using SPSS version 13.

3. Results and discussion

Ion source temperature (I. S. T), interface temperature (I. T) and scanning speed (S. S) were set at five different levels (i.e. 200 °C, 225 °C, 250 °C, 275 °C, 300 °C), four distinctive stages (i.e. 250 °C, 275 °C, 300 °C, 325 °C) and five altered points (500 u sec-1, 1,000 u sec-1, 5,000 u sec-1, 10,000 u sec-1, 20,000 u sec-1). Therefore, there are 100 combinations of I.S.T, I.T and S.S (i.e. 4×5×5 = 100) included in the statistical analysis.

3.1 Pearson correlation analysis

To find the relationship between peak area of PAHs and GC-MS tuning parameters (i.e. I.S.T, I.T and S.S), the Pearson correlation coefficients were computed and displayed in Table 3. Some comparatively strong negative correlations were observed between two continuous variables, namely S.S. and high molecular weight (HMW) PAHs. For instance, R-values of S.S. vs Benzo[a]pyrene, S.S. vs Ind, S.S. vs D[a,h]A were -0.77, -0.76, -0.75 respectively. As discussed earlier in Section 1, the relatively fast scanning speed can dramatically cause a significant drop of instrumental sensitivity. Therefore, ASSP and FASST were developed for resolving this technical problem. By carefully controlling the kinetic energy of positively charged ions travelling inside the mass spectrometry, one can reduce its passing time and thus allowing more ions to enter the quadrupole filter. However, if the kinetic energy is not high enough to shorten the passing time of positively charged ions inside the quadrupole, a major drop of instrumental sensitivity can be detected as previously mentioned. Some positive correlations between S.S. vs Phe (R = 0.61), An (R = 0.65), deut-Fl (R = 0.63) proved that ASSP and FASST seem to work fairly well on LMW PAHs (see Table 3). This can be explained through a competition between the kinetic energy of positively charged ions and the voltage scanning speed. It appears that ASSP and FASST successfully resolved the sensitivity drop problems for LMW PAHs. In addition, neither I.S.T. nor I.T. play a significant role on the variation of PAHs' sensitivities.

3.2 Cluster analysis

Cluster analysis (CA), also called segmentation analysis or taxonomy analysis, seeks to identify homogeneous subgroups of cases in a population. That is, cluster analysis seeks to identify a set of groups, which both minimize within-group variation and maximize between-group variation. In this study, CA was conducted using SPSS 13.0 for Windows. CA techniques may be hierarchical (i.e. the resultant classification has an increasing number of nested classes) or non-hierarchical (i.e. k-means clustering). Hierarchical clustering allows users to select a definition of distance, then select a linking method of forming clusters, then determine how many clusters best suit the data. Hierarchical clustering methods do not require pre-set knowledge of the number of groups.

	I.S.T.*	**I.T.****	**S.S.*****	**Ph**	**An**	**deut-Fl**	**Fluo**	**Pyr**	**11H-B[a]F**	**11H-B[b]F**	**B[a]A**	**Chry**	**B[b]F**	**B[k]F**	**B[e]P**	**B[a]P**	**Ind**	**D[a,h]A**	**B[g,h]P**
I.S.T.	1																		
I.T.	-1.4E-17	1																	
S.S.	2.4E-17	1.76E-17	1																
Ph	0.289	0.160	**0.610**	1															
An	0.284	0.113	**0.647**	**0.984**	1														
deut-Fl	0.189	0.117	**0.632**	**0.979**	**0.991**	1													
Fluo	0.457	0.208	0.356	**0.914**	**0.901**	**0.877**	1												
Pyr	0.454	0.198	0.338	**0.904**	**0.889**	**0.866**	**0.989**	1											
11H-B[a]F	0.485	0.243	-0.188	0.509	0.465	0.435	**0.783**	**0.798**	1										
11H-B[b]F	0.456	0.251	-0.202	0.477	0.430	0.396	**0.756**	**0.769**	**0.975**	1									
B[a]A	0.315	0.240	-0.687	-0.121	-0.179	-0.207	0.239	0.262	**0.754**	**0.777**	1								
Chry	0.397	0.262	-0.574	0.040	-0.012	-0.047	0.393	0.415	**0.846**	**0.856**	**0.965**	1							
B[b]F	0.302	0.204	**-0.777**	-0.333	-0.388	-0.419	0.025	0.046	0.598	**0.630**	**0.955**	**0.898**	1						
B[k]F	0.242	0.220	**-0.746**	-0.341	-0.393	-0.426	0.006	0.029	0.574	**0.601**	**0.941**	**0.895**	**0.951**	1					
B[e]P	0.268	0.219	**-0.779**	-0.346	-0.401	-0.432	0.011	0.034	0.592	**0.621**	**0.961**	**0.906**	**0.986**	**0.977**	1				
B[a]P	0.306	0.194	**-0.766**	-0.324	-0.378	-0.413	0.037	0.060	**0.610**	**0.634**	**0.964**	**0.914**	**0.987**	**0.974**	**0.993**	1			
Ind	0.313	0.200	**-0.756**	-0.309	-0.365	-0.400	0.052	0.070	**0.618**	**0.641**	**0.953**	**0.911**	**0.974**	**0.959**	**0.978**	**0.980**	1		
D[a,h]A	0.391	0.183	**-0.753**	-0.319	-0.372	-0.413	0.045	0.065	**0.612**	**0.632**	**0.948**	**0.906**	**0.982**	**0.962**	**0.983**	**0.988**	**0.981**	1	
B[g,h]P	-0.022	0.185	**-0.621**	-0.195	-0.250	-0.245	0.077	0.085	0.517	0.544	**0.763**	**0.724**	**0.733**	**0.730**	**0.748**	**0.738**	**0.825**	**0.713**	1
deut-Per	0.257	0.181	**-0.795**	-0.400	-0.452	-0.482	-0.044	-0.022	0.552	0.576	**0.943**	**0.883**	**0.981**	**0.976**	**0.990**	**0.989**	**0.978**	**0.983**	0.740

Table 3. The Pearson correlation coefficients of PAH's peak area (PA)

There are three general approaches to clustering groups of data, namely "Hierarchical Cluster Analysis", "K-means Cluster Analysis" and "Two-Step Cluster Analysis". In both K-means clustering and two-step clustering, researchers have to specify the number of clusters in advance then calculate how to assign cases to the K clusters. Furthermore, these two clustering techniques require a very large scale of data set (e.g. n > 1,000). On the contrary, Hierarchical Cluster Analysis (HCA) is appropriate for smaller samples (e.g. n < 200) and can be carried out without any data pre-treatment. It is also important to note that there are two types of Hierarchical Clustering namely Agglomerative Hierarchical Clustering and Divisive Hierarchical Clustering.

In agglomerative hierarchical clustering every case is initially considered as a cluster then the two cases with the lowest distance (i.e. highest similarity) are combined into a cluster. The case with the lowest distance to either of the first two is considered next. If that third case is closer to a fourth case than it is to either of the first two, the third and fourth cases become the second two-case cluster, if not, the third case is added to the first cluster. The process is repeated, adding cases to existing clusters, creating new clusters, or combining clusters to get to the desired final number of clusters. In contrast, the divisive clustering works in the opposite direction, starting with all cases in one large cluster. Since the

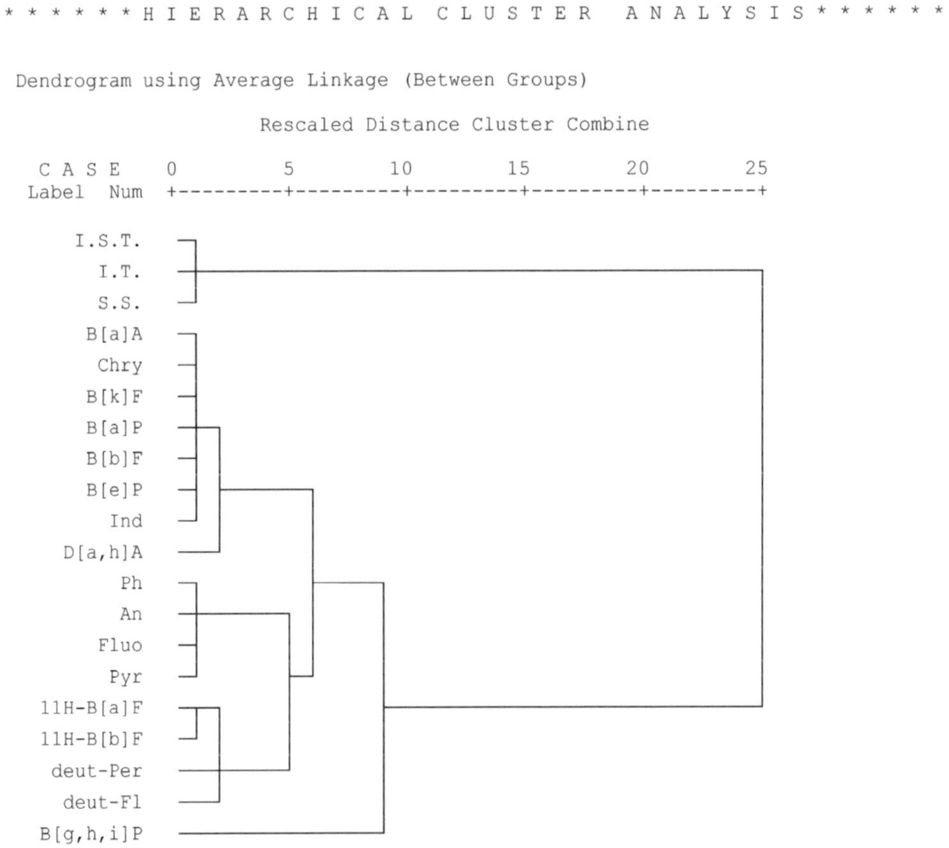

Fig. 1. Dendrogram using average linkage between group of PAH's peak area, I.S.T, I.T and S.S.

objective of this section is to clusterize the wide range of GC-MS tuning parameters coupled with RRFs of PAHs with relatively small sample numbers, the agglomerative hierarchical cluster analysis is probably the most suitable for this purpose. HCA was carried out using the multivariate data analysis software "SPSS version 13". To obtain more information on the affinities of GC-MS tuning parameters and PAHs' peak area, HCA had been conducted on the 20 variables (i.e. 17 PAHs and 3 GC-MS tuning parameters). The results demonstrated in the dendrogram (Fig. 1) distinguished the 20 individual parameters into two major clusters. The first cluster consists mainly of I.S.T, I.T and S.S, which can be considered as a domination of GC-MS tuning parameters.

The second major cluster can be subdivided into three sub-clusters. The first sub-cluster consists of B[a]A, Chry, B[k]F, Benzo[a]pyrene, B[b]F, B[e]P, Ind, D[a,h]A, which were all composed of four to five benzene rings. Therefore, this sub-cluster represents the typical marks of HMW PAHs. The second sub-cluster contains LMW PAHs (Ph, An, Fluo, Pyr). The third sub-cluster composed of 11H-B[a]F, 11H-B[b]F, deut-Per and deut-Fl. It is interesting to note that these congeners have different chemical structures from those of other PAHs. Therefore, the third cluster can be considered as an indicative of mixing of unique structures of PAHs.

Since dendrogram provides a visual accounting of how closely connected one parameter is to another, the more properties of two parameters have in common, the closer they are associated. Figure 1 demonstrates that, for this dendrogram of four clusters, there is a fairly close-knit group in the first major cluster (i.e. I.S.T, I.T and S.S) and the first sub-cluster of second major cluster (i.e. B[a]A, Chry, B[k]F, Benzo[a]pyrene, B[b]F, B[e]P, Ind, D[a,h]A). B[g,h,i]P is the outsider in this dendrogram, while the second sub-cluster of second major cluster (i.e. Ph, An, Fluo, Pyr) takes an intermediary position. Similar to previous results obtained by using Pearson correlation analysis, S.S. is still the most influential GC-MS tuning parameters governing the instrumental sensitivity of PAHs. Apart from B[g,h,i]P, the four-to-five aromatic ring PAHs are closely associated with the fluctuations of S.S. emphasizing that one should carefully select the suitable scanning speed in order to minimize the sacrifice of instrumental sensitivity.

3.3 Multiple linear regression analysis

Table 4 displays the standardized coefficients of multiple linear regression analysis (β-value) of PAH's relative response factors (RRFs). It is remarkable to note that the moderately strong negative β-values (i.e. -0.78 ~ -0.83) and fairly strong positive β-values (i.e. 0.68 ~ 0.76) were observed with S.S as displayed in Fig 4. These results are consistent with previous findings computed by using Pearson correlation analysis and hierarchical cluster analysis. On the contrary, the comparatively weak β-values (i.e. -0.78 ~ -0.83) were detected with I.T. as displayed in Fig 3. It is well known that too high interface temperature can considerably degrade the chemical structure of target compound and hence reducing the instrumental sensitivity. On the other hand, if the interface temperature was too low, the cold condensation will occur and conclusively undermining the instrumental sensitivity of PAHs. Interestingly, there are no obvious signs of either "thermal decomposition" or "cold condensation" which can appreciably degrade the instrumental sensitivity of PAHs. In addition, I.S.T. show a fairly strong positive β-values with Ph (0.38), An (0.70), Ind (0.40) and D[a,h]A (0.72) as illustrated in Fig. 4.

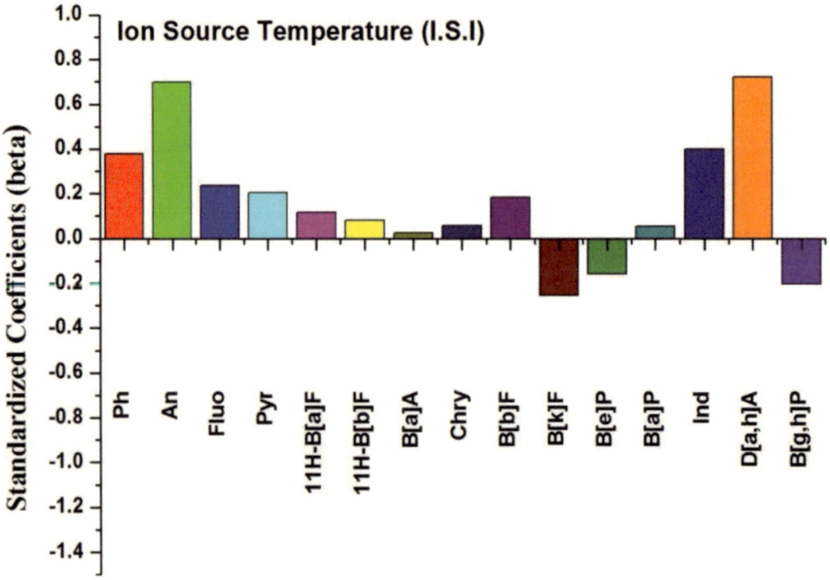

Fig. 2. The standardized coefficients of multiple linear regression analysis (MLRA) of PAH's relative response factors (RRFs) (Ion Source Temperature)

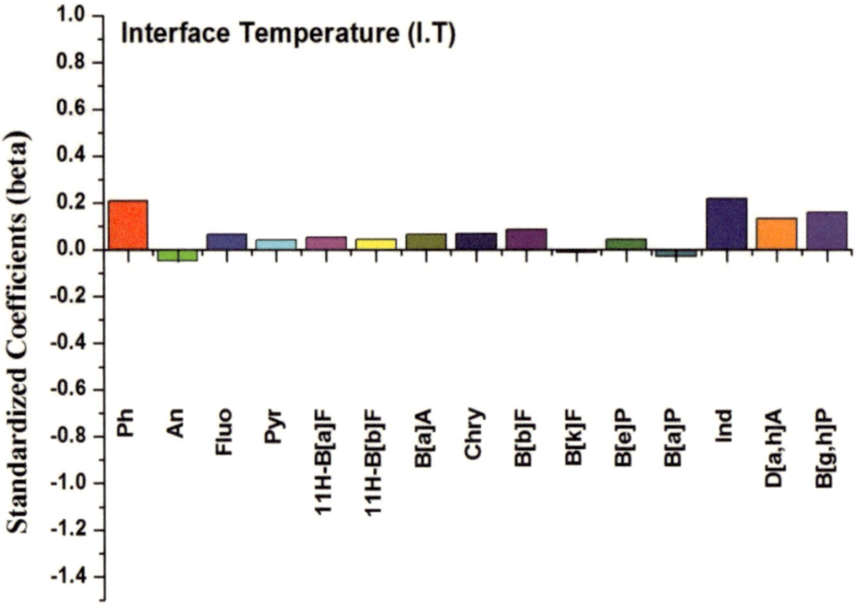

Fig. 3. The standardized coefficients of multiple linear regression analysis (MLRA) of PAH's relative response factors (RRFs) (Interface Temperature)

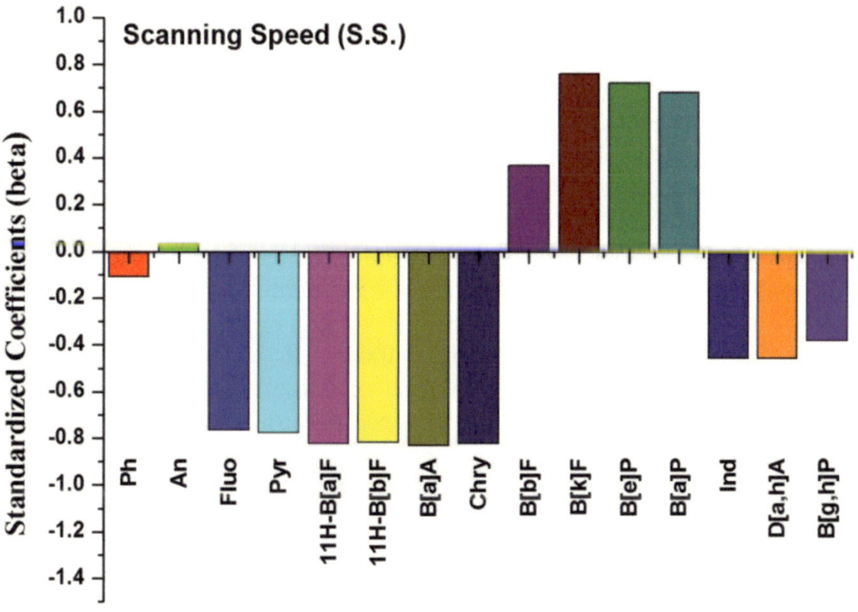

Fig. 4. The standardized coefficients of multiple linear regression analysis (MLRA) of PAH's relative response factors (RRFs) (Scanning Speed)

	Ph	An	Fluo	Pyr	11H-B[a]F	11H-B[b]F	B[a]A	Chry	B[b]F	B[k]F	B[e]P	B[a]P	Ind	D [a,h]A	B [g,h]P
I.S.T.*	0.379	**0.698**	0.237	0.204	0.117	0.080	0.024	0.057	0.184	-0.255	-0.158	0.055	0.401	**0.721**	-0.203
I.T.**	0.209	-0.047	0.067	0.040	0.053	0.044	0.066	0.069	0.086	-0.013	0.043	-0.031	0.219	0.134	0.160
S.S.***	-0.110	0.030	**-0.766**	**-0.775**	**-0.824**	**-0.818**	**-0.832**	**-0.822**	0.367	**0.760**	**0.721**	**0.682**	-0.457	-0.458	-0.381

*I.S.T: Ion Source Temperature, **I.T: Interface Temperature, ***S.S: Scanning Speed

Table 4. The standardized coefficients of multiple linear regression analysis (MLRA) of PAH's relative response factors (RRFs)

3.4 Principal Component Analysis

PCA as the multivariate analytical tool is employed to reduce a set of original variables (i.e. GC-MS tuning parameters and PAHs' RRFs) and to extract a small number of latent factors (principal components, PCs) for analyzing relationships among the observed variables. Data submitted for the analysis were arranged in a matrix, where each column corresponds to one parameter and each row represents the number of data. Data matrixes were evaluated through PCA allowing the summarized data to be further analyzed and plotted. Table 5 displays the principal component patterns for Varimax rotated components of I.S.T., I.T., S.S. and PAHs'RRFs. In order to enable further interpretation of potential correlations between GC-MS tuning parameters and PAHs' RRFs, a PCA model with five significant PCs, each representing 54.08 %, 14.91 %, 9.42 %, 5.61 % and 4.72 % of the variance, thus accounting for 88.74 % of the total variation in the data, was calculated. The first component (PC1) shows considerable strong negative correlation coefficients on five aromatic ring PAHs (i.e. B[k]F, B[e]P, Benzo[a]pyrene) and strong positive correlation coefficients on three-to-four aromatic ring PAHs (i.e. Fluo, Pyr, 11H-B[a]F, B[a]A, Chry). These results confirm that GC-MS tuning parameters do significantly affect these two classes of PAHs in a totally different way. The strong negative correlation coefficient (R = -0.87) detected in S.S. reflected the fact that higher scanning velocity reduced numbers of positively charged ions of HMW PAHs and thus decreased the instrumental sensitivity of five aromatic compounds. The second component (PC2) has higher loadings for I.S.T. (R = 0.89), Ph (R = 0.61), An (R = 0.87) and D[a,h]A (R = 0.65), which could be due to the fact that higher ion source temperature enhanced the signal of three ring aromatic compound (i.e. Ph, An) and five ring aromatic PAHs (i.e. D[a,h]A). The third component (PC3), the fourth component (PC4) and the fifth component (PC5) shows high loading of Ind (R = 0.81), B[b]F (R = 0.77) and I.T. (R = 0.91) respectively. Since I.T. showed the highest R-value in PC5, it appears reasonable to conclude that interface temperature plays a minor role in fluctuations of PAHs' signals detected by the GCMS-QP2010 Ultra. In addition, some HMW PAHs such as Ind and B[b]F appears to insensitive to the variations of GC-MS tuning parameters.

	Principal Component				
	PC1	**PC2**	**PC3**	**PC4**	**PC5**
I.S.T.	0.034	**0.886**	0.017	0.279	-0.107
I.T.	0.024	-0.036	0.096	0.105	**0.906**
S.S.	**-0.866**	0.096	-0.269	-0.142	0.034
Ph	0.093	**0.610**	0.207	-0.177	0.448
An	0.148	**0.869**	-0.146	-0.136	-0.043
Fluo	**0.926**	0.296	0.072	-0.084	0.091
Pyr	**0.936**	0.259	0.046	-0.083	0.073
11H-B[a]F	**0.965**	0.159	0.099	-0.084	0.070
11H-B[b]F	**0.962**	0.134	0.080	-0.077	0.081
B[a]A	**0.969**	0.070	0.043	-0.095	0.104
Chry	**0.965**	0.104	0.050	-0.113	0.099
B[b]F	-0.499	0.080	-0.028	**0.772**	0.116
B[k]F	**-0.840**	-0.151	-0.242	-0.171	0.109
B[e]P	**-0.844**	-0.116	-0.192	0.175	0.156
B[a]P	**-0.821**	0.104	-0.160	0.290	0.071
Ind	0.316	0.396	**0.810**	0.186	0.136
D[a,h]A	0.452	**0.646**	0.249	0.456	0.061
B[g,h]P	0.166	-0.185	0.937	-0.115	0.084
% of Variance	54.083	14.913	9.420	5.613	4.724

Table 5. Principal component analysis of PAH's relative response factors (RRFs)

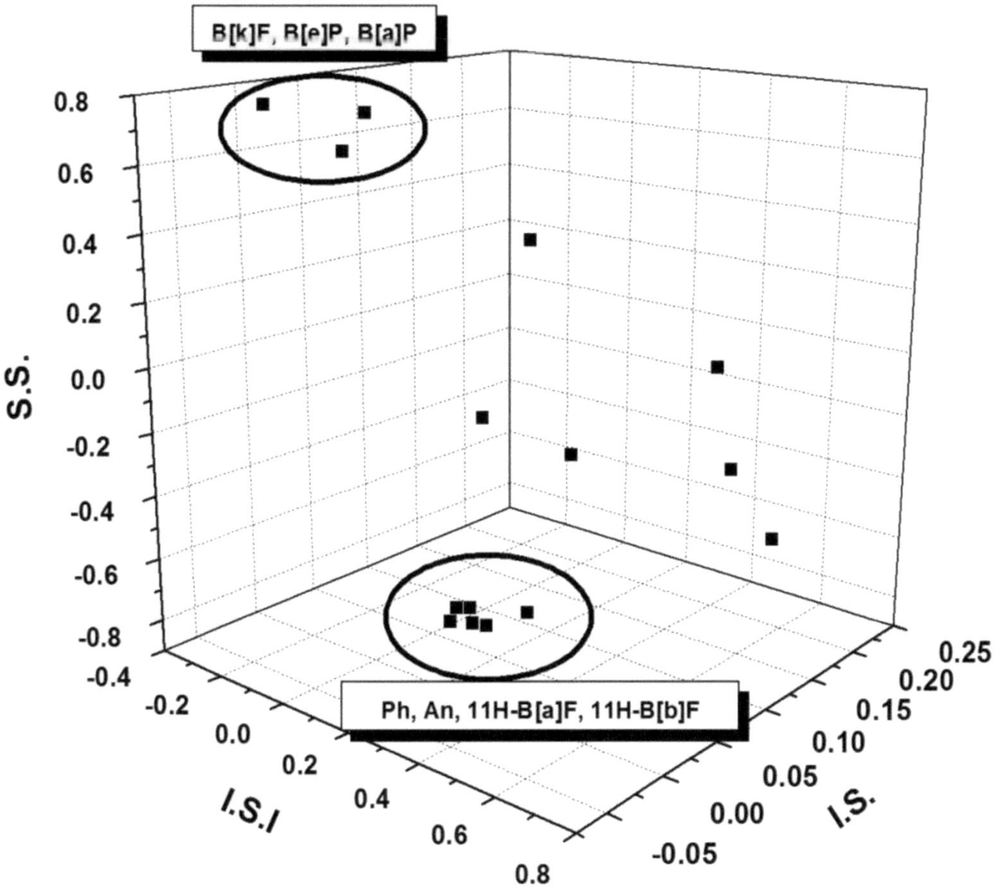

Fig. 5. Three dimensional plots of standardized coefficients of I.S.I, I.S. and S.S.

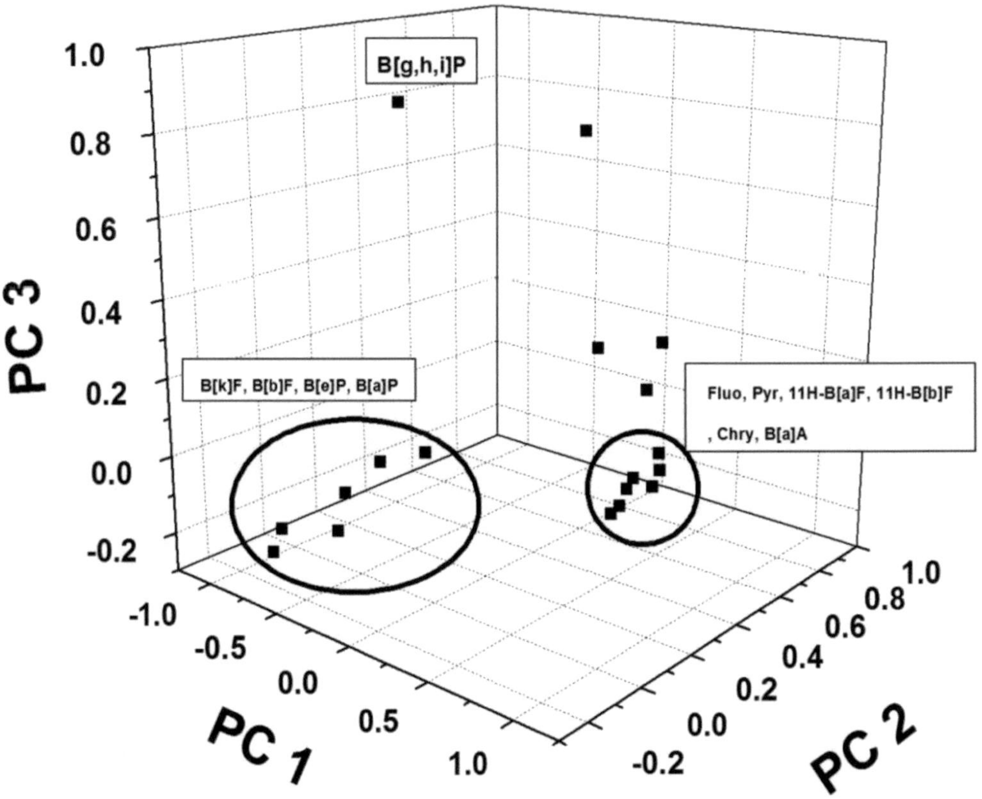

Fig. 6. Three dimensional plots of correlation coefficients of PC1, PC2 and PC3

4. Conclusions

This study is the first, to our best knowledge, to be presented for the quantitative approach to investigate the effects of "ion source temperature", "interface temperature" and "scanning speed" on variations of RRFs of 15 PAHs determined by Shimadzu GCMS-QP2010 Ultra. In fact, the impact of "ion source temperature" and "interface temperature" affect only a few highly volatile PAHs, whereas the effects of "scanning speed" significantly cause the variations of instrumental sensitivities of PAHs. Despite of various factors deteriorating sensitivity of RRFs, the statistical analysis reflect the considerable contribution of scanning speed on relative sensitivities of PAHs in both positive and negative directions. Since the reliability of analytical results are highly dependent on the "sensitivity" of RRFs, it is of great importance for the analyzer to ensure the correct selection of "scanning speed" suitable for the target compounds.

5. Acknowledgement

The authors would like to thank Bara Scientific Co., Ltd and NIDA Center for Research and Development on Natural Disaster Managment, School of Social and Environmental

Development, National Institute of Development Administration (NIDA) for the continued financial support of this work.

6. References

Bernstein, M.P., Sandford, S.A. and Allamandola, L.J. (1999). Life's far-flung raw materials. *Sci. Amer.* 281 1, pp. 42–49.

Blake, D.F. and Jenniskens, P. (2001). The ice of life. *Sci. Amer.* 285 (2): 44–50.

Bossert, I.P. and Bartha, R. (1986). *Bull. Environ. Contam. Toxicol.*, 37: 490-495.

Ehrenfreund, P (1999) Molecules on a space odyssey. *Science* 283: 1123–1124.

EPA Appendix A to 40 CFR, Part 423–126 Priority Pollutants Available from: http://www.epa.gov/region01/npdes/permits/generic/prioritypollutants.pdf, (2003).

Gohlke, R. S. (1959). Time-of-Flight Mass Spectrometry and Gas-Liquid Partition Chromatography. *Analytical Chemistry* 31 (4): 535. doi:10.1021/ac50164a024.

Gohlke, R and McLafferty, F. W. (1993). Early gas chromatography/mass spectrometry. *Journal of the American Society for Mass Spectrometry* 4 (5): 367. doi:10.1016/1044-0305(93)85001-E.

Harvey, G. R (1991) *Polycyclic aromatic hydrocarbons*, Cambridge University Press, Cambridge, USA.

International Agency for Research on Cancer (1983) IARC Monographs on the evaluation of the carcinogenic risk of chemicals to man, Vol.32: Polycyclic aromatic hydrocarbons, Part 1: Chemical, environmental and experimental data. IARC, Lyon, France.

International Agency for Research on Cancer (1984) IARC Monographs on the evaluation of the carcinogenic risk of chemicals to man, Vol.34: Polycyclic aromatic hydrocarbons, Part 3: Industrial exposures in aluminium production, coal gasification, coke production, and iron and steel founding. IARC, Lyon, France.

International Agency for Research on Cancer (1985) IARC Monographs on the evaluation of the carcinogenic risk of chemicals to man, Bitumens, coal-tars and derived products, shale-oils and soots. *Monnograph No 35*, IARC, Lyon, France.

International Agency for Research on Cancer (1987) IARC Monographs on the evaluation of the carcinogenic risk of chemicals to man, Supplement, IARC, Lyon, France.

Jie, F and Kai-Xiong, W. (2007). Multiresidual Analysis of Organochlorine Pesticides, Polychlorinated Biphenyls and Polycyclic Aromatic Hydrocarbons in Marine Shellfishes by Gas Chromatography-Ion Trap Mass Spectrometry. *Chinese J Anal Chem* 35 (11): 1607-1613.

Mackay, D., Callcot, D (1998) Partitioning and physical properties of PAHs. In: Neilson, A.H. (Ed.), The Handbook of Environmental Chemistry, vol. 3, Part J. PAHs and Related Compounds. Springer, Berlin, pp. 325–346.

Pongpiachan, S., Bualert, S., Sompongchaiyakul, P and Kositanont, C. (2009). Factors affecting sensitivity and stability of polycyclic aromatic hydrocarbons. *Journal of Analytical Letters* 42 (13): 2106-2130.

Simoneit, B.R.T. and Fetzer, J.C (1996) High molecular weight polycyclic aromatic hydrocarbons in hydrothermal petroleums from the Gulf of California and Northeast Pacific Ocean. *Org. Geochem* 24: 1065–1077.

Simoneit, B.R.T. (1998). Biomarker PAHs in the environment. In: Neilson, A. and Hutzinger, O., Editors, *The Handbook of Environmental Chemistry*, Springer Verlag, Berlin: 175–221.

Wild, S.R. and Jones, K.C. (1995) Polynuclear Aromatic Hydrocarbons in the United Kingdon Environment: A Preliminary Source Inventory and Budget. *Environ. Pollut.* 88: 91-108.

Part 2

Selected Biomedical Applications of Gas Chromatography Techniques

7

Analysis of Toxicants by Gas Chromatography

Sukesh Narayan Sinha[1,*] and V. K. Bhatnagar[2]
[1]National Institute of Nutrition (ICMR), Jamia-Osmania, P.O-Hyderabad,
[2]Principal and Dean, Subharti Institute of Technology & Engineering, Swami Vivekanand Subharti University, Meerut, India

1. Introduction

Chromatography is an analytical technique which has been used for isolation, purification and separation of organic and inorganic compounds including qualitative and quantitative estimation of compounds. Basically, there are two types of chromatography. One is liquid chromatography and other is Gas chromatography. The chromatography was discovered in 1906 and gas chromatography was developed by Tshett in 1950s. Generally, in GC the mobile phases are gases such as helium (He), or Nitrogen (N_2). GC depends upon temperature programming and boiling point of the compounds. The volatile organic compounds (VOCs), poly-aromatic hydrocarbon and pesticides have been analysed by gas chromatography-mass spectrometric technique. In this chapter, we present a review of existing analytical methodology for the environmental and biological monitoring of exposure to toxicants. In an effort to provide a more concise exploration of existing toxicants bio-monitoring methodology that is relevant today. In addition, as much of the toxicants analysis work involving effective dose measurements is its infancy (adduct measurements) or do not provide sensitive indicators of exposure, All these methods are in use but some form of chromatography, but the detection systems range from simple UV absorbance detection to sophisticated mass spectrometric analyses. These methods possess limits of detection (LODs) that span a wide range; some are suitable for only occupational or forensic applications while those with LODs near or lower than the low-µg/ l are useful for detecting incidental environmental exposures. In addition, these methods have been used to measure toxicants and/or their metabolites in a variety of matrices including urine, serum, breast milk, saliva, and post-partum meconium.

2. Gas chromatography equipped with different detectors for analysis of Arial pollutants

Generally, five types of detectors are coupled with GC (GC-ECD, GC-NPD, GC-FID, GC-PID and GC-MS). The analysis of compounds on GC depends upon nature of compound,

* Corresponding Author

physical and chemical properties of compound. Gas chromatography coupled with electron capture detector (GC-ECD) is used for qualitative as well as quantitative estimation of organochlorine compounds especially organochlorine pesticides. Gas chromatography coupled with nitrogen phosphorous detector (GC-NPD) is used for quantification of nitrogen and phosphorous containing compounds, especially organophosphate compounds. Gas chromatography coupled with flame ionization detector (GC-FID) is used for quantification of volatile organic compounds. Gas chromatography mass spectrometer (GC-MS) is used for quantitative as well as qualitative estimation of non-polar compounds. This technique is very useful for structural confirmation as well as molecular weight determination of compounds. Now, mass spectrometer is used for measurement of molecular weight as well as structural estimation and quantification of compounds. Various types of MS may be used for analysis of protein, peptide, mapping of oligonucleotides, pesticides and air pollutants. MS is also used in industrial as well as pharmaceutical application in drug discovery, poly-aromatic hydrocarbons, poly-chlorinated biphenyl, clinical screening, haemoglobin analysis and drug testing). GC-MS is now a well established routine technique where the MS is viewed as another detector and used for analysis of air samples. GC-MS (eg. High resolution GC-MS (Electron impact (EI), chemical ionization (CI) and negative chemical ionization (NCI) (Ballschmiter et al., 1992; Fischer et al., 1992) techniques were used for the determination of relevant environment parameters (eg. PO, SH_20, KOW) in the mixture of organic compound. This technique was also used to study the chemistry of incomplete combustion and high temperature pyrolisis. The emission from the technical burning process, including the organic emissions of automobile and their occurrence in environmental samples were studied for the groups of polychlorinated, dibenzofuran, biphenyls, diphenyl ethers, chlorophenols and microbial degradation of polychlorinated pesticides and arenes like the dibenzo-p-dioxins, dibenzofurans and naphthalines. A combined GC-MS technique is used in USA for the Qualitative and quantitative analysis of methyl terbutyl ether (MTBE) and benzene in gasoline (Star et al., 2000; Quach et al., 1998). The GC-MS experiments demonstrate the use of internal standards in the method to improve precision, standard deviation, and ion extraction/monitoring for the measurement of specific highly volatile organic compounds in air samples. The thermally desorbed compounds were analysed by GC/FID (flame ionization detector)/MS utilising a GC oven program ranging from 2^0C (to prevent column plugging by ice) to 160^0C and separation on a 30 m x 0.32 mm DB5-MS capillary column. VOCs are identified on the basis of retention time and mass spectral search, where as quantification is based on FID response which should be calibrated weekly with VOC standards (Hodgson et al., 1989; Reidel,1996). The FID detector is more sensitive than MS detector for analysis of VOC's. Structural characterization of PAC by combined GC-MS and GC-FTIR provides unambiguous identification of the isomers in iomex samples.

GC-MS was used to quantify triphenyl hydroquinol THP (Abdulrahman (2001) in environmental samples. Analysis of these data shows that there is no statistical difference between TPH values quantified with GC-MS and those derived from conventional TPH methods with GC/FID. Further, the GC-MS technique used in the study is readily adaptable to most environmental laboratories for analysis of volatile and semi-volatile compounds (EPA analysis ie 8260 and 8270 method respectively, EPA 845 (Robert, (1993). The implications of these results are 1) additional information (e.g. PAH) is often required when conducting a risk based assessment which can be derived from pre-existing site assessment data, thereby decreasing the cost and time required to obtain the additional samples and analytical information, and (2) unique TPH distributions can be critically evaluated with mass spectra in order to ascertain the nature and potential source of the compounds present in the TPH mixtures.

A gas chromatography–mass spectroscopic method in electron ionization (EI) mode with MS/MS ion preparation using helium at flow rate 1 ml min^{-1} as carrier gas on DB-5 capillary column (30 m × 0.25 mm i.d. film thickness 0.25 µm) has been developed for the determination of benzene in indoor air. The detection limit for benzene was 0.002 µg ml^{-1} with S/N: 4 (S: 66, N: 14). The benzene concentration for cooks during cooking time in indoor kitchen using dung fuel was 114.1 μgm^{-3} while it was 6.6 μgm^{-3} for open type kitchen. The benzene concentration was significantly higher ($p < 0.01$) in indoor kitchen with respect to open type kitchen using dung fuels. The wood fuel produces 36.5 μgm^{-3} of benzene in indoor kitchen. The concentration of benzene in indoor kitchen using wood fuel was significantly ($p < 0.01$) lower in comparison to dung fuel. This method may be helpful for environmental analytical chemist dealing with GC–MS in confirmation and quantification of benzene in environmental samples with health risk exposure assessment (Sinha et al., 2005).

The benzene and toluene was quantified using the GC-MS/ MS technique from indoor air. The geometrical mean (GM) of benzene exposure for cooks during cooking hours in an indoor kitchen using mixed fuel was 75.3 µg/m^3 (with partition) and 63.206 µg/m^3 (without partition), while the exposure was 1.7 µg/m^3 for open type. The benzene exposure was significantly higher ($p < 0.05$) in an indoor kitchen with respect to open type using mixed fuels. Concentration of benzene (114.1 µg/m^3) for cooks in an indoor kitchen with partition using dung fuel was significantly higher in comparison to open type kitchen. (Sinha et al., 2005). The toxic compounds are toluene, xylene, ethylbenzene, 2-4dimethyl heptanes ,cyclohexane propanol, 1-(methyl ethyl) cyclohexane, propyl benzene, 1-ethyl-3-methyl benzene, heptyl cyclohexane, 1-methyl-3propyl benzene, diethyl benzene, 4-ethyl octane, naphthalene and 3-methyl decane were present in thinner. All these compounds were confirmed using retention time parameter, molecular ion peak and other characteristic peak using National Institute of Science Technology (NIST) library search. The percentage composition of thinner has been also explained (Sinha and Zadi 2008). A GC-MS method is also very useful for measurement of main urinary metabolites of benzene, namely phenol, catechol,hydroquinone, 1,2,4-trihydroxybenzene, t,t-muconic acid and S-phenylmercapturic acid. Measurement of urinary benzene was performed via headspace solid phase microextraction of 0.5 mL of urine specimens followed by GC-MS. A number of methods including GC-FID and GC-MS are available for analysis of benzene from ambient environment but the more sensitive procedure of Pre-concentration on charcoal followed by GC-MS analysis is preferred. GC-MS technique being used for the qualitative and quantitative analysis for methyl tertiary butyl ether (MTBE) and benzene in gasoline. In this method, the internal standard was used to improve precision, standard deviation and ion extraction/ monitoring for the measurement of specific highly volatile organic compounds in air pollution samples (Carmichael et al., 1990). A carbotrap tube (2 mm) was used to determine volatile organic compounds in ambient air. Such compounds were desorbed and thermally analyzed with GC-MS With the use of GC-MS technique about 54 toxic hydrocarbons were quantified in the ambient air of Tehran. Polyurethane foam (PUF) cartridge samples were analyzed for dioxins and furans as per EPA test method TO-9 (Keen and Doug project No 46310001131). This gas chromatography high resolution mass spectrometric (GC-HRMS) method was also used for the analysis of poly chlorinated dibenzofurans in ambient air.

The gas chromatography-electron ionization with single ion monitoring (GC-EI/SIM), gas chromatography-negative chemical ionization with single ion monitoring (GCNCI/SIM), and gas chromatography-negative chemical ionization with single reaction monitoring (GC-NCI/SRM) method was developed for the determination of pesticides in air sample extracts at concentrations <100-μgL^{-1}. In general, GC-NCI/SIM provided the lowest method detection limits (MDLs commonly 2.5–10 μgL^{-1}), while GC-NCI/SRM was used for confirmation of parathion-ethyl, tokuthion, carbofenothion. But, as per the reported method GC-EI/ SRM at concentration <100- μgL^{-1} was not suitable for most pesticides. GC-EI/SIM was more prone to interference issues than negative chemical ionization (NCI) methods, but of fixed good sensitivity (MDLs 1–10 μgL^{-1}) for pesticides with poor NCI response (organophosphates (Ops): sulfotep, phorate, aspon, ethion, and OCs: alachlor,aldrin, perthane, and DDE, DDD, DDT) (Martinez et al., 2004).

3. Chromatography equipped with different detector for analysis of pesticides in food samples

GC–MS/MS method was used for the analysis of organochlorine pesticides (OCs) residues (α-HCH, β-HCH, γ-HCH, δ-HCH, p,p-DDE, p,p-DDD, p,p-DDT, α-endosulfan, β-endosulfan, and endosulfan sulphate) in carbonated drinks (Miller and Miline 2008). Furthermore, a method was reported (Paya et al., 2007) for the analysis of pesticide residues using a quick, cheap, effective, rugged, and safe (QuEChERS) multi-residue method in combination with gas chromatography and tandem mass spectrometric detection (GC–MS/MS). A mixture of 38 pesticides was quantitatively recovered from spiked lemon, raisins, wheat and flour using GC–MS/MS, while 42 pesticides were recovered from oranges, red wine, red grapes, raisins and wheat flour using GC–MS/MS for determination (Paya et al., 2007). Several GC methods have also been developed for pesticide residual analysis in different food (Kumari and Kathpal 2009; Kumari et al., 2006; kumara et al., 2002) commodities (e.g., vegetables, fruits and other products of food).

A sensitive method for the quantification of 11 pesticides in sugar samples to the μgkg^{-1} level has been developed. These pesticides are often used in an agricultural context. A simple solvent extraction followed by selective analysis using a gas chromatography–mass spectrometric method was used. This method was accurate (> 99%) as it possesses limits of detection in the 0.1-ug kg^{-1} range, and the coefficients of variations are less than 15% at the low μgkg^{-1} end of the method's linear range. The percent recovery of all the pesticides at the lowest levels of detection ranges from 82% to 104%. This method was used for the quantification of pesticides in sugar samples collected from different factory outlets from different parts of India. In this study, 27 refined sugar samples were analysed in which one sample showed a detectable level of the chlorpyrifos. This study showed that Indian sugar is free from the commonly-used pesticides at the low μgg^{-1} levels (Sinha et al., 2011).

Pesticide	Matrices	Sample preparation	Instru-ment used	[a]LOD ranged (μgkg^{-1})	[b]LOQ ranged (μgkg^{-1})	% Recovery ranged (%mean)	[c]RSD ranged (%)	Refe-rences
OPs	vegetables	Solvent extraction	GC-PID	3 to10	NR	75.2% to 111.5%,	2.8% to 10.4%.	Li et al., 2010
OPs	vegetables	solvent extraction with SPE	CEI with quantum dot fluore-scence detection	50 to 180, and 100-30000	NR	88.7-96.1%		Chen et al., 2010)
pesticides	Fruit+ vegetable	QuEChERS	LC-MS/MS	NR	NR	70% to 120%,	>20%	Camino-Sánchez et al., 2010)
pesticides	olives	solid-phase dispersion and QuEChERS	LC-MS/MS	10	NR	70-120%		Gilbert-López et al., 2010
Pesticides (OCs + Ops)	vegetable	QuEChERS	GC-NPD	1 to 9	NR	70.22% to 96.32%	>15%	Srivastava et al., 2010
pesticides	fruits and vegetables	QuEChERS	LP-GC/TOFMS	NR	NR	70-120%	< 20%	Koesukwiwat et al., 2010
Pesticides	vegetables	NR	GC-MS	NR	NR	NR	NR	Osman et al., 2010
Pesticides	vegetables	QuEChERS	LC-MS/MS	NR	NR	NR	NR	Kmellár et al., 2010
Ops and carba-mates	total diet	QuEChERS	LC-MS/MS	NR	200	70-120%	< 20%	Chung et al., 2010
pesticides	vegetables	Liquid extraction	HPLC	0.5-3.0	NR	77.8-98.2%	NR	Lin et al., 2010
pesticides	Vegetable + fruits	Solid phase extraction	UPLC-MS/MS	1-2000	NR	70-120%	NR	Kamel et al., 2010

NR = Not reported
LOD= Limit of detection
LOQ= Limit of quantification
RSD= Relative standard deviation

Table 1. Different reported methods in food matrices.

The reported method for the analysis of pesticides in food samples is tabulated in Table-1. However, recently, a method has been reported (Li Wu, Chen & Zhang 2010) detailing the percentage recoveries, RSD, and LOD in vegetable sample. The reported RSD for all the pesticides used in this study ranged from 2.8% to 10.4% and the recovery ranged from 75% to 11%. The reported LOD in vegetable samples ranged from 3 to 10µgkg^{-1} for all pesticides. Also, an LC-MS/MS method has been used for the analysis of pesticides in fruit and vegetable samples (Camino sanchez et al., 2010). In this method, QuEChERS was used for the extraction. The reported mean percentage recoveries mostly ranged between 70% and 120%, and RSD were generally below 20% at the lower spiked level. The CEI with quantum dot fluorescence detector was used for analysis of vegetable samples (Chen & Fung 2010). The reported LOD ranged from 50 to 180µgkg^{-1}, while the reported recovery ranged between 88% and 96%. Additionally, the LC-MS/MS was used by different researcher (Gilbert-Lopez et al., 2010; Kmellar, Abranko, Fodor & Lehotay 2010; Chun & Chan 2010) for analysis of pesticides in olives, vegetables and total diet samples. In this method, QuEChERS was used for the extraction. The reported mean percentage recoveries mostly ranged between 70% and 120%, and RSD were below 20%. The reported LOD was 10µgkg^{-1}. The GC-MS method was used for analysis of pesticides in fruits and vegetables samples (Koesukwiwat, Lehotay, Miao & Lepipatpiboon, 2010; Osman, Al-Humaid, Al-Rehiayani & Al-Redhalman, 2010). The simultaneous identification and quantification of pesticide residues in fruits, milk and vegetables were reported (Kamel, Qian, Kolbe & Stafford, 2010). This method was based on solid phase extraction followed by UPLC-MS/MS chromatographic separation and a full-scan mass spectrometric detection. The recovery for all pesticides ranged from 70% to 120% with LOD ranged from 1 to 2000µgkg^{-1}. The HPLC method was also used for analysis of pesticides in vegetables samples (Lin et al., 2010). In this method liquid extraction was used. In the vegetable samples, the observed recoveries ranged from 77% to 99% and LOD ranged from 0.5 to 3µgkg^{-1} while GC-NPD was used for analysis of vegetable samples reported mean percentage recoveries mostly ranged between 70% and 96%, and RSD were generally below 15% (Srivastava, Trivedi, Srivastava, Lohani & Srivastava 2010). The LOD values ranged from 1 to 9µgkg^{-1}. However it should be noted that HPLC and GC-NPD are not confirmatory techniques.

4. Gas chromatography equipped with different detector for analysis of pesticides in biological samples

A sensitive GC–MS method in MS/MS ion preparation was developed for quantitative estimation and qualitative determination of chlorpyrifos in human blood samples. Dissociation energy effects on ion formation of chlorpyrifos and sensitivity of this analytical method was well demonstrated. Chlorpyrifos was extracted using methanol/hexane mixture from 0.2 ml human blood, deactivated with saturated acidic salt solution. The extract was then re-concentrated and analyzed by electron impact (EI) MS/MS gas chromatography-mass spectrometer. The MS/MS spectra of chlorpyrifos ion were recorded on different dissociation energy (30–100 V) to establish the structural confirmation and to demonstrate the effect of this energy on sensitivity, S/N ratio and detection limit for quantification of chlorpyrifos, which is reported for the first time. At different exciting amplitude (30–100 V), different behaviors of base peak, sensitivity, S/N ratio and detection limit of this method were observed for quantification of chlorpyrifos. The mass spectra recorded at dissociation energy <70 V, in between 70–80 V and >80 V correspond to the m/z 314 (100%), m/z 286 (100%) and m/z 258 (100%), respectively. The sensitivity, signal to noise ratio and detection limit of this method increased on 95 V at m/z 258. Therefore, m/z 258 was used for the quantification of chlorpyrifos. The detection limit for quantification was 0.1 ng/ml with *S/N*: 2 in human blood. The linear calibration curve with the correlation coefficient ($r > 0.99$) was obtained. The percentage recoveries from 95.33% to 107.67% were observed for chlorpyrifos from human blood. The blood samples were collected at different time intervals. The concentration of chlorpyrifos in Poisoning case was 3300, 3000, 2200, 1000, 600, 300 ng/ml on day 1, 3, 6, 8, 10 and 12, respectively. On the 12th day of exposure of chlorpyrifos, 90.9% reduction in concentration was observed. On day 14th, the chlorpyrifos was not detected in the blood of the same subject. Thus, the present study is useful for detection of chlorpyrifos in critical care practices and also provides tremendous selectivity advantages due to matrix elimination in the parent ion isolation step in blood sample analysis for chlorpyrifos (Sinha et al., 2006). The effect of dissociation energy on ion formation and sensitivity of triazofos in blood samples was studied. Six millilitres hexane was used for the extraction of triazofos from 2mL serum samples. The extract was reconstituted in 1mL hexane and analyzed by GC-MS/MS in electron impact MS/MS mode. The structure, ion formations, nature of base peak and fragmentation schemes were correlated with the different dissociation energies. The new ion was obtained at mass to charge ratio 161 (100%), which was the characteristic ion peak of triazofos. On using different exciting amplitudes, different behaviours of fragmentation schemes were obtained. The effect of dissociation energy on sensitivity of the analyte was also demonstrated. The mass spectra recorded at different exciting amplitudes <50V in between 50-60V and >60V correspond to the m/z 161 (100%), 77 (100%) and 119 (100%), respectively. The maximum sensitivity of analyte in blood sample was obtained on using 50V dissociation energy. Additionally, the effect of current on sensitivity of the method was also demonstrated. In all conditions, the new characteristic ion at m/z 161 was obtained and used for quantification of triazofos in blood samples with maximum sensitivity. The limit of detection and quantification was 0.351 and 1.17 $ngmL^{-1}$, respectively, with 99% accuracy. The observed correlation coefficient was 0.995. The inter-day percentage recoveries from 83.9% to 111% were obtained below 9.38 percentage RSD. This present method gives combined picture of confirmation and quantification of triazofos in critical care practices and also provides tremendous selectivity advantages due to matrix elimination in the parent ion isolation step in blood sample analysis for triazofos in poisoning emergency cases (Sinha 2010). A number of families in a rural area of Jabalpur District (Madhya Pradesh), India, were affected by repeated episodes of convulsive illness over a period of three weeks. The aim of this investigation was to determine the cause of the illness. The investigation included a house-to-house survey, interviews of affected families, discussions with treating physicians, and examination of hospital records. Endosulfan poisoning was suspected as many villagers were using empty pesticide containers for food storage. To confirm this, our team collected blood and food samples, which were transported to the laboratory and analyzed with GC-ECD. Thirty-six persons of all age groups had illness of varying severity over a period of three weeks. In the first week, due to superstitions and lack of treatment, three children died. In the second week, symptomatic treatment of affected persons in a district hospital led to recovery but recurrence of convulsive episodes occurred after the return home. In the third week, 10 people were again hospitalized in a teaching hospital. Investigations carried out in this hospital ruled out infective etiology but no facilities were available for chemical analysis. All persons responded to symptomatic treatment. The presence of endosulfan in blood and food samples was confirmed by GCMS. One of the food items (Laddu) prepared from wheat flour was found to contain 676 ppm of α-endosulfan. Contamination of wheat grains or flour with endosulfan and its consumption over a period of time was the most likely cause of repeated episodes of convulsions, but the exact reason for this contamination could not be determined. This report highlights the unsafe disposal of pesticide containers by illiterate farm workers, superstitions leading to delay in treatment, and susceptibility of children to endosulfan (Dewan et al., 2004). Mass spectrometry method has been used with capillary gas chromatography in electron ionization mode to identify the presence of α, β and endosulfan sulphate in human serum samples. The fragmentation ions were obtained due to ring cleavages, rearrangement and remote charge process. Solvent extraction procedure was used for isolation of compounds from serum sample. A detection limit as low as 100 ppb for α, β and endosulfan sulphate could be easily achieved for confirmation. For endosulfan and endosulfan sulphate the m/z values were obtained at intervals of M^{+2}. The m/z values of α, β endosulfan are identical but they differed in retention time (Sinha et al., 2004). Therefore, the spectra reported might serve as reference spectra for identification of different type of organo-chlorine pesticide in human serum samples. The GC-MS has been employed to quantitatively detect at low ppb levels of α and β endosulfan in human serum, urine and liver but this failed to separate α, β-isomers. GC equipped with a mass detector was used to measure the levels of endosulfan in human blood at ppb level.

Although some OC pesticide metabolites are monitored in urine, they are most commonly measured as the intact pesticide and/or its metabolite in whole blood, serum, plasma, or other lipid-rich matrices. These methods and the specific pesticides measured are outlined in Table 2. Although some pesticide metabolites are monitored in urine, they are most commonly measured as the intact pesticide and/or its metabolite in, whole blood, serum, plasma, or other lipid-rich matrices. Typically, serum or plasma is extracted using a liquid partitioning or solid-phase extraction (SPE) and the extract is analyzed using capillary gas chromatography (GC) with electron-capture detection (ECD) (Table-2). These methods are reliable and use affordable instrumentation. However, GC–ECD analyses have a higher potential for detecting interfering components than do more selective analysis techniques. Other methods for analysis of serum extracts include mass selective detection (MSD aka MS) [Table-2] and high-resolution mass spectrometry with isotope dilution quantification. These analyses are typically more selective and sensitive than GC–ECD analyses; however, the high cost of instrumentation and isotopically labeled standards and the complex operation and maintenance of these instruments often preclude their routine use in most laboratories.

Typically, serum or plasma is extracted using a liquid partitioning or solid-phase extraction (SPE) and the extract is analyzed using capillary gas chromatography (GC) with electron-capture detection (ECD) (Table-2). A sensitive and accurate analytical method was developed for quantifying 29 contemporary pesticides in human serum or plasma. These pesticides include organophosphates, carbamates, chloroacetanilides, and synthetic pyrethroids among others and include pesticides used in agricultural and residential settings. This method employs a simple solid-phase extraction followed by a highly selective analysis using isotope dilution gas chromatography–high-resolution mass spectrometry. This method is very accurate, has limits of detection in the low pg/g range and coefficients of variation of typically less than 20% at the low pg/g end of the method linear range (Barr et al., 2002).

Method	Analytes[a]	Matrix	Extraction	Analytical system	I.S. type	Recovery[e] (%)	LOD (µg/g)	RSD (%)
Frenzel, 2000	15	Whole blood	Kieselguhr SPE	GC-MS	None	97	30–40	7
Rohrig, 2000	1-6	Breast milk	SPME	GC-ECD	None	b	Low	b
Ward, 2000	1-10	Serum, breast	SPE	GC-HRMS	Isotope dilution	60–80	0.07–0.26	<20
Najam, 1999	1-3, 5-12, 13, 17-18	Serum	Solvent extraction, Silica / Florosil cleanup	GC-ECD	Surrogates	39–126	0.15–0.5	7-32
Pauwels, 1999	f	Serum	C_{18} SPE, acid wash	GC-MS	$^{13}C_{12}$ PCB 149	48–140	Low	b
Lino, 1998	f	Serum	Florisil SPE	GC-ECD	f	>84	1–37 µg/l	<!9
Luo, 1997	1,2	Serum	*n*-Hexane	GC-ECD	b	93–106	b	b
Walisze-wski, 1982	1, 2,4,6	Adipose	Light petroleum, acid wash	GC-ECD	b	91–99	0.01 µg/kg	<10
Brock, 1996	1-11	Serum	C_{18} SPE, Florosil	GC-ECD	Surrogates	63–80	0.08–0.66 µg/l	0.7–5.9[c]
Noren, 1996	MeSO - DDE	Breast milk	Liquid–gel partitioning, absorption/ gel permeation chromatography	GC-HRMS	Surrogates	80–97	0.01–0.05 ng/g lipid	4–14
Gill, 1996	1–9, 11–13, 15+ others	Serum	Solvent extraction /SPE	GC-MS	Surrogates	60–110	b	b
Guardino, 1996	1, 2	Blood	C_{18} SPE	GC-ECD;	None	b	b	b
Minelli, 1996	1, 2, 4–6	Serum	Serum–silica suspension, hexane/ acetone, alumina cleanup	GC-ECD	b	80–99	< 1 mg/l	< 15
Gallelli, 1995	f	Adipose liver	Light petroleum, Florosil	GC-ECD	f	f	f	f

Method	Analytes[a]	Matrix	Extraction	Analytical system	I.S. type	Recovery[e] (%)	LOD (µg/g)	RSD (%)
Prapamontol, 1991	1-2, 4-7, 10-13	Milk	Ethyl acetate / acetone /methanol, C_{18}SPE	GC-ECD	Surrogates	90-110	0.5-2.5 µg/l	≤16
Burse, 1990	1-3, 5-10, 12	Serum	Hexane/ ether, Florosil	GC-ECD	Surrogates	48-122	<1 µg/l	7-23
Saady, 1990	1-2, 4, 6, 7-10	Serum	C_{18} SPE	GC-ECD	Surrogate	70-75	0.1-0.7 µg/l	4-25
Liao, 1988	1-5, 11	Adipose	solvent, Florosil	GC-MS	Surrogat	72-120	b	b
Mardones 1999	14	Urine	Acid hydrolysis, on-line cleanup SFE	MEKC-UV	b	58-103	1-12 µg/l	3-7
LeBel, 1983	f	f	Acetone/ hexane, gel permeation,. dichloromethane/ cyclohexane	GC-ECD	B	>80	b	b
Bristol, 1982	3, 5, 7-9, 11	f	f	GC-ECD[d]	Surrogates	35-99	Low µg/ l	3-20
Tessari, 1980	1-3, 7-10, 13	Breast milk	ACN/ hexane, Florosil	GC-ECD	b	68-90	0.5-30 µg/l	b
Strassman 1977	1-13	Breast milk	Solvent, Florosil	GC-ECD[d]	b	b	10-100 µg/l	b
Martinez, 1998	15	Urine	SPE	GC-MS-MS	Dieldrin	>89	0.006-0.018 µg/l	9-13
Nigg, 1991	17	Urine	Oxidation, solvent	GC (detector not noted)	None	b	1	b
Angerer, 1981	14	Urine	Acid hydrolysis, derivatization	GC-ECD	f	87-119	5-20 µg/l	4-10
Holler, 1989	14	Urine	Acid hydrolysis, derivatization	GC-MS-MS	Isotope analogues/surrogates	> 50	1	c
Hill, 1995	14	Urine	Enzyme hydrolysis, chlorobutane / ether; derivatization	GC-MS-MS	Stable isotope analogues	b	1-2 µg/l	21-24

Method	Analytes[a]	Matrix	Extraction	Analytical system	I.S. type	Recovery[e] (%)	LOD (µg/g)	RSD (%)
Heleni Tsoukali et al., 2005	malathion, parathion, methyl parathion and diazinon	biological samples	SPME	GC-NPD	Fenitrothion	b	2 to 55 ng/ml	b
Zlatković M, et al., 2010	dimethoate, diazinon, malathion and malaoxon	biological samples	solid-phase extraction	LC-MS	b	90-99	0.007-0.07 mg/l	b
John H et al., 2010	dimethoate	biological samples	Deproteinization by precipitation and extensive dilution	LC-MS/MS	b	90-115	0.12-0.24 µg/l	5-14
Paula Proença et al., 2005	imidacloprid	biological samples	b	LC-MS	b	86	0.002 µg/mL	5.9
Gallardo et al., 2006	dimethoate	biological samples	SPME	GC-MS	ethion	1.24 and 0.50%.	50-500 ng/ml	
Saito et al., 2008		biological samples	SPME	HS-GC-MS	b	b	0.25 µg/mL	12.6
Bichon et al 2008	Fipronil	biological samples	SPE	GC-MS	b	b	0.05-0.73 pg/ml	b
Zhou et al., 2007	pentachlorophenol	biological samples	HS-SPME	GC-MS	b	b	0.02 ng/ml	12.6

I.S.= Internal standard; LOD = limit of detection; RSD = relative standard deviation; SPME = solid-phase micro-extraction; SPE = solid-phase extraction; SFE = supercritical fluid extraction; GC-ECD = gas chromatography electron capture detection; GC–MS= gas chromatography–mass spectrometry, GC–HRMS=gas chromatography–high-resolution mass spectrometry;. a 1= *p, p*-DDT; 2= *p, p*-DDE; 3= hexachlorobenzene; 4= a-hexachlorocyclohexane; 5= b-hexachlorocyclohexane; 6= g-hexachlorocyclohexane; 7= heptachlor epoxide; 8= oxychlordane; 9= *trans*-nonachlor; 10= dieldrin; 11= aldrin; 12= endrin; 13= mirex; 14= lindane and/or metabolites (chlorinated phenols); 15= endosulfan and/or metabolites; 16= methylsulfonyl-DDE; 17= *o, p*-DDT; 18=*cis*-nonachlor. b Not given. c Standard error about the mean. d GC–MS used for confirmation of positive samples. e Recovery refers only to the absolute recovery from extraction or isolation of the analyte. f Unable to obtain full article. Details taken from abstract. Missing details may be available in full article.

Table 2. Methods for analysis of organochlorine pesticides in human matrices

5. Conclusion

In this chapter, we present a review of existing analytical methodology for the biological and environmental monitoring of exposure to pesticides has been presented. A critical assessment of the existing methodology has been done and other areas which need more research have been identified. An effort has been made to use to measure toxicants. There are many detectors which are equipped with Gas Chromatography like Mass, electron capture detector, flame ionization detector, phosphorous nitrogen detector etc. But, in one modern context the mass detector is the best detector which is very useful for confirmation and quantification of all types of compounds. The electron capture detector is only suitable for quantification of organo-chlorine compounds but it is not confirmatory technique. The nitrogen phosphorous detector is suitable for quantification of nitrogen and phosphorous compounds. Other types of detectors except mass detector are not suitable as confirmatory technique.

6. References

A.R., Najam, M.P., Korver, C.C., Williams, V.W., Burse, & L.L., Needham, (1999). Analysis of mixture of polychlorinated biphenyl and chlorinated pesticides in human serum by column fraction and dual-column capillary gas chromatography with electron capture detection. J. AOAC Int. 82, 177-185.

Abdulrahman, R. B. (2001). Distribution of VOC in ambient air of Tehran , Arch Env Health, 56, 380-383,

Angerer, J., Heinrich, R., & Laudehr, H., (1981). Occupational exposure to hexachlorocyclohexane. V. Gas chromatographic determination of monohydroxychlorobenzenes (chlorophenols) in urine. Int Arch Occup Environ Health. 48(4), 319-24

Arrebola, F.J., Martinez Vidal, J.L., & Fernandez-Gutierrez , A., (1999). Excretion study of endosulfan in urine of a pest control operator. Toxicology Letters, 107, 15-20.

Ballschmiter, R., Bacher, A., Mennel, R.C., Fischer, U., Richle, Swerev, M., (1992). The determination of chlorinated biphenyls, chlorinated dibenzodioxins and chlorinated dibenzofurans by GC-MS. J high Resolution chromatography 15, 260-270. Barcelona. Hum Toxicol 6, 397-400

Barr, D. B., Barr, J. R., Maggio, V. L., Whitehead, R. D., Jr., Sadowski, M. A., Whyatt, R., et al. (2002). A multi-analyte method for the quantification of contemporary pesticides in human serum and plasma using high-resolution mass spectrometry. Journal of Chromatography B, 778, 99–111.

Bichon, E., Richard, C.A., & Le Bizec, B., (2008). Development and validation of a method for fipronil residue determination in ovine plasma using 96-well plate solid-phase extraction and gas chromatography-tandem mass spectrometry, J Chromatogr A. 2008 1201(1), 91-9.

Bristol, D.W., Crist, H.L., Lewis, R.G., MacLeod, K.E., Sovocool, G.W., (1982). Chemical analysis of human blood for assessment of environmental exposure to semivolatile organochlorine chemical contaminants, J Anal Toxicol. 6(6), 269-75.

Brock, J.W., Burse, V.W., Ashley, D.L., Najam, A.R., Green, V.E., Korver, M.P., Powell, M.K., Hodge, C.C., & Needham, L. L., (1996). An improved analysis for chlorinated pesticides and polychlorinated biphenyls (PCBs) in human and bovine sera using solid-phase extraction. J Anal Toxicol 20, 528–536.

Burse, V.W., Head, S.L., Korver, M.P., McClure, P.C., Donahue, J.F., & Needham, L.L., (1990). Determination of selected organochlorine pesticides and polychlorinated biphenyls in human serum, J Anal Toxicol. 1990 May-Jun;14(3):137-42

C. Mardones, A., Rios, M., & Valcarcel, (1999). Determination of chlorophenols in human urine based on the integration of on-line automated clean-up and preconcentration unit with micellar electrokinetic chromatography, Electrophoresis, 20, 2922-2929.

Camino-Sánchez, F.J., Zafra-Gómez, A., Oliver-Rodríguez, B., Ballesteros, O., Navalón, A., Crovetto, G., Vílchez, J.L., (2010) UNE-EN ISO/IEC 17025:2005-accredited method for the determination of pesticide residues in fruit and vegetable samples by LC-MS/MS. Food Addit Contam Part A Chem Anal Control Expo Risk Assess. 27(11), 1532-44

Carmichael, P.L., Jacob, J., Grimmer, G., Phillips, D.H. (1990). Analysis of the PAH content of petrol and diesel engine lubricating oils and determination of DNA adductsin topically treated mice by 32 P-post-labelling, Carcinogenesis, 11, 2025-2032

Chen, Q. Fung, Y., (2010). Capillary electrophoresis with immobilized quantum dot fluorescence detection for rapid determination of organophosphorus pesticides in vegetables. Electrophoresis, 31 (18), 3107-3114.

Chung, S.W., Chan, B.T., (2010). Validation and use of a fast sample preparation method and liquid chromatography-tandem mass spectrometry in analysis of ultra-trace levels of 98 organophosphorus pesticide and carbamate residues in a total diet study involving diversified food types, J Chromatography A, 1217(29), 4815-24.

Dewan, A., Bhatnagar, V. K., Mathur, M. L., Chakma, T., Kashyap, R., Sadhu, G. H., et al. (2004). Repeated episodes of endosulfan poisoning. Journal of Toxicology –Clinical Toxicology, 42, 1–7.

E. V., Minelli & M. L., Ribeiro, (1996). Quantitative method for the determination of organochlorine pesticides in serum. *J. Anal. Toxicol.*, 20(1), 23-26.

Fischer, R. C., Witlinger, R., Ballschmiter, K., (1992). Retention index based vapour pressure estimation for PCB by GC, Fresenius. J of Analytical Chemistry 342, 421-425

Frenzel, T. Sochor, H., Speer K., & and Uihlein, M., (2000). Rapid multi-method for verification and determination of toxic pesticides in whole blood by means of capillary GC-MS. *J. Anal. Toxicol.*, 24(5), 365-371.

Frenzel, T., Sochor, H., Speer, K., & Uihlein, M., (2000) Rapid multi-method for verification and determination of toxic pesticides in whole blood by means of capillary GC-MS. *J. Anal. Toxicol.*, 24(5), 365-371

Gallardo, E., Barroso, M., Margalho, C., Cruz, A., Vieira, D. N., & López-Rivadulla, M., (2006). Solid-phase microextraction for gas chromatographic/mass spectrometric analysis of dimethoate in human biological samples, Rapid Communications in Mass Spectrometry 20(5), 865–869.

Gallelli, G., Mangini, S.,& Gerbino, C.,(1995). Organochlorine residues in humans adipose and hepatic tissues from autopsy sources in northern Italy. J. Toxicol. Environ Health 46, 293-300.

Gomez-Catalan, J. J., To-Figueras, J., Planas, M., Rodamilans, J., Corbella, (1987).

Guardino, X., Serra, C., Obiols, J., Rosell, M.G., Beren guer, M.J., Lopez, F., & Brosa, J., (1996). Determination of DDT and related compounds in blood samples from agricultural workers J. Chromatogr. A 719 , 141-147.

Heleni, T., Georgios, T., Nikolaos, R., & Ifigeneia G., (2005). Solid phase microextraction gas chromatographic analysis of organophosphorus pesticides in biological samples , Journal of Chromatography B, 822 (1-2)194-200.

Hill, R.H., Shealy Jr, D.B., Head, S.L., Williams, C.C., Bailey, S.L., Gregg, M., Baker, S.E., & Needham, L.L., (1995). Determination of Pesticide Metabolites in Human Urine Using an Isotope Dilution Technique and Tandem Mass Spectrometry J. Anal. Toxicol. 19, 323-329.

Hodgson AT. Grimmer JR. Application of multisorbent sampling technique for investigations of volatile organic compounds buildings, In: Design and protocol for monitoring of indoor air-quality. ASTM STP 1002, NL Nagota and JP Harper (Eds.) 1989, 244-256

Holler J.S., Fast, D.M., Hill, R.H., Jr, Cardinali, F.L., Todd, G.D., McCraw, J.M., Bailey, S.L., & Needham, L.L., (1989). Quantification of selected herbicides and chlorinated phenols in urine by using gas chromatography/mass spectrometry/mass spectrometry. J Anal Toxicol. 13(3),152-7.

John, H., Eddleston, M., Clutton, R.E., Worek, F., & Thiermann, H., (2010). Simultaneous quantification of the organophosphorus pesticides dimethoate and omethoate in porcine plasma and urine by LC-ESI-MS/MS and flow-injection-ESI-MS/MS, J Chromatogr B Analyt Technol Biomed Life Sci. 878(17-18), 1234-45

Kamel, A., Qian, Y., Kolbe, E., Stafford, C., (2010). Development and validation of a multiresidue method for the determination of neonicotinoid and macrocyclic lactone pesticide residues in milk, fruits, and vegetables by ultra-performance liquid chromatography/MS/MS. J AOAC International, 93(2), 389-99.

Keenan, F., Doug, C., Vegetation clearance study work plan ordinance and explosive. Prepared by US Dept. of the Army Sacrament District Corps of Engineers, 1325, J Street, Sacramento, California 95814-2922. HLLA project no. 46310001131

Kmellár, B., Abrankó, L., Fodor, P., Lehotay, S.J., (2010), Routine approach to qualitatively screening 300 pesticides and quantification of those frequently detected in fruit and vegetables using liquid chromatography tandem mass spectrometry (LC-MS/MS). Food Addit Contam Part A Chem Anal Control Expo Risk Assess. 27(10),1415-30.

Koesukwiwat , U., Lehotay, S.J., Miao, S., Leepipatpiboon, N., (2010) High throughput analysis of 150 pesticides in fruits and vegetables using QuEChERS and low-pressure gas chromatography-time-of-flight mass spectrometry. J Chromatography A, 1217(43), 6692-703.

Kumari, B., & Kathpal, T. S. (2009). Monitoring of pesticide residues in vegetarian diet. Environmental Monitoring Assessment, 151, 19–26.

Kumari, B., Madan, V. K., & Kathpal, T. S. (2006). Monitoring of pesticide residues in fruits. Environmental Monitoring and Assessment, 123, 407–412.

Kumari, B., Madan, V. K., Kumar, R., & Kathpal, T. S. (2002). Monitoring of seasonal vegetables for pesticide residues. Environmental Monitoring and Assessment, 74, 263–270.

LeBel G.L., & Williams, D.T., (1983). Determination of organic phosphate triesters in human adipose tissue. J Assoc Off Anal Chem. 66(3), 691-9

Li, L., Wu, C., Chen, J., Zhang, S., Ye, Y., (2010). Determination of organophosphorus pesticide residues in vegetables by gas chromatography using back-flush technique. Se Pu. July 28 (7), 724-8

Liao, W., Smith, W.D., Chiang, T.C., & Williams, L.R., (1988). Rapid, low-cost cleanup procedure for determination of semi-volatile organic compounds in human and bovine adipose tissues , J Assoc Off Anal Chem.71(4), 742-7

Lin, X., Chen, X., Huo, X., Yu, Z., Bi, K., Li, Q., (2010). Dispersive liquid-liquid microextraction coupled with high-performance liquid chromatography-diode array detection for the determination of N-methyl carbamate pesticides in vegetables. journal of Separation Science. PMID: 21137098

Lino, C.M., Azzolini, C.B., Nunes, D.S., Silva, J.M., & da Silveira, M.I., (1998). Methods for the determination of organochlorine pesticide residues in human serum J. Chromatogr. B 716 , 147-152.

Luo, X.W., Foo, S.C., & Ong, H.Y., (1997). Serum DDT and DDE levels in Singapore general population Sci. Total Environ. 208 , 97-104.

Martínez Vidal, J.L., Arrebola, F.J., Fernández-Gutiérrez, A., & Rams, M.A., (1998). Determination of endosulfan and its metabolites in human urine using gas chromatography-tandem mass spectrometry, J Chromatogr B Biomed Sci Appl. 719(1-2), 71-8.

Martinez, J. L. V., Arrebola, F. J., Garrido, A. F., Martinez, J. F., & Mateu-Sanchez, M. (2004). Validation of a gas chromatographic–tandem mass spectrometric method for analysis of pesticide residues in six food commodities. Selection of a reference matrix for calibration. Chromatographia, 59(5–6), 321–327.

Miller, K. D., & Milne, P. (2008). Determination of low-level pesticides residues (0.5 lg/L) in soft drinks and sport drinks by gas chromatography with mass spectrometry: Collaborative study. Journal of AOAC International, 91(1), 202–236.

Najam, A.R., Korver, M.P., Williams, C.C., Burse, V.W., & Needham, L L., (1999) Analysis of a mixture of polychlorinated biphenyls and chlorinated pesticides in human serum by column fractionation and dual-column capillary gas chromatography with electron capture detection. J. AOAC Int. 82, 177-185.

Nigg, H. N., Stamper, J. H., Deshmukh S. N., & Queen, R. M., (1991). 4,4'-Dichlorobenzilic acid urinary excretion by dicofol pesticide applicators, Chemosphere. 22, (3-4) , 365-373.

Norén, K., Lundén, A., Pettersson, E., & Bergman, A., (1996). Methylsulfonyl metabolites of PCBs and DDE in human milk in Sweden, 1972-1992. Environ Health Perspect. 104(7), 766–772

Osman, K.A., Al-Humaid, A. M., Al-Rehiayani, S.M., Al-Redhaiman, K.N., (2010). Monitoring of pesticide residues in vegetables marketed in Al-Qassim region, Saudi Arabia. Ecotoxicol Environ Saf. 73(6), 1433-9.

Paula, P., Helena, T., Fernando, C., João, P., Paula ,V., Monsanto, E., Marques, P., & Duarte N. V., (2005). Two fatal intoxication cases with imidacloprid: LC/MS analysis, Forensic Science International. Volume 153(1) ,75-80

Pauwels, A., Wells, D., Covaci, A., & Schepens, P.J.C. (1999). Improved sample preparation method for selected persistent organochlorine pollutants in human serum using solidphase disk extraction with gas chromatographic analysis. J Chromatogr B 723:117-125.

Paya, P., Anastassiades, M., Mack, D., Sigalova, I., Tasdelen, B., Oliva, J., et al. (2007). Analysis of pesticide residues using the Quick Easy Cheap Effective Rugged and Safe (QuEChERS) pesticide multiresidue method in combination with gas and liquid chromatography and tandem mass spectrometric detection. Analytical Bioanalytical Chemistry, 389, 1697–1714.

Pentachlorophenol and hexachlorobenzene in serum and urine of the population of Prapamontol T., & Stevenson, D., (1991). Rapid method for the determination of organochlorine pesticides in milk. J Chromatogr 552:249-257.

Quach, D.T.I.,. Ciszkouzski, N. A. P., Fienlayson, J., Barbara, A., (1998). New GC-MS experiment for the undergraduate instrumental laboratory in environmental chemistry: methyl -t-butyl ether and benzene in gasoline. J Chem Educ 75. 595-599.

Reidel, K. (1996). Determination of benzene and alkylated biomass in ambient and exhaled air by microwave desorption coupled with GC-MS. J of Chromatography A 719, 383-389.

Robert H. TPH measurements; the advantage of using GC-MS. 1995: PhD. Dissrtation. ENTRIX INC JHON MAC MURPHY ZYMAX ENVIR TECHNOLOGY INC.

Rohrig, L., Meisch, H.U., & Fresenius, J., (2000). Application of solid phase micro extraction for the rapid analysis of chlorinated organics in breast milk Anal. Chem. 366 , 106-111

Saady J.J., & Poklis A., (1990). Determination of chlorinated hydrocarbon pesticides by solid-phase extraction and capillary GC with electron capture detection J Anal Toxicol. 1990 Sep-Oct;14(5):301-4

Saito, T., Morita, S., Motojyuku, M., Akieda, K., Otsuka, H., Yamamoto, I., & Inokuchi, S., (2008). Determination of metaldehyde in human serum by headspace solid-phase microextraction and gas chromatography-mass spectrometry, J Chromatogr B Analyt Technol Biomed Life Sci. 875(2), 573-6.

Sinha SN. Kulkarni, P.K. Desai, N. M., Shah, S. H., Patel, G.M., Mansuri, M.M., Parih, D. J., Saiyed, H.N. (2005). Gas chromatographic-mass spectroscopic determination of benzene in Indoor air during the use of biomass fuels in cooking time, J of chromatography A 315-319.

Sinha SN. Kulkarni, P.K. Shah, S. H., Desai, N. M., Patel, G.M., Mansuri, M.M., Saiyed, H.N. (2006). Environmental monitoring of benzene and toluene produced in indoor air due to combustion of solid biomass fuels, Science of the Total Environment 357, 280-287.

Sinha, S. N. (2010). Effect of dissociation energy: Signal to noise ratio on ion formation and sensitivity of analytical method for quantification and confirmation of triazofos in blood samples using gas chromatography–mass spectrometer (GC–MS/MS). International Journal of Mass Spectrometry, 296, 47–52.

Sinha, S. N., Bhatnagar, V.K., Doctor, P., Toteja, G.S., Agnihotri, N.P., Kalra R.L., A novel method for pesticide analysis in refined sugar samples using a gas chromatography-mass spectrometer (GC-MS/MS) and simple solvent extraction method, Food Chemistry 126 (2011) 379-386

Sinha, S. N., Pal, R., Dewan, A., Mansuri, M. M., & Saiyed, H. N. (2006). Effect of dissociation energy on ion formation and sensitivity of an analytical method for determination of chlorpyrifos in human blood, using gas chromatography-masss pectrometer (GC-MS in MS/MS). International Journal of Mass Spectrometry, 253, 48-57.

Sinha, S. N., Patel, T. S., Desai, N. M., Mansuri, M. M., Dewan, A., & Saiyed, H. N. (2004). GC-MS study of endosulfan in biological samples. Asian Journal of Chemistry, 16(3-4), 1685-1690.

Sinha, S. N., Zadi, S. S. (2008). Gas Chromatography-Mass spectral analysis of paint thinner, Asian Journal of Chemistry 20(8), 6365—6368.

Srivastava, A.K., Trivedi, P., Srivastava, M.K., Lohani, M., & Srivastava, L.P. (2011). Monitoring of pesticide residues in market basket samples of vegetable from Lucknow City, India: QuEChERS method. Environmental Monitoring and Assessment, 176(1-4, 465-472.

Star J. M., Rone, J.R., Metnacche, M.G., Henderson, R. F. (2000). Development of methods for analysis of biomarkers of Benzene exposure using GC/MS and MALDI. The Toxicologist. Presented at the Annual meeting of the society of Toxicology, Philadelphia, 54, 310.

Strassman S.C., & Kutz, F.W., (1977). InseAicide residues in human milk from Arkansas and Mississippi,1 973-1974.Pestic Monit J , 10(4),1 30-133.

Tessari, J.D., & Savage, E.P., (1980). Gas-liquid chromatographic determination of organochlorine pesticides and polychlorinated biphenyls in human milk. J Assoc Off Anal Chem. 1980 Jul;63(4):736-41.

U. S., Gill, H. M., Schwartz, B. Wheatley, (1996). Development of a method for the analysis of PCB congeners and organochlorine pesticides in blood/serum. Chemosphere Volume: 32 (6), 1055-1076.

Waliszewski, S.M., & Szymczynski, G.A., (1982). Simple, low-cost method for determination of selected chlorinated pesticides in fat samples J. Assoc. Off. Anal. Chem. 65, 677-9.

Ward, E.M., Schulte, P., Grajewski, B., Andersen, A., Patterson Jr., D.G., Turner, W., Jellum, E., Deddens, J.A., Fried- land, J., Roeleveld, N., Waters, M., Butler, M.A., DiPietro, E., & Needham, L.L., (2000). Serum organochlorine levels and breast cancer: a nested case-control study of Norwegian women. Cancer Epidemiol. Biomarkers Prev. 9, 1357-1367.

Ward, E.M., Schulte, P., Grajewski, B., Andersen, A., Patterson, D.G., Jr, Turner, W., Jellum, E., Deddens, J.A., Friedland, J., Roeleveld, N., Waters, M., Butler, M.A., DiPietro, E., & Needham, L.L., (2000). Serum organochlorine levels and breast cancer: a nested case-control study of Norwegian women. Cancer Epidemiol Biomarkers Prev. 9(12), 1357-67.

Zhou, Y., Jiang, Q., Peng, Q., Xuan, D., & Qu, W., (2007). Development of a solid phase microextraction-gas chromatography-mass spectrometry method for the determination of pentachlorophenol in human plasma using experimental design. Chemosphere. 2007 70(2), 256-62.

Zlatković, M., Jovanović, M., Djordjević, D., & Vucinić, S., (2010). Rapid simultaneous determination of organophosphorus pesticides in human serum and urine by liquid chromatography-mass spectrometry, Vojnosanit Pregl. 67(9),717-22.

8

Acetone Response with Exercise Intensity

Tetsuo Ohkuwa[1,*], Toshiaki Funada[1] and Takao Tsuda[2]
[1]*Department of Materials Science and Engineering, Nagoya Institute of Technology, Showa-ku, Nagoya,*
[2]*Pico-Device Co., Nagoya Ikourenkei Incubator, Chikusa-ku, Nagoya, Japan*

1. Introduction

Blood samples are commonly used to explain the mechanism of metabolism during exercise and diagnosis of certain diseases. However, blood sampling is invasive and usually accompanied by problems such as loss of blood, emotional stress and discomfort for volunteers. The collection and analysis of expired air and skin gas has a number of advantages compared to that of blood, and can be performed by non-invasive and painless procedure. These non-invasive volatile gas monitoring is attractive since they can be repeated frequently without any risk and their sampling does not require skilled medical staff. Especially, skin gases which emanated from human skin which enables collection from hands, arms, fingers and other local areas of the human bodies (Tsuda et al. 2011). Volatile organic compounds can be produced anywhere in the body and may reflect physiologic or pathologic biochemical processes. These substrates are transported via the bloodstream and exhaled through the lung (Schubert et al. 2004). Analysis of volatile organic compounds enable the observation of biochemical process in the body in noninvasive manner.

It well known that free fatty acid and glucose are the major energy fuels under usual circumstances. Ketones such as β-hydroxybutyrate, acetoacetate and acetone are generated in the liver via decarboxylation of excess Acetyl-CoA (Miekisch et al. (2004), mainly from the oxidation of fatty acids, and are exported to peripheral tissues, such as the brain, heart, kidney and skeletal muscle for use as energy fuels (Mitchell et al. 1995; Laffel et al. 1999). Acetone is one of the most abundant compounds in human breath (Miekisch et al. 2004). Acetone is mainly generated from non-enzymatic, decarboxylation of acetoacetate (Laffel et al. 1999; Owen et al. 1982). Ketones in blood increased gradually with respect to the length of fasting periods in diabetics and obese people (Tassopoulos et al. 1969; Reichard et al. 1979). Owen et al. (1982) demonstrated that in diabetic patients, plasma acetone concentration is significantly related to breath acetone concentration. Acetone in expired air increased after high-fat ketogenic diet treatment (Musa-Veloso et al. 2002). The relationship between acetone concentration in plasma and breath has been well established (Naioth et al. 2002). Yamane et al. (2006) reported that skin gas acetone concentrations of patients with diabetes were significantly higher than those of the control subjects. Turner et al. (2008) demonstrated that there is a clear relationship between the level of breath acetone and

---

* Corresponding Author

acetone released from the skin. This result suggested that emission rates of acetone from skin can be used to estimate the blood acetone level. The acetone concentrations in both skin and breath gases are potentially useful as an indicator of β-oxidation of fatty acid and ketosis (Naitoh et al. 2002; Miekish et al. 2004; Musa-Veloso et al. 2002), but less is known as to the effects of exercise on acetone concentration of skin gas and exhaled acetone.

Senthilmohan et al. (2000) reported that running exercise increased breath acetone concentration compared to the basal level. However, in their study exercise intensity was not clarified. Sasaki et al. (2011) reported that acetone concentration in expired air increased during graded and prolonged exercise such as walking or running. We have reported that acetone concentrations emanated from skin gas and in expired air increased with high exercise intensities of cycle exercise (Yamai et al. 2009) and hand-grip exercise (Mori et al. 2008).

2. Methods

2.1 Procedure of exercise

The participants for hand-grip and cycle exercise were 7 and 8 healthy males, respectively. When this study was started, the first step was to ascertain the reproducibility of the results for skin gas acetone concentration test during hand-grip exercise. First the skin gas acetone was collected for four times in one participant. In hand-grip exercise, the participants performed a dynamic hand-grip exercise of three different types of exercise during 60 sec (Exercise 1, 2 and 3). Exercise 1 was performed at 20 kg with one contraction per two sec. Exercise 2 was 30kg with one contraction per three sec. Exercise 3 was 10kg with one contraction per sec. In cycle exercise, the workloads were 360 (1.0kg), 720 (2.0kg), and 990 (2.75kg) kgm/min, and each stage was 5 min in duration. A pedaling frequency of 60 rpm was maintained.

2.2 Collection of expired air and skin gases

The expired air was collected in Douglas bag, and skin gas was collected in the sampling bag at rest, during and after cycle and hand-grip exercises. In skin gas sampling, the right hand was cleaned, and wiped with paper before skin gas collection. After the hand was inserted into the sampling bag, which was fixed to the elbow (cycle exercise) or the middle of the upper arm (the center point between olecranon and acromion) (Fig 1-A; cycle exercise; Fig 1-B; hand-grip exercise). All the air in the bag was sucked out with a glass syringe; The gas in the bag was replaced with 600ml nitrogen gas. The skin gas collected for 1 min, and measured acetone concentration by gas chromatography.

2.3 Analytical procedure and conditions

The sampling gas was analyzed with a cold trap gas chromatographic system (Yamane et al. 2006; Nose et al. 2006). The sample was automatically sucked into the stainless-steel tube from loop cooled with liquid nitrogen. After the sample injection valve was rotated from the trapping position to the injection position, the trap tubing was heated directly to aid thermal desorption of the acetones in the sample. Acetone concentration in gas samples was detected by a gas chromatography with a flame ionization (FID).

The concentration of acetone in expired air and skin gases was measured using modified methods of Yamane et al. (2006). Acetone was detected with a flame ionization detector

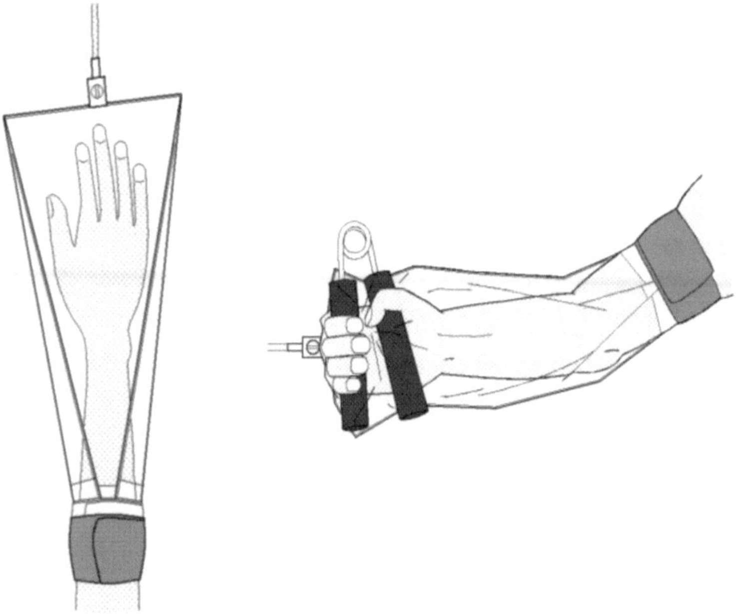

Fig. 1. Skin gas sampling methods (A: cycle exercise, B: hand-grip exercise).

(FID) by a Type GC-14B gas chromatograph (Shimadzu, Kyoto, Japan). The separation conditions for measuring the acetone were as follows: a Porapack Q capillary column (Type G-950: 1.2 mm internal diameter and 5.8 m long, Chemical Evaluation and Research Institute, Tokyo, Japan); the injection and detection temperature was 150℃; the column temperature was 100℃; and the retention time was 2.3 min.

3. Results

3.1 Hand-grip exercise

Table 1 demonstrated the mean (M) and standard deviation (S.D.) of skin gas acetone concentration at rest, during exercise, after exercise, and each coefficient of variation (C.V.) when one subject performed the same exercise four times. The C.V. of at rest, during exercise and recovery were 10.98, 9.52, 9.73, 8.67 and 8.49, respectively. Fig.2 demonstrates changes in acetone emanated from skin gas when one subject performed the same hand-grip exercise four times. The skin gas acetone concentration significantly increased during exercise compared to resting level, and decreased to almost the same as basal level immediately after 1.0-2.0 min during recovery.

	Rest	Exercise	Recovery		
			1.0-2.0 min	3.0-4.0 min	5.0-6.0 min
Mean (ppm)	0.082	0.147	0.113	0.105	0.106
SD	0.009	0.014	0.011	0.0091	0.009
C.V. (%)	10.98	9.52	9.73	8.67	8.49

Table 1. Mean, SD and CV of skin gas acetone concentration at rest, during exercise and recovery.

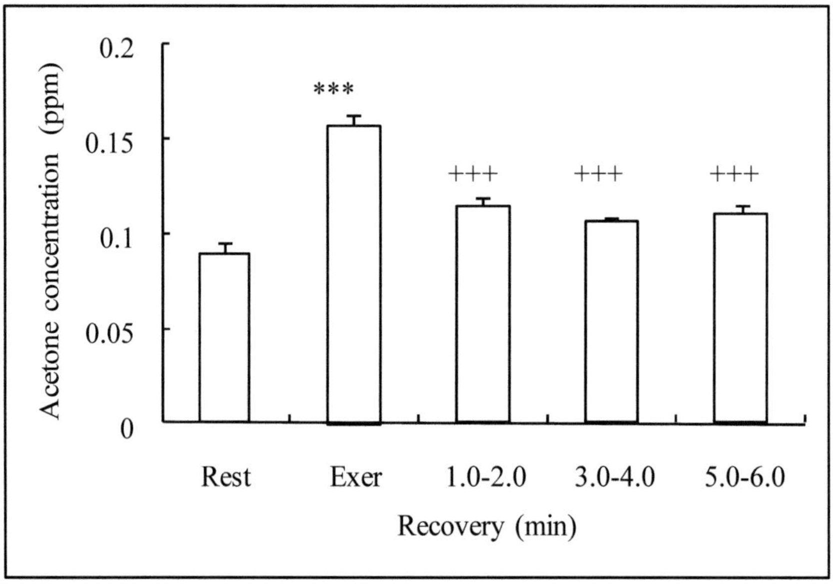

Fig. 2. Skin gas acetone concentration at rest, during hand-grip exercise and recovery. One subject performed the same hand-grip exercise four times. Values are M±SEM. **p<0.01, significant difference compared to rest ; ++p<0.01 significant difference compared to exercise.

Acetone concentration in skin gas during hand-grip exercise 2 (30kg ×20 times) was significantly higher than basal level (Figure 3). Although skin gas acetone during exercise 1 (20kg×30 times) and exercise 3 (10kg×60 times) increased, significant difference was not found. No significant difference was found in skin gas acetone concentration during hand-grip exercise among exercise 1, 2, and 3.

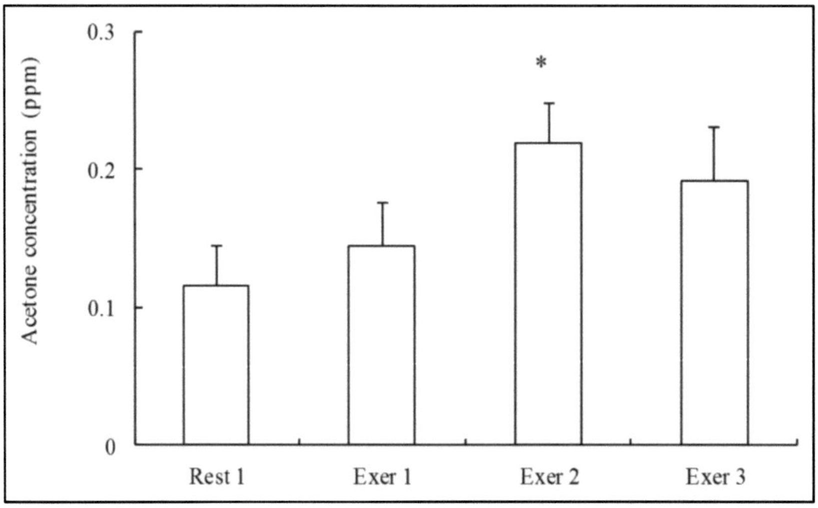

Fig. 3. Comparison of skin-gas acetone concentration among exercise 1, 2 and 3. Values are M±SEM. *p<0.05, significant difference compared to rest.

3.2 Cycle exercise

Figure 4 represents the changes in acetone concentration in expired air at the basal level, 360, 720 and 990 kgm/min during cycle exercise and recovery of 5 min after the 990 kgm/min exercise. Acetone concentration of expired air tend to increase with an increase in exercise intensity. The acetone concentration of expired air at 990 kgm/min significantly increased compared with the basal level ($p<0.05$). Figure 5 shows the changes in acetone excretion in expired air at the basal level, during cycle exercise and recovery. The acetone

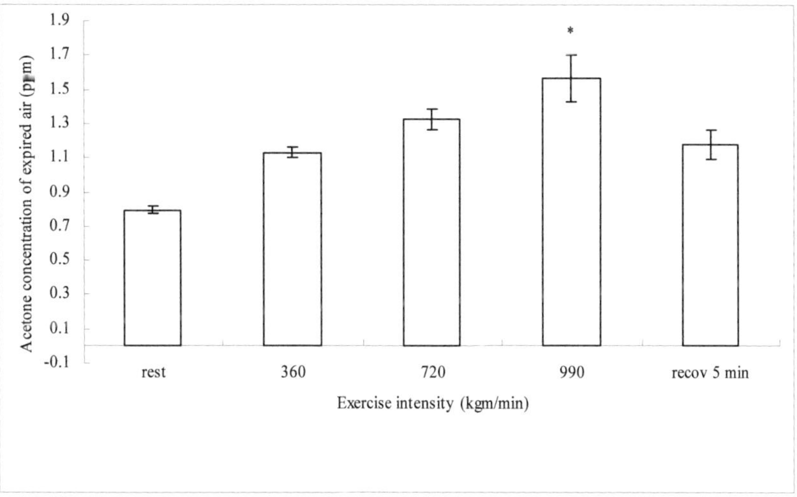

Fig. 4. Changes in acetone concentration in expired air at the basal level, during cycle exercise and recovery.*$p<0.05$ significant difference compared to rest.

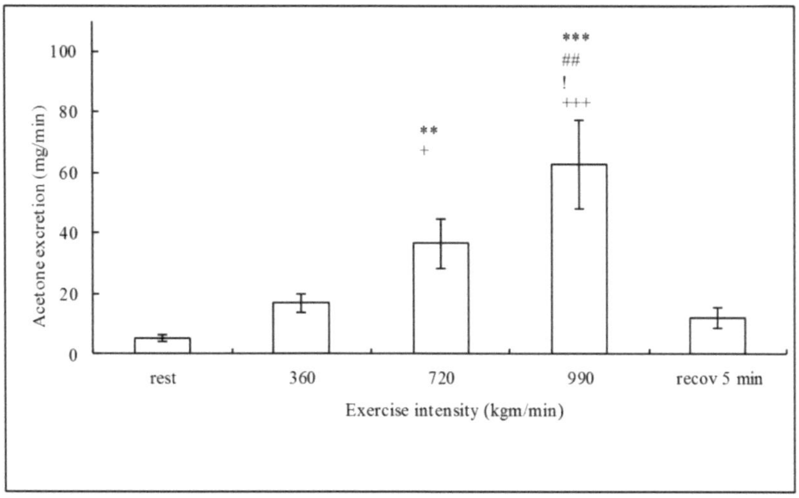

Fig. 5. Changes in acetone excretion in expired air at the basal level, during exercise and recovery. *** $p<0.001$, **$p<0.01$ significant difference compared with resting value. ## $p<0.01$ significant difference compared to 360 kgm/min, ! $p<0.05$ significant difference compared to 720 kgm/min, +++ $p<0.001$, + $p<0.05$ significant difference compared to 5 min after exercise.

excretion at 720 and 990 kgm/min significantly increased compared with the basal level ($p<0.05$). As shown in Figure 6, the skin gas acetone concentration at 990 kgm/min significantly increased compared with the basal level and 360 kgm/min ($p<0.05$). Acetone concentration in expired air was 4-fold greater than skin gas at rest and 3-fold greater during exercise ($p<0.01$). There is a significant relationship between skin gas acetone concentration with expired air ($r=0.752$, $p<0.01$, Fig.7).

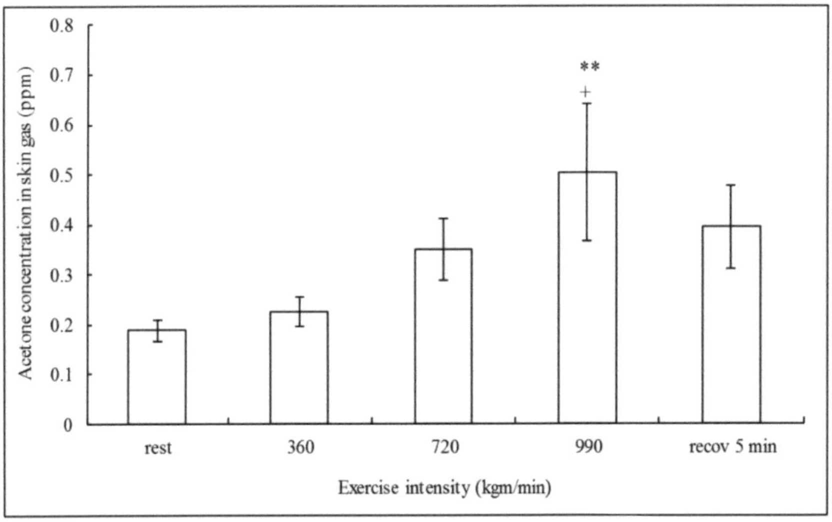

Fig. 6. Changes in skin gas acetone concentration at the basal level, during exercise and recovery. ** $p<0.01$ significant difference compared to rest. # $p<0.05$ significant difference compared to 360kgm/min.

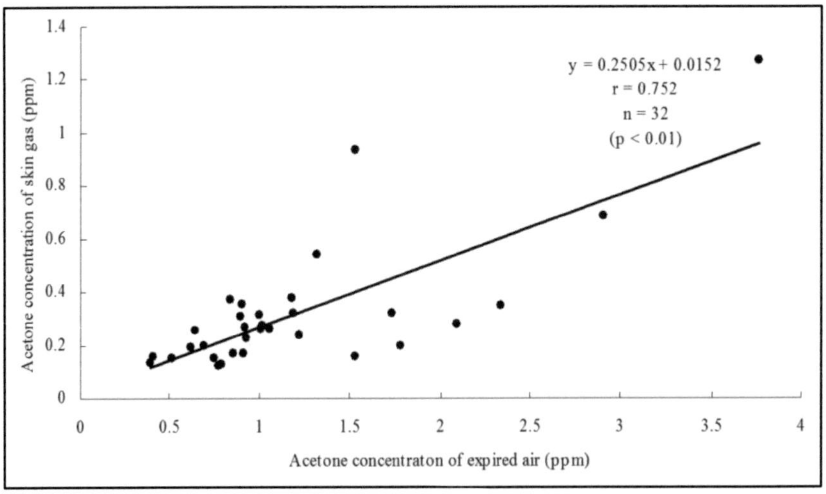

Fig. 7. The relationship in acetone concentration between in expired air and skin gas.

4. Discussion

Acetone is drived from spontaneous, non-enzymatic, and decarboxylation of acetoacetate when there is an insufficient supply of blood glucose (Owen et al. 1982). The relationship between acetone concentration in blood and breath has been well established (Owen et al. 1982). Previously, it has been demonstrated that skin gas acetone is correlated with breath acetone (Naitoh et al. 2002). Yamane et al (2006) also reported that skin acetone concentrations of patients with diabetes were significantly higher than those of the control subject. Naito al. (2002) reported that concentrations of acetone in both skin gas and breath increased according to the length of fasting periods, and also demonstrated that there is a good relationship between skin and breath acetone concentration.

It is a well known fact that ketogenesis is enhanced in diabetes, fasting and exercise (Féry et al. 1983; Balasse and Féry (1989); Wahren et al. 1984). Balasse and Féry (1989) reported that plasma ketone bodies increased after treadmill running at the intensity of 40% ~ 60% maximal aerobic capacity for 120 min compared with before exercise, and also observed that the plasma ketone bodies was higher in 60% maximal power compared to that of 40 % maximal power.

The C.V. of skin gas acetone concentration during hand-grip exercise was 9.52, when skin gas acetone measured four times in one subject. These results indicate that reasonable reproducibility is obtained in measurements of skin gas acetone concentration during hand-grip exercise and that determination of skin gas acetone is highly reliable.

The Exercise 2 of the hand-grip exercise (30kg ×20 times) showed an increase in acetone level of skin gas compared to basal level. The exercise 1 (20kg×30 times) and 3 (10kg×60 times) tend to increase skin gas acetone, but the increase was not significant. Although the total work is the same between Exercise 1,2 and 3, acetone concentration in Exercise 2 was increased compared to rest. The results of our study coincidently is comparable with the result from previous study (Balasse and Féry 1989). These experiments suggest that acetone concentration in blood and skin gas are related to intensity of exercise.

Senthilmohan et al. (2000) have demonstrated that acetone concentration of expired air increased in running exercise (Senthilmohan et al. 2000), but was not clear whether the acetone concentrations of skin gas increased with an elevation of exercise intensity. Acetone concentration of expired air and skin gas at the intensity of 990kgm/min has significantly increased compared to the basal level (Fig. 5 and 6). These results basically agreed with those of other studies (Balasse and Féry 1989; Féry et al.1983; Féry et al.1986). Sasaki et al. (2011) demonstrated in the graded exercise test that acetone level in exhaled air began to increase at the intensity of 39.6 % of maximal oxygen uptake. From these results, it is clear that acetone concentration of expired and skin gases increased with elevation of exercise intensity.

FFA and ketone bodies in blood increase during prolonged exercise and increase in mobilization results in elevation of utilization (Loy et al. 1986; Ravussin et al. 1986; Sasaki et al. 2011). The acetone concentration correlates with the concentration of blood ß-hydroxybutyrate (Reichard et al. 1979; Yamane et al. 2006). The kinetic responses of ketone bodies to exercise is very complex. The increase in ketogenesis is related to the rise in FFA levels and to an increase in the ketogenic capacity of the liver (Balasse and Féry 1989). Exercise is known to increase the hepatic conversion of FFA to ketones in control group (Wahren et al. 1984). However, Wahren et al. (1975) have shown that uptake of ketone bodies by the leg muscle was of minor importance during bicycle exercise in control subject. Wahren et al. (1984) suggested that the rate of hepatic ketogenesis can be influenced by the rate of lipolysis (FFA release from adipose tissue) and the blood flow to the liver, and by the fractional extraction of FFA by the liver and an increase in the proportion of FFA converted to ketones. Furthermore, the increase of plasma catecholamines in response to the exercise may have contributed to the augmented fractional extraction of FFA by the splanchnic bed, which was observed in the control in response to exercise (Wahren et al. 1984). Dynamic hand-grip exercise increased norepinephrine (Costa et al. 2001). The increase of skin gas acetone might have been caused by norepinephrine secretion, which was increased during dynamic hand-grip exercise.

5. Acknowledgement

It was concluded in this report that acetone concentration in expired air and skin gas is directly related to exercise intensity. This human skin gas project was supported by the Aichi Science and Technology Foundation.

6. References

Balasse EO, Féry F. Keton body production and disposal: effects of fasting, diabetes, and exercise. Diabetes Metab Rev 1989; 5:247-270.

Costa F, Christensen NJ, Farley G, Biaggioni I. NO modulates norepinephrine release in human skeletal muscle: implications for neural preconditioning. Am J Physiol. Regul Integr Comp Physiol 2001; 280: R1494-R1498.

Féry F, Blasse EO. Ketone body turnover during and after exercise in overnight-fasted and starved humans. Am J Physiol 1983; 245: E318-E325.

Féry F, Blasse EO. Response of ketone body metabolism to exercise during transition from postabsorptive to fasted state. Am J Physiol 1986; 250: E495-E501.

Laffel L. Ketone bodies: a review of physiology, pathophysiology and application of monitoring to diabetes. Diabetes Metab Res Rev 1999; 15: 412-426.

Loy SF, Conlee RK, Winder WW, Nelson AG, Arnall DA, Fisher AG: Effect of 24-h fast on cycling endurance time at two different intensities. J Appl Physiol 1986; 61: 654-659.

Miekisch W, Schubert JK, Noeldge-Schomburg GFE. Diagnostic potential of breath analysis-focus on volatile organic compounds. Clinica Chimica Acta 2004; 347: 25-39.

Mitchell GA, Kassovska-Bratinova S, Boukaftane Y, Robert MF, Wang SP, Ashmarina L, Lambert M, Lapierre P, Potier E. Medical aspects of ketone body metabolism. Clin Invest Med 1995; 18: 193-216.

Mori K, Funada T, Kikuchi M, Ohkuwa T, Itoh H, Yamazaki Y. Tsuda T. Influence of dynamic hand-grip exercise on acetone in gas emanating from human skin. Redox Rep 2008; 13:139-142.

Musa-Veloso K, Likhodii SS, Cunnane SC. Breath acetone is a reliable indicator of ketosis in adults consuming ketogenic meals. Am Clin Nutr 2002; 76: 65-70.

Naitoh K, Inai Y, Hirabayashi T, Tsuda T. Direct temperature-controlled trapping system and its use for the gas chromatographic determination of organic vapor released from human skin. Anal Chem 2000; 72: 2797-2801.

Naitoh K, Tsuda T, Nose K, Kondo T, Takasu A, Hirabayashi T. New measurement of hydrogen gas and acetone vapor in gases emanating from human skin. Instrum Sci Technol. 2002; 3: 267-280.

Nose K, Ueda H, Ohkuwa T, Kondo T, Araki S, Ohtani H, Tsuda T. Identification and assessment of carbon monoxide in gas emanated from human skin Chromatography 2006; 27: 63-65.

Owen OE, Trapp VE, Skutches CL, Mozzoli MA, Hoeldtke RD, Boden G, Reichard GA. Acetone metabolism during diabetic ketoacidosis. Diabetes. 1982; 31: 242-248.

Ravussin E, Bogardus C, Scheidegger K, Lagrange B, Horton ED, Horton ES. Effect of elevated FFA on carbohydrate and lipid oxidation during prolonged exercise in humans. J Appl Physiol 1986; 60: 893-900.

Reichard GA, Haff AC, Skutches CL, Paul P, Holroyde CP, Owen OE. Plasma acetone metabolism in the fasting human. J Clin Invest 1979; 63: 619-626.

Sasaki H, Ishikawa S, Ueda H, Kimura Y. Acetone response during graded and prolonged exercise. Yoshikawa T, Naito Y (eds): Gas Biology Research in Clinical Practice. Basel, Karger, 2011,pp119-124.

Schubert JK, Miekisch W, Geiger K, Noldge-Schomburg GFE. Breath analysis in critically ill patients: potential and limitations. Expert Rev Mol Diagn 2004; 4: 619-629.

Senthilmohan ST, Milligan DB, McEwan MJ, Freeman CG, Wilson PF. Quantitative analysis of trace gases of breath during exercise using the new SIFT-MS technique. Redox Rep 2000; 5: 151-153.

Tassopoulos CN, Barnett D, Fraser TR. Breath-acetone and blood-sugar measurements in diabetes. Lancet 1969; 28: 1282-1286.

Tsuda T, Ohkuwa T, Itoh H. Findings of skin gases and their possibilities in healthcare monitoring. Yoshikawa T, Naitoh Y (eds): Gas Biology Research in Clinical Practice. Basel, Karger, 2011,pp125-132.

Turner C, Parekh B, Walton C, Spanel P, Smith D, Evans M. An exploratory comparative study of volatile compounds in exhaled breath and emitted by skin using selected ion flow tube mass spectrometry. Rapid Commun Mass Spectrom 2008; 22: 526-532.

Wahren J, Hagenfeldt L, Fehlig P. Splanchnic and leg exchange of glucose, amino acids, and free fatty acids during exercise in diabetes mellitus. *J Clin Invest* 1975; 55: 1303-1314.

Wahren J., Sato Y., Ostman J., Hagenfeldt L., Felig P. Turnover and splanchnic metabolism of free fatty acids and ketones in insulin-dependent diabetics at rest and in response to exercise. J Clin Invest 1984; 73: 1367-1376.

Yamai K, Ohkuwa T, Ito H, Yamazaki Y, Tsuda T. Influence of cycle exercise on acetone in expired air and skin gas. Redox Rep. 2009; 14: 285-289.

Yamane N, Tsuda T, Nose K, Yamamoto A, Ishiguro H, Kondo T. Relationship between skin acetone and blood β-hydroxybutyrate concentrations in diabetes. Clin Chimi Acta 2006; 365: 325-329.

9

Indoor Air Monitoring of Volatile Organic Compounds and Evaluation of Their Emission from Various Building Materials and Common Products by Gas Chromatography-Mass Spectrometry

Hiroyuki Kataoka[1], Yasuhiro Ohashi[2], Tomoko Mamiya[1], Kaori Nami[1], Keita Saito[1], Kurie Ohcho[1] and Tomoko Takigawa[3]

[1]School of Pharmacy, Shujitsu University, Nishigawara, Okayama,
[2]Department of Health Chemistry, Okayama University Graduate School of Medicine, Dentistry and Pharmaceutical Sciences, Tsushima, Okayama,
[3]Department of Public Health, Okayama University Graduate School of Medicine, Dentistry and Pharmaceutical Sciences, Shikata, Okayama,
Japan

1. Introduction

In recent years, increased numbers of people entering modern buildings complain of various symptoms such as dry mucous membranes and skin; irritation of eyes, nose, and throat; chest tightness; headache; and mental fatigue (Kirkeskov et al., 2009). These nonspecific health problems related to indoor environments are caused by volatile organic compounds (VOCs) emitted from various sources such as building materials (Haghighat et al., 2002; Lee et al., 2005; Claeson et al., 2007; Nicolle et al., 2008; Han et al., 2010; Jia et al., 2010), household materials (Kwon et al. 2008), and combusted materials (Liu et al., 2003; Ye, 2008; Fromme et al., 2009; Kabir & Kim, 2011). VOCs are widely used in many household products and are emitted by paints (Afshari et al., 2003; Wieslander & Norbäck, 2010; Chang et al., 2011), adhesives (Wilke et al., 2004), waxes, solvents, detergents, woods (Jensen et al., 2001; Kirkeskov et al., 2009), and items containing them, including carpets (Katsoyiannis et al., 2008), vinyl flooring (Cox et al., 2001 and 2002), air-conditioners (Tham et al., 2004), newspapers (Caselli et al. 2009), printers and photocopiers (Lee et al., 2006). VOCs emitted by these materials can be classified as primary or secondary. Primary emissions are emissions of non-bound or free VOCs within building materials; these are generally low molecular weight compounds utilized in additives, solvents and unreacted raw materials like monomers. Secondary emissions refer to VOCs that were originally chemically or physically bound, and are usually generated following decomposition, oxidation, chain scission, sorption processes, maintenance, or microbial action, followed by their emission (Pedersen et al., 2003; Lee et al., 2005; Wady & Larsson, 2005; Araki et al., 2009 and 2010).

Indoor air quality (Tumbiolo et al., 2005; Salthammer, 2011) has been assessed in various environments, including non-residential buildings (Abbritti & Muzi, 2006; Bruno et al., 2008; Barro et al., 2009; Massolo et al., 2010), residences (Son et al., 2003; Hippelein, 2004; Sax et al., 2004; Ohura et al., 2006; Yamaguchi et al., 2006; Dodson et al., 2009; Liu et al., 2008; Takigawa et al., 2010; Logue et al., 2011), schools (Adgate et al., 2004a; Sohn et al., 2009), hospitals (Takigawa et al., 2004), stores and restaurants (Vainiotalo et al., 2008; Loh et al., 2009). VOCs are regarded as one of the main causes of "sick building syndrome (SBS)" (Harada et al., 2007; Glas et al., 2008; Takeda et al., 2009), and exposure to high concentrations of VOCs can lead to adverse health effects such as acute and chronic respiratory effects, functional alterations of the central nervous system, mucous and dermal irritations, chromosome aberrations, and cancer (Boeglin et al., 2006; Rumchev et al., 2007; Sarigiannis et al., 2011; Zhou et al., 2011). SBS is a serious problem in Japan, and the Ministry of Health, Labour and Welfare (MHLW) of Japan (2002) has advised that total VOC (TVOC) be limited to 400 $\mu g/m^3$. This TVOC value, however, was not based on the possible effects of long-term exposure on chronic toxicity. Furthermore, air concentrations of VOCs are generally lower in the home than in the workplace (Larroque et al., 2006; LeBouf et al., 2010), and symptoms related to these low indoor VOC levels and their emission sources are not sufficiently clear. To systematically evaluate the relationship between indoor air pollution and human exposure to VOCs (Gokhale et al., 2008; Delgado-Saborit et al., 2011), it is important to measure VOCs in indoor environments, to assess their possible sources and to determine the source strengths of VOCs to which humans are exposed during working, commuting and rest times. In this chapter, we describe a sensitive and reliable method for the simultaneous determination of VOCs by gas chromatography-mass spectrometry (GC-MS). Using this method, we measured the VOC levels in indoor air of a new building, and we characterized the VOCs emitted from various building materials and common household products.

2. Experimental

2.1 Reagents

A 1 mg/mL standard solution of 39 VOCs (Table 1) in carbon disulfide (CS_2) was purchased from Kanto Kagaku (Tokyo, Japan). All other chemicals were of analytical-reagent grade.

2.2 Gas chromatography-mass spectrometry

GC-MS analysis was performed using a Shimadzu Model QP-2010 gas chromatograph-mass spectrometer in conjunction with a GCMS solution Ver.2 workstation. A fused-silica capillary column of cross-linked DB-1 (J&W, Folsom, CA, USA: 60 m × 0.25 mm i.d., 1.0 µm film thickness) was used. The GC operating conditions included injection and detector temperatures of 260°C; a column temperature of 40°C for 10 min, increasing to 280°C at 8°C/min; an inlet helium carrier gas flow rate of 1.0 mL/min maintained with an electronic pressure controller; and a split ratio of 10:1. The electro impact (EI)-MS conditions included an ion-source temperature of 200°C; ionizing voltage of 70 eV; and selected ion monitoring (SIM) mode detection for each compound in each time fraction. Selected ions and peak numbers of each VOC are shown in Table 1. The 39 VOCs were separated into 8 functional groups (A-H), and the results obtained by an average of duplicate analyses were reported as the total concentrations of target VOCs in each group.

Table 1. VOCs used in this study

Peak	Retention time (min)	Selected ion (m/z)	VOCs	Group[1)]	Peak	Retention time (min)	Selected ion (m/z)	VOCs	Group[1)]
1	12-13.5	61	Ethyl acetate	E	20		91	*m*-Xylene + *p*-Xylene	A
2		57	*n*-Hexane	B	21	22.5-25	104	Styrene	A
3		83	Chloroform	C	22		91	*o*-Xylene	A
4		62	1,2-Dichloroethane	C	23		43	*n*-Nonane	B
5		43	2,4-Dimethyl-pentane	B	24		93	α-Pinene	H
6	13.5-16	97	1,1,1-Trichloroethane	C	25		105	1,2,3-Trimethyl-benzene	A
7		56	*n*-Butanol	D	26	25-28.5	43	*n*-Decane	B
8		78	Benzene	A	27		146	*p*-Dichloro-benzene	C
9		117	Carbon tetrachloride	C	28		105	1,2,4-Trimethyl-benzene	A
10		63	1,2-Dichloro-propane	C	29		68	Limonene	H
11		57	2,2,4-Trimethyl-pentane	B	30		41	*n*-Nonanal	F
12	16-19	43	*n*-Heptane	B	31	28.5-30.5	43	*n*-Undecane	B
13		43	Methyli-sobutylketone	G	32		119	1,2,4,5-Tetramethyl-benzene	A
14		91	Toluene	A	33		43	*n*-Decanal	F
15	19-22.5	129	Chlorodibromometh ane	C	34		43	*n*-Dodecane	B
16		43	Butyl acetate	E	35	30.5-40	43	*n*-Tridecane	B
17		43	*n*-Octane	B	36		43	*n*-Tetra-decane	B
18		166	Tetrachloro-ethylene	C	37		57	*n*-Penta-decane	B
19	22.5-25	91	Ethylbenzene	A	38		57	*n*-Hexa-decane	B

[1)] A: aromatic hydrocarbon, B: aliphatic hydrocarbon, C: halocarbon, D: alcohol, E: ester, F: aldehyde, G: ketone, H: terpene.

2.3 Sampling and analysis of indoor air VOCs

Indoor air quality in 13 rooms in a newly built hospital was assessed by active air sampling and VOC analysis before the hospital was opened (in March) and after one year later (in May). In addition, indoor air VOC monitoring was performed in another newly constructed hospital and in a newly constructed school (in March). This new hospital was built without using adhesives in all floors and walls. The mean room temperature and relative humidity of the rooms were 14°C and 65%, respectively, in March and 25.0°C and 42%, respectively, in May. Active sampling was performed using charcoal sorbent tubes (glass tubes with two sections, 130 mg in front and 65 mg in back; Shibata Kagaku, Tokyo, Japan) and a sampling pump (SP-208 Dual, GL Science Inc., Tokyo, Japan), using the standard method of the MHLW. To enable measuring maximum indoor chemical concentrations, sampling was performed in a room that had been closed for more than 5 h following ventilation. From when ventilation occurred to sampling, all doors of built-in furniture in the room were open. In the center of the room (more than 1 m from the wall and 1.2-1.5 m above the floor), VOCs were collected from air onto charcoal sorbent tubes in duplicate, at a flow-rate of 0.2 L/min for 0.5 h in newly constructed building (before occupation) and at a flow-rate of 6 L/h for 24 h after occupation for one year. As controls, VOCs in the air were also trapped outside, 2-5 m from the building and 1.2-1.5 m above the ground. All samples were sealed in a container with an activated carbon bed, stored in an insulated container, and shipped to our laboratory. The front charcoal sorbent was desorbed with 1 mL of CS_2 by shaking and standing for 1 h. After centrifugation at 3000 rpm for 1 min, the supernatant CS_2 solution was transferred to an autosampler vial, and 1 µL of this solution was injected into the GC-MS system. Outlines of indoor air sampling and the analytical procedure are illustrated in Fig. 1.

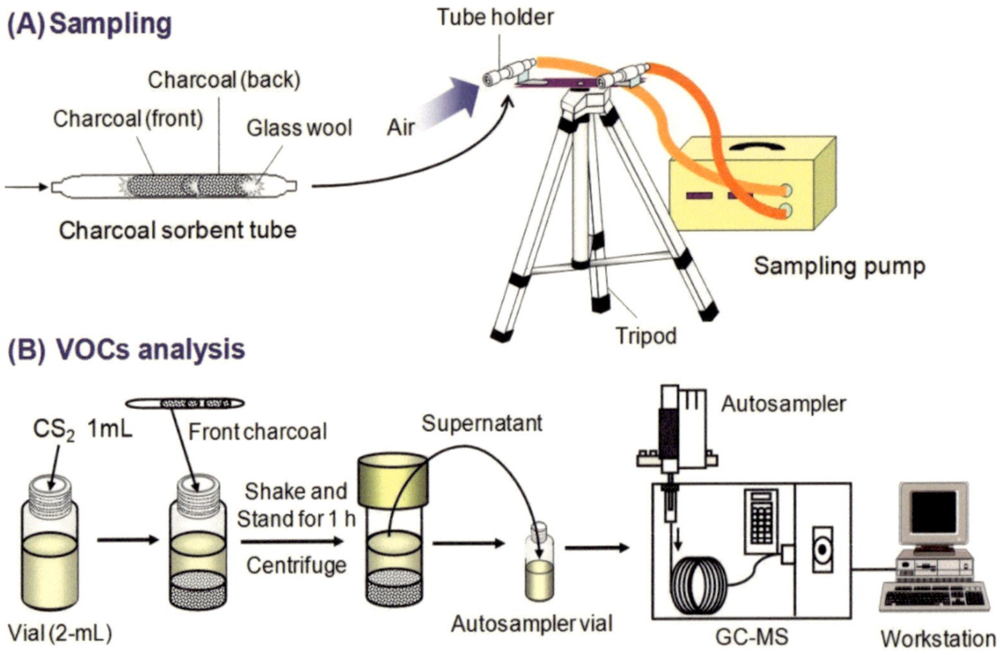

Fig. 1. Outline of indoor air sampling and VOC analysis.

2.4 Sampling and analysis of VOCs emitted from various building materials and common products

VOC emission tests were performed on 16 building materials and 31 common household products, including school supplies, purchased from a local market. Some photographs of these materials are shown in Fig. 2. Woods and hard plastic products were sawn and the sawdust was used for emission tests. Other dry materials, such as carpet, wall paper, newspaper and soft plastic products, were cut finely with scissors or a knife, and the cut pieces were used for emission tests. Wet materials, such as paint, wax, shampoo, glue, paste and ink, were used directly for emission test. Fifteen g of each material were placed in a cleaned small chamber (500-mL volume), and the emitted VOCs were collected onto charcoal sorbent tubes (Shibata Kagaku) by absorption of headspace air using an air sampling pump at a flow-rate of 500 mL/min for 6 h. The adsorbed VOCs on charcoal sorbent were desorbed with 1 mL of CS_2 as described in section 2.3 and analyzed by GC-MS. The VOCs emitted by each material were reported per 180 L. An outline of the emission test is illustrated in Fig. 3.

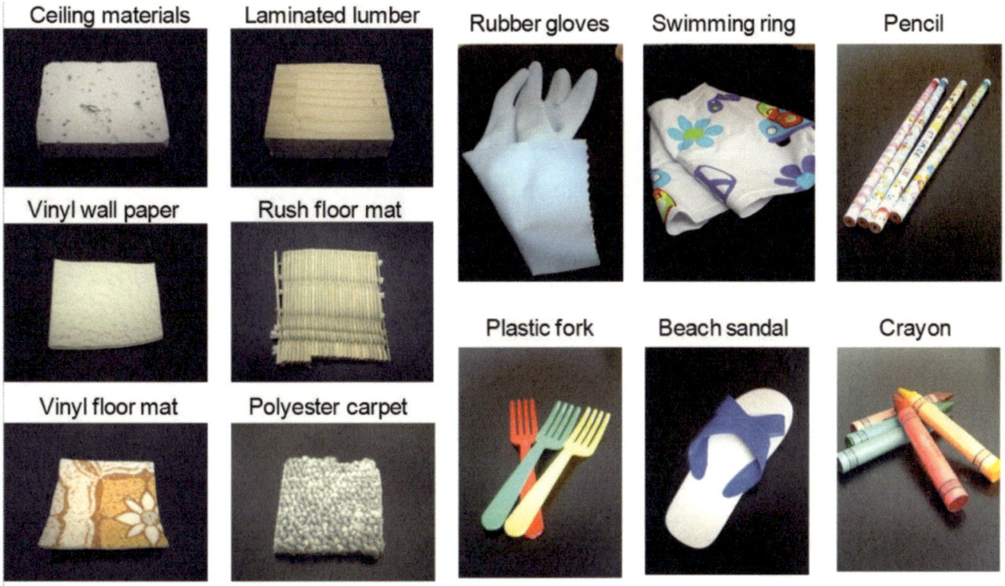

Fig. 2. Several building materials and common products used for emission test.

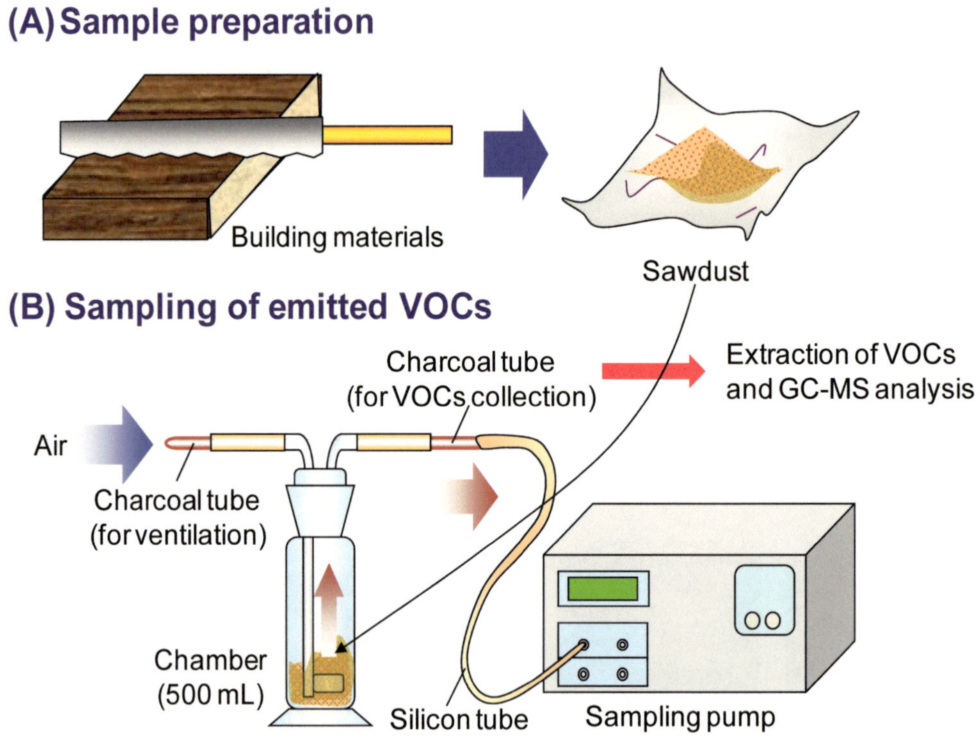

Fig. 3. Outline of emission test.

3. Results and discussion

3.1 GC-MS analysis of VOCs

Mass spectra of VOCs were confirmed by scan mode detection. Although molecular ion peaks of each VOC were observed, the base ion peaks shown in Table 1 were selected for SIM mode detection. The GC-MS method was selective and sensitive, with all 39 VOCs separated on a DB-1 capillary column within 40 min. A typical total ion chromatogram of the VOCs is shown in Fig. 4. The calibration curve for each VOC was linear ($r>0.9992$) in a range from 0.1 to 10 µg/mL CS_2, and the limits of detection (LOD) that gave a signal-to-noise ratio of 3 were 0.4-13.4 ng/mL CS_2 (Table 2).

VOCs	Range [1] (µg/mL CS_2)	Correlation coefficient	LOD [2] (ng/mL CS_2)	LOQ [3] (µg/m^3 for 30 min)
Ethyl acetate	0.1-2.0	0.9993	13.4	21.1
n-Hexane	0.1-2.0	0.9996	4.4	11.5
Chloroform	0.1-2.0	0.9995	4.5	31.3
1,2-Dichloroethane	0.1-2.0	0.9994	2.3	34.7
2,4-Dimethylpentane	0.1-2.0	0.9995	0.7	10.9
1,1,1-Trichloroethane	0.1-2.0	0.9994	8.0	35.7
n-Butanol	0.1-2.0	0.9994	12.4	28.5
Benzene	0.1-2.0	0.9998	2.0	19.4
Carbon tetrachloride	0.1-2.0	0.9992	3.9	44.8
1,2-Dichloropropane	0.1-2.0	0.9992	0.5	25.0
2,2,4-Trimethylpentane	0.1-2.0	0.9996	0.9	11.7
n-Heptane	0.1-2.0	0.9995	0.2	13.7
Methylisobutylketone	0.1-2.0	0.9998	1.3	17.3
Toluene	0.1-10	0.9994	0.4	24.6
Chlorodibromomethane	0.1-2.0	0.9997	1.3	87.5
Butyl acetate	0.1-2.0	0.9997	1.1	19.5
n-Octane	0.1-2.0	0.9994	1.6	15.6
Tetrachloroethylene	0.1-2.0	0.9996	1.8	50.5
Ethylbenzene	0.1-10	0.9996	0.6	27.9
m-Xylene + *p*-Xylene	0.1-10	0.9997	0.7	18.0
Styrene	0.1-2.0	0.9997	0.6	30.4
o-Xylene	0.1-10	0.9997	0.4	29.2
n-Nonane	0.1-2.0	0.9998	1.5	16.1
a-Pinene	0.1-2.0	0.9998	0.8	27.3
1,2,3-Trimethylbenzene	0.1-2.0	0.9998	0.4	31.3
n-Decane	0.1-2.0	0.9996	1.4	13.2
p-Dichlorobenzene	0.1-2.0	0.9998	0.5	42.6
1,2,4-Trimethylbenzene	0.1-2.0	0.9998	0.5	38.6
Limonene	0.1-2.0	0.9998	1.1	25.6
n-Nonanal	0.1-2.0	0.9998	2.8	31.0
n-Undecane	0.1-2.0	0.9995	1.5	17.9
1,2,4,5-Tetramethylbenzene	0.1-2.0	0.9998	0.6	34.7
n-Decanal	0.1-2.0	0.9994	3.9	32.0
n-Dodecane	0.1-2.0	0.9994	1.0	16.7
n-Tridecane	0.1-2.0	0.9992	2.1	17.2
n-Tetradecane	0.1-2.0	0.9994	1.4	18.0
n-Pentadecane	0.1-2.0	0.9996	2.4	19.5
n-Hexadecane	0.1-2.0	0.9995	2.3	20.2

[1] Range 0.1-2.0 µg/mL (n=12), range 0.1-10 µg/mL (n=18); [2] S/N=3; [3] S/N=10.

Table 2. Linearity of calibration, limits of detection and limits of quantitation of target VOCs

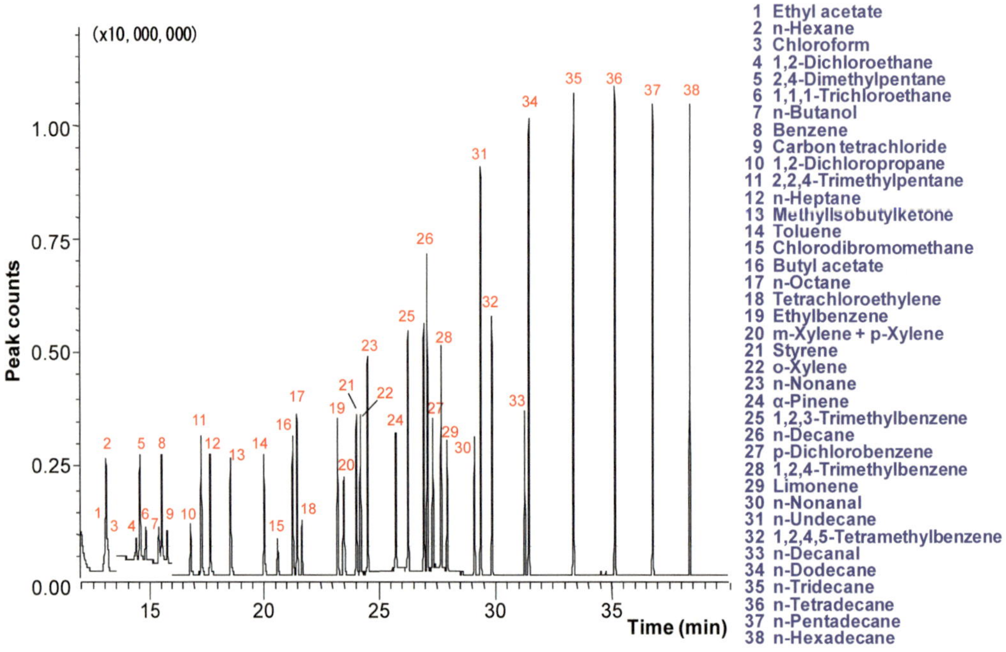

Fig. 4. Typical total ion chromatogram obtained from standard VOCs including 1 µg/mL of each compound.

3.2 Indoor air monitoring of VOCs in newly built buildings

Indoor air VOCs were easily trapped onto charcoal sorbent by the MHLW standard method, with limits of quantitation (LOQ) of VOCs being 10.9-87.5 µg/m^3 for 30 min (Table 2). Using this method, we measured the indoor air VOC concentrations in 13 rooms in a newly built, 10 story hospital before occupation and after occupation for one year (Tables 3 and 4). VOC levels varied depending on the presence of indoor building materials, such as paint and furniture. VOCs were not detected, in air sampling obtained once daily from one site outside the hospital. Prior to the building being occupied, aromatic hydrocarbons (toluene, xylenes and ethylbenzene), aliphatic hydrocarbons (mainly *n*-hexane) and esters (ethyl acetate and butyl acetate) were detected with TVOC concentrations exceeding the recommended maximum concentration (400 µg/m^3) in 12 of 13 rooms (Fig. 5). Particularly, toluene was detected in all rooms and its concentration exceeded the MHLW recommended maximum concentration (260 µg/m^3) in 12 rooms. One year after occupation, however, the TVOC concentrations in the same rooms were below 80 µg/m^3, and the indoor levels of toluene and *n*-hexane decreased dramatically, to about 1/100 and 1/60, respectively, of their previous values. **Table 3.** Indoor air VOC amounts in 13 rooms of newly built hospital prior to occupation.

VOCs	VOC amounts in indoor air (μg/ m³) [1]/ sampling at 0.2 L/ min for 30 min												
	Guidance room	Doctor office	Radiograph room	10F Lounge	6A Nurse station	6B Nurse station	7A Nurse station	8A Nurse station	8B Nurse station	9A Nurse station	9B Nurse station	10A Nurse station	10B Nurse station
Ethyl acetate	ND[2]	ND	255	ND	ND	ND	ND	ND	ND	ND	ND	ND	ND
n-Hexane	ND	263	ND	ND	ND	111	ND	ND	ND	ND	ND	362	417
Chloroform	ND	ND	ND	ND	ND	ND	ND	ND	ND	ND	ND	ND	ND
1,2-Dichloroethane	ND	ND	ND	ND	ND	ND	ND	ND	ND	ND	ND	ND	ND
2,4-Dimethylpentane	ND	ND	ND	ND	ND	ND	ND	ND	ND	ND	ND	ND	ND
1,1,1-Trichloroethane	ND	ND	ND	ND	ND	ND	ND	ND	ND	ND	ND	ND	ND
n-Butanol	ND	ND	ND	ND	ND	ND	ND	ND	ND	ND	ND	ND	ND
Benzene	ND	ND	ND	ND	ND	ND	ND	ND	ND	ND	ND	ND	ND
Carbon tetrachloride	ND	ND	ND	ND	ND	ND	ND	ND	ND	ND	ND	ND	ND
1,2-Dichloropropane	ND	ND	ND	ND	ND	ND	ND	ND	ND	ND	ND	ND	ND
2,2,4-Trimethylpentane	ND	ND	ND	ND	ND	ND	ND	ND	ND	ND	ND	ND	ND
n-Heptane	ND	ND	ND	ND	ND	ND	ND	ND	ND	ND	ND	ND	ND
Methylisobutylketone	ND	ND	ND	ND	ND	ND	ND	ND	ND	ND	ND	ND	ND
Toluene	218	682	605	640	355	785	1051	420	1733	847	934	850	1494
Chlorodibromomethane	ND	ND	ND	ND	ND	ND	ND	ND	ND	ND	ND	ND	ND
Butyl acetate	ND	ND	ND	ND	618	272	ND	ND	ND	ND	ND	ND	ND
n-Octane	ND	ND	ND	ND	ND	ND	ND	ND	ND	ND	ND	ND	ND
Tetrachloroethylene	ND	ND	ND	ND	ND	ND	ND	ND	ND	ND	ND	ND	ND
Ethylbenzene	ND	186	ND	ND	712	317	160	ND	ND	ND	ND	ND	ND
m-Xylene + *p*-Xylene	ND	ND	ND	ND	744	476	460	ND	ND	ND	ND	ND	ND
Styrene	ND	ND	ND	ND	ND	ND	ND	ND	ND	ND	ND	ND	ND
o-Xylene	ND	ND	ND	ND	250	ND	ND	ND	ND	ND	ND	ND	ND
n-Nonane	ND	ND	ND	ND	ND	ND	ND	ND	ND	ND	ND	ND	ND
α-Pinene	ND	ND	ND	ND	ND	ND	ND	ND	ND	ND	ND	ND	ND
1,2,3-Trimethylbenzene	ND	ND	ND	ND	ND	ND	ND	ND	ND	ND	ND	ND	ND
n-Decane	ND	ND	ND	ND	ND	ND	ND	ND	ND	ND	ND	ND	ND
p-Dichlorobenzene	ND	ND	ND	ND	ND	ND	ND	ND	ND	ND	ND	ND	ND
1,2,4-Trimethylbenzene	ND	ND	ND	ND	ND	ND	ND	ND	ND	ND	ND	ND	ND
Limonene	ND	ND	ND	ND	ND	ND	ND	ND	ND	ND	ND	ND	ND
n-Nonanal	ND	ND	ND	ND	ND	ND	ND	ND	ND	ND	ND	ND	ND
n-Undecane	ND	ND	ND	ND	ND	ND	ND	ND	ND	ND	ND	ND	ND
1,2,4,5-Tetramethylbenzene	ND	ND	ND	ND	ND	ND	ND	ND	ND	ND	ND	ND	ND
n-Decanal	ND	ND	ND	ND	ND	ND	ND	ND	ND	ND	ND	ND	ND
n-Dodecane	ND	ND	ND	ND	ND	ND	ND	ND	ND	ND	ND	ND	ND
n-Tridecane	ND	ND	ND	ND	ND	ND	ND	ND	ND	ND	ND	ND	ND
n-Tetradecane	ND	ND	ND	ND	ND	ND	ND	ND	ND	ND	ND	ND	ND
n-Pentadecane	ND	ND	ND	ND	ND	ND	ND	ND	ND	ND	ND	ND	ND
n-Hexadecane	ND	ND	ND	ND	ND	ND	ND	ND	ND	ND	ND	ND	ND

[1] Mean of duplicate analysis; [2] Not detectable.

Table 3. Indoor air VOC amounts in 13 rooms of newly built hospital prior to occupation.

VOCs	VOC amounts in indoor air (μg/ m³) [1]/ sampling at 0.2 L/ min for 30 min												
	Guidance room	Doctor office	Radiograph room	10F Lounge	6A Nurse station	6B Nurse station	7A Nurse station	8A Nurse station	8B Nurse station	9A Nurse station	9B Nurse station	10A Nurse station	10B Nurse station
Ethyl acetate	9.0	6.9	7.1	11.7	5.32	11.5	10.8	2.1	ND	3.5	3.1	6.1	8.5
n-Hexane	2.5	ND	1.9	6.0	ND	47.4	22.7	ND	ND	2.3	1.0	1.9	5.9
Chloroform	ND[2]	ND	ND	ND	ND	ND	ND	ND	ND	ND	ND	ND	ND
1,2-Dichloroethane	ND	ND	ND	ND	ND	ND	ND	ND	ND	ND	ND	ND	ND
2,4-Dimethylpentane	ND	ND	0.9	0.8	ND	3.0	1.8	ND	ND	0.5	ND	0.8	1.0
1,1,1-Trichloroethane	ND	ND	ND	ND	ND	ND	ND	ND	ND	ND	ND	ND	ND
n-Butanol	4.2	ND	ND	ND	ND	ND	ND	ND	ND	ND	ND	ND	ND
Benzene	2.7	ND	2.0	2.3	ND	ND	ND	1.2	0.9	1.1	1.1	1.1	2.2
Carbon tetrachloride	ND	ND	ND	ND	ND	ND	ND	ND	ND	ND	ND	ND	ND
1,2-Dichloropropane	ND	ND	ND	ND	ND	ND	ND	ND	ND	ND	ND	ND	ND
2,2,4-Trimethylpentane	ND	ND	ND	ND	ND	ND	ND	ND	ND	ND	ND	ND	ND
n-Heptane	1.1	ND	ND	1.3	ND	ND	ND	ND	ND	ND	ND	ND	1.2
Methylisobutylketone	2.2	2.0	ND	1.6	1.6	1.8	2.1	ND	ND	ND	ND	ND	ND
Toluene	12.9	8.8	8.5	9.8	4.7	8.7	10.0	5.3	5.6	4.6	3.8	7.5	15.6
Chlorodibromomethane	ND	ND	ND	ND	ND	ND	ND	ND	ND	ND	ND	ND	ND
Butyl acetate	2.1	2.8	1.4	1.9	0.7	1.4	1.4	ND	ND	ND	ND	1.5	2.2
n-Octane	ND	ND	ND	1.4	ND	ND	ND	ND	ND	ND	ND	ND	1.3
Tetrachloroethylene	ND	ND	ND	ND	ND	ND	ND	ND	ND	ND	ND	ND	ND
Ethylbenzene	3.4	2.3	3.0	3.3	ND	2.0	1.8	1.3	1.9	1.1	1.5	1.7	4.3
m-Xylene + *p*-Xylene	3.9	3.8	2.5	3.2	ND	1.4	2.1	2.1	2.2	1.7	1.9	2.4	4.6
Styrene	ND	ND	ND	ND	ND	ND	ND	ND	ND	ND	ND	ND	ND
o-Xylene	ND	ND	ND	ND	ND	ND	ND	ND	ND	ND	ND	ND	1.8
n-Nonane	ND	ND	ND	ND	ND	ND	ND	ND	ND	ND	ND	ND	ND
α-Pinene	ND	ND	ND	ND	ND	ND	ND	ND	ND	ND	ND	ND	ND
1,2,3-Trimethylbenzene	ND	ND	ND	ND	ND	ND	ND	ND	ND	ND	ND	ND	ND
n-Decane	ND	ND	ND	ND	ND	ND	ND	ND	ND	ND	ND	ND	1.1
p-Dichlorobenzene	ND	ND	ND	ND	ND	ND	ND	ND	ND	ND	ND	ND	ND
1,2,4-Trimethylbenzene	ND	ND	ND	ND	ND	ND	ND	ND	ND	ND	ND	ND	ND
Limonene	ND	ND	ND	ND	ND	ND	ND	ND	ND	ND	ND	ND	ND
n-Nonanal	ND	ND	ND	ND	ND	ND	ND	ND	ND	ND	ND	ND	ND
n-Undecane	ND	ND	ND	1.5	ND	ND	ND	ND	ND	ND	ND	ND	ND
1,2,4,5-Tetramethylbenzene	ND	ND	ND	ND	ND	ND	ND	ND	ND	ND	ND	ND	ND
n-Decanal	ND	ND	ND	ND	ND	ND	ND	ND	ND	ND	ND	ND	ND
n-Dodecane	ND	ND	ND	ND	ND	ND	ND	ND	ND	ND	ND	ND	ND
n-Tridecane	ND	ND	ND	ND	ND	ND	ND	ND	ND	ND	ND	ND	ND
n-Tetradecane	ND	ND	ND	ND	ND	ND	ND	ND	ND	ND	ND	ND	ND
n-Pentadecane	ND	ND	ND	ND	ND	ND	ND	ND	ND	ND	ND	ND	ND
n-Hexadecane	ND	ND	ND	ND	ND	ND	ND	ND	ND	ND	ND	ND	ND

[1] Mean of duplicate analysis; [2] Not detectable.

Table 4. Indoor air VOC amounts in 13 rooms of newly built hospital after occupation for one year.

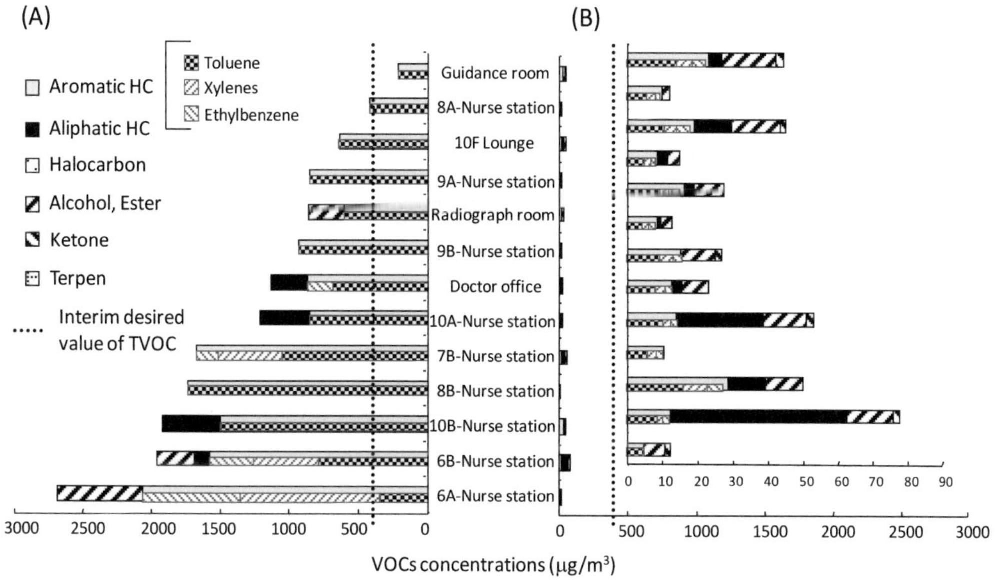

Fig. 5. Comparison of indoor air VOC amounts in 13 rooms of a newly built hospital (A) prior to occupation and (B) after occupation for one year. Air sampling: (A) 0.2 L/min × 30 min and (B) 6 L/h × 24 h.

We also evaluated the relationships among environmental, personal, and occupational factors and changes in the subjective health symptoms in 214 hospital employees (Takigawa et al., 2004). Multiple logistic regression analysis was applied to select variables significantly associated with subjective symptoms that can be induced by SBS. Subjective symptoms of deterioration in the skin, eyes, ears, throat, chest, central nervous system, autonomic system, musculoskeletal system, and digestive system among employees were associated mainly with gender differences and high TVOC concentrations (>1200 μg/m^3). These findings suggest the importance of reducing indoor air VOCs in new buildings to protect employees from the risks of indoor environment-related adverse health effects.

Indoor air VOCs were also measured in unoccupied new buildings, including another newly built hospital that attempted to reduce SBS by not using adhesives in all floors and walls. As shown in Fig. 6A, VOCs were not detected in any rooms or corridors of this hospital. In contrast, TVOC concentrations exceeded the recommended maximum value (400 μg/m^3) in 4 of 10 rooms of a newly built school (Fig. 6B), whereas VOCs were not detected in the other 4 rooms. In 4 rooms, the concentrations of toluene were high, and exceeding the guideline value (260 μg/m^3) of the MHLW. Furthermore, relatively high concentrations of esters (ethyl acetate and butyl acetate) were detected in 4 rooms.

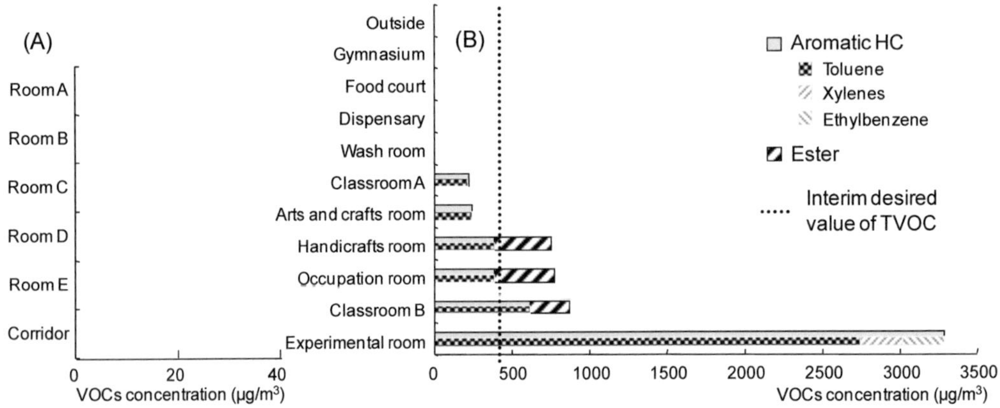

Fig. 6. Indoor air VOC concentrations in rooms of the unoccupied new buildings, (A) a newly built hospital designed to prevent SBS and (B) a newly built school. Air sampling: 0.2 L/min × 30 min

The occurrence and concentrations of VOCs in indoor environments can be affected by outdoor atmospheric conditions, indoor sources, indoor volume, human activities, chemical reactions, ventilation rates, and seasonal factors (Son et al., 2003; Schlink et al., 2004; Massolo et al., 2010). Indoor VOC concentrations have decreased recently in Japan and may be easily reduced by sufficient ventilation and SBS measures. However, measurement of VOC exposure in households with children (Adgate et al., 2004a, b; Sohn et al., 2009) suggested a significant association between VOC exposure and respiratory symptoms such as childhood asthma (Khalequzzaman et al., 2007; Hulin et al., 2010). These findings indicate the necessity of frequent monitoring of VOC exposure in children.

3.3 Emission of VOCs from various building materials and household products

Although various VOCs were detected in newly constructed buildings, they were not detected in the building that took measures to avoid SBS. Therefore, to determine the causal relationship between VOC exposure and SBS onset, it is important to determine the types of building materials and household products that emit VOCs, and the type and quality of VOCs emitted. We therefore collected the VOCs emitted by 16 building materials and 31 household products by a small chamber sampling method (Fig. 3). These emitted VOCs were quantitatively collected onto charcoal sorbent tubes at a flow-rate of 0.5 L/min for 6 h and analyzed by GC-MS.

While there was little emission of VOCs from rush floor mats and ceiling board materials, toluene, chloroform, ethyl acetate, and *n*-hexane were detected in wood chipboard, vinyl wall paper, and vinyl floor mats (Table 5 and Fig. 7). These VOCs may have originated from adhesives and painting materials, which are used to manufacture these products. We found that water-based paints emitted significant amounts of toluene, xylenes, *n*-butanol and high-molecular weight aliphatic hydrocarbons. These quantities emitted may depend on the thickness of the paint layer (Afshari et al., 2003). Some components in these emissions are also highly reactive and may contribute to the health damage.

VOCs	Amounts of VOCs from sample (μg/ 180 L/ 15 g) [1]/ sampling at 0.5 L/ min for 6 h															
	Ceiling board	Gypsum board	Laminated lumber	Plywood	Wood chipboard	Wall paper	Vinyl wall paper (A)	Vinyl wall paper (B)	Rush floor mats	Vinyl floor mats (A)	Vinyl floor mats (B)	Polyester carpets	Floor wax (A)	Floor wax (B)	Adhesive	Water-based paint
Ethyl acetate	ND [2]	1.07	ND	ND	19.59	ND	ND	ND	ND	ND	ND	ND	ND	ND	ND	ND
n-Hexane	ND	1.91	ND	1.17	ND	5.18	0.91	ND	ND	1.51	1.56	1.18	N.D.	5.74	0.56	ND
Chloroform	ND	1.32	ND	1.37	ND	ND	2.80	0.91	ND	ND	ND	3.06	2.35	ND	1.23	ND
1,2-Dichloroethane	ND	ND	ND	ND	ND	ND	ND	ND	ND	ND	ND	ND	ND	ND	ND	ND
2,4-Dimethylpentane	ND	ND	ND	ND	ND	ND	ND	ND	ND	ND	ND	ND	ND	ND	ND	ND
1,1,1-Trichloroethane	ND	ND	ND	ND	ND	ND	ND	ND	ND	ND	ND	ND	ND	ND	ND	ND
n-Butanol	ND	ND	ND	ND	ND	ND	ND	ND	ND	ND	ND	ND	ND	ND	ND	176.55
Benzene	ND	ND	ND	ND	ND	ND	ND	ND	ND	0.28	ND	ND	ND	ND	ND	ND
Carbon tetrachloride	ND	ND	ND	ND	ND	ND	ND	ND	ND	ND	ND	ND	ND	ND	ND	ND
1,2-Dichloropropane	ND	ND	ND	ND	ND	ND	ND	ND	ND	ND	ND	ND	ND	ND	ND	ND
2,2,4-Trimethylpentane	ND	ND	ND	ND	ND	ND	ND	ND	ND	ND	ND	ND	ND	ND	ND	ND
n-Heptane	ND	ND	ND	ND	ND	ND	ND	0.46	ND	ND	ND	ND	ND	ND	ND	8.67
Methylisobutylketone	ND	ND	ND	ND	ND	ND	ND	ND	ND	ND	ND	ND	ND	ND	ND	73.50
Toluene	ND	ND	ND	197	13.91	ND	1.32	0.71	ND	0.93	1.49	0.54	ND	ND	ND	649.64
Chlorodibromomethane	ND	ND	ND	ND	ND	ND	ND	ND	ND	ND	ND	ND	ND	ND	ND	ND
Butyl acetate	ND	ND	ND	ND	ND	ND	ND	ND	ND	ND	ND	ND	ND	ND	ND	ND
n-Octane	ND	ND	ND	ND	ND	ND	ND	ND	ND	ND	ND	ND	ND	ND	ND	ND
Tetrachloroethylene	ND	ND	ND	ND	ND	ND	ND	ND	ND	ND	ND	ND	ND	ND	ND	ND
Ethylbenzene	ND	ND	ND	ND	ND	ND	ND	ND	ND	ND	ND	ND	ND	ND	ND	67.94
m-Xylene + *p*-Xylene	ND	ND	ND	ND	ND	ND	ND	ND	ND	ND	ND	ND	ND	ND	ND	58.89
Styrene	ND	ND	ND	ND	ND	ND	ND	ND	ND	ND	ND	ND	ND	ND	ND	ND
o-Xylene	ND	ND	ND	ND	ND	ND	ND	ND	ND	ND	1.31	ND	ND	ND	ND	23.68
n-Nonane	ND	ND	ND	ND	ND	ND	ND	8.94	ND	ND	ND	ND	ND	ND	ND	27.08
α-Pinene	ND	ND	ND	ND	ND	ND	ND	ND	ND	ND	1.81	9.66	ND	ND	ND	ND
1,2,3-Trimethylbenzene	ND	ND	ND	ND	ND	ND	ND	ND	ND	ND	ND	ND	ND	ND	ND	ND
n-Decane	ND	ND	ND	ND	ND	ND	ND	32.21	ND	ND	ND	ND	ND	ND	ND	88.04
p-Dichlorobenzene	ND	ND	ND	ND	ND	ND	ND	ND	ND	ND	ND	ND	ND	ND	ND	ND
1,2,4-Trimethylbenzene	ND	ND	ND	ND	ND	ND	ND	ND	ND	ND	ND	ND	ND	ND	ND	ND
Limonene	ND	ND	ND	ND	ND	ND	ND	ND	ND	ND	ND	ND	ND	ND	ND	ND
n-Nonanal	ND	ND	ND	ND	ND	ND	ND	ND	ND	ND	ND	ND	ND	ND	ND	ND
n-Undecane	ND	ND	ND	ND	ND	ND	ND	6.38	ND	ND	ND	ND	ND	ND	ND	ND
1,2,4,5-Tetramethylbenzene	ND	ND	ND	ND	ND	ND	ND	4.11	ND	ND	ND	ND	ND	ND	ND	ND
n-Decanal	ND	ND	ND	ND	ND	ND	ND	ND	ND	ND	ND	ND	ND	ND	ND	ND
n-Dodecane	ND	ND	ND	ND	ND	ND	ND	ND	ND	ND	ND	ND	1.41	ND	ND	ND
n-Tridecane	ND	ND	ND	ND	ND	ND	ND	ND	ND	ND	1.68	ND	ND	9.44	ND	ND
n-Tetradecane	ND	ND	ND	ND	ND	ND	ND	1.25	ND	ND	ND	ND	1.56	18.62	ND	ND
n-Pentadecane	ND	ND	ND	ND	ND	ND	ND	ND	ND	ND	ND	ND	ND	27.35	ND	ND
n-Hexadecane	ND	ND	ND	ND	ND	ND	ND	ND	ND	ND	ND	ND	ND	27.59	ND	ND

[1] Mean of duplicate analysis; [2] Not detectable.

Table 5. Amounts of VOCs emitted from various building materials.

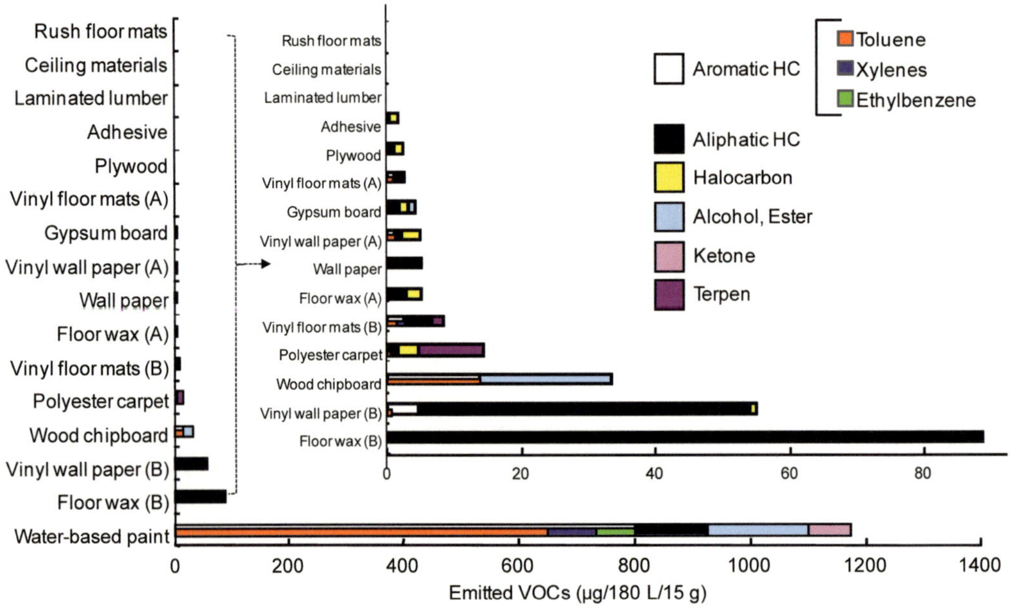

Fig. 7. VOCs emitted from various building materials.

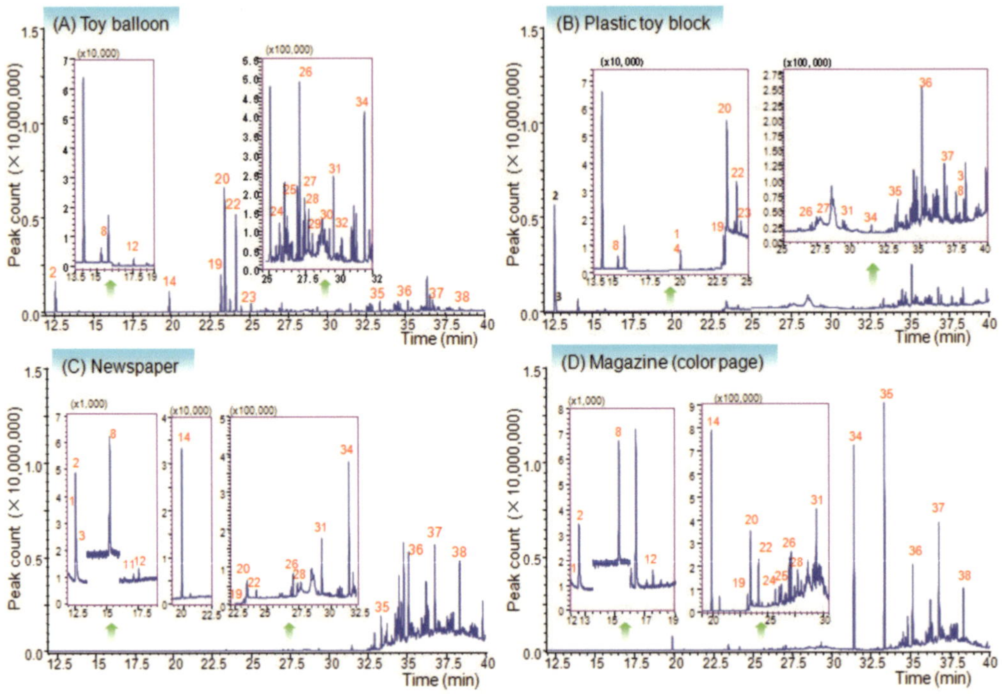

Fig. 8. Typical total ion chromatograms of VOCs emitted from some common products. Peak numbers appear in Fig. 4.

Typical total ion chromatograms of VOCs emitted from some household products are shown in Fig. 8. High concentrations of *n*-hexane, toluene, ethylbenzene and xylenes were detected in a toy rubber balloon (Table 6 and Fig. 9). In addition to toluene, *n*-hexane and chloroform, high concentrations of high-molecular weight aliphatic hydrocarbons were also detected from printed materials such as newspapers and magazines. These are doubtless the main sources of indoor air VOCs at newspaper stands, printing shops, and bookstores (Lee et al., 2006; Barro et al., 2008; Caselli et al., 2009). Evidence has indicated a close relationship between occupational VOC exposure and adverse health effects on workers in the printing industry and in copy centers (Yu et al., 2004). Furthermore, various VOCs were detected in school supplies, including clay, India ink, paint, crayons, glue, and pencils printed with colored paint (Table 7 and Fig. 10). Particularly, paint coating materials are recognized as a major source of VOC exposures (Zhang & Niu, 2002).

These findings may provide semiquantitative estimations of inhalation exposure to VOCs in indoor environments and may allow the selection of safer household products. In particular, the emissions from school supplies are of importance, because they affect the health of children.

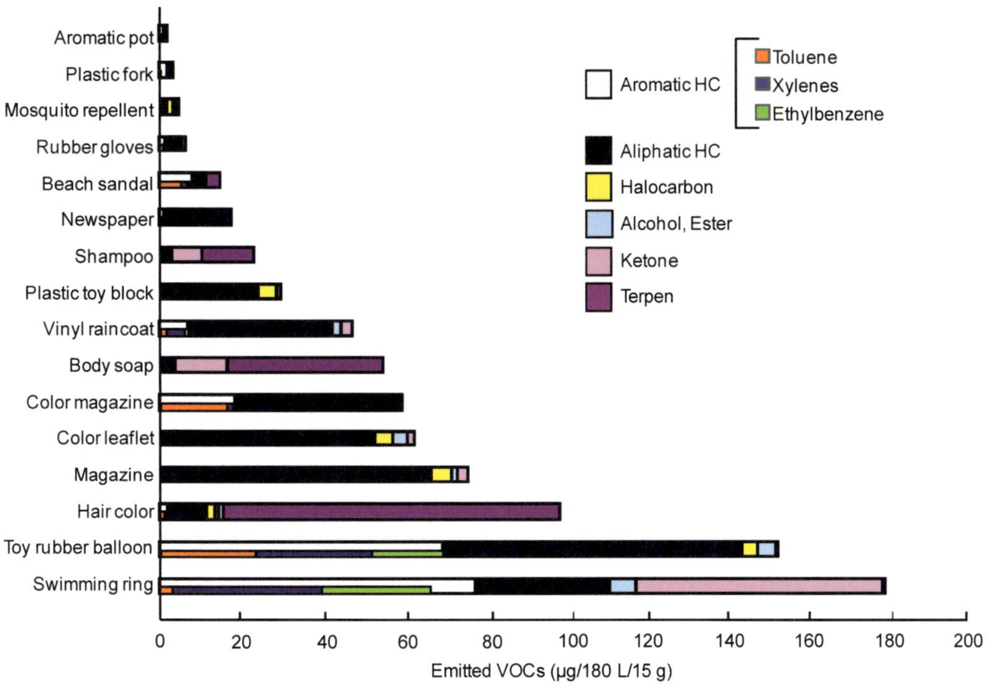

Fig. 9. VOCs emitted from various common products.

VOCs	Amounts of VOCs from sample (μg/ 180 L/ 15 g) [1]/ sampling at 0.5 L/ min for 6 h															
	Newspaper	Magazine	Color magazine	Color leaflet	Plastic toy block	Rubber balloon	Swimming ring	Beach sandal	Vinyl raincoat	Plastic fork	Rubber gloves	Mosquito repellent	Aromatic pot	Shampoo	Body soap	Hair color
Ethyl acetate	0.19	1.87	0.04	2.37	1.03	3.83	1.10	0.07	1.50	0.08	0.28	0.06	0.09	0.07	ND	0.34
n-Hexane	0.06	39.27	0.06	47.71	23.63	70.68	23.59	1.43	33.94	0.72	2.88	0.70	0.28	0.82	0.63	2.73
Chloroform	0.01	4.96	ND	4.39	4.56	3.71	ND	ND	ND	ND	ND	ND	ND	0.12	0.10	0.19
1,2-Dichloroethane	ND [2]	0.19	ND	0.01	ND	ND	ND	ND	ND	ND	ND	ND	ND	ND	ND	ND
2,4-Dimethylpentane	ND	0.39	ND	0.27	ND	ND	ND	ND	ND	ND	ND	ND	ND	ND	ND	ND
1,1,1-Trichloroethane	ND	0.28	ND	0.01	ND	ND	ND	ND	ND	ND	ND	ND	ND	ND	ND	ND
n-Butanol	0.07	1.24	0.03	1.24	ND	0.04	5.57	0.20	0.08	0.00	0.00	0.09	ND	0.04	ND	0.50
Benzene	0.04	0.11	0.06	0.07	0.01	0.01	0.23	0.13	0.31	0.07	0.09	0.08	0.08	0.07	0.07	0.19
Carbon tetrachloride	ND	0.23	ND	0.02	0.04	0.03	0.05	0.01	0.05	ND	ND	ND	ND	0.02	ND	1.84
1,2-Dichloropropane	ND	0.19	ND	ND	ND	0.01	ND	ND	ND	ND	ND	ND	ND	ND	ND	ND
2,2,4-Trimethylpentane	0.01	0.19	0.01	ND	ND	0.01	0.02	ND	ND	ND	ND	ND	ND	ND	ND	ND
n-Heptane	ND	0.18	0.01	0.01	ND	0.06	0.03	ND	0.04	ND	ND	ND	ND	ND	ND	0.08
Methylisobutylketone	ND	0.26	ND	0.01	ND	0.03	57.65	ND	0.06	ND	0.04	ND	0.24	ND	ND	0.09
Toluene	0.64	0.18	16.23	0.05	0.12	23.51	2.86	5.38	1.28	0.05	0.74	0.04	0.10	0.17	0.09	0.93
Chlorodibromomethane	ND	0.12	ND	ND	ND	ND	ND	ND	ND	ND	ND	ND	ND	ND	ND	ND
Butyl acetate	ND	0.29	ND	0.01	ND	0.34	ND	0.39	0.67	0.05	0.52	0.16	0.10	0.02	0.02	0.05
n-Octane	ND	0.14	0.02	0.01	ND	0.27	ND	0.04	0.05	ND	ND	ND	ND	ND	ND	0.23
Tetrachloroethylene	ND	0.07	ND	ND	ND	0.27	ND	ND	ND	ND	ND	ND	ND	ND	ND	ND
Ethylbenzene	0.08	0.11	0.6	0.05	0.08	17.54	26.52	0.34	1.01	0.18	0.20	0.07	0.10	0.07	0.04	0.28
m-Xylene + *p*-Xylene	0.13	0.21	0.99	0.08	0.1	19.93	27.04	1.12	2.71	0.19	0.38	0.11	0.22	0.11	0.08	0.44
Styrene	ND	0.14	0.01	ND	0.01	0.17	0.06	0.03	0.01	1.38	0.01	ND	0.02	0.02	ND	0.01
o-Xylene	0.04	0.11	0.34	0.02	0.03	8.55	10.31	0.33	1.63	0.05	0.11	0.04	0.05	0.03	0.02	0.12
n-Nonane	0.01	0.01	0.03	0.01	0.01	0.56	0.69	0.08	0.05	ND	0.01	ND	0.11	ND	0.01	0.06
α-Pinene	ND	0.09	0.17	ND	ND	0.24	0.13	0.09	0.27	0.01	0.16	0.19	0.28	1.02	0.33	10.54
1,2,3-Trimethylbenzene	0.01	0.11	0.11	0.02	ND	0.1	4.70	0.06	0.05	ND	0.02	0.01	0.16	ND	ND	0.02
n-Decane	0.08	0.05	0.31	0.02	0.01	0.66	2.61	0.19	0.65	0.01	0.03	0.04	ND	0.02	0.04	0.06
p-Dichlorobenzene	ND	0.33	0.01	0.19	0.03	0.24	0.05	0.08	0.12	0.01	0.06	1.47	ND	0.02	0.02	0.07
1,2,4-Trimethylbenzene	0.02	0.14	0.1	0.06	ND	0.12	4.99	0.12	0.04	ND	0.02	0.02	ND	0.05	0.05	0.02
Limonene	0.01	0.15	0.1	0.01	0.01	0.17	0.12	3.76	0.10	0.03	0.48	0.84	ND	11.80	38.65	72.48
n-Nonanal	0.03	2.84	0.52	1.80	ND	0.06	2.12	0.22	2.39	0.02	0.02	0.03	ND	7.27	12.97	1.16
n-Undecane	0.18	0.49	0.39	0.10	0.01	0.25	3.53	0.21	0.18	0.01	0.02	0.06	ND	0.03	1.40	0.17
1,2,4,5-Tetramethylbenzene	0.01	0.13	0.02	0.03	ND	0.06	1.82	0.04	0.02	ND	0.02	0.01	ND	0.03	0.03	0.01
n-Decanal	ND	0.40	0.03	0.14	ND	ND	2.04	0.22	0.09	ND	ND	0.11	ND	0.29	0.16	0.28
n-Dodecane	0.4	2.55	12.18	0.06	0.01	0.44	1.91	0.25	0.98	0.02	0.03	0.08	ND	0.39	0.53	4.63
n-Tridecane	1.83	1.09	13.62	0.14	0.04	0.61	0.35	0.13	0.12	0.02	0.03	0.17	ND	0.63	0.22	0.28
n-Tetradecane	4.85	5.62	4.56	2.20	0.21	0.57	0.08	0.20	0.49	0.06	0.07	0.26	ND	0.11	0.24	0.97
n-Pentadecane	5.15	14.62	6.77	1.49	0.09	0.25	0.02	0.08	0.05	0.03	0.04	0.08	ND	0.04	0.08	0.23
n-Hexadecane	4.05	4.04	2.96	0.85	0.07	0.11	0.02	0.04	0.08	0.02	0.03	0.05	ND	0.03	0.06	0.10

[1] Mean of duplicate analysis; [2] Not detectable.

Table 6. Amounts of VOCs emitted from various household products.

VOCs	Amounts of VOCs from sample (μg/ 180 L/ 15 g) [1]/ sampling at 0.5 L/ min for 6 h														
	Pencil	Fluid paste	Glue	Adhesive tape	Clay	Rubber band	Indian ink	Stamp ink	Paint (white)	Paint (black)	Paint (red)	Crayon (black)	Crayon (blue)	Crayon (yellow)	Crayon (red)
Ethyl acetate	10.84	0.07	11.13	1.42	1.50	2.31	0.30	ND	0.33	0.27	1.88	0.08	0.07	0.19	3.13
n-Hexane	1.27	0.66	110.16	15.07	33.94	70.52	9.48	0.53	2.25	2.42	36.17	3.12	2.78	6.14	37.33
Chloroform	7.47	ND	0.44	0.03	ND	5.58	4.71	0.10	1.27	1.63	ND	ND	ND	ND	16.17
1,2-Dichloroethane	0.05	ND	0.01	ND	ND	0.02	0.02	0.00	0.01	0.01	ND	ND	ND	ND	0.04
2,4-Dimethylpentane	0.01	ND	ND	0.02	ND	0.02	ND	ND	ND	ND	ND	ND	ND	ND	0.01
1,1,1-Trichloroethane	0.01	ND	0.01	ND	ND	0.01	0.01	ND	ND	ND	ND	ND	ND	ND	0.01
n-Butanol	206.77	ND	0.08	ND	0.08	0.04	ND	0.44	0.07	4.11	0.10	ND	ND	ND	0.44
Benzene	0.08	0.07	0.08	0.01	0.31	0.14	0.08	0.06	0.06	0.09	0.16	0.08	0.08	0.11	0.21
Carbon tetrachloride	0.02	ND	0.02	0.01	0.05	0.04	0.04	ND	0.01	0.01	ND	ND	ND	ND	0.02
1,2-Dichloropropane	ND [2]	ND	ND	ND	ND	ND	ND	ND	ND	ND	ND	ND	ND	ND	ND
2,2,4-Trimethylpentane	ND	ND	0.01	0.03	ND	0.01	ND	ND	ND	ND	ND	ND	ND	ND	ND
n-Heptane	0.01	ND	0.35	8.16	0.04	0.04	ND	ND	ND	0.05	ND	ND	0.04	ND	0.02
Methylisobutylketone	0.08	ND	ND	0.29	0.06	0.06	0.01	ND	ND	0.01	ND	ND	ND	ND	0.01
Toluene	14.81	0.03	0.14	62.35	1.28	1.65	0.12	0.03	0.05	0.10	0.37	0.16	0.15	0.20	0.53
Chlorodibromomethane	ND	ND	ND	ND	ND	ND	ND	ND	ND	ND	ND	ND	ND	ND	ND
Butyl acetate	83.91	0.33	ND	0.68	0.67	0.14	ND	0.54	0.02	0.02	0.03	0.02	0.02	0.04	0.03
n-Octane	84.34	ND	ND	0.64	0.05	0.02	ND	ND	ND	ND	ND	0.03	0.10	0.03	0.05
Tetrachloroethylene	ND	ND	ND	ND	ND	0.01	ND	ND	ND	ND	ND	ND	ND	ND	ND
Ethylbenzene	6.03	0.05	0.05	0.32	1.01	1.11	0.08	0.04	0.04	0.05	0.27	0.13	0.12	0.16	0.31
m-Xylene + *p*-Xylene	7.15	0.07	0.08	0.98	2.71	1.53	0.08	0.07	0.06	0.08	0.39	0.28	0.30	0.32	0.55
Styrene	0.01	ND	ND	ND	0.01	0.02	ND	ND	ND	ND	0.01	ND	ND	ND	ND
o-Xylene	3.10	0.02	0.02	0.01	1.63	0.41	0.02	0.02	0.02	0.03	0.11	0.11	0.13	0.13	0.17
n-Nonane	0.02	ND	ND	ND	0.05	0.03	ND	ND	0.01	0.01	0.01	0.62	0.95	0.69	0.60
α-Pinene	0.03	ND	0.02	ND	0.27	0.09	0.36	ND	ND	0.01	0.03	0.02	0.03	0.03	0.03
1,2,3-Trimethylbenzene	0.19	ND	ND	ND	0.05	0.09	ND	ND	ND	ND	0.01	0.45	0.59	0.49	0.32
n-Decane	0.13	0.01	0.01	ND	0.65	0.10	0.01	0.01	0.01	0.02	0.02	3.85	4.60	4.41	3.08
p-Dichlorobenzene	0.01	ND	0.01	ND	0.12	0.13	ND	0.01	ND	ND	0.02	0.09	0.09	0.16	0.10
1,2,4-Trimethylbenzene	2.23	ND	ND	ND	0.04	0.24	ND	ND	ND	ND	0.01	0.53	0.66	0.58	0.37
Limonene	0.02	0.02	0.04	ND	0.10	0.21	0.02	0.03	ND	0.04	0.02	0.05	0.07	0.06	0.05
n-Nonanal	1.69	ND	5.67	ND	2.39	0.02	0.01	ND	0.01	0.02	0.02	0.55	0.59	0.54	0.24
n-Undecane	0.15	0.01	0.02	ND	0.18	0.16	0.01	0.01	0.01	0.01	0.02	4.32	5.50	5.49	3.58
1,2,4,5-Tetramethylbenzene	2.89	ND	ND	ND	0.02	0.01	ND	ND	ND	ND	ND	0.23	0.29	0.28	0.16
n-Decanal	0.12	ND	0.11	ND	0.09	0.02	1.70	ND	ND	ND	0.05	0.12	ND	9.10	2.80
n-Dodecane	0.08	0.01	0.02	ND	0.98	0.15	0.22	0.01	0.01	0.01	0.02	1.91	2.83	2.80	1.85
n-Tridecane	0.06	0.01	0.05	ND	0.12	0.11	2.08	0.02	0.02	0.04	0.04	0.43	0.70	0.71	0.41
n-Tetradecane	0.08	0.04	0.08	ND	0.49	0.13	0.05	0.04	0.05	0.09	0.09	0.77	1.16	1.36	0.77
n-Pentadecane	0.06	0.02	0.04	ND	0.05	0.07	0.03	0.03	0.04	0.04	0.05	0.18	0.24	0.32	0.16
n-Hexadecane	0.03	0.02	0.03	ND	0.08	0.04	0.02	0.02	0.03	0.03	0.05	0.34	0.43	0.72	0.30

[1] Mean of duplicate analysis; [2] Not detectable.

Table 7. Amounts of VOCs emitted from various school items.

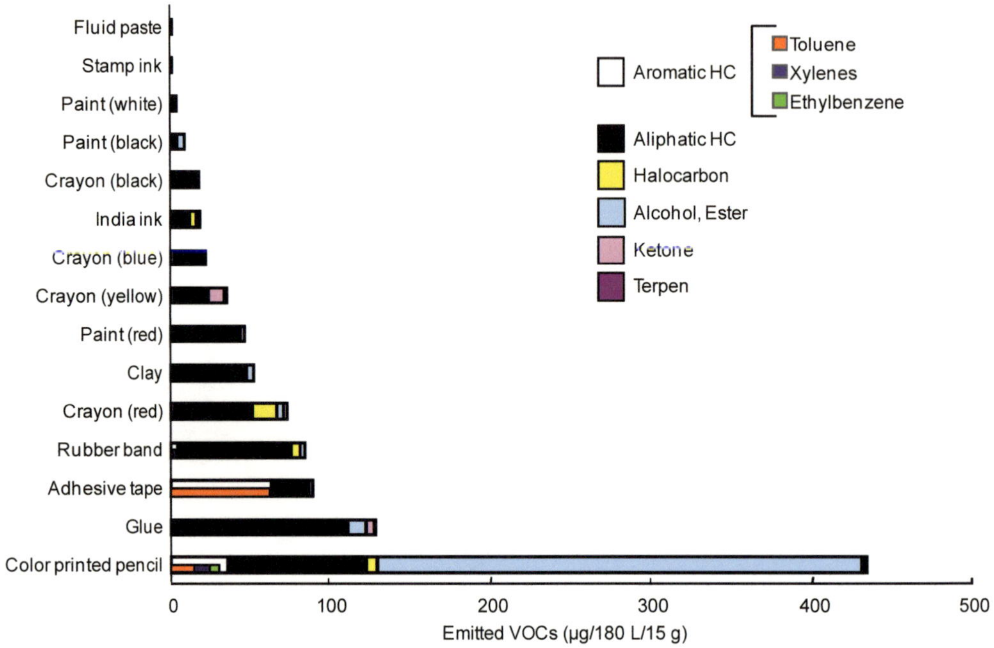

Fig. 10. VOCs emitted from various school items.

4. Conclusion

This chapter provides an analytical method for the determination of VOCs in environmental air samples by GC-MS. This GC-MS method is convenient and reliable, and is useful in evaluating indoor air quality and the sources of VOCs emitted in indoor environments. Indoor air VOC levels in newly constructed buildings exceeded those set by the MHLW. Since humans spend most of their lives indoors, it is necessary to minimize exposure to VOCs affecting human health. Furthermore, we found that various building materials and household products were emission sources of VOCs. Indoor VOC levels associated with these sources can be reduced by increasing outdoor air ventilation, but this entails increased costs in building construction, operation, and energy (Cox et al., 2010). Low VOC-emitting materials are being developed and are used more widely in buildings to help achieve healthier and more productive indoor environments. While VOC-exposure from household products is less than that from building materials, children hypersensitive to these chemicals may be at high risk from directly touching toys and school supplies. Sufficient assessment of the hazards and risks of indoor environments and the regulation of indoor air pollutants such as VOCs are necessary to protect human health, especially children and people who are sensitive to these chemicals. Finally, we hope that this chapter will be beneficial and informative for scientists and students studying environmental pollution and related research fields.

5. Acknowledgment

We are extremely grateful to the late professor Shohei Kira for his valuable instruction. This work was supported by The Ministry of Health, Labour and Walfare of Japan Health Science Research Grant for Research on Environmental Health, The Science Research Promotion Fund, and a Grant-in-Aid for Basic Scientific Research, The Promotion and Mutual Aid Corporation for Private Schools of Japan, and The Yakumo Foundation for Environmental Science.

6. References

Abbritti G. & Muzi G. (2006). Indoor air quality and health in offices and other non-industrial working environments, *Med. Lav.* Vol. 97(2): 410-417.

Adgate J.L., Church T.R., Ryan A.D., Ramachandran G., Fredrickson A.L., Stock T.H., Morandi M.T. & Sexton K. (2004a). Outdoor, indoor, and personal exposure to VOCs in children, *Environ. Health Perspect.* Vol. 112(14): 1386-1392.

Adgate J.L., Eberly L.E., Stroebel C., Pellizzari E.D. & Sexton K. (2004b). Personal, indoor, and outdoor VOC exposures in a probability sample of children, *J. Expo. Anal. Environ. Epidemiol.* Vol. 14: S4-S13.

Afshari A., Lundgren B. & Ekberg L.E. (2003). Comparison of three small chamber test methods for the measurement of VOC emission rates from paint, *Indoor Air* Vol. 13(2):156-165.

Araki A., Eitaki Y., Kawai T., Kanazawa A., Takeda M. & Kishi R. (2009). Diffusive sampling and measurement of microbial volatile organic compounds in indoor air, *Indoor Air* Vol. 19(5): 421-432.

Araki A., Kawai T., Eitaki Y., Kanazawa A., Morimoto K., Nakayama K., Shibata E., Tanaka M., Takigawa T., Yoshimura T., Chikara H., Saijo Y. & Kishi R. (2010). Relationship between selected indoor volatile organic compounds, so-called microbial VOC, and the prevalence of mucous membrane symptoms in single family homes, *Sci. Total Environ.* Vol. 408(10): 2208-2215.

Barro R., Regueiro J., Llompart M. & Garcia-Jares C. (2009). Analysis of industrial contaminants in indoor air: part 1. Volatile organic compounds, carbonyl compounds, polycyclic aromatic hydrocarbons and polychlorinated biphenyls, *J. Chromatogr. A* Vol. 1216(3): 540-566.

Boeglin M.L., Wessels D. & Henshel D. (2006). An investigation of the relationship between air emissions of volatile organic compounds and the incidence of cancer in Indiana countries, *Environ. Res.* Vol. 100: 242-254.

Bruno P., Caselli M., de Gennaro G., Iacobellis S. & Tutino M. (2008). Monitoring of volatile organic compounds in non-residential indoor environments, *Indoor Air* Vol. 18(3): 250-256.

Caselli M., de Gennaro G., Saracino M.R. & Tutino M. (2009). Indoor contaminants from newspapers: VOCs emissions in newspaper stands, *Environ. Res.* Vol. 109(2): 149-157.

Chang Y.M., Hu W.H., Fang W.B., Chen S.S., Chang C.T. & Ching H.W. (2011). A study on dynamic volatile organic compound emission characterization of water-based paints, *J. Air Waste Manag. Assoc.* Vol. 61(1): 35-45.

Claeson A.-S., Sandstrom M. & Sunesson A.-L. (2007). Volatile organic compounds (VOCs) emitted from materials collected from buildings affected by microorganisms, *J. Environ. Monit.* Vol. 9: 240-245.

Cox S.S., Little J.C. & Hodgson A.T. (2001). Measuring concentrations of volatile organic compounds in vinyl flooring, *J. Air Waste Manag. Assoc.* Vol. 51(8): 1195-1201.

Cox S.S., Little J.C. & Hodgson A.T. (2002). Predicting the emission rate of volatile organic compounds from vinyl flooring, *Environ. Sci. Technol.* Vol. 36(4): 709-714.

Cox S.S., Liu Z., Little J.C., Howard-Reed C., Nabinger S.J. & Persily A. (2010). Diffusion-controlled reference material for VOC emissions testing: proof of concept, *Indoor Air* Vol. 20(5): 424-433.

Delgado-Saborit J.M., Aquilina N.J., Meddings C., Baker S. & Harrison R.M. (2011). Relationship of personal exposure to volatile organic compounds to home, work and fixed site outdoor concentrations, *Sci. Total Environ.* Vol. 409(3): 478-488.

Dodson R.E., Levy J.I., Houseman E.A., Spengler J.D. & Bennett D.H. (2009). Evaluating methods for predicting indoor residential volatile organic compound concentration distributions, *J. Expo. Sci. Environ. Epidemiol.* Vol. 19(7): 682-693.

Fromme H., Dietrich S., Heitmann D., Dressel H., Diemer J., Schulz T., Jörres R.A., Berlin K. & Völkel W. (2009). Indoor air contamination during a waterpipe (narghile) smoking session, *Food Chem. Toxicol.* Vol. 47(7): 1636-1641.

Glas B., Stenberg B., Stenlund H. & Sunesson A.L. (2008). A novel approach to evaluation of adsorbents for sampling indoor volatile organic compounds associated with symptom reports, *J. Environ. Monit.* Vol. 10(11): 1297-1303.

Gokhale S., Kohajda T. & Schlink U. (2008). Source apportionment of human personal exposure to volatile organic compounds in homes, offices and outdoors by chemical mass balance and genetic algorithm receptor models, *Sci. Total Environ.* Vol. 407(1): 122-138.

Haghighat F., Lee C.S. & Ghaly W.S. (2002). Measurement of diffusion coefficients of VOCs for building materials: review and development of a calculation procedure, *Indoor Air* Vol. 12(2): 81-91.

Han K.H., Zhang J.S., Wargocki P., Knudsen H.N. & Guo B. (2010). Determination of material emission signatures by PTR-MS and their correlations with odor assessments by human subjects, *Indoor Air* Vol. 20(4): 341-354.

Harada K., Hara K., Wei C.N., Ohmori S., Matsushita O. & Ueda A. (2007). Case study of volatile organic compounds in indoor air of a house before and after repair where sick building syndrome occurred, *Int. J. Immunopathol. Pharmacol.* Vol. 20(2 Suppl 2): 69-74.

Hippelein M. (2004). Background concentrations of individual and total volatile organic compounds in residential indoor air of Schleswig-Holstein, Germany, *J. Environ. Monit.* Vol. 6(7): 745-752.

Hulin M., Caillaud D. & Annesi-Maesano I. (2010). Indoor air pollution and childhood asthma: variations between urban and rural areas, *Indoor Air* Vol. 20(6): 502-514.

Jensen L.K., Larsen A., Mølhave L., Hansen M.K. & Knudsen B. (2001). Health evaluation of volatile organic compound (VOC) emissions from wood and wood-based materials, *Arch. Environ. Health.* Vol. 56(5): 419-432.

Jia C., Batterman S., Godwin C., Charles S. & Chin J.Y. (2010). Sources and migration of volatile organic compounds in mixed-use buildings, *Indoor Air* Vol. 20(5): 357-369.

Kabir E. & Kim K.H. (2011). An investigation on hazardous and odorous pollutant emission during cooking activities, *J. Hazard. Mater.* Vol. 188(1-3): 443-454.

Katsoyiannis A., Leva P. & Kotzias D. (2008). VOC and carbonyl emissions from carpets: a comparative study using four types of environmental chambers, *J. Hazard Mater.* Vol. 152(2): 669-676.

Khalequzzaman M., Kamijima M., Sakai K., Chowdhury N.A., Hamajima N. & Nakajima T. (2007). Indoor air pollution and its impact on children under five years old in Bangladesh, *Indoor Air* Vol. 17(4): 297-304.

Kirkeskov L., Witterseh T., Funch L.W., Kristiansen E., Mølhave L., Hansen M.K. & Knudsen B.B. (2009). Health evaluation of volatile organic compound (VOC) emission from exotic wood products, *Indoor Air* Vol. 19(1): 45-57.

Kwon K.D., Jo W.K., Lim H.J. & Jeong W.S. (2008). Volatile pollutants emitted from selected liquid household products, *Environ. Sci. Pollut. Res. Int.* Vol. 15(6): 521-526.

Larroque V., Desauziers V. & Mocho P. (2006). Development of a solid phase microextraction (SPME) method for the sampling of VOC traces in indoor air, *J. Environ. Monit.* Vol. 8: 106-111.

LeBouf R.F., Casteel C. & Rossner A. (2010). Evaluation of an air sampling technique for assessing low-level volatile organic compounds in indoor environments, *J. Air Waste Manag. Assoc.* Vol. 60(2): 156-162.

Lee C.S., Haghighat F. & Ghaly W.S. (2005). A study on VOC source and sink behavior in porous building materials - analytical model development and assessment, *Indoor Air* Vol. 15(3): 183-196.

Lee C.W., Dai Y.T., Chien C.H. & Hsu D.J. (2006). Characteristics and health impacts of volatile organic compounds in photocopy centers, *Environ. Res.* 100(2): 139-149.

Liu J., Guo Y. & Pan X. (2008). Study of the current status and factors that influence indoor air pollution in 138 houses in the urban area in Xi'an, *Ann. N. Y. Acad Sci.* Vol. 1140: 246-255.

Liu W., Zhang J., Hashim J.H., Jalaudin J., Hashim Z. & Goldstein B.D. (2003). Mosquito coil emissions and health implications, *Environ. Health Perspect.* Vol. 111(12): 1454-1460.

Logue J.M., McKone T.E., Sherman M.H. & Singer B.C. (2011). Hazard assessment of chemical air contaminants measured in residences, *Indoor Air* Vol. 21(2): 92-109.

Loh M.M., Houseman E.A., Levy J.I., Spengler J.D. & Bennett D.H. (2009). Contribution to volatile organic compound exposures from time spent in stores and restaurants and bars, *J. Expo. Sci. Environ. Epidemiol.* Vol. 19(7): 660-673.

Massolo L., Rehwagen M., Porta A., Ronco A., Herbarth O. & Mueller A. (2010). Indoor-outdoor distribution and risk assessment of volatile organic compounds in the atmosphere of industrial and urban areas, *Environ. Toxicol.* Vol. 25(4): 339-349.

Ministry of Health, Labour and Walfare of Japan (2002) Committee on sick house syndrome: indoor air pollution progress report no. 4. Ministry of Health, Labour and Walfare of Japan, Tokyo.

Nicolle J., Desauziers V. & Mocho P. (2008). Solid phase microextraction sampling for a rapid and simple on-site evaluation of volatile organic compounds emitted from building materials, *J. Chromatogr. A* Vol. 1208(1-2): 10-15.

Ohura T., Amagai T., Senga Y. & Fusaya M. (2006). Organic air pollutants inside and outside residences in Shimizu, Japan: levels, sources and risks, *Sci. Total Environ.* Vol. 366(2-3): 485-499.

Pedersen E.K., Bjørseth O., Syversen T. & Mathiesen M. (2003). A screening assessment of emissions of volatile organic compounds and particles from heated indoor dust samples, *Indoor Air* Vol. 13(2): 106-117.

Ribes A., Carrera G., Gallego E., Roca X., Berenguer M.A. & Guardino X. (2007). Development and validation of a method for air-quality and nuisance odors monitoring of volatile organic compounds using multi-sorbent adsorption and gas chromatography/mass spectrometry thermal desorption system, *J. Chromatogr. A* Vol. 1140(1-2): 44-55.

Rumchev K., Brown H. & Spickett J. (2007). Volatile organic compounds: do they present a risk to our health?, *Rev. Environ. Health* Vol. 22(1): 39-55.

Salthammer T. (2011). Critical evaluation of approaches in setting indoor air quality guidelines and reference values, Chemosphere Vol. 82(11): 1507-1517.

Sarigiannis D.A., Karakitsios S.P., Gotti A., Liakos I.L. & Katsoyiannis A. (2011). Exposure to major volatile organic compounds and carbonyls in European indoor environments and associated health risk, *Environ. Int.* Vol. 37(4): 743-765.

Sax S.N., Bennett D.H., Chillrud S.N., Kinney P.L. & Spengler J.D. (2004). Differences in source emission rates of volatile organic compounds in inner-city residences of New York City and Los Angeles, *J. Expo. Anal. Environ. Epidemiol.* Vol. 14: S95-S109.

Schlink U., Rehwagen M., Damm M., Richter M., Borte M. & Herbarth O. (2004). Seasonal cycle of indoor-VOCs: comparison of apartments and cites. *Atmos. Environ.* Vol. 38: 1181-1190.

Serrano-Trespalacios P.I., Ryan L. & Spengler J.D. (2004). Ambient, indoor and personal exposure relationship of volatile organic compounds in Mexico City Metropolitan area, *J. Expo. Anal. Environ. Epidemiol.* Vol. 14: S118-S132.

Sohn J., Yang W., Kim J., Son B. & Park J. (2009). Indoor air quality investigation according to age of the school buildings in Korea, *J. Environ. Manage.* Vol. 90(1): 348-354.

Son B., Breysse P. & Yang W. (2003). Volatile organic compounds concentrations in residential indoor and outdoor and its personal exposure in Korea, *Environ. Int.* Vol. 29(1): 79-85.

Takeda M., Saijo Y., Yuasa M., Kanazawa A., Araki A. & Kishi R. (2009). Relationship between sick building syndrome and indoor environmental factors in newly built Japanese dwellings, *Int. Arch. Occup. Environ. Health* Vol. 82(5): 583-593.

Takigawa T., Horike T., Ohashi Y., Kataoka H., Wang, Da-H. & Kira S. (2004). Were volatile organic compounds the inducing factors for subjective symptoms of employees working in newly constructed hospitals?, *Environ. Toxicol.* 19: 280-290.

Takigawa T., Wang B.L., Saijo Y., Morimoto K., Nakayama K., Tanaka M., Shibata E., Yoshimura T., Chikara H., Ogino K. & Kishi R. (2010). Relationship between indoor chemical concentrations and subjective symptoms associated with sick building syndrome in newly built houses in Japan, *Int. Arch. Occup. Environ. Health* Vol. 83(2): 225-235.

Tham K.W., Zuraimi M.S. & Sekhar S.C. (2004). Emission modelling and validation of VOCs' source strengths in air-conditioned office premises, *Environ. Int.* Vol. 30(8): 1075-1088.

Tumbiolo S., Gal J.F., Maria P.C. & Zerbinati O. (2005). SPME sampling of BTEX before GC/MS analysis: examples of outdoor and indoor air quality measurements in public and private sites, *Ann. Chim.* Vol. 95(11-12): 757-766.

Vainiotalo S., Väänänen V. & Vaaranrinta R. (2008). Measurement of 16 volatile organic compounds in restaurant air contaminated with environmental tobacco smoke, *Environ. Res.* Vol. 108(3): 280-288.

Wady L. & Larsson L. (2005). Determination of microbial volatile organic compounds adsorbed on house dust particles and gypsum board using SPME/GC-MS, *Indoor Air* Vol. 15 Suppl 9: 27-32.

Wieslander G. & Norbäck D. (2010). Ocular symptoms, tear film stability, nasal patency, and biomarkers in nasal lavage in indoor painters in relation to emissions from water-based paint, *Int. Arch. Occup. Environ. Health* Vol. 83(7): 733-741.

Wilke O., Jann O. & Brödner D. (2004). VOC- and SVOC-emissions from adhesives, floor coverings and complete floor structures, *Indoor Air* Vol. 14 Suppl 8: 98-107.

Yamaguchi T., Nakajima D., Ezoe Y., Fujimaki H., Shimada Y., Kozawa K., Arashidani K. & Goto S. (2006). Measurement of volatile organic compounds (VOCs) in new residential buildings and VOCs behavior over time, *J. UOEH.* Vol. 28(1): 13-27.

Ye Q. (2008). Development of solid-phase microextraction followed by gas chromatography-mass spectrometry for rapid analysis of volatile organic chemicals in mainstream cigarette smoke, *J. Chromatogr. A* Vol. 1213(2): 239-244.

Yu I.T.S., Lee N.L., Zhang X.H., Chen W.O., Jam Y.T. & Wong T.W. (2004). Occupational exposure to mixtures of organic solvents increases the risk of neurological symptoms among printing workers in Hong Kong, *J. Occup. Environ. Med.* Vol. 46: 323-330.

Zhang L.Z. & Niu J.L. (2002). Mass transfer of volatile organic compounds from painting material in a standard field and laboratory emission cell, *Int. J. Heat & Mass Transfer* Vol. 46: 2415-2423.

Zhou J., You Y., Bai Z., Hu Y., Zhang J. & Zhang N. (2011). Health risk assessment of personal inhalation exposure to volatile organic compounds in Tianjin, China, *Sci. Total Environ.* Vol. 409(3): 452-459.

10

Gas Chromatography in the Analysis of Compounds Released from Wood into Wine

Maria João B. Cabrita[1], Raquel Garcia[1], Nuno Martins[1],
Marco D.R. Gomes da Silva[2] and Ana M. Costa Freitas[1]
[1]*School of Science and Technology, Department of Plant Science, Institute of Agricultural and Environmental Science ICAAM Mediterranean, University of Évora, Évora,*
[2]*REQUIMTE, Department of Chemistry, Faculty of Science and Technology, New University of Lisbon, Campus Caparica,*
Portugal

1. Introduction

Wood has been used in alcoholic beverages for centuries, mainly as material for containers used for alcoholic beverages aging. Recently OIV (Organisation International de la Vigne et du Vin) approved the use of chips (Resolution oeno 3/2005) and staves as alternatives for barrels. These practices are being rapidly spread among winemakers. The increased used of these alternatives are mainly related to low investments, similar sensorial results obtained in shorter time, simplicity of use and the possibility of avoiding contamination and off-flavours, too-often related to aged or contaminated barrels.

Besides oak, other woods are being looked at for enological purposes, such as acacia, cherry, chestnut and mulberry. Their characteristics are commonly compared to oak. In the past, chestnut (*Castanea sativa*) was widely used in the Mediterranean area, because of its availability and its cheap price. Chestnut wood has higher porosity than oak. Cherry wood (*Prunus avium*) has high porosity and oxygen permeation, and is usually used for short aging times. Acacia wood (*Robinia pseudoacacia*) is hard, with low porosity. Mulberry wood (*Morus alba* and *Morus nigra*) is tender and elastic, with medium porosity, and is characterized by a low release of compounds. The lack of properties for cooperage is now overcome by their possible use as staves or chips.

The aim of this work is to present an overview on volatile and semi-volatile composition of different kind of wood with oenological interest. Within this purpose, this work will be focused on a bibliographic review of the most used chromatographic methods for characterization of volatile and semi-volatile compounds, including also a brief description of the most common reported sample preparation methods for chromatographic analysis.

The composition of woods volatile fraction depends on the botanical species and geographic origin. Prior to its final use a toasting step is needed. This toasting process induces the formation of a great number of volatile and odoriferous compounds, and it is the main reason for significant differences among non-toasted and toasted wood. Volatiles and semivolatiles compounds from wood belong to several different chemical families: lactones, terpenes, norisoprenoids, aldehydes, ketones, alcohols, phenols and esters.

Because of the odourific impact of some volatiles, a number of studies dealt with their characterization in wood-aged beverages and woods itself. Volatile phenols such as guaiacol and eugenol, phenolic aldehydes such as vanillin and syringaldehyde, oak lactones, and furanic aldehydes (furfural and derivatives) have been described as the main contributors to the sensory fingerprinting of aged alcoholic beverages.

Most of the volatile compounds are formed during open-air seasoning and toasting phases of wood processing. Several furanic aldehydes and ketones come from the thermo degradation of celluloses and hemicelluloses. This is the case of hydroxy-methylfurfural (HMF) (from cellulose-derived glucose) and furfural arising from pentoses produced by partial hydrolysis of hemicelluloses. These latter compounds are responsible for almond and toasty odours. Thermal degradation of lignin determines the formation of methoxylated volatile phenols (*i.e.*, guaiacyl and syringyl derivatives), phenolic ketones, and phenolic aldehydes contributing to smoked or spiced and vanilla aromas, respectively.

To study volatiles in woods or in alcoholic beverages some preparation methods must be considered. The volatile fraction from woods can be studied from three different points of view: the volatiles existing in wood itself, the volatiles that wood can release into synthetic medium miming alcoholic beverages and the study of alcoholic beverages like wines and spirits after aging with wood contact. In practical terms, that implies using sample preparation methods for solid and liquid matrices.

Sample preparation for gas chromatography (GC) analysis of volatile compounds from woods can be performed by pressurized liquid extraction, a process that combines temperature and pressure with liquid solvent to achieve extracts rapidly and efficiently. This technique is available in an automated or a manual version as accelerated solvent extraction (ASE), and it is being currently used as a good alternative for the solid-liquid extraction used previously for studying volatiles from solid matrices. Most of the "oldest" sample preparation methods involved Soxhlet extraction with dichloromethane or methanol or simultaneous distillation-extraction procedures.

The sample preparation procedures for GC analysis of volatile compounds in wines or hydro-alcoholic mixtures in general, are usually done by LLE (liquid-liquid extraction), SPE (solid-phase extraction) or SPME (solid-phase microextraction). But, more environmental friendly sample preparation techniques like SBSE (stir bar sorptive extraction) can also be employed with good results when thermal desorption is used instead of retro-extraction, for subsequent chromatographic analysis.

The chromatographic technique more suitable for these analyses is gas chromatography-mass spectrometry (GC/MS), as the mass spectrometer allows a more powerful tool for compounds identification. When MS is coupled with two-dimensional (2D) chromatographic systems, broader capabilities can be open in order to fully characterized wine matrices. An olfactometric approach is also mandatory in order to identify chromatogram zones where peak identification should take place. Consequently gas chromatography-olfactometry (GC-O) emerges as an important technique to achieve this purpose. GC-O analysis has been widely used to identify odour-active compounds or to screen the odour volatile composition in wines. This technique allows to obtain relatively simple olfactograms and to establish a hierarchy of the most important odorants according to their potential sensory impact. In general it implies the use of headspace technique as sample preparation method, such as purge-and-trap (P&T) or solid-phase microextraction (HS-SPME) systems.

2. Composition and biosynthesis of the volatile fraction of woods: Effect of botanical species and toasting

For centuries, wood has been used in wine technology, either as containers for transportation either for aging wine and spirits. Over the years, many different tree species have been used in cooperage, but oak and chestnut woods proved to be the best ones to manufacture barrels and therefore the most suitable ones for wine and spirits aging.

The use of barrels is, however, expensive since barrels can only be used few times for wine aging. Besides the above mentioned issue, barrels can often be the cause of wine contaminations and off-flavours, if sanitation procedures are not correctly applied.

All woods are composed of cellulose, lignin, ash-forming minerals, and extractives formed into a cellular structure. The characteristics and amounts of these compounds and differences in cellular structures give to each wood its specific characteristics. Some are more flexible than others, some are easier to work, some are harder or softer, heavier or lighter.

Extractives from different wood species comprise several substances which belong to an extremely wide range of chemical families that are characteristics of each wood. Thus, extractives allow the identification of wood species, contributing also for the determination of other wood properties, namely permeability, density, hardness and compressibility.

Besides botanical species, geographical origin also contributes to the final chemical composition of the extractives. Technological treatments during cooperage such as seasoning and toasting give the final characteristics to the barrel (Chatonnet & Dubourdieu, 1998; Cadahía et al., 2001; Doussot et al., 2002; Pérez-Prieto et al., 2002; Cadahía et al., 2003).

Wood physical properties play a significant role in cooperage. Usually, oak is the most used specie due to its unique physical and chemical nature. Physically, oak has strength since its wide radial rays give strength when shaped for a cask. Oak is also a "pure wood" as opposed to pine or rubber trees which contain resin canals that can transfer strong flavours to maturing beverages.

Several studies have been conducted in order to evaluate the accumulation of oak volatile compounds in wines and spirits and how different factors affect the final concentration of these compounds both in barrels (Puech, 1987; Sefton et al., 1993; Yokotsuka et al., 1994; Waterhouse & Towey, 1994; Piggot et al., 1995; Escalona et al., 2002; Ferreras et al., 2002; Pérez-Prieto et al., 2002, 2003; Netto et al., 2003; Madrera et al., 2003; Garde-Cerdán et al., 2002, 2004; Gómez-Plaza et al., 2004; Jarauta et al., 2005; Garde Cerdán & Ancín-Azpilicueta, 2006 b; Caldeira et al., 2006 a,b; Jiménez-Moreno & Ancín-Azpilicueta, 2007; Jiménez-Moreno et al., 2007; Frangipane et al., 2007) and with oak chips (Pérez-Coello et al., 2000; Arapitsas et al., 2004; Caldeira et al., 2004; Guchu et al., 2006; Frangipane et al., 2007; Bautista-Ortin et al., 2008; Rodríguez-Bencomo et al., 2008, 2009; Rodríguez- Rodríguez & Gómez-Plaza, 2011).

Other studies were made in model solutions to avoid matrix influence (Singleton & Draper, 1961; Ancín-Azpilicueta et al., 2004; Fernández de Simón et al., 2010 a; Rodríguez-Rodríguez & Gómez-Plaza, 2011) or in wood itself (Vichi et al., 2007; Natali et al., 2006; Díaz-Maroto et al., 2008).

More recently some studies have been made with other woods than oak (Flamini, et al., 2007; de Rosso et al., 2009; Fernández de Simón et al., 2009; Caldeira et al., 2010) as they becoming increaseling interesting.

Ever since OIV approved the use of alternatives for barrels, like chips and staves, researchers are looking into other kind of woods such as acacia, cherry and mulberry, which were abandoned in the past due to the lack of cooperage properties.

2.1 Effect of botanical species

The main botanical species used in cooperage is oak (genus *Quercus*) and in a lesser extension, chestnut (genus *Castanea*). The oak species more used in cooperage are *Quercus alba* (American oak) growing in different areas of the United States, *Quercus patraea* Liebl. (sessili oak), *Quercus Robur* L. (pedunculate oak) and *Quercus pyrenaica* growing in Europe. French oaks are the most widespread wood for barrels, especially those from Allier and Limousin regions. But oaks from Hungary, Russia, Spain, Romania and Portugal are also becoming more and more attractive.

Many researchers are now dedicated to the study of chemical compounds in different kind of woods, although oak is still the most studied one. Whenever a new wood is tested it is compared with the oak effects in wines and spirits. Oak belonging to different species can be significantly different regarding volatile composition. Jordão et al. (2006) concluded that *Quercus pyrenaica* released significantly more volatile compounds than *Quercus patraea,* in a study with SPME in hydro-alcoholic oak wood solutions. Fernández de Simón et al., (2010 a) concluded that *Quercus pyrenaica* chips, althought similar to other species, have some odourific particularities, such as high levels of furanic compounds, eugenol, furaneol and *cis*-whiskylactone and low levels of vanillin. Several authors pointed out that β-methyl-γ-octalactone (whisky lactone), particularly the *cis* isomer, can be used to differentiate American from French oaks (Masson et al., 1995; Waterhouse & Towey, 1994; Masson et al., 1997; Chatonnet & Dubourdieu, 1998; Pérez-Coello et al, 1999). Several norisoprenoid compounds were found in American oak, but were almost absent in European oaks (Sefton et al., 1990). Same wood species from French oak and East European oaks can be distinguished by their contents of eugenol, 2-phenylethanol and aromatic aldehydes, mostly vanillin and syringaldehyde (Prida & Puech 2006).

The chemical compounds released from acacia, chestnut, cherry, mulberry and oak untoasted woods into alcoholic extracts, shows that each wood has a different and characteristic profile (Flamini et al., 2007, de Rosso et al., 2009, Fernández de Simon et al., 2009). Acacia had significant aromatic aldehydes, particularly vanillin, syringaldehyde and dihydroxy-benzaldehyde but no eugenol (de Rosso et al., 2009) or methoxyeugenol (Flamini et al., 2007) were found. Chestnut and oak showed the highest content of volatile compounds namely vanillin, eugenol, methoxyeugenol, syringaldehyde, α-terpineol, and oak presented high amouts of *cis*- and *trans*-β-methyl-γ-octalactones. In cherry, several aromatic compounds were found although in low abundance, but mulberry was the poorest, with small amounts of eugenol and absence of methoxyeugenol (de Rosso et al., 2009). Cherry is also characterized by methoxyphenols, particularly, high content of trimethoxyphenol (Flamini et al., 2007). Chestnut and oak woods are known to release significant amounts of eugenol and methoxyeugenol into wines. High amounts of vanillin are released from chestnut and high levels of syringaldehyde are released from acacia and oak woods.

2.2 Effect of toasting level

Oak wood chemical composition mainly depends on the species, its provenience and the various treatments that wood undergoes in cooperage, such as seasoning, region of seasoning and toasting (Marco et al., 1994; Chatonnet & Dubourdieu, 1998; Cadahía et al., 2001; Doussot et al., 2002; Pérez-Prieto et al., 2002; Cadahía et al., 2003).

Seasoning prevents the wood from shrinking after barrel production while firing is applied to stabilise the curved shape of the barrel (van Jaarsveld et al., 2009) and both these steps play a crucial role in wood flavour development.

Wood seasoning in cooperage is usually performed under natural conditions in open air during a variable period of time (18 to 36 months), but artificial seasoning can also be performed. However, natural seasoning has a more positive effect on the odourific profile of wood. During this process the wood volatile compounds profiles, which include lactones, phenolic aldehydes or volatile phenols, show significant differences.

Nevertheless, among all variables that can influence the impact of woods in wine or spirits sensory changes, heating is the most important. Toasting has a significant influence on wood's chemical compounds, modifying both, the quantity and the quality of the extractable substances (Cutzach et al., 1997; Chatonnet et al., 1999). The toasting process drastically enhances the gain in volatile compounds arising from the thermal degradation of oak wood (Cutzach et al., 1997; Chatonnet et al., 1999; Doussot et al., 2002).

When heat is applied to wood during toasting process, chemical bonds are disrupted within biopolymers such as cellulose, hemicellulose, lignin, polysaccharides, polyphenols and lipids, resulting in degradation or compositional changes by pyrolysis and thermolysis (Fernández de Simón et al., 2009; van Jaarsveld et al., 2009), which induce a notable modification of wood chemical composition.

Volatile phenols, phenolic aldehydes, phenyl ketones and some phenyl alcohols are mainly formed from lignin thermodegradation. In particular, high levels of mono and dimethoxylated phenols, benzoic and cinnamic aldehydes were identified in toasted wood.

Heat degradation of polyosides leads to the production of furanic aldehydes, pentacyclic and hexacyclic ketones. Lactones, formed from wood lipids, increase their concentration at the beginning of toasting, although they can be destroyed by a lengthy toasting process (Giménez-Martínez et al., 1996; Chatonnet et al., 1999; Cadahía et al., 2003). This thermo degradation process leads to the formation of several compounds that can be transferred to wine and spirits during aging or maturation, hence, extensive information about the volatile composition of wood used in cooperage would be of great interest to the wine industry. Table 1 resumes information regarding main volatile compounds formed from wood biopolymer during toasting process.

Common name	IUPAC name	m/z fragment	LRI Apolar column	LRI Polar column	Aroma notes
DERIVED FROM POLYSACHARIDES					
Furanic aldeydes					
Furfural	2-Furancarboxaldehyde	95,**96**	834[(a)] 811[(h)]	1444[(b)]	Bread, sweet almond [(i)], slightly toasty, caramel[(h)]
5-Methylfurfural	5-Methyl-2-furancarboxaldehyde	109,**110**	965[(a)]	1551[(b)]	Almond, caramel[(i)], spicy, toasty[(h)]
5-Hydroxy methylfurfural	5-Hydroxymethyl-2-furancarboxaldehyde	**97**,126	1235[(a)]	2466[(b)]	Odourless[(n)]
Maltol	3-Hydroxy-2-methyl-4*H*-pyran-4-one	71,**126**	1111[(a)]	1938[(b)]	Toasty[(n)]
Furaneol	4-Hydroxy-2,5-dimethyl-3(2*H*)-furanone	85,**128**	1083[(a)]	2013[(b)]	Caramel-like[(o)]
Acids					
Acetic acid	Ethanoic acid	**43**,45,60	602[(g)]	1464[(f)]	vinegar[(f)]
DERIVED FROM LIGNIN AND POLYPHENOLS					
Volatile phenols					
Eugenol	2-Methoxy-4-(prop-2-enyl) phenol	77,103, **164**	1359[(a)]	2139[(b)]	Clove,honey [(i)] spicy, cinnamon[(h)]
Isoeugenol (*cis* and *trans*)	*cis-* and *trans*-1-Methoxy-4-(prop-2-enyl) phenol	77,149, **164**	1408[(a)] 1451[(a)]	2226[(b)] 2314[(b)]	Spicy [(m)] clove,woody[(h)]
Phenol	Hydroxybenzene	66,94	983[(a)]	1978[(b)]	Ink[(n)]
3,4-Dimethylphenol	1-Hydroxy-3,4-dimethylbenzene	**107**	1193[(a)]	2192[(b)]	
o-Cresol	2-Methylphenol	**107**/108	1059[(a)]	1980[(b)]	Leather, spicy[(m)]
p-Cresol	4-Methylphenol	**107**/108	1079[(a)]	2056[(b)]	
m-Cresol	3-Methylphenol	**107**/108	1086[(a)]	2064[(b)]	
Guaiacol	2-Methoxyphenol	109/**124**	1089[(a)]	1833[(b)]	Smoke,sweet, medicine [(i)]
4-Methylguaiacol	4-Methyl-2-methoxyphenol	123/**138**	1191[(a)]	1928[(b)]	Spicy,phenolic, lightly green[(h)]
4-Ethylguaiacol	4-Ethyl-2-methoxyphenol	**137**/152	1274[(a)]	2002[(b)]	Phenolic[(i,m)] leather[(m)] smoked[(i)]
4-Propylguaiacol	4-Propyl-2-methoxyphenol	**137**/166	1461[(a)]	2083[(b)]	Leather animal[(m)]

Common name	IUPAC name	m/z fragment	LRI Apolar column	LRI Polar column	Aroma notes
4-Vinylguaiacol	4-Vinyl-2-metoxyphenol	135/**150**	1314(a)	2165(b)	Clove (i)
Syringol	2,6-Dimethoxyphenol	139/**154**	1353(a)	2237(b)	Smoke, burned wood(f)
4-Methylsyringol	4-Methyl-2,6-dimetoxyphenol	168	1449(a)	2322(b)	Smoke,burned, flowery(f)
4-Ethylsyringol	4-Ethyl-2,6-dimethoxyphenol	167	1528(a)	2381(b)	
4-Allylsyringol	4-Allyl-2,6-dimethoxyphenol	194	1605(a)	2511(b)	Spicy smoky(m)
4-Propylsyringol	4-Propyl-2,6-dimethoxyphenol	167	1612(a)	2452(b)	
4-Ethylphenol	1-Ethyl-4-hydroxybenzene	**107**,122		2201(f)	Animal, horse, stable(f)
2-Phenylethanol	Hydroxyethylbenzene	91	1113(a)	1888(b)	Floral, roses(h)
Phenolic aldeydes					
Vanillin	4-Hydroxy-3-methoxybenzaldehyde	**151**/152	1399(a)	2518(b)	Vanilla (i)
Benzaldehyde	Phenylmethanal	77,**106**	962(a)	1493(b)	Bitter almonds (h)
Conyferaldehyde	3-Methoxy-4-hydroxycinnamaldehyde	178	1747(a)	3096(b)	Vanilla, woody(n)
Syringaldehyde	4-Hydroxy-3,5-dimethoxybenzaldehyde	181,**182**	1643(a)	2904(b)	Vanilla (i)
Sinapaldehyde	3,5-Dimethoxy-4-hydroxycinnamaldehyde	208	2002(a)	3458(b)	Vanilla(n)
Phenolic esters					
Ethyl vanillate	4-Hydroxy-3-methoxy benzoic acid, ethyl ester	**151**, 167,196	1648(g)		Flower, vanilla,fruit, sweet, (i)
Methyl vanillate	4-Hydroxy-3-methoxy benzoic acid, methyl ester	**151**,182	1518(a)	2565(b)	Caramel, butterscotch, vanilla (i)
Phenyl ketones					
Acetovanillone	1-(4-Hydroxy-3,5-dimethoxyphenyl)-ethanone	**151**/166	1487(a)	2595(b)	Vanilla(i)
Propiovanillone	1-(4-Hydroxy-3,5-dimethoxyphenyl)-propanone	**151**/180	1501(a)	2661(b)	Vanilla(n)
Butyrovanillone	1-(4-Hydroxy-3-methoxyphenyl)-butanone	**151**/194	1590(a)	2770(b)	Caramel, sweet, Buttery (i)

Common name	IUPAC name	m/z fragment	LRI Apolar column	LRI Polar column	Aroma notes
Acetosyringone	2-(4-Hydroxy-3-methoxyphenyl)-acetaldehyde	**181**/196	1744(a)	2953(b)	
Propiosyringone	1-(4-Hydroxy-3,5-dimethoxyphenyl-) propanone	**181**/210	1753(a)	3010(b)	Vanilla(n)
Isoacetosyringone	2-(4-Hydroxy-3,5-dimethoxyphenyl)-acetaldehyde	167	1712(a)	2927(b)	
Isopropiosyringone	1-(4-Hydroxy-3,5-dimethoxyphenyl)-2-propanone	167	1785(a)	2979(b)	
Alcohols					
Coniferyl alcohol	4-(1-*trans*)-3-Hydroxy-prop-1-enyl-2-methoxyphenol	137	1745(a)	3213(b)	
Benzyl alcohol	Hydroxy methylbenzene	79,108		1879(p)	Sweet, floral (m)
		DERIVED FROM LIPIDS			
Lactones					
β-Methyl-γ-octalactone (*cis*)	*cis*-4-Methyl-5-butyldihydro-2(3*H*)-furanone	99	1325(a)	1928(b)	Sweet, coconut (j) woody (k,l)
β-Methyl-γ-octalactone (*trans*)	*trans*-4-Methyl-5-butyldihydro-2(3*H*)-furanone	99	1292(a)	1861(b)	Sweet, coconut (j) woody (k,l)
γ-Butyrolactone	Dihydro-2(3*H*)-furanone	86	913(a)	1593(b)	Cheese (m)
Acids					
Propionic acid	Propanoic acid	45,**74**	668(g)		Fruity, floral(q)
Butyric acid	Butanoic acid	**60**,73,88	827(g)	1627(d)	Sweaty, cheesy unpleasent(h)
3-Methyl butyric acid	3-Methylbutanoic acid		834(h)	1675(r)	Cheesy, sweaty(h)
Valeric acid	Pentanoic acid	**60**,73,101			Strawberry(q)
Caproic acid	Hexanoic acid	**60**,73	982(g)	1865(d)	Faintly cheesy sweaty(h)
Caprylic acid	Octanoic acid	**60**,73,144	1163(h)		Sweaty, penetrating(h)
Capric acid	Decanoic acid	**60**,73,172			Rancid(p)
Lauric acid	Dodecanoic acid	60,73,200			Soap(f)
Myristic acid	Tetradecanoic acid	43,60,**73**			Coconut oil
Palmitic acid	Hexadecanoic acid	**43**,60,73		2820(d)	

Common name	IUPAC name	m/z fragment	LRI Apolar column	LRI Polar column	Aroma notes
DERIVED FROM CAROTENOIDS					
Norisoprenoids					
β-Ionone	4-(2,6,6-Trimethyl-cyclohex-1-enyl)-but-3-en-2-one	43, **177**,192		1614(c)	Violet (m)
3-Oxo-*α*-ionol	9-Hydroxymegastigma-4,7-dien-3-one	108	1648(a)	2518(b) 2658(d) 1937(c)	
Blumenol C	4-(3-Hydroxybutyl)-3,5,5-trimethyl-cyclohex-2-en-1-one	41,**43**,108		2002(c)	
Blumenol A (vomifoliol)	6,9-Dihydroxymegastigma-4,7-dien-3-one	189,**207**, 224		2180(c)	
Vitispirane	2,6,6-Trimethyl-10-methyliden-1-oxospiro[4,5]dec-7-ene		1286(e)	1327(c)	Floral, fruity, earthy, woody
3-Oxo-retro-*α*-ionol	9-hidroximegastigma-4,6-dien-3-one (*cis* and *trans*)			2001(c) 2081(c) 2797(d)	

(a)DB5 column, Fernandez de Simon et al., 2009 ; (b)Carbowax column, Fernandez de Simon et al., 2009; (c)DB1701 (medium polar) column, Sefton et al., 1990; (d)Stabilwax column, Natali et al., 2006 (e)HP5 column, Jordão et al.,2006; (f)DBwax column, Caldeira et al.,2008; (g)HP5 column, Vichi et al., 2007; (h) SPB1 column, Diaz-Maroto et al., 2008; (i)Rodriguez-Bencomo et al., 2009; (j)Mosedale & Puech, 1998; (k)Piggott et al., 1995; (l)Garde-Cerdán & Ancín-Azpilicueta , 2006a (m)Sáenz-Navajas et al., 2010; (n) Togores, 2004; (o)Gomes da Silva & Chaves das Neves, 1999; (p) Zhao et al, 2011 (q) Cormier et al. 1991; (r)Barata et al., 2011

Table 1. Main volatile compounds formed from wood biopolymer during toasting process. LRI denotes linear retention indexes; *m/z* denotes mass fragment ions and base peak is presented in bold.

The intensity and length of the applied heat affects the production of compounds during macromolecules degradation and define the toasting levels. Designations like untoasted, light, medium and heavy toast are common, but there is no industry standard for toast level.

According to Vivas et al. (1991) the quality and quantity of each volatile compound are strongly related with toasting intensity, but particular characteristics of each species can also determine the rate of modification during toasting process. For instance, when comparing volatile composition of *Q. pyrenaica* and *Q. petraea* wood chips, the increase amount of compounds with the toasting process was less evident in the Portuguese oak (Jordão et al., 2006). These authors pointed out that physical properties and structure of wood may influence heat conduction and reactions upon heating. Caldeira et al., (2006 b) also found that as toasting intensity increases, concentration of the majority of volatile compounds found in wood matrix rises. In several oak species, as well as in Portuguese chestnut, increasing toasting level led to an increase of furanic aldehydes, volatile phenols, 4-hydroxy-but-2-en-lactone and vanillin (Caldeira et al., 2006 b)

It is also relevant that the response of a wood to a particular seasoning and toasting condition is determined by the size of the wood piece, as it affects their structural properties, and hence, the flavour characteristics. Moreover, each piece size shows different extraction kinetics when in contact with wines (Fernández de Simón et al., 2010 b).

Indeed, when heat is applied to wood, a depolymerization of the lignin takes place producing phenolic aldehydes. Therefore, their concentration increases with toast level. Toasting also produce the cleavage of $\alpha-\beta$ bonds of cinnamic aldehydes and their thermo-oxidation and thermo-decarboxylation, leading to the formation of dimethoxyphenyl units such as syringol, and in heavy toast, to simple phenols as phenol and *o*-cresol (Nonier et al., 2006).

Toasting is such an important parameter that the characteristic profile of a toasted wood is completely different from the untoasted wood. The particular characteristics of macromolecules, lignin, cellulose, hemicellulose and lipids in each wood have a great influence on the volatiles composition of toasted woods. Fernández de Simón et al., (2009) concluded that toasting led to an increase in almost all the compounds when compared to untoasted woods. These increases were more evident in acacia, chestnut and ash woods, concerning lignin, lipids and carbohydrate derivatives. Cherry and ash woods were found to be richer than toasted oak in lignin derivatives, but much poorer in lipid and carbohydrate derivatives.

3. Analytical methodologies for quantification and identification of volatiles compounds from woods

3.1 Sample preparation methods

Wine aroma is a very complex matrix comprising an enormous variety of compounds of many chemical families. Not all of them are odour active compounds, which imply the necessity of a target strategy in order to isolate the compounds of interest from the aqueous/alcoholic matrix.

Frequently, prior to analysis, samples are submitted to a preliminary preparation step, including isolation and concentration of the target compounds. The more used extraction and enrichment technique are liquid-liquid extraction (LLE), solid-phase extraction (SPE) and headspace, comprising the static/equilibration method (HS) and the dynamic method (purge & trap - P&T). These techniques have been recently reviewed by Costa Freitas et al. The quality of the subsequent analysis and results, depend largely on the isolation procedures. Different preparation methodologies might affect the final extract composition for the very same matrix. Nowadays researchers are focusing their attention not only to the extraction efficiency, of the methods used, but also to sustainable methods and thus more environmental friendly. Considering this, miniaturization has been applied and micro-LLE methods, such as dispersive liquid-liquid extraction (DLLME) (Fariña et al., 2007) or pressurized solvent extraction (Natali et al., 2006, Vichi et al., 2007), solid phase micro extraction methods (SPME) (Cai et al., 2009) and single drop microextraction (SDME)

(Martendal et al., 2007; Sillero et al., 2011) arise as prominent techniques for isolation of compounds in several liquid and solid matrices. Special attention has been given to methods which preclude the use of solvents, such as SPME (Bozalongo et al., 2007) and stir bar sorptive extraction (SBSE) (Marín et al., 2005; Tredoux et al., 2008; Rojas et al., 2009). These methods increase, in the case of gas chromatography (GC), the quality of the resulting chromatogram, allowing the detection of several volatile compounds that would, otherwise, be hidden under the solvent peak. As a definition one might state that preparation methods should be reproducible, and should provide an extract, as much as possible, similar to the original matrix.

Extracts obtained after exhaustive extraction techniques, which include solvent extraction and distillation, normally do not reflect the real matrix aroma profile, since they isolate compounds according, simultaneously, to solubility in the used solvent and volatility, considering the matrix properties (e.g. water or alcohol content). These methods provide a profile that reflects the volatile and semi-volatile composition of matrix rather then "real" aroma compounds of the matrix. Moreover, literature also agree, that only a small fraction of the isolated compounds can be correlated with sensory notes of the matrices (Plutowska & Wardencki, 2008). Consequently HS-based methods are currently the most used ones, since they allow the target isolation of the released volatiles from the matrix.

3.1.2 Extraction methods using solvents

In the analysis of wine aroma, the most used solvent extraction methods are LLE (Ortega et al., 2001; Garde-Cerdán & Ancín-Azpilicueta, 2006b), micro-wave assisted extraction (MAE) (Serradilla & de Castro, 2011), ultrasound assisted extraction (UAE) (Cocito et al., 1995; Fernandes et al., 2003), PLE (Natali et al., 2006; Vichi et al., 2007) and SPE (Morales et al., 2004; Campo et al 2007, Weldegergis et al., 2011). The type and quantity of the used organic solvent varies according to the methodology. In order to perform a classical LLE, eventually, a generous amount of solvent is used, and that is the reason why MAE, UAE and PLE became more popular, since the total amount of solvent can be significantly reduced. Pre-fractionation of the extract can also be subsequently performed (Fernandes et al, 2003) in order to simplify the extract for chromatographic analysis. Nevertheless, in all techniques, the obtained organic extract must be concentrated in order to be chromatographically analyzed. If gas chromatography hyphenated with olfactometry (GC-O) is going to be used, the organic extract should be rinsed with aqueous solutions, at different pH, in order to eliminate the presence of non-volatile compounds which are also extracted during the isolation process (Plutowska & Wardencki, 2008). The final extract has to be free of compounds which can produce artifacts during injection into the hot injector of the GC instrument, such as amino acids or fatty acids. The latter present a prolonged odor in the olfactometric port which can falsify the results of the analytes eluting subsequently. Thus a clean-up with resins should be performed. Before analysis a concentration step is also needed, in order to evaporate the excess of solvent prior to the chromatographic analysis. With this necessary step one can simultaneously promote the evaporation of the more volatile compounds and, eventually, odor components degradation, through oxidation processes, since air contact is almost unavoidable. Thus the final extract will not reproduce the original aroma matrix.

3.1.3 Solvent-less extraction methods

Extraction methods precluding the use of solvents have the main advantage of, in a single step, promoting extraction, isolation and concentration of the target compounds, without any other manipulation. At the same time, since it relies in the headspace of a sample, they will, indeed, mirroring the original sample and thus representing the true olfactory perception. This technique can be used in static or dynamic modes, or at a limit, the sample itself can be submitted to direct analysis without any previous treatment. In static mode, after reaching equilibrium, although not necessarily, sampling takes place and, normally, the more volatile compounds are successfully detected. Less volatile compounds, on the other, and depending on their individual vapour pressure, might be also detected. To overcome this drawback, dynamic sampling can be used. Here, purge & trap techniques, in which the compounds of interest are trapped in a suitable adsorbent are the most used. After sampling the trapped compounds can be directly desorbed, by means of a thermal desorber injector, or an extraction with a suitable solvent, before analysis, can be performed. Except for this latter case, the sample is completely lost after analysis, and no repeats are possible. However, no solvent peak is obtained, allowing detection of the more volatile components that, otherwise, could be co-eluting with the solvent peak. Other drawbacks of the dynamic mode should also be mentioned, namely the impact of the chemical nature of the adsorbent, on the trapped compounds. The most used trap materials are Tenax TA, Porapak Q and Lichrolut EN, since their affinity for ethanol is small (Escudero et al., 2007; Weldegergis et al, 2011). After the introduction of SPME (Belardi & Pawliszyn, 1989; Arthur & Pawliszyn, 1990), most of the HS methods started to use this method to perform aroma analysis. The obvious advantages of SPME are the fact that it does not involve critical sample manipulations. SPME is a simple and clean extraction method, comprising in a single step, all the necessary steps mandatory in aroma analysis: extraction, isolation and concentration (Rocha et al., 2007). In this technique, depending on the fibre physico-chemical properties (chemical nature and thickness), target analysis is, somehow, possible. However if a wide screening of the sample is aimed, more than one type of fibre should be tested. SPME demands a careful optimization of the experimental conditions in order to be used for qualitative and quantitative studies. The efficiency, accuracy and precision of the extraction methodology is directly dependent on extraction time, sample agitation, pH adjustment, salting out, sample and/or headspace volume, temperature of operation, adsorption on container walls and desorption conditions (Pawliszyn, 1997) besides fibre coating and thickness.

3.2 Chromatographic methods

The contact between wines and wood during the fermentation or storage process is a common practice in wine making. In order to improve the organoleptic characteristics of wine, storage in wood based barrels or the addition of wood chips is used in their natural form or after seasoning and thermal treatment by toasting or charring. The wood types for enological purposes comprises cherry, acacia, chestnut, mulberry, and especially, oak. Together with changes in colour and taste, the aroma of the resulting wine product is strongly affected (Garde-Cerdán & Ancín-Azpilicueta, 2006a). The porous nature of wood, promotes oxidation reactions, forming a wide variety of volatile compounds, otherwise not present in the final product. Some compounds are also directly extracted into the wine during the long lasting contact between the wood and the wine. Together with endogenous volatile compounds, they contribute to the final wine bouquet. The most representative compounds, which reflect the wood influence in wine aroma, are furfural derivates, such as furfural, 5-methylfurfural and furfuryl alcohol, phenolic aldehydes and ketones, such as vanillin and acetovanillone and a large variety of volatile phenols, such as guaiacol, 4-ethylphenol, 4-ethylguaiacol, 4-vinylphenol, 4-vinylguaiacol and eugenol and also phenolic acids (Tredoux et al., 2008; Weldergegis et al., 2011). The two oak lactones isomers, *cis*- and *trans*-β-methyl-γ-octalactone and γ-nonalactone are also present, especially if wood has been submitted to a previous seasoning or toasting process (Carrillo et al., 2006). Because they possess a very low sensory threshold, even in very low concentration, they strongly influence the final aroma. Liquid chromatographic (LC) methods, like capillary electrophoresis (CE) (Sádecká & Polonský, 2000), high performance liquid chromatography (HPLC) and ultra performance liquid chromatography (UPLC) are suitable methods to be used in order to separate, detect and quantify the above mentioned chemical classes, as recently reviewed (Cabrita et al., 2008; Kalili & de Villiers, 2011). The strong aromatic ring chromophore allows ease and sensitive detection by ultra-violet-visible (UV) detection devices or by mass spectrometry (MS). UPLC also reached better limits of detection (LOD) in the low ppb range, ten times higher for phenol compounds and furanic derivatives, when compared with HPLC methods (Chinnici et al., 2007). Considering the long-term storage in charred barrels, the presence of polycyclic aromatic hydrocarbons (PAHs) in aged alcoholic beverages has been proposed as an emerging issue. Although the transference rate, to the hydro-alcoholic mixture, of these high hydrophobic compounds is expected to be low, LODs less then 1 ppb are achievable when LC techniques are coupled with fluorimetric detection. Since the majority of the above mentioned compound classes are semi-volatile compounds, together with the endogenous compounds present in the wine, wine becomes an even more complex matrix to be analyzed as a whole. Therefore the analytical methodologies should be sufficiently powerful and sensitive to separate and detect hundreds of compounds of quite different concentrations and volatilities. The advent of gas chromatography (GC) significantly enhanced the systematic study of wine aroma, and its use, particularly when combined with mass spectrometry (MS), contributed significantly to the knowledge of wine volatile composition. Extracts from acacia, chestnut, cherry, mulberry, and oak wood, used in barrels for wine and spirits aging, have already been studied by GC/MS positive ion chemical ionization (PICI). These method, using methane as reagent gas, produce a high yield of the protonated molecular ion of volatile phenols, and allows compounds identification to be confirmed by collision-induced-dissociation (CID) experiments on $[M+H]^+$ species. MS/MS fragmentation patterns can be studied with standard compounds giving more accurate results (Flamini et al., 2007).

Considering that the human nose is still the best detector for aroma active compounds, the hyphenation of GC with olfactometry (GC-O) is mandatory, when sensorial activity, of individual compounds, and the knowledge of the relations between the perceived odour and chemical composition, of the volatile fraction, is searched. In GC-O detection is performed by an educated panel of persons and the qualitative and quantitative evaluation of the odour is carried out for each single compound eluting from the chromatographic column. For a given compound a relationship is established between its sensorial activity at a particular concentration (if above the threshold of sensory detection) and its smell, as well as the determination of the time period of sensory activity and the intensity of the odor. This is possible if in parallel to a conventional detector, e.g. flame ionization detector (FID), MS or even a flame photometric detection (FPD) for sulfur speciation (Chin et al., 2011), an olfactometric port is attached to the instrument in order to determine the compound odor. The splitted flow carries the compounds simultaneously to both detection devices, allowing the comparison between each obtained signals. The mass structural information obtained by MS spectra allows the identification of odour active compounds. The quantitative evaluation of the intensity of the odours can be divided in three groups: detection frequency (where NIF - Nasal Impact Frequency values or SNIF - Surface of Nasal Impact Frequency are measured), dilution to threshold (CHARM method - Combined Hedonic Aroma Response Measurement and AEDA - Aroma Extract Dilution Analysis) and finally direct intensity methods (where the intensity of the stimuli and its duration are measured – OSME - Olfactometry by Finger Span Method). All these methods were carefully revised by Plutowska & Wardencki (2008).

It is easily recognizable that GC is the most appropriate instrumental approach, for wine aroma analysis. However, the enormous quantity of peaks obtained together with the similarity of retention factors of many related components means that component overlap will be the general expectation, meaning that complete separation may be largely unachievable for a given extract. In order to improve analysis, new separation technologies should be embraced. It is desirable to increase peak capacity and thus increasing the probability of locating a greater number of baseline resolved components within the chromatographic run. Together with new stationary phases (more thermal and chemically stable and enantioselective) and faster detection systems, this will largely enhance results quality. Multidimensional chromatographic (MD) methods are emerging as important alternatives to analyze such complex matrix, as wines. Whether using the heart-cutting mode (MD-GC) (Campo et al., 2007; Schmarr et al., 2007; Pons et al., 2008), enantio-MDGC (Darriet et al, 2001; Culleré et al., 2009) or comprehensive two-dimensional GC (2D-GC) approach (GC × GC) (Ryan et al., 2005; Rocha et al., 2007; Torres et al., 2011; Weldegergis et al., 2011), MD methods allow the combination of two or more independent, or nearly independent, separation steps, increasing significantly the separation power of the corresponding one-dimensional (1D-GC) techniques and, therefore the physical separation of compounds in complex samples. When coupled with a preparative fraction collection device (selectable 1D/2D GC–O/MS with PFC) one can perform the enrichment of trace compounds with an high odourific impact (Ochiai & Sasamoto, 2011) in order to eventually proceed to further identification by NMR (Marriott et al., 2009). In order to overcome matrix complexity, a further dimensionality can be coupled to achieve separation efficiency. Methoxypyrazines were analyzed by LC-MDGC/MS (Schmarr et al., 2010a), nevertheless the complexity and price of the instrumental set-up encourages more research in this domain. 2D-LC × LC methods (Cacciola et al., 2007, Dugo et al., 2008; Cesla et al. 2009) are also replacing the traditional 1D-LC separation modes, especially in what phenolic volatile concerns. This technique improves the separation of non-volatile polyphenols (Kivilompolo et al., 2008), and allows ease switch between LC and LC × LC during a single analysis providing a better separation target, only, to the most complex part of the chromatogram.

Additionally to the enhanced separations achieved by 2D methods, comprehensive GC and LC also allow an easy comparison between samples, since the obtained chromatograms can be easily inspected for compound markers in quality control, certification and fraud detection. In fact, the probability of correctly determining a particular sample substance is increased, by the correlation of two or more independent data sets, in complex sample analysis. To generate these two sets of data, analysis has to be performed, individually, in two orthogonal columns phases, e.g. non-polar and a polar column for the case or GC × GC or e.g. exclusion and a reversed-phase column in LC × LC. Although this practice improves the quantitative data for key target components, it is often difficult, and may be tedious, to correlate the retention times for all of the peaks from the individual chromatograms. In any 2D chromatogram retention time data from the two independent stationary phases are immediately available. Thus a 2D chromatogram contains much more information than two independent 1D chromatograms. 2D analysis offers a genuine opportunity to characterize individual components based on retention time data alone, provided that high reproducibility of retention times can be achieved, reducing and eventually precluding, the requirement of MS detection for many samples. The reliability of peak position co-ordinates opens the possibility of fingerprinting comparison or statistical treatment in order to characterize the effects of distinct types of wood, the same wood from different origins, or even aging time (Cardeal et al., 2008; Vaz-Freire, 2009; Schmarr et al., 2010 b; Schmarr & Bernhardt, 2010).

4. Conclusion

Each wood species has a characteristic profile, regarding their volatile composition. Heating during processing of wood to be used in cooperage, specially toasting, has a great influence on biopolymers degradation, thus the resulting compounds. The volatile profile of an untoasted wood can be different from the same wood after toasting. Increasing toasting levels generally leads to an increase of volatile compounds but some can also suffer degradation. Compounds released from wood into wine or spirits depends on several factors such as type of wood, size of alternatives (*e.g.* chips and staves), time of contact, but also depends on the chemical characteristics of the beverage, specially the alcohol content.

Recent development in 2D analytical approaches, together with more sensitive and fast mass spectrometers, allow to gain insight in the volatile fraction of the wine/wood interaction system, aiming wine quality.

5. Acknowledgment

Authors wish to thank Fundação para a Ciência e Tecnologia, Ministério da Ciência, Tecnologia e Ensino Superior and Programa Operacional Ciência e Inovação for financial support PTDC/QUI-QUI/100672/2008.

6. References

Ancín-Azpilicueta, C.; Garde-Cerdán T.; Torrea, D. & Jimenez, N. (2004). Extraction of volatile compounds in model wine from different oak woods: Effect of SO_2. *Food Research International,* Vol. 37, pp. 375-383.

Arapitsas, P.; Antonopoulos, A.; Stefanou, E. & Dourtoglou, V.G.(2004). Artificial aging of wines using oak chips. *Food Chemistry*, Vol.4, pp. 563–570

Arthur, C. & Pawliszyn, J. (1990). Solid Phase Microextraction with Thermal Desorption Using Fused Silica Optical Fibers. *Analytical Chemistry*, Vol.62, pp. 2145-2148

Barata, A.; Campo, E; Malfeito-Ferreira, M.; Loureiro, V; Cacho, J. & Ferreira, V. (2011). Analytical and Sensorial Characterization of the Aroma of Wines Produced with Sour Rotten Grapes Using GC-O and GC-MS: Identification of Key Aroma Compounds. *Journal of Agricultural and Food Chemistry*, Vol. 59, pp. 2543–2553

Bautista-Ortín, A.B.; Lencina, A.G.; Cano-López, M.; Pardo-Minguez, F.; López-Roca, J.M. & Gómez-Plaza, E. (2008). The use of oak chips during the ageing of a red wine in stainless steel tanks or used barrels: Effect of the contact time and size of the oak chips on aroma compounds. *Australian Journal of Grape and Wine Research*, Vol. 14, pp. 63-70.

Belardi, R. P. & Pawliszyn, J. B. (1989). The Application of Chemically Modified Fused Silica Fibers in the extraction of Organics from Water Matrix Samples and their Rapid Transfer to capillary columns. *Water Pollution Research Journal of Canada*, Vol.24, pp. 179-189

Bozalongo, R.; Carrillo, J. D.; Torroba, M. A. F. & Tena, M. T. (2007). Analysis of French and American oak chips with different toasting degrees by headspace solid-phase microextraction-gas chromatography–mass spectrometry. *Journal of Chromatography A*, Vol.1173, pp. 10–17

Cabrita, M.J.; Torres, M^a.; Palma, V.; Alves, E.; Patão, R.; Costa Freitas, A. M^a. (2008). Impact of Malolactic Fermentation on Low Molecular Weight Phenolic Compounds. *Talanta*, Vol. 74, pp. 1281–1286

Cacciola, F.; Jandera, P.; Hajduá, Z.; Cĕsla, P. & Mondello, L. (2007). Comprehensive two-dimensional liquid chromatography with parallel gradients for separation of phenolic and flavone antioxidants. *Journal of Chromatography A*, Vol.1149, pp. 73–87

Cadahía, E.; Fernández de Simón, B.; & Jalocha, J. (2003). Volatile compounds in Spanish, French, and American oak woods after natural seasoning and toasting. *Journal of Agricultural and Food Chemistry*, Vol. 51, pp. 5923–5932.

Cadahía, E.; Muñoz, L.; Fernández de Simón, B. & García-Vallejo, M. C. (2001). Changes in low molecular weight phenolic compounds in Spanish, French, and American oak woods during natural seasoning and toasting. *Journal of Agricultural and Food Chemistry*, Vol. 49, pp. 1790–1798.

Cai, L.; Koziel, J. A.; Dharmadhikarib, M. & van Leeuwenc, J. (H). (2009). Rapid determination of *trans*-resveratrol in red wine by solid-phase microextraction with on-fiber derivatization and multidimensional gas chromatography–mass spectrometry. *Journal of Chromatography A*, Vol.1216, pp. 281–287

Caldeira, I.; Anjos, O.; Portal, V.; Belchior, A.P. & Canas, S. (2010). Sensory and chemical modifications of wine-brandy aged with chestnut and oak wood fragments in comparison to wooden barrels. *Analytica Chimica Acta*, Vol. 660, pp. 43–52

Caldeira, I.; Bruno de Sousa, R.; Belchior, A.P. & Clímaco, M.C. (2008). A sensory and chemical approach to the aroma of wooden aged Lourinha wine brandy. *Journal of Viticulture and Enology*, Vol.23, pp.97-110

Caldeira, I.; Pereira, R.; Clímaco, M.C.; Belchior, A.P. & Bruno de Sousa, R. (2004). Improved method for extraction of aroma compounds in aged brandies and aqueous alcoholic wood extracts using ultrasound. *Analytica Chimica Acta*, Vol. 513, pp. 125–134

Caldeira, I.; Clímaco, M.C.; Bruno de Sousa, R. & Belchior A.P. (2006 b) Volatile composition of oak and chestnut woods used in brandy ageing: modification induced by heat treatment. *Journal of Food Engineering*, Vol. 76, pp. 202-211

Caldeira, I.; Mateus, A.M. & Belchior A.P. (2006 a). Flavour and odour profile modifications during the first five years of Lourinhâ brandy maturation on different wooden barrels. *Analytica Chimica Acta*, Vol. 563, pp. 265-273

Campo, E.; Cacho, J. & Ferreira, V. (2007). Solid phase extraction, multidimensional gas chromatography mass spectrometry determination of four novel aroma powerful ethyl esters. Assessment of their occurrence and importance in wine and other alcoholic beverages. *Journal of Chromatography A*, Vol.1140, pp. 180–188

Cardeal, Z. L.; de Souza, P. P.; Gomes da Silva, M. D. R. & Marriott, P. J. (2008). Comprehensive two-dimensional gas chromatography for fingerprint pattern recognition in *cachaça* production. *Talanta* Vol.74, pp. 793–799

Carrillo, J.D.; Garrido-López, A. & Tena, M.T. (2006) Determination of volatile oak compounds in wine by headspace solid-phase microextraction and gas chromatography-mass spectrometry *Journal of Chromatography A*, Vol. 1102, pp. 25-36

Cĕsla, P. Hájek, T. & Jandera, P. (2009). Optimization of two-dimensional gradient liquid chromatography separations. *Journal of Chromatography A*, Vol. 1216, pp. 3443–3457

Chatonnet, P. & Dubourdieu, D. (1998) Comparative study of the characteristics of American white oak (*Quercus alba*) and European oak (*Quercus petraea* and *Q. robur*) for production of barrels used in barrel aging of wines. *American Journal of Enology and Viticulture*, Vol 49, pp. 79-85.

Chatonnet, P. ; Cutzach, I. ; Pons, M. & Dubourdieu, D. (1999). Monitoring toasting intensity of barrels by chromatographic analysis of volatile compounds from toasted oak wood. *Journal of Agricultural and Food Chemistry*, Vol. 47, pp. 4310–4318.

Chin, S. T.; Eyres, G. T. & Marriott, P. J. (2011). Identification of potent odourants in wine and brewed coffee using gas chromatography-olfactometry and comprehensive two-dimensional gas chromatography. *Journal of Chromatography A. Vol. 1218, pp.7487-7498*

Chinnici, F.; Natali, N.; Spinabelli, U. & Riponi, C. (2007). Presence of polycyclic aromatic hydrocarbons in woody chips used as adjuvant in wines, vinegars and distillates *LWT*, Vol. 40, pp. 1587–1592

Cormier, F.; Raymond, Y. ; Champagne, C. P. & Morin, A. (1991). Analysis of odor-active volatiles from Pseudomonas fragi grown in milk. *Journal of Agricultural and Food Chemistry*, Vol. 39, pp. 159–161.

Costa Freitas, A. M.; Gomes da Silva, M. D. R. & Cabrita M. J. (accepted). Extraction Techniques and Applications: Food and Beverage. Sampling techniques for the determination of volatile components in grape juice, wine and alcoholic beverages", in *Comprehensive Sampling and Sample Preparation*, J. Pawliszyn (Editor), Elsevier

Cocito, C.; Gaetano, G. & Delli, C. (1995). Rapid extraction of aroma compounds in must and wine by means of ultrasound. *Food Chemistry*, Vol.52, pp. 311-320

Culleré, L.; Escudero, A.; Campo, E.; Cacho, J. & Ferreira V. (2009). Multidimensional gas chromatography–mass spectrometry determination of 3-alkyl-2-methoxypyrazines in wine and must. A comparison of solid-phase extraction and headspace solid-phase extraction methods. *Journal of Chromatography A*, Vol.1216, pp. 4040–4045

Cutzach, I. ; Chatonnet, P. ; Henry, R. & Dubourdieu, D. (1997). Identification of volatile compounds with a "toasty" aroma in heated oak used in barrelmaking. *Journal of Agricultural and Food Chemistry*, Vol. 45, pp. 2217–2224.

Darriet, P.; Lamy, S.; La Guerche, S.; Pons, M.; Dubourdieu, D.; Blancard, D.; Steliopoulos, P. & Mosandl, A. (2001). Stereodifferentiation of geosmin in wine *European Food Research & Technology*, Vol.213, pp. 122–125

de Rosso, M.; Cancian, D.; Panighel, A.; Vedova, A.D. & Flamini, R. (2009). Chemical composition released from five different woods used to make barrels for aging wines and spirits: volatile compounds and polyphenols. *Wood Science Technology*, Vol 43, pp. 375-385

Díaz-Maroto, M.C.; Guchu, E.; Castro-Vásquez, L.; de Torres, C. & Pérez-Coello, M.S. (2008) Aroma-active compounds of American, French, Hungarian and Russian oak woods, studied by GC-MS and GC-O. *Flavour and Fragance Journal*, Vol.23, pp. 93-98

Doussot, F. ; De Jeso, B. ; Quideau, S. & Pardon, P. (2002). Extractives content in cooperage oak wood during natural seasoning and toasting; influence of tree species, geographic location, and single-tree effect. *Journal of Agricultural and Food Chemistry*, Vol. 50, pp. 5955–5961.

Dugo, P.; Cacciola, F. Herrero, M.; Donato, P. & Mondello, L. (2008). Use of partially porous column as second dimension in comprehensive two-dimensional system for analysis of polyphenolic antioxidants. *Journal of Separation Science*, Vol.31, pp. 3297–3308

Escalona, H.; Birkmyre, L.; Piggot, J.R. & Paterson, A. (2002) Effect of maturation in small oak casks on the Volatility of red wine aroma compounds. *Analytica Chimica Acta*, Vol. 458, pp. 45-54

Escudero, A.; Campo, E.; Fariña, L.; Cacho, J. & Ferreira, V. (2007). Analytical Characterization of the Aroma of Five Premium Red Wines. Insights into the Role of Odor Families and the Concept of Fruitiness of Wines. *Journal of Agricultural and Food Chemistry*, Vol.55, pp. 4501–4510

Fariña, L.; Boido, E.; Carrau, F. & Dellacassa, E. (2007). Determination of volatile phenols in red wines by dispersive liquid–liquid microextraction and gas chromatography–mass spectrometry detection. *Journal of Chromatography A*, Vol.1157, pp. 46-50

Fernandes, L.; Relva, A. M.; Gomes da Silva, M. D. R. & Costa Freitas, A. M. (2003). Different multidimensional chromatographic approaches applied to the study of wine malolactic fermentation. *Journal of Chromatography A*, Vol. 995, pp. 161-169

Fernández de Simón, B.; Muiño, I. & Cadahía, E. (2010a). Characterization of volatile constituints in commercial oak wood chips. *Journal of Agricultural and Food Chemistry,* Vol. 58, pp. 9587-9596

Fernández de Simón, B.; Cadahía, E.; del Álamo, M. & Nevares, I. (2010b) Effect of size, seasoning and toasting in the volatile compounds in toasted oak wood and in a red wine treated with them. *Analytica Chimica Acta,* Vol. 660 pp. 211-220

Fernández de Simón, B.; Esteruelas, E.; Muñoz, A.M.; Cadahía, E. & Sanz, M. (2009). Volatile Compounds in Acacia, Chestnut, Cherry, Ash, and Oak Woods, with a View to Their Use in Cooperage. *Journal of Agricultural and Food Chemistry,* Vol. 57,pp. 3217-3227

Ferreras, D.; Fernández, E. & Falqué, E. (2002) Note: Effects of oak wood on the aromatic composition of *Vitis vinifera* L. var. *treixadura* wines. *Food Science and Technology International,* Vol. 8, pp. 343-349

Flamini, R.; Vedova, A.D.; Cancian, D.; Panighel, A. & de Rosso, M. (2007) GC/MS-positive ion chemical ionization and MS/MS study of volatile benzene compounds in five different woods used in barrel making. *Journal of Mass Spectrometry,* Vol.42, pp.641-646

Frangipane M.Y.; de Santis, D. & Ceccarelli, A. (2007) Influence of oak woods of different geographical origins on quality of wines aged in barriques and using oak chips. *Food Chemistry,* Vol.103, pp. 46-54

Garde-Cerdán, T., & Ancín-Azpilicueta, C. (2006a). Review of quality factors on wine ageing in oak barrels. *Trends in Food Science &Technology,* Vol.17, pp. 438-447

Garde-Cerdán, T., & Ancín-Azpilicueta, C. (2006b). Effect of oak barrel type on the volatile composition of wine: Storage time optimization. *Lebensmittel-Wissenschaft und-Technologie,*Vol. 39, pp. 199-205

Garde-Cerdán, T.; Rodríguez-Mozaz, S. & Ancín-Azpilicueta, C. (2002). Volatile composition of aged wine in used barrels of French oak and of American oak. *Food Research International,* Vol. 35, pp. 603-610

Garde-Cerdán, T.; Torrea-Goñi, D. & Ancín-Azpilicueta, C. (2004). Acumulation of volatile compounds during ageing of two red wines with different composition. *Journal of Food Engineering,* Vol. 65, pp. 349-356

Giménez-Martínez, R.; López-García de la Serrana, H.; Villalón-Mir, M.: Quesada-Granados, J. & López-Martínez, M. C. (1996). Influence of wood heat treatment, temperature and maceration time on vanillin, syringaldehyde, and gallic acid contents in oak word and wine spirit mixtures. *American Journal of Enology and Viticulture,* Vol. 47, pp. 441-446.

Gomes da Silva, M. D. R. & Chaves das Neves, H. J. (1999). Complementary Use of Hyphenated Purge-and-Trap Gas Chromatography Techniques and Sensory Analysis in the Aroma Profiling of Strawberries (Fragaria ananassa). *Journal of Agricultural and Food Chemistry,* Vol. 47, pp. 4568-4573.

Gómez-Plaza, E.; Pérez-Prieto, L.J.; Fernández-Fernández, J.I. & López-Roca, J. M. (2004). The effect of successive uses of oak barrels on the extraction of oak-related volatile compounds from wine. *International Journal of Food Science & Technology*, Vol. 39, pp. 1069-1078.

Guchu, E.; Díaz-Maroto, M.C.; Pérez-Coello, M.S.; González-Viñas, M.A.; Cabezudo Ibáñez, M.D. (2006) Volatile composition and sensory characteristics of Chardonnay wines treated with American and Hungarian oak chips. *Food Chemistry*, Vol. 99, pp. 350–359

Jarauta, I.; Cacho, J. & Ferreira, V. (2005). Concurrent phenomena contributing to the formation of the aroma of wine during aging in oak wood: An analytical study. *Journal of Agricultural and Food Chemistry*, Vol. 53, pp. 4166-4177

Jiménez-Moreno, N. & Ancín-Azpilicueta, C. (2007). Binding of oak volatile compounds by wine lees during simulation of wine ageing. *Lebensmittel-Wissenschaft und-Technologie*,Vol. 40, pp. 619-624.

Jiménez-Moreno, N.; González-Marco, A. & Ancín-Azpilicueta, C. (2007). Influence of wine turbidity on the accumulation of volatile compounds from the oak barrels. *Journal of Agricultural and Food Chemistry*, Vol. 55, pp. 6244-6251.

Jordão, A.M.; Ricardo da Silva, J.; Laureano, O.; Adams, A.; Demyttenaere, J.; Verhé, R. & De Kimpe, N. (2006). Volatile composition analysis by solid-phase microextraction applied to oak wood used in cooperage (*Quercus pyrenaica* and *Quercus Petraea*): effect of botanical species and toasting process. *Journal of Wood Science*, Vol. 52, pp. 514-521

Kalili, K. M. & de Villiers, A. (2011). Recent developments in the HPLC separation of phenolic compounds. *Journal of Separation Science*, Vol.34, pp. 854–876

Kivilompolo, M.; Obůrka, V. & Hyötyläinen, T. (2008). Comprehensive two-dimensional liquid chromatography in the analysis of antioxidant phenolic compounds in wines and juices. *Analytical & Bioanalytical Chemistry*, Vol.391, pp. 373–380

Madrera, R.R.; Gomis, D.B. & Alonso J.J.M. (2003). Influence of Distillation System, Oak Wood Type, and Aging Time on Volatile Compounds of Cider Brandy. *Journal of Agricultural and Food Chemistry*, Vol. 51, pp. 5709-5714

Marco, J.; Artajona, J.; Larrechi, M. S. & Rius, F. X. (1994). Relationship between geographical origin and chemical composition of wood for oak barrels. *American Journal of Enology and Viticulture*, Vol. 45, pp. 192–200.

Marín, J.; Zalacain, A.; De Miguel, C.; Alonso, G. L. & Salinas, M. R. (2005). Stir bar sorptive extraction for the determination of volatile compounds in oak-aged wines. *Journal of Chromatography A*, Vol.1098, pp. 1–6

Marriott, P. J.; Eyres, G. T. & Dufour, J. -P. (2009). Emerging Opportunities for Flavor Analysis through Hyphenated Gas Chromatography, *Journal of Agricultutal and Food Chemistry*, Vol.57, pp. 9962–9971

Martendal, E.; Budziak, D. & Carasek, E. (2007). Application of fractional factorial experimental and Box-Behnken designs for optimization of single-drop microextraction of 2,4,6-trichloroanisole and 2,4,6-tribromoanisole from wine samples. *Journal of Chromatography A*, Vol. 1148, pp. 131–136

Masson, G.; Guichard, E.; Fournier, N. & Puech, J. L. (1995). Stereoisomers of β-methyl-γ-octalactone. 2. Contents in the wood of French (*Quercus robur* and *Quercus petraea*) and American (*Quercus alba*) oaks. *American Journal of Enology and Viticulture,* Vol. 46, pp. 424- 428.

Masson, G.; Guichard, E.; Fournier, N. & Puech, J. L. (1997) The β-methyl-γ-octalactone stereoisomer contents of European and American oak. Applicable to wines and spirits. *Journal des Sciences et Techniques de la Tonnellerie,* Vol. 3, pp. 9-15.

Morales, M.L.; Benitez, B. & Troncoso, A. M. (2004). Accelerated aging of wine vinegars with oak chips: evaluation of wood flavour compounds. *Food Chemistry,* Vol.88, pp. 305-315

Mosedale, J.R. & Puech, J.-L. (1998) Wood maturation of distelled beverages. *Trends in Food Science & Technology,* Vol. 9, pp. 95-101

Natali, N.; Chinnici, F. & Riponi, C. (2006). Characterization of Volatiles in Extracts from Oak Chips Obtained by Accelerated Solvent Extraction (ASE). *Journal of Agricultural Food Chemistry,* Vol.54, pp. 8190-8198

Netto, C.C.; Moreira, R.F.A. & De Maria, C.A.B. (2003). Note: Volatile profile from Caninha aged in oak (*Quercus* sp) and balsam (*Myroxylon,* sp) barrels. *Food science and Technology International,* Vol. 9, pp. 59-64

Nonier, M.F. ;Vivas, N. ; Vivas de Gaulejac, N. ; Absalon, C.; Soulie, Ph & Fouquet, E. (2006) Pyrolysis-gas chromatography/mass spectrometry of Quercus sp. wood. Application to structural elucidation of macromolecules and aromatic profiles of different species, *Journal of Analytical and Applied Pyrolysis,* Vol. 75, pp. 181–193.

Ochiai. N & Sasamoto, K. (2011). Selectable one-dimensional or two-dimensional gas chromatography–olfactometry/mass spectrometry with preparative fraction collection for analysis of ultra-trace amounts of odor compounds. *Journal of Chromatography A,* Vol.1218, pp. 3180–3185

Ortega, C.; Lopez, R.; Cacho, J. & Ferreira, V. (2001). Fast analysis of important wine volatile compounds: Development and validation of a new method based on gas chromatographic-flame ionization detection analysis of dichloromethane microextracts. *Journal of Chromatography,* Vol.923, pp. 205–214.

Pawliszyn, J. (1997). *Solid Phase Microextraction: Theory and Practice.* Wiley-VCH, Inc., New York, USA, 242 pp..

Pérez-Coello, M. S.; Sanz, J. & Cabezudo, D. (1999) Determination of volatile compounds in hydroalcoholic extracts of French and American oak wood. *American Journal of Enology and Viticulture,* Vol. 50 pp. 162- 165

Pérez-Coello, M.S.; Sánchez, M. A.; García, E.; González-Viñas, M.A.; Sanz, J. &. Cabezudo, M.D. (2000). Fermentation of White Wines in the Presence of Wood Chips of American and French Oak. *Journal of Agricultural and Food Chemistry,* Vol. 48, pp. 885-889

Pérez-Prieto, L.J.; Lopez-Roca, J.M.; Martínez-Cutillas, A.; Pardo-Minguez, F. & Gómez-Plaza, E. (2003). Extraction and formation dynamic of oak-related volatile compounds from different volume barrels to wine and their behavior during bottle storage. *Journal of Agricultural and Food Chemistry,* Vol. 51, pp. 5444-5449.

Pérez-Prieto, L.J.; López-Roca, J.M.; Martínez-Cutillas, A.; Pardo-Minguez, F. & Gómez-Plaza, E. (2002). Maturing wine in oak barrels. Effects of origin, volume, and age of the barrel on the wine volatile composition. *Journal of Agricultural and Food Chemistry*, Vol. 50, pp. 3272-3276.

Piggott, J.R. ; Conner, J.M. & Paterson, A. (1995). *Flavour development in whisky maturation*. in Food Flavors: Generation, Analysis and Process Influence, G. Charalambous (Ed.), pp. 1731 - 1751

Plutowska, B & Wardencki, W. (2008). Application of gas chromatography–olfactometry (GC–O) in analysis and quality assessment of alcoholic beverages – A review, *Food Chem*istry, Vol.107, pp. 449–463

Pons, A.; Lavigne, V.; Eric, F.; Darriet, P. & Dubourdieu, D. (2008). Identification of Volatile Compounds Responsible for Prune Aroma in Prematurely Aged Red Wines. *Journal of Agricultural Food Chemistry*, Vol.56, pp. 5285–5290

Prida, A. & Puech, J.L. (2006) Influence of Geographical Origin and Botanical Species on the Content of Extractives in American, French, and East European Oak Woods. *Journal of Agricultural and Food Chem*istry, Vol. 54, pp. 8115-8126

Puech, J.L. (1987). Extraction of phenolic compounds from oak wood in model solutions and evolution of aromatic aldehydes in wine aged in oak barrels. *American Journal of Enology and Viticulture*, Vol. 38, pp. 236-238

Rocha, S. M.; Coelho, E.; Zrostlíková, J. Delgadillo, I & Coimbra, M. A. (2007). Comprehensive two-dimensional gas chromatography with time-of-flight mass spectrometry of monoterpenoids as a powerful tool for grape origin traceability. *Journal of Chromatography A*, Vol.1161, pp. 292–299

Rodríguez-Bencomo, J.J.; Ortega-Heras, M.; Pérez-Magariño, S. & Gónzalez-Huerta (2009). Volatile compounds of red wines macerated with Spanish, American, and French oak chips. *Journal of Agricultural and Food Chemistry*, Vol. 57, pp. 6383-6391

Rodríguez-Bencomo, J.J.; Ortega-Heras, M.; Pérez-Magariño, S.; Gónzalez-Huerta, C. & Gónzalez-Sanjosé, M.L. (2008). Importance of chip selection and elaboration process on the aromatic composition of finished wines. *Journal of Agricultural and Food Chemistry*, Vol. 56, pp. 5102-5111

Rodríguez-Rodríguez, P. & Gómez-Plaza, E. (2011). Differences in the Extraction of Volatile Compounds from Oak Chips in Wine and Model Solutions *American Journal of Enology and Viticulture*, Vol. 62, pp. 127-132

Rojas, F. S.; Ojeda, C. B. & , Pavón, J. M. C. (2009). A Review of Stir Bar Sorptive Extraction. *Chromatographia*, Vol. 69, pp. S79-S94

Ryan, D.; Watkins, P.; Smith, J.; Allen, M. & Marriott, P. (2005). Analysis of methoxypyrazines in wine using headspace solid phase microextraction with isotope dilution and comprehensive two-dimensional gas chromatography. *Journal of Separation Science*, Vol.28, pp. 1075–1082

Sádecká, J. & Polonský. J. (2000). Electrophoretic methods in the analysis of beverages. *Journal of Chromatography A*, Vol. 880, pp.243–279

Sáenz-Navajas M-P; Campo, E.; Fernández-Zurbano, P.; Valentin, D. & Ferreira, V. (2010). An assessment of the effects of wine volatiles on the perception of taste and astringency in wine. *Food Chemistry,* Vol. 121, pp. 1139–1149.

Schmar, H. -G; Ganß, S.; Koschinski, S.; Fischer, U; Riehle, C.; Kinnart, J.; Potouridis, T. & Kutyrev, M. (2010 a). Pitfalls encountered during quantitative determination of 3-alkyl-2-methoxypyrazines in grape must and wine using gas chromatography-mass spectrometry with stable isotope dilution analysis. Comprehensive two-dimensional gas chromatography-mass spectrometry and on-line liquid chromatography-multidimensional gas chromatography-mass spectrometry as potential loopholes. *Journal of Chromatography A,* Vol. 217, pp. 6769–6777

Schmarr, H.-G & Bernhardt, J. (2010). Profiling analysis of volatile compounds from fruits using comprehensivetwo-dimensional gas chromatography and image processing techniques. *Journal of Chromatography A,* Vol.1217, pp. 565–574

Schmarr, H.-G.; Ganß, S.; Sang, W. & Potouridis, T. (2007). Analysis of 2-aminoacetophenone in wine using a stable isotope dilution assay and multidimensional gas chromatography-mass spectrometry. *Journal of Chromatography A,* Vol.1150, pp. 78–84

Schmarr, H.-G; Bernhardt, J.; Fischer, U.; Stephan, A.; Müller, P. & Durner, D. (2010 b). Two-dimensional gas chromatographic profiling as a tool for a rapid screening of the changes in volatile composition occurring due to microoxygenation of red wines. *Analytica Chimica Acta,* Vol.672, pp. 114–123

Sefton, M.; Francis, I.; Williams, P. (1990). Volatile norisoprenoid compounds as constituents of oak woods used in wine and spirit maturation. *Journal of Agricultural and Food Chemistry,* Vol 38, pp. 2045-2049

Sefton, M.A., Spillman, P.J.; Pocock, K.F.; Francis, I.L. & Williams, P.J. (1993). The influence of oak origin, seasoning, and other industry practices on the sensory characteristics and composition of oak extracts and barrel-aged white wines. *Australian Grapegrower & Winemaker,* Vol. 355, pp. 17-25

Serradilla, J. A. P. & de Castro, M. D. L. (2011). Microwave-assisted extraction of phenolic compounds from wine lees and spray-drying of the extract. *Food Chemistry,* Vol.124, pp. 1652–1659

Sillero, I. M.; Herrador, E. A.; Cárdenas, S. & Valcárcel, M. (2011). Determination of 2,4,6-tricholoroanisole in water and wine samples by ionic liquid-based single-drop microextraction and ion mobility spectrometry. *Analytica Chimica Acta,* Vol.702, pp. 199- 204

Singleton, V.L., & Draper, D.E. (1961). Wood chips and wine treatment: The nature of aqueous alcohol extracts. *American Journal of Enology and Viticulture,* Vol.12, pp.152-158

Togores, J. H. (2004). *Tratado de enologia Tomo 2.* Mundiprensa Ed.

Torres, M.P.; Cabrita, M.J.; Gomes da Silva, M. D. R.; Palma, V. & Costa Freitas, A. Mª. (2011).The Impact of the Malolactic Fermentation in the Volatile Composition of Trincadeira Wine Variety. *Journal of Food Biochemistry,* Vol. 35, pp. 898-913.

Tredoux, A.; de Villiers, A.; Májek, P.; Lynen, F.; Crouch, A. & Sandra, P. (2008). Stir Bar Sorptive Extraction Combined with GC-MS Analysis and Chemometric Methods for the Classification of South African Wines According to the Volatile Composition, *Journal of Agricultural and. Food Chemistry,* Vol. 56, pp. 4286–4296

van Jaarsveld, F.P.; Hattingh, S. & Minnaar, P. (2009). Rapid induction of ageing character in brandy products - Part III. Influence of toasting. *South African Journal for Enology & Viticulture,* Vol. 30, pp. 24-37

Vaz-Freire, L. T.; Gomes da Silva, M. D. R. & Costa Freitas, A. M. (2009). Comprehensive two-dimensional gas chromatography for fingerprint pattern recognition in olive oils produced by two different techniques in Portuguese olive varieties *Galega Vulgar, Cobrançosa* e *Carrasquenha. Analytica Chimica Acta,* Vol.633, pp. 263–270

Vichi, S.; Santini, C.; Natali, N.; Riponi, C.; Tamames, E. L. & Buxaderas, S. (2007). Volatile and semi-volatile components of oak wood chips analysed by Accelerated Solvent Extraction (ASE) coupled to gas chromatography–mass spectrometry (GC–MS). *Food Chemistry,* Vol.102, pp. 1260–1269

Vivas, N. Glories, Y.; Doneche, B & Gueho, E. (1991). Observation on the oak wood microflora (Quercus sp) during its natural air drying. Annales des Sciences Naturelles- Botanique et Biologie Vegetale, Vol. 11, pp. 149-153

Waterhouse, A.L., & Towey, J.P. (1994). Oak lactone isomer ration distinguishes between wine fermented in American and French oak barrels. *Journal of Agricultural and Food Chemistry,* Vol. 42, pp. 1971-1974

Weldegergis, B. T.; Croucha, A. M.; Górecki, T. & de Villiers, A. (2011). Solid phase extraction in combination with comprehensive two-dimensional gas chromatography coupled to time-of-flight mass spectrometry for the detailed investigation of volatiles in South African red wines. *Analytica Chimica Acta,* Vol.701, pp. 98- 111

Yokotsuka, K; Matsunaga, M. & Singleton, V.L (1994). Comparision of composition of Koshu white wines fermented in oak barrels and plastic tanks. *American Journal of Enology and Viticulture,* Vol. 45, pp. 11-16

Zhao, Y.P.; Wang, L.; Li, J. M.; Pei G. R. & Liu, Q. S. (2011) Comparison of Volatile Compounds in Two Brandies Using HS-SPME Coupled with GC-O, GC-MS and Sensory Evaluation- *South African Journal of Enology and Viticulture,* Vol. 32, pp. 9-20

Part 3

Selected Application of Gas Chromatography in Life Sciences

11

GC Analysis of Volatiles and Other Products from Biomass Torrefaction Process

Jaya Shankar Tumuluru[1,*], Shahab Sokhansanj[2], Christopher T. Wright[1] and Timothy Kremer[3]

[1]*Biofuels and Renewable Energies Technologies, Idaho National Laboratory, Idaho Falls,*

[2]*Bioenergy Resource and Engineering Systems Group, Environmental Sciences Division, Oak Ridge National Laboratory, Oak Ridge,*

[3]*Chemical and Biomolecular Engineering Department, The Ohio State University, Columbus, USA*

1. Introduction

Gas chromatography (GC) is a common method used to analyze gases produced during various chemical processes. Torrefaction, for example, is a method for pretreating biomass to make it more suitable for bioenergy applications that uses GC to characterize products formed during the process. During torrefaction, biomass is heated in an inert environment to temperatures ranging between 200–300°C. Torrefaction causes biomass to lose low-energy condensables (liquids) and non-condensable volatiles, initially in gas form, thereby making biomass more energy dense.

The condensable volatiles are divided into three subgroups: reaction water, organics, and lipids. The first subgroup is reaction water, which contains free and bound water released from the biomass by evaporation. The second subgroup consists of organics (in liquid form) that are mainly produced during devolatilization, depolymerization and carbonization reactions in the biomass. The final subgroup, lipids, consists of compounds that are present in the original biomass, such as waxes, fatty acids, and non-condensable gases—mainly CO_2, CO, and small amounts of methane. Knowing the composition of volatiles produced in the torrefaction temperature range can shed light on the raw material adaption process, process control, process behavior and operation, energy process and energy optimization, and green chemical production.

In general, GC with mass spectroscopy is used for both condensable and non-condensable gases. GC configuration plays an important role in accurately identifying the compounds in these gases. A combination of different detectors—like thermal, flame, and photo ionization detectors—combined with mass spectrometers are used for profiling both condensable and non-condensable gases.

---

* Corresponding Author

2. Gas Chromatography

Chromatography is an important analytical tool that allows for the separation of components in a gas mixture. GC is a common type of chromatography used to separate and analyze compounds that can be vaporized without decomposition. Typical uses of GC include testing the purity of a particular substance or separating the different components and relative amounts of different components of a mixture. GC can also be used to prepare pure compounds from a mixture (Pavia et al., 2006). In GC, the mobile phase (or "moving phase") is a carrier gas, usually an inert gas such as helium or an unreactive gas such as nitrogen. The stationary phase is a microscopic layer of liquid or polymer on an inert solid substrate inside a glass or metal tube called a column. The gases are analyzed as they interact with the column walls, which have been coated with different stationary phases. This coating results in the compounds eluting at different times, called the retention time for each compound. These compounds are then further analyzed by comparison with calibrated standard gases (Pavia et al., 2006).

The working principle of GC is similar to that of fractional distillation, as both processes use temperature to separate the components. The only difference is that distillation is used in large-scale applications, where GC is only used in smaller-scale applications (Pavia et al., 2006). Because of its mechanisms for analysis, GC is also sometimes known as vapor-phase chromatography (VPC), or gas–liquid partition chromatography (GLPC) (Pavia et al., 2006).

3. GC components

Figure 1 shows the typical components of GC system: the sample injector, column, detector, and carrier gases.

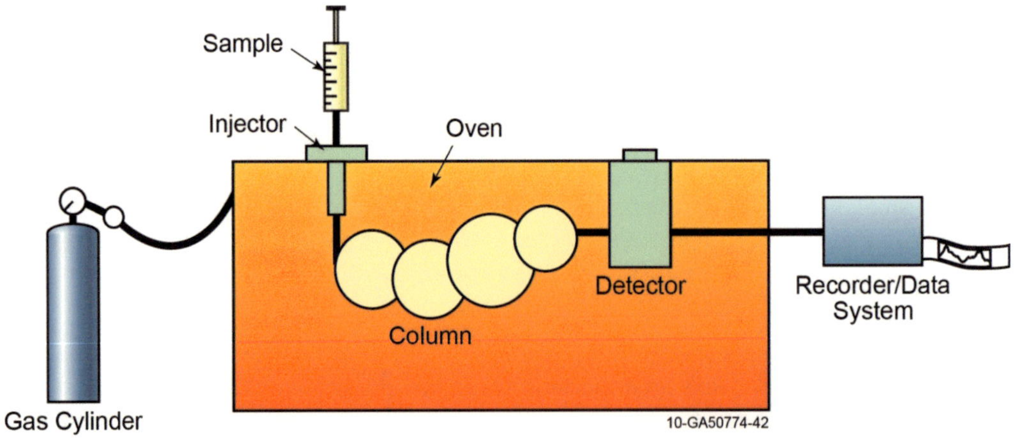

Fig. 1. Schematic view of gas chromatography (Punrattanasin and Spada, 2011)

3.1 Types of samplers

Manual insertion of a sample in a GC system for analysis is possible, but is no longer preferred for reasons of accuracy. In fact, manual injection is avoided where there is a need for high accuracy and precise results. A better option is an auto-sampler that injects samples automatically into the inlets of the GC system. Automatic injection helps achieve consistency and reproducibility and reduces analysis time (Pavia et al., 2006). The several types of automatic samplers are:

- Liquid injection
- Static head-space by syringe technology
- Dynamic head-space by transfer-line technology
- Solid phase microextraction (SPME).

3.2 Types of inlets

An inlet is hardware attached to the column head that introduces the samples into the continuous flow carrier gas. Common types of inlets are split/splitless (S/SL), on-column, programmed temperature vaporization (PTV) injectors, gas switching valve, and purge and trap systems.

An S/SL injector introduces a sample into a heated chamber via a syringe through a septum, and the heat helps volatilization of the sample and sample matrix. In split mode, the sample/carrier gas is exhausted into the chamber through the split vent, and this is preferred when the sample has high analyte concentrations (>0.1%). In the case of the splitless mode, the valve opens after a pre-set time to purge heavier elements in order to prevent the contamination of the system. An on-column inlet is a simple component through which the sample is introduced directly into the column without heat. A PTV injector introduces a sample into the liner at a controlled injection rate. The temperature of the liner is usually slightly below the boiling point of the solvent so that the solvent is continuously evaporated and vented through the split line. PTV injectors are particularly suited to in-liner derivatisation due to the flexibility of control over parameters such as injection volume (Bosboom et al., 1996), carrier gas flow and liner temperature (Poy et al., 1981). A gas-source inlet (or gas-switching valve) injects gaseous samples into the collection bottles through a six-port switching valve. Upon switching, the contents of the sample loop are inserted into the carrier gas stream. Finally, the purge-and-trap system purges insoluble volatile chemicals from the matrix by bubbling inert gas through an aqueous sample. The volatiles are 'trapped' on an adsorbent column (also known as a trap or concentrator) at ambient temperature. The trap is then heated, and the volatiles are directed into the carrier gas stream (Pavia et al., 2006; Kaufmann, 1997).

3.3 Types of detectors

A number of detectors may be selected in gas chromatography, based on the sample to be separated. The most common are the flame ionization detector (FID) and the thermal conductivity detector (TCD). These detectors are sensitive to a wide range of components and work for wide-ranging concentrations. TCDs are essentially universal and can be used to detect any component other than the carrier gas so long as the thermal conductivities are different from those of the carrier gas at a specific detector temperature. TCD has a detection limit of ~100 ppm (Goorts, 2008).

FIDs are more sensitive than TCDs and have a detection limit of ~1 ppm (Goorts, 2008), but their major limitation is the number of molecules they can successfully detect. FIDs can only detect combustible carbon atoms; therefore, FIDs are primarily used for detecting organic molecules. FIDs cannot detect H_2O or CO_2, so they are undesirable for some applications (Goorts, 2008). As TCD analysis is non-destructive, it can be operated in-series before an FID, which is destructive, thus providing complementary detection of the same analyte (Goorts, 2008).

Other detectors used for specific types of substances or narrower concentration ranges are

- Catalytic combustion detector (CCD), which measures combustible hydrocarbons and hydrogen
- Discharge ionization detector (DID), which uses a high-voltage electric discharge to produce ions
- Dry electrolytic conductivity detector (DELCD), which uses an air phase and high temperature (v. Coulsen) to measure chlorinated compounds
- Electron capture detector (ECD), which uses a radioactive Beta particle (electron) source to measure the degree of electron capture
- Flame photometric detector (FPD)
- Hall electrolytic conductivity detector (ElCD)
- Helium ionization detector (HID)
- Nitrogen phosphorus detector (NPD)
- Infrared detector (IRD)
- Mass selective detector (MSD)
- Photo-ionization detector (PID)
- Pulsed discharge ionization detector (PDD)
- Thermal energy (conductivity) analyzer/detector (TEA/TCD)
- Thermionic ionization detector (TID).

The various detectors, support gases, selectivity, detectability and dynamic range are given in Table 1 (SHU, 2011).

Some gas chromatographs are connected to a mass spectrometer (MS), nuclear magnetic resonance (NMR) spectrometer, or infrared (IR) spectrophotometer that acts as the detector. These combinations are known as GC-MS, GC-MS-NMR, and GC-MS-NMR-IR. However, it is very rare to use the NMR and IR along with GC. The most commonly used is GC-MS (Scott, 2003).

4. Carrier gas

The carrier gas plays an important role in GC analysis and can vary depending on the GC used. Carrier gas must be dry, free of oxygen, and chemically inert. Typical carrier gases include helium, nitrogen, argon, hydrogen, and air. The choice of carrier gas (mobile phase) is important, with hydrogen providing the best separation. Helium is most commonly used as a carrier gas because it has larger flow rates, is non-flammable, and can work with many detectors. The type of carrier gas used depends on the following factors: (a) detector used; (b) sample's composition; (c) safety and availability (i.e., hydrogen is flammable, high-purity helium can be difficult to obtain in some areas of the world); and (d) purity of the carrier gas (where high pure gases of 99.995% are selected when high sensitivities in the measurement are required) (Guiochon and Guillemin, 1988).

Detector	Type	Support Gases	Selectivity	Detectability	Dynamic Range
Flame ionization (FID)	Mass flow	Hydrogen and air	Most organic compounds	100 pg	10^7
Thermal conductivity (TCD)	Concentration	Reference	Universal	1 ng	10^7
Electron capture (ECD)	Concentration	Make-up	Halides, nitrates, nitriles, peroxides, anhydrides, and organic metals	50 fg	10^5
Nitrogen phosphorous	Mass flow	Hydrogen and air	Nitrogen, phosphorous	10 pg	10^6
Flame photometric (FPD)	Mass flow	Hydrogen and air, possibly oxygen	Sulphur, phosphorous, tin, boron, arsenic, germanium, selenium, and chromium	100 pg	10^3
Photo-ionization (PID)	Concentration	Make-up	Alaphatics, aromatics, ketones, esters, aldehydes, amines, hetrocyclics, organosulphurs, some organometallics	2 pg	10^7
Hall electrolytic conductivity	Mass flow	Hydrogen, oxygen	Halides, nitrogen, nitroamine, sulphur		
Note: pg: picogram; ng: nanogram, and fg: femtogram					

Table 1. Various detectors and their detectability

4.1 GC methods for sample analysis

An analysis method comprises conditions in which a GC operates for a specific analysis. Developing the method involves finding the conditions required for the analysis of a specific sample. The process conditions that can be used to generate the method file are (a) inlet temperature, (b) detector temperature, (c) column temperature, (d) temperature program, (e) carrier gases and their flow rates, (f) the column's stationary phase, (g) column diameter and length, (h) inlet type and flow rates, and (i) sample size and injection technique. These operating conditions depend on the detector selected and the compounds to be analyzed (Grob et al., 2004).

4.2 Gas Chromatography-Mass Spectrometry (GC-MS)

GC-MS uses two techniques that are combined into a single method for analyzing mixtures of chemicals. Gas chromatography separates the components of a mixture, and mass spectroscopy characterizes each of the components individually. Combining the two techniques helps to analyze the samples both qualitatively and quantitatively. As the sample is injected into the chromatograph, the sample mixture gets separated into individual components due to different flow rates. This results in quantitative analysis of the components, along with a mass spectrum of each component. Applications of GC-MS include drug detection, fire investigation, environmental analysis, explosives investigation, and identification of unknown samples. Strengths of GC/MS analysis are (a) identification of organic components from complex mixtures, (b) quantitative analysis, and (c) determination of traces of organic contamination. A major limitation of GC/MS is that the sample must be volatile or capable of derivatization (Hübschmann, 2008).

5. Volatile analysis during biomass torrefaction

5.1 Biomass as an energy source

Utilization of biomass for energy can help reduce world dependence on fossil fuels, reduce the impact of greenhouse gasses on the environment, and help meet targets established in the Kyoto Protocol (UN, 1998). Biomass usage as a renewable energy resource is increasing as it is considered carbon neutral (carbon dioxide released is already part of the carbon cycle). Commercial limitations of biomass are its low bulk density, high moisture content, hydrophilic nature, and low calorific value (Arias et al., 2008). Due to its low energy content compared to fossil fuels, it may be necessary to use high volumes of biomass. This results in storage, transportation, and feed-handling problems at bioenergy conversion facilities. These drawbacks have led to the development of new technologies to successfully use biomass for fuel applications. One way to overcome these limitations is pretreating the biomass. There are different pre-treatment methods, including thermal, mechanical, and chemical. Torrefaction is a thermal method that significantly improves the physical and chemical properties of biomass. These compositional changes make torrefaction a promising pre-treatment method for thermochemical conversion and combustion applications (Tumuluru et al., 2011).

5.2 Biomass components

Fig. 2 indicates the low-molecular and macromolecular weight substances in biomass (Mohan et al., 2006). The various components of the biomass are significantly influenced by thermal treatments (Tumuluru et al., 2011).

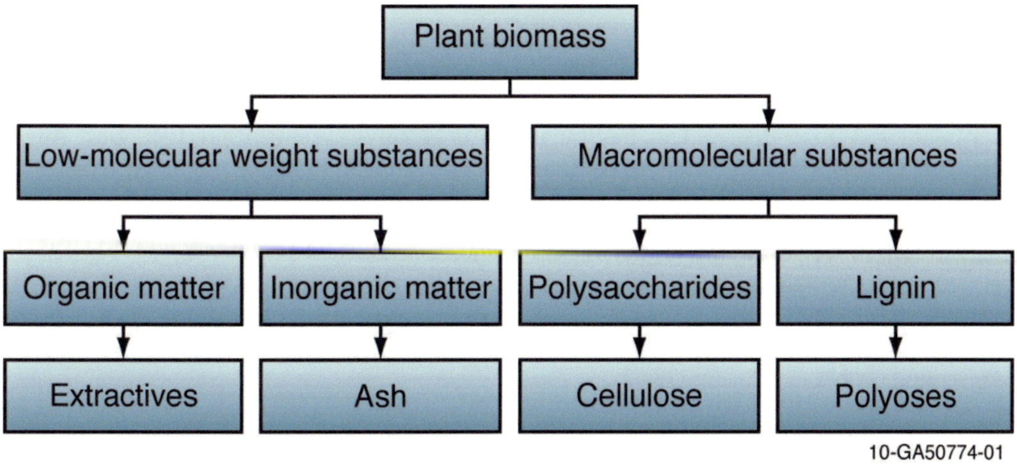

Fig. 2. Plant biomass composition (Mohan et al. 2006)

5.2.1 Cellulose

Cellulose is a high-molecular-weight polymer which provides structural rigidity to the plants. Cellulose degradation begins at 240°C, resulting in anhydrous cellulose and levoglucosan polymer, and makes up the fibers in wood and levoglucosan (Mohan et al., 2006). The crystalline structure resists thermal depolymerization during thermal treatment to a greater degree than is seen in unstructured hemicelluloses.

5.2.2 Hemicellulose

Hemicelluloses are branched polymers which account for about 25-35 wt% in biomass. Thermal degradation of hemicelluloses takes place from 130-260°C, with the majority of weight loss occurring above 180°C (Demibras, 2009; Mohan et al., 2006). Degradation of hemicellulose results in emission of volatiles.

5.2.3 Lignin

Lignin is an amorphous, highly branched, cross-linked macromolecular polyphenolic resin which is covalently linked to hemicellulose and cross-linked with different plant polysaccharides. These linkages give mechanical strength to the cell wall. It is relatively hydrophobic and aromatic in nature. Lignin decomposes when heated above 280°C, producing phenols due to cleavage of ether bonds (Demibras, 2009; Mohan et al., 2006). Lignin converts into char at temperatures >300°C.

5.2.4 Extractables

The other organic extractables present in biomass are fats, waxes, alkaloids, proteins, phenolics, simple sugars, pectins, mucilages, gums, resins, terpenes, starches, glycosides, saponins, and essential oils (Mohan et al., 2006).

6. Torrefaction process

Torrefaction is a process of heating biomass slowly in an inert atmosphere to a maximum temperature of 300°C (Fonseca et al., 1998) and has been defined as the partially controlled and isothermal pyrolysis of biomass occurring in a temperature range of 200–300°C (Bergman and Kiel, 2005). The treatment yields a solid, uniform product with lower moisture and higher energy content compared to raw biomass (Sadaka and Negi, 2009).

The initial heating of biomass during torrefaction removes unbound water. Heating above 160°C removes the bound water due to the thermo-condensation process (Zanzi et al., 2002). Increasing the temperatures to 180–270°C initiates the decomposition of hemicellulose. At temperatures >280°C, the process is completely exothermic and results in significant increase in the production of CO_2, phenols, acetic acid, and other higher hydrocarbons (Zanzi et al., 2002).

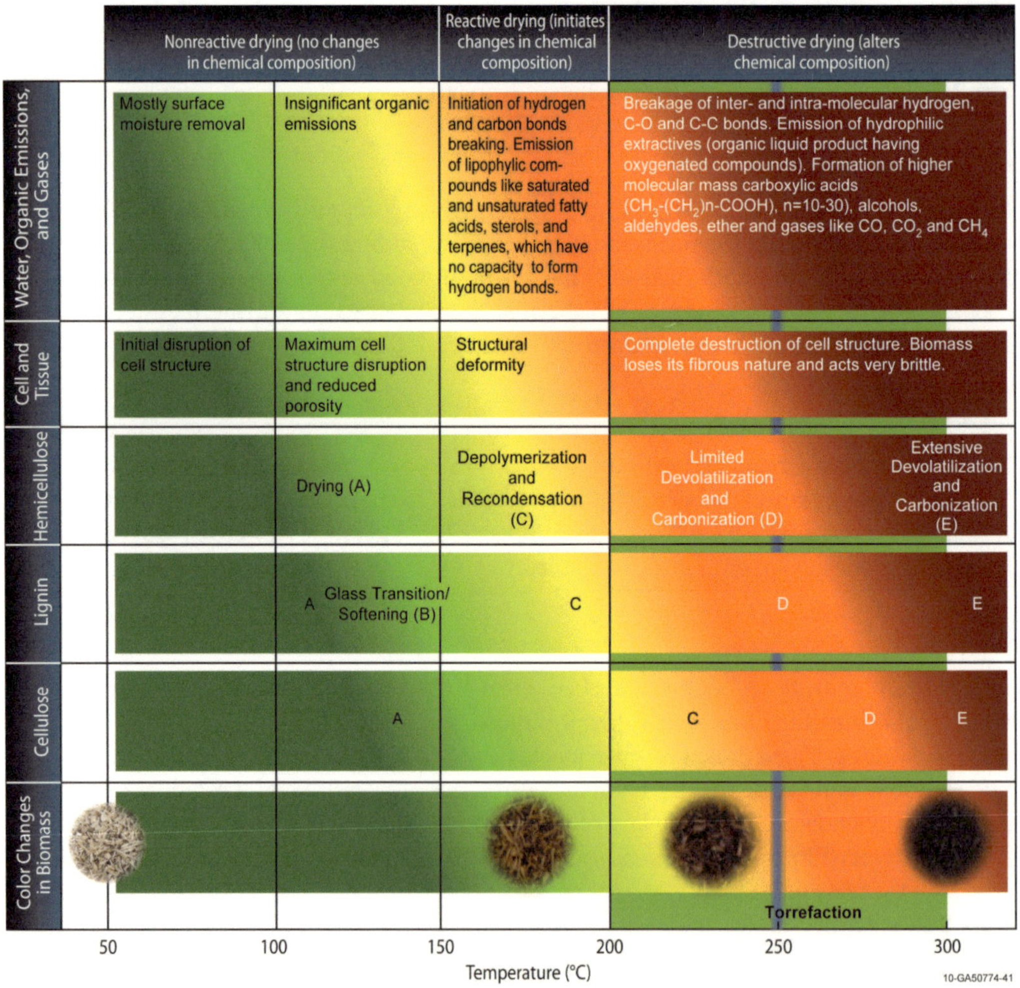

Fig. 3. Structural, chemical, and color changes in biomass at different drying temperatures (Tumuluru et al., 2011)

6.1 Biomass reactions

The physiochemical and structural changes in biomass at different temperature regimes is given in Fig. 3 (Tumuluru et al., 2011). The figure indicates that at higher temperature regimes, the drying is more destructive in terms of breakage of inter- and intra-molecular, hydrogen, C-O, and C-C bonds and leads to emission of some hydrophilic, oxygenated compounds. In addition, these changes lead to formation of higher-molecular-mass carboxylic acids, aldehydes, ethers and some gases like carbon monoxide, carbon dioxide and methane.

6.2 Torrefaction gas composition

The three main products produced during torrefaction are a brown/dark-color solid; a condensable liquid, including mostly moisture, acetic acid, and other oxygenates; and non-condensable gases—mainly CO_2, CO, and small amounts of methane. The last two products can be represented by volatiles. Fig. 4 (Bergman et al., 2005) gives an overview of the torrefaction products, classified based on their state at room temperature, which can be,

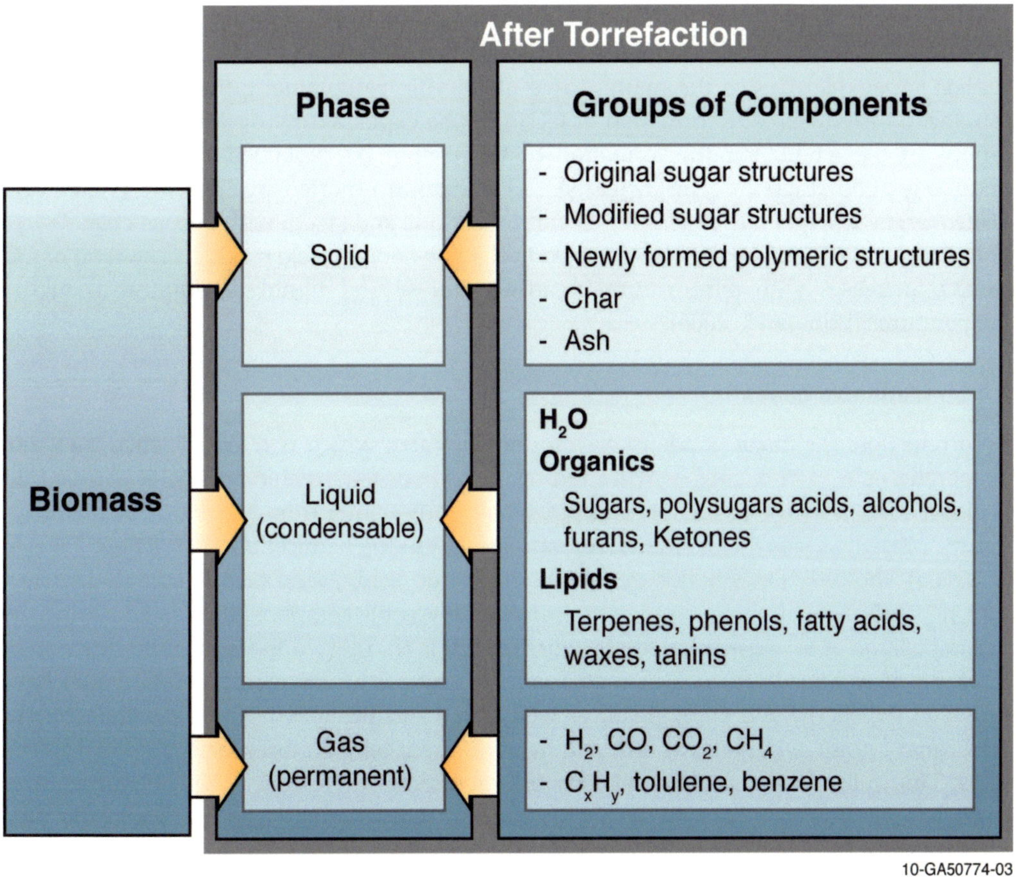

Fig. 4. Products formed during the torrefaction of biomass (Bergman et al. 2005)

solid, liquid, or gas. The solid phase consists of a chaotic structure of the original sugar structures and reaction products. The gas phase includes the gases that are considered permanent gases, but also light aromatic components such as benzene and toluene. The condensables, or liquids, can be divided into three subgroups which include water, organics, and lipids, as shown in Fig. 4. One subgroup is reaction water as a product of thermal decomposition. The liquid also contains the free and bound water that has been released from the biomass by evaporation and dehydration reactions. The organics subgroup (in liquid form) consists of organics that are mainly produced during devolatilization and carbonization. Finally, the lipids are a group of compounds that are present in the original biomass. This subgroup contains compounds such as waxes and fatty acids. The type and amount of the gases that come out depend on the raw material type and torrefaction process conditions, including process temperature and residence time (Tumuluru et al., 2011).

Currently, there is much emphasis on understanding the torrefaction gas composition, as its energy content plays a major role in improving the overall efficiency. Some researchers like Bergman et al. (2005, 2005a), Prins et al. (2006), and Deng et al. (2009) have investigated the torrefaction gas composition.

6.3 Non-condensable gases

Carbon monoxide (CO) is the main source of calorific value among the non-condensable torrefaction products. The formation of CO_2 may be explained by decarboxylation of acid groups in the wood, but the formation of CO cannot be explained by dehydration or decarboxylation reactions. The increased CO formation reported in literature (White and Dietenberger, 2001) is due to reaction of carbon dioxide and steam with porous char. Traces of hydrogen and methane are also detected in the non-condensable product. The ratio of CO to CO_2 increases with temperature because cellulose and lignin decompose at higher temperatures (Prins et al., 2006).

6.4 Condensable gases

In torrefaction, the major condensable product is water, which is released when moisture evaporates as well as during dehydration reactions between organic molecules. Acetic acid is also a condensable torrefaction product, mainly originating from acetoxy- and methoxy-groups present as side chains in xylose units present in the hemicelluloses fraction. Prins et al. (2006) showed that smaller quantities of formic acid, lactic acid, furfural, hydroxy acetone, and traces of phenol are also present in the volatile component liberated during the decomposition of biomass. As a result, more energy is transferred to the volatiles fraction in the form of combustibles such as methanol and acetic acid. Bridgeman et al. (2008) have provided the energy and mass yields, as well as the volatiles lost during the torrefaction of reed canary grass, wheat straw, and willow at different temperatures, as shown in Table 2. Table 3 indicates the physiochemical composition of pine and torrefied pine at temperatures from 240–290°C. With the increase in torrefaction temperature, fixed carbon in the pine increased while volatiles and moisture content decreased. Tables 3 and 4 indicate that, in the torrefaction range (200–300°C), there is significant change in the volatile concentration within the biomass.

	Temperature (K)			
	503	**523**	**543**	**563**
Reed Canary Grass				
Mass yield (daf)	92.6	84.0	72.0	61.5
Energy yield (daf)	93.5	86.6	77.1	69.0
Volatiles (daf)	7.4	16.0	28.0	38.5
Wheat Straw				
Mass yield (daf)	91.0	82.6	71.5	55.1
Energy yield (daf)	93.5	86.2	78.2	65.8
Volatiles (daf)	9.0	17.4	28.5	44.9
Willow				
Mass yield (daf)	95.1	89.6	79.8	72.0
Energy yield (daf)	96.5	92.7	85.8	79.2
Volatiles (daf)	4.9	10.4	20.2	28
Note: daf: dry ash free basis				

Table 2. Energy and mass yields and volatiles lost during torrefaction of reed canary grass, wheat straw, and willow at different temperatures (Bridgeman et al., 2008)

T[a],°C[b]	P[c]	240	250	260	270	290
Fixed carbon, %	20.64	23.55	25.59	25.69	29.38	35.39
Elementary analysis **C, %** **O, %**	50.98 42.80	51.14 42.70	51.93 42.18	53.78 40.66	53.57 40.67	58.08 36.40
Pentosans, %	9.61	5.93	5.90	3.10	2.54	1.40
Lignin, %	22.84	24.90	28.72	33.44	39.23	53.47
Extractable, %[d]	14.67	8.19	14.09	19.35	16.49	17.98
Moisture, %[e]	10.80	5.66	4.08	3.96	3.76	3.88
Yield, %	-	86.2	81.8	75.7	66.4	48.8

a. in each case, the mean result given was obtained from a minimum of four different experiments
b. roasting time: 30 minutes
c. native pine
d. neutral-solvent extractable (ethanol, benzene, boiling water)
e. powder samples left at the laboratory atmosphere still had a constant humidity

Table 3. Physiochemical analysis of pine and torrefied pine

6.5 Significance of torrefaction gas

The knowledge of the composition of volatiles produced in the temperature range of torrefaction is a topic of interest for understanding of

- Raw material adaption to the process
- Process control
- Process behavior and operation
- Energy process and energy optimization
- Production of green chemicals.

6.5.1 Improving torrefaction process efficiency

Volatiles produced during torrefaction have relatively low calorific values, but the gas is still flammable. If the gases produced during torrefaction are flammable, they may be combusted and used to heat the torrefaction reactor in a recycle loop as suggested by Bergman et al. (2005), as shown in Fig. 5. The torrefaction gas composition must be studied further to find out if the gas is combustible, if autothermal operation of the reactor is possible, or if an external fuel is required, as well as what torrefaction temperature and residence times give flammable gases and the best efficiencies for the process. Using gas chromatography to identify and quantify compounds in these gases can reveal what portion of the gas is flammable, what flame temperature is required to ignite all the flammable components, and theoretically what heating value the gas would have. With this information, the potential use of these gases as a heat source for the reactor can be further evaluated.

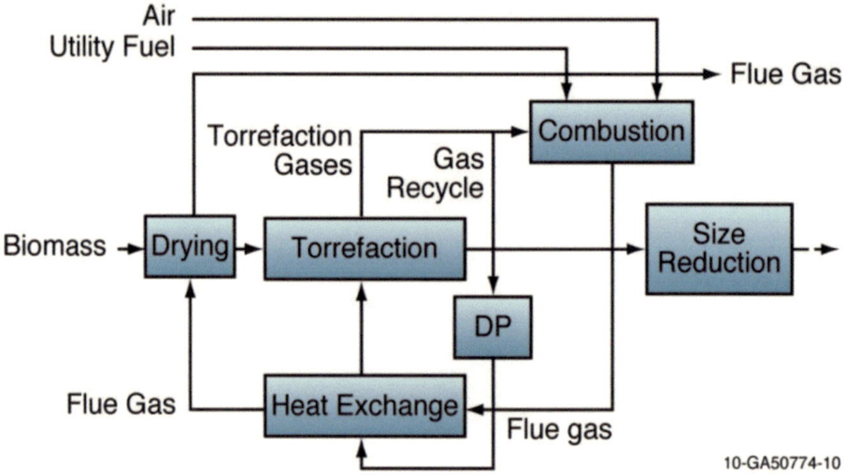

Fig. 5. Combined drying and torrefaction process developed by the Energy Research Centre of the Netherlands (ECN).

6.5.2 Chemical production

Other uses of the condensable fraction of torrefaction gases are for the production of chemicals like concentrated acetic acid, formic acid, methanol, and furfural. The yield of these chemicals depends greatly on torrefaction parameters, including temperature, biomass type, and residence time. In a study of the torrefaction of willow, acetic acid was found to be as much as 5% of the total yield. Formic acid and furfural were as much as 4%, and methanol was at just over 1% (Prins, 2006).

Acetic acid is used mainly to derive more specialized compounds. The largest use of acetic acid is in making vinyl acetate monomer (VAM), which is used in the production of paints, adhesives, paper coatings, and textile finishes. VAM is also used in the production of ethylene vinyl alcohol (EVOH) polymers, which are used in food packing films, plastic bottles, and automobile gasoline tanks. Acetic acid is also used in creating purified terephthalic acid, which is used in making bottle resins and polyester fiber. Another derivative is ethyl acetate, used as a solvent in oil-based lacquers and enamels, including polyurethane finishes and in inks and adhesives. Another large use of acetic acid is in deriving acetate esters, which are used in a wide variety of paints, inks, and other coatings, as well as in many other chemical processes (ICIS News).

Formic acid is a chemical intermediate in the production of various other chemicals and pharmaceuticals, such as caffeine, enzymes, antibiotics, artificial sweeteners, plant protection agents, PVC-plasticizers, and rubber antioxidants. Formic acid is used directly in dyeing, pickling, and tanning processes for leather, the coagulation of rubber, and in various cleaning products (BASF).

Furfural and its derivatives can be used as a replacement for crude-oil-based organics used in industry, such as flavoring compounds, furane resin, surface coatings, pharmaceuticals, mortar, polymers, adhesives, and moulds. Tetrahydrofurfuryl alcohol and tetrahydrofuran (THF) are also used to create specialty chemicals in a wide range of chemical synthesis (Win, 2011).

7. Gas chromatography methods for torrefaction gases

Various GC configurations are used for the analysis of pyrolysis gases that can also be used to analyze gases from the torrefaction processes. GC systems used for pyrolysis should be reviewed because those volatiles and permanent gases produced during pyrolysis are likewise produced in the torrefaction process. The authors feel that with less or no modification, these GC systems can be adequately adapted for torrefaction-gas analysis.

In the study of Prins et al. (2006b), the volatile products are split into a liquid and gas phase in a cold trap at –5°C. Liquid products collected in the cold trap were diluted with 2-butanol because not all the products dissolved in water. The diluted liquid was analyzed with high-performance liquid chromatography (HPLC) using a Chrompack Organic Acids column with detection based on refraction index. The composition of the non-condensable gas was analyzed with a Varian Micro GC with a Poraplot Q and a Molsieve column.

In the study of Deng et al. (2009), torrefaction products were removed from the hot zone to minimize the secondary reactions between the liquid and char and to maximize the solid yield. A two-necked flask immersed in liquid nitrogen was used as a trapping system for the condensable gases. Non-condensable gases went through a filter to remove the carbon soot before entering an infrared gas analysis. The gas composition and concentration were recorded continuously throughout the heating process. Finally, the weight of the biochar and the amount of liquids obtained were measured.

In Bridgeman's (2008) study, a Nicolet Magma-IR Auxiliary Equipment Module (AEM), connected to a Stanton Redcroft Simultaneous Analyser STA-780 Series, was used to perform torrefaction experiments at laboratory scale while simultaneously analyzing the volatile pyrolysis products for corresponding thermogravemetric analyzer (TGA) data.

Bergman et al. (2005) used two impinger bottles immersed in a cold bath (<5°C) for collection of the condensables during the torrefaction process. Torrefaction gas is sucked through the bottles at a pre-specified rate so that yields can be determined. A polar solvent (water or ethanol) is used in the bottles. Tests showed that two bottles removed all the condensables from the gas phase. Some aerosols are formed by quenching (going from ~200°C to ~0°C). Quantification of water was found to be unreliable due to aerosol formation and because reaction water mixes with physically bound water. It is unclear whether lipids are collected efficiently because the components of the lipid subgroup were not quantified afterwards. Analysis of the product was done with a combination of GC-FID, GC-MS, and ion chromatography. These quantified products are shown in Table 4.

2-furaldehyde (furfural)	1-Hydroxy-2-butanone
5-methyl-2-furaldehyde	Eugenol
Acetaldehyde	Isoeuganol
2(5H)-Furanone	Propionic acid
Ethylene-glycol diacetate	3,4,5 trimethoxytoluene
Ethanol	Furan-2-methanol
Pyrrole	Furan-3-methanol
Methanol	Methylacetate
Methylformiate	Hydroacetone
Cyclohexanone	Butanone
Propanal	2-butenal
Pyrrole-2-carboxaldehyde	Phenol
2-methoxyphenol	3-methoxyphenol
4-methoxyphenol	2,6 dimethoxyphenol
Acetone	3-methoxypyridine
4-methoxypyridine	2-methoxypyridine
Acetic acid	Formic acid

Table 4. Quantified condensable products produced during torrefaction of wood

Permanent gases, as shown in Table 5, are also measured using an online measurement procedure. This comprises micro-GC, by which the composition of the dry gas is determined. Before the gas is charged to the micro-GC, water is removed in a cold trap because the presence of the water disturbs the measurement.

O_2	C_2H_2
Ar	C_2H_4
CO_2	C_2H_6
CO	Benzene
N_2	Toluene
H_2	Xylenes
CH_4	

Table 5. Non-condensable gases quantified during torrefaction of wood

In Pommer et al.'s 2011 studies on torrefaction of different biomass sources using GC-MS, the most abundant compounds were identified due to the decomposition of hemicellulose, cellulose, and lignin, as shown in Table 6.

4-Allyl-syringol	Formaldehyde
Acetaldehyde	Formic acid
Acetic acid	Furaldehyde
Carbon monoxide	(3H)-Furan-2-one
Carbon dioxide	2-Furfuralalcohol
Conifer alcohol	Hydroxyacetaldehyde
Eugenol	Methanol
Propanal-2-one	Sinapaldehyde
Pyranone	Synapyl alcohol
Syringaldehyde	Vanillin
4-Vinyl-syringol	Water

Table 6. GC-MS analysis of abundant compounds produced due to the torrefaction of biomass

In addition, the analysis of biomass tar can also be used for reference. Two methods for the sampling and analysis of tar produced from wood pyrolysis were compared in the study of Dufour et al. (2007). A Clarus 500 GC system (Perkin-Elmer) coupled to a Clarus 500 MS quadrupole MS system (Perkin-Elmer) was used for the analysis. The gas chromatograph was equipped with an electronically controlled split/splitless injection port. GC was carried out on a 5% diphenyl-/95% dimethylpolysiloxane fused-silica capillary column (Elite-5 ms, 60 m × 0.25 mm, 0.25 μm film thickness; Perkin-Elmer). Helium (Alphagaz 2, Air Liquide) was used as the carrier gas, with a constant flow of 1.2 mL/min. The first method used a conventional cold-trapping technique in solvent-filled impingers followed by liquid injection. The second employed a new application of multi-bed solid-phase adsorbent (SPA) tubes, followed by thermal desorption (TD). Both methods are based on GC coupled with MS. Quantification was performed with a well reproducible GC–MS method with three internal deuterated standards. The main compounds sampled by impingers with methanol or 2-propanol and SPA tubes are listed in Table 7.

Barrefors and Petersson (1995) used GC and GC–MS for volatiles produced during the pyrolysis of wheat straw. Mahinpey et al. (2009) have carried out GC and GC–MS analysis of bio-oil, biogas, and biochar from pressurized pyrolysis of wheat straw using a tubular reactor. The GC system used was a gas chromatograph (Agilent Technologies 6890 N) equipped with two columns (Porapak; molecular sieve), flame ionization detection (FID), and thermal conductivity detection (TCD). Standard gas mixtures were used for quantitative calibration, and argon was used as the carrier gas.

Benzene	1,2,4-Trimethylbenzene; 1-methylethylbenzene
Acetic acid	3,3-Dimethylphenol
Pyridine	Phenol
Pyrrole	2- or 3-Methylstyrene
Toluene	Benzofuran
Ethylbenzene	Indene
1,3-Dimethylbenzene	2- or 3- or 4-Methylphenol
Ethynylbenzene	5-Methylpyrimidine
o-Xylene	2- or 3- or 4-Methylphenol
1- or 2-Propenylbenzene	7- or 2-Methyl-benzofuran; 3-phenyl-2-propenal
7- or 2-Methyl-benzofuran	3-Phenylfuran; 1- or 2-naphthol
4,7-Dimethylbenzofuran; 1-hydroxy-5,8-dihydronaphthalene	1- or 2-Naphthol
2-Methylnaphthalene	7- or 2-Methyl-benzofuran or 3-phenyl-2-propenal
1-Methylnaphthalene	Diethenylbenzene
Fluorene	3-Phenyl-1,2-butadiene; 1-, 2-methylindene
4-Methyldibenzofuran; 9H-xanthene	3-Phenyl-1,2-butadiene; 1-, 2-methylindene
1H-phenalene	Biphenyl
2-, 3-, or 4-Methylbiphenyl	Acenaphthene
1-Phenylmethylene-1H-indene; 1,9-dihydropyrene	2-Ethenylnaphthalene
Phenanthrene	Acenaphthylene
Anthracene	2-Methylanthracene
1-Phenylmethylene-1H-indene; 1,9-dihydropyrene	6H-Cyclobuta[J,K]phenanthrene; 8,9-dihydrocyclopenta [D,E,F]phenanthrene
Dibenzofuran	1- or 2-Phenylnaphthalene
Acenaphthenone	Fluoranthene
1H phenalene	11H-benzo[A] or [B]fluorene
1-Methylpyrene	11H-benzo[A] or [B]fluorene
Pyrene	

Table 7. Main identified compounds sampled with SPA tube

The GC–MS system consisted of an Agilent HP6890GC coupled to a quadrupole mass spectrometer (Agilent 5973 Network MSD). Electron ionization (EI) was used with an ion source temperature of 230°C and an interface temperature of 280°C, with EI spectra obtained at 70 eV. In EI, the instrument was used in SCAN mode initially to confirm the identity of the 18 compounds and then in selected ion monitoring (SIM) mode for quantitative analysis. The GC system was equipped with a split/splitless inlet with a splitless sleeve containing carbofrit (4 mm i.d., 6.5 × 78.5 mm, Restek). The injector temperature was 225°C. A LEAP Technologies autosampler with a 10 µL syringe was used for injections of 1 µL at a rate of 10 µL s^{-1}. The analytical column was DB-5 ms, 5% polydiphenyl-95% dimethylpolysiloxane, 30

× 0.25 mm i.d., and 0.25 µm film thickness (J&W Scientific). The carrier gas was helium (UHP) at a constant flow of 1.0 mL min^{-1}. The oven temperature program had an initial temperature of 65°C, which was held for 4 min, rising by 10°C/min to 120°C, 15°C/min to 170°C, 3°C/min to 200°C, and finally 30°C/min to 300°C, where it was held for 5 min for a total run time of 31.17 minutes. This temperature program was selected to provide adequate separation of most of the 18 compounds of interest, as shown in the total ion chromatogram (TIC) for a standard and sample. Only two compounds coeluted (pyrocatechol and phenol 2-methoxy 4-methyl), and the remaining compounds are provided in Table 8.

Phenol	4-methylcatechol	3,5-dimethoxy-4'-hydroxy acetophenone
Methylhydroquinone	2,6-dimethoxyphenol	Syringaldehyde
Guajacol	Eugenol	2-methoxy -4-propylphenol
4-ethyl phenol	Vanillin	phenol 2 -methoxy 4-methyl
pyrocatechol	Isoeugenol	3-methoxycatechol
4'-hydroxy-3'-methoxyacetophenone	4-ethyl guaiacol	2-propanone, 1-(4-hydroxy-3-methoxyphenol

Table 8. GC and GC−MS analysis of gases produced during pyrolysis of wheat straw

In the studies conducted by Bacidan et al. (2007), product distribution and gas release in pyrolysis of thermally thick biomass residues samples have used micro GC and FTIR analyzer for gases like CO_2, CO, CH_4, H_2, C_2H_2, C_2H_6, and C_2H_4. The FTIR analysis of the gases was performed with a Bomem 9100 analyser (sampling line and cell heated at 176 °C with a volume of 5l and an optical path length of 6.4 m). The instrument is equipped with a DTGS detector at the maximum resolution of 1 cm^{-1}. The gas samples were also quantified online using a Varian CP-4900 micro-GC equipped with two TCD detectors and a double injector connected to two columns: (1) a CP-PoraPLOT Q column (10 m length, 0.25 mm inner diameter and 10 µm film thickness produced by Varian, Inc.) to separate and quantify CO_2, CH_4, C_2H_2 + C_2H_4 (not separated), and C_2H_6 ; and (2) a CP-MolSieve 5 Å PLOT column (20 m length, 0.25 mm inner diameter and 30 µm film thickness produced by Varian, Inc.) to analyse H_2, O_2, CH_4, CO, and N_2. Helium and argon were used, respectively, as carrier gases in the two columns.

Lappas et al. (2002) focused on identifying the composition of the gases produced during biomass pyrolysis, where an online connected GC (HP 5890) with two detectors (TCD and FID) was used, along with a split/splitless injector and four columns (OV 101, Molecular sieve 5A, Porapaq N, and a capillary Poraplot Al2O3/KCl). The composition of the regenerator gas exit stream (i.e., flue gases) was determined with a GC (HP 5890) detector (TCD, Molecular sieve 5A and Porapaq N columns), and the hydrocarbons, CO and CO_2 were analyzed.

Qing et al. (2011) analyzed the gases and volatiles produced during the pyrolysis of wheat straw. They used a Shimazu QP2010 Plus GC-MS, and the GC column type was a DB-Wax fused silica capillary column (30 m × 0.25 mm; i.e., film thickness = 0.25 µm).

And finally, Wang et al. (2011) concentrated on the pyrolysis of pine sawdust by using a GC-MS (Finnigan Trace MS). The column used in these studies was a GC-MS with a capillary column DB-1301 (30 m x 0.25mm; i.e., film thickness = 0.25 µm). Helium was the carrier gas, with a constant flow of 0.5 ml/min. The MS was operated in electron ionization mode with a 70 eV ionization potential, and an m/z range from 30 to 500 was scanned. The peaks were identified based on the computer matching of the mass spectra with the National Institute of Standards and Technology (NIST) library. The summary of the systems used for analysis of torrefaction and pyrolysis gas analysis is given in Table 9.

S. No	**GC configuration**	**Process**	**Volatiles and permanent gases**		**Reference**
			Volatiles	Gases	
1	HPLC using Chrompack organic acids column for volatiles. Varian micro gas chromatography with Porapolt Q and Molsieve column for non condensable gases	Torrefaction	Water, acetic, lactic acid, formic acid, furfural, hydroxyl acetone, phenol, methonal	CO, CO_2	Prins et al. (2006b)
2	Infrared gas analysis (Gasboard-5110, Wuhan, China).	Torrefaction	Moisture, acetic acid and other oxygenates	CO_2, CO and small amounts of methane	Deng et al. (2009)
3	TGA-FTIR (A Nicolet Magma-IR AEM connected to a Stanton Redcroft Simultaneous Analyser STA-780 Series)	Torrrefaction	Volatiles	CO_2, CO	Bridgeman et al. (2008)
5	GC–MS (A Clarus 500 GC system (Perkin-Elmer, Shelton, CT, USA) coupled to a Clarus 500 MS quadrupole MS system (Perkin-Elmer)	Torrefaction	Benzene, acetic acid, pyridine, pyrole, toluene, ethylbenzene, 1,3-dimethylbenzene,etc		Dufour et al. (2007)
6	GC-FID and GC-MS	Torrefaction	2-furaldehyde (furfural), 1-Hydroxy-2-butanone, acetic acid, formic acid etc.	O_2, Ar, CO_2, H_2, CO	Bergman et al. (2005)
7	GC-MS	Torrefaction	Acetic acid, formic acid, formaldehyde, vanillin, methanol, acetaldehyde etc.	CO, CO_2	Pommer et al. (2011)

S. No	**GC configuration**	**Process**	**Volatiles and permanent gases**		**Reference**
8	GC-MS	Pyrolysis	Acetic acid, formic acid, formaldehyde etc.		Barrefors and Petersson (1995)
9	GC-MS (Agilent HP6890GC coupled to a quadrupole mass spectrometer) (Agilent 5973 Network MSD) GC (Agilent Technologies 6890 N) equipped with two columns (Porapak; Molecular sieve), flame ionization detection (FID), and thermal conductivity detection (TCD)	Pyrolysis	Phenol, vanillin, acetophenone, syringaldehyde etc.	CO, CO_2 etc	Mahinpey et al. (2009)
10	FTIR (Bomem 9100 analyser) and Micro GC	Pyrolysis	Hydrocarbons	CO, CO_2	Bacidan et al. (2007)
11	Gas chromatograph-GC (HP 5890) with two detectors (TCD and FID)	Pyrolysis		CO, CO_2	Lappas et al. (2002)
12	Shimazu QP2010 Plus GC-MS. GC Column Type: A DB-Wax fused silica capillary column (30 m × 0.25 mm, i.e., film thickness = 0.25 μm)	Pyrolysis			Qing and Wu (2011)
13	GC-MS: The column used in GC-MS was a capillary column DB-1301 (30mx0.25mm i.d., 0.25μm film thickness) and MS was operated in electron ionization mode with a 70 eV ionization potential.	Pyrolysis	Higher hydrocarbons	CO, CO_2	Wang et al. (2011)

Table 9. GC systems for torrefaction and pyrolysis gas analysis

8. Conclusions

Torrefaction of biomass produces a product that has a higher energy content compared to that of raw biomass. During torrefaction, biomass loses moisture as well as some condensable and non-condensable gases. The condensable gases include water, organics, and lipids. The liquid contains the free and bound water that has been released from the biomass by evaporation. The organics subgroup (in liquid form) consists of organics that are mainly produced during devolatilization, depolymerization, and carbonization reactions in the biomass. The knowledge of the composition of volatiles produced in the temperature range of torrefaction is a topic of interest as it helps in the raw material adaption process, process control, process behavior and operation, energy process and energy optimization, and in the production of green chemicals. Lipids are a group of compounds that are present in the original biomass and contain compounds such as waxes and fatty acids. The non-condensable gases are CO_2, CO, and small amounts of methane. GC configuration plays an important role in the accurate identification of the compounds. A combination of different detectors, such as thermal, flame, and photo ionization detectors, in combination with a mass spectrometer, are used for the profiling of both condensable and non-condensable gases. It can be concluded that a GC-MS can help to analyze most of the volatiles emitted from biomass during torrefaction as well as a micro GC for the analysis of permanent gases like CO, CO_2, and CH_4.

9. Acknowledgements

The authors acknowledge Leslie Park Ovard for her valuable contribution to the manuscript and Barney Hadden, Ph.D, Gordon Holt, Lisa Plaster, and Allen Haroldsen from INL's R&D Publication Support Service for editorial and graphics support. This work is supported by the U.S. Department of Energy, under DOE Idaho Operations Office Contract DE-AC07-05ID14517. Accordingly, the U. S. Government retains and the publisher, by accepting the article for publication, acknowledges that the U. S. Government retains a nonexclusive, paid-up, irrevocable, world-wide license to publish or reproduce the published form of this manuscript, or allow others to do so, for U. S. Government purposes.

10. Disclaimer

11. References

Arias, B.R., Pevida, C.G., Fermoso, J.D., Plaza, M.G., Rubiera, F.G., and Pis Martinez, J.J. (2008). Influence of Torrefaction on the Grindability and Reactivity of Woody

Biomass, *Fuel Processing Technology*, Vol.89, No.2, (February 2008), pp. 169–175, ISSN 0378-3820

Barrefors, G., and Petersson, G. (1995). Assessment by gc and gc-ms of Volatile, *Journal of Chromatography A*, Vol.710, No.1, (1 September 1995), pp. 71-77, ISSN 0021-9673

BASF: The Chemical Company, Formic Acid (webpage). http://www2.basf.us/specialtyintermediates/formic_acid.html

Becidan, M., Skreiberg, Ø., and Hustad, J.E. (2007). Products Distribution and Gas Release in Pyrolysis of Thermally Thick Biomass Residues Samples, *Journal of Analytical and Applied Pyrolysis*, Vol.78, No.1, (January 2007), pp. 207–213, ISSN 0165-2370

Bergman, P.C.A., and Keil, J.H.A. (2005). Torrefaction for Biomass Upgrading, *14th European Biomass Conference & Exhibition*, Paris, France, 17–21 October 2005

Bergman P.C.A., Boersma, A.R., Kiel, J.H.A., Prins, M.J., Ptasinski, K.J., and Janssen, F.J.J.G. (2005a). Torrefied Biomass for Entrained-flow Gasification of Biomass, ECN Report #ECN-C--05-026

Bergman, P.C.A., Boersma, A.R., Zwart, R.W.R., and Kiel, J.H.A. (2005). Torrefaction for Biomass Co-firing in Existing Coal-fired Power Stations: BIOCOAL, ECN Report #ECN-C-05-013

Bosboom, J. C., Janssen, H. G., Mol, H. G. J., and Cameras, C. A. (1996). Large-volume Injection in Capillary Gas Chromatography using a Programmed-temperature Vaporizing Injector in the On-column or Solvent-vent Injection Model, *Journal of Chromotography A*, Vol.724, No.1–2, (16 February 1996), pp. 384-391, ISSN 0021-9673

Bourgois, J.P., and Guyonnet, R. (1988). Characterization and Analysis of Torrefied Wood, *Wood Science and Technology*, Vol.22, No.2, (June 1988), pp. 143–155, ISSN 0043-7719

Bourgois, J.P., and Doat, J. (1984). Torrefied Wood from Temperate and Tropical Species: Advantages and Prospects, In: *Bioenergy 84, Proceedings of the International Conference on Bioenergy, Goteborg, Sweden, 15–21 June, 1984*, H. Egneus, A. Ellegard, P. O'Keefe, and L. Kristofferson (Eds.), Elsevier Applied Science Publishers, pp. 153–159, ISBN 978-085-3343-51-6

Bridgeman, T.G., Jones, J.M., Shield, I., and Williams, P.T. (2008). Torrefaction of Reed Canary Grass, Wheat Straw, and Willow to Enhance Solid Fuel Qualities and Combustion Properties, *Fuel*, Vol.87, No.6, (May 2008), pp. 844–856, ISSN 0016-2361

Demirbas, A. (2009). Pyrolysis Mechanisms of Biomass Materials, *Energy Sources, Part A: Recovery, Utilization, and Environmental Effects*, Vol.31, No.13, (13 May 2009), pp. 1186–1193, ISSN 1556-7230

Deng, J., Wang, G., Kuang, J., Zhang, Y., and Luo, Y. (2009). Pretreatment of Agricultural Residues for Co-Gasification Via Torrefaction, *Journal of Analytical and Applied Pyrolysis*, Vol.86, No.2, (November 2009), pp. 331–337, ISSN 0165-2370

Dufour A., Girods, P., Masson, E., Normand, S., Rogaume, Y., and Zoulalian, A. (2007). Comparison of Two Methods of Measuring Wood Pyrolysis Tar, *Journal of Chromatography A*, Vol.1164, No.1–2 (14 September 2007), pp. 240–247, ISSN 0021-9673

Fonseca Felfli, F., Luengo, C.A., Bezzon, G., and Beaton Soler, P. (1998). Bench Unit for Biomass Residues Torrefaction, *Biomass for Energy and Industry, Proceedings of the International Conference, Würzburg, Germany, 8-11 June 1998,* H. Kopetz, T. Weber, W. Palz, P. Chartier, and G.L. Ferrero (Eds.), C.A.R.M.E.N., Rimpar, Germany, pp. 1593–1595

Goorts, M.P. (2008). Applying Gas Chromatography to Analyze the Composition and Tar Content of the Product Gas of a Biomass Gasifier, Eindhoven University of Technology, Master's Thesis, Report Number WVT 2008.24, Available from http://w3.wtb.tue.nl/fileadmin/wtb/ct-pdfs/Master_Theses/Final_report_martijn_goorts.pdf

Grob, R.L., and Barry, E.F. (2004). *Modern Practice of Gas Chromatography (4th Ed.),* John Wiley & Sons, Inc., Hoboken, NJ, ISBN 0-471-22983-0

Guiochon, G., and Guillemin, C.L. (1988). *Quantitative Gas Chromatography for Laboratory Analyses and On-Line Process Control, (Journal of Chromatography Library,* No.42), Elsevier Ltd., Amsterdam, the Netherlands, ISBN 978-044-4428-57-8

Hübschmann, H-J. (2008). *Handbook of GC/MS: Fundamentals and Applications,* Wiley-VCH, Weinheim, Germany, ISBN 3-527-31427-X

ICIS, Acetic Acid Uses and Market Data (webpage). http://www.icis.com/v2/chemicals/9074779/acetic-acid/uses.html

ICIS, Methanol Uses and Market Data (webpage). http://www.icis.com/v2/chemicals/9076035/methanol/uses.html

IEA Bioenergy, Task 34 - Pyrolysis of Biomass, Leader: Doug Elliott (webpage). http://www.pyne.co.uk/

Kaufmann, A. (1997). Determination of polar analytes in aqueous matrices by purge and trap, *Journal of High Resolution Chromatography,* Vol.20, No.1, (January 1997), pp. 10–16, ISSN 0935-6304

Lappas, A., Samolada, M., Latridis, D., Voutetakis, S., and Vasalos, I. (2002). Biomass Pyrolysis in a Circulating Fluid Bed Reactor for the Production of Fuels and Chemicals, *Fuel,* Vol.81, No.16, (1 November 2002), pp. 2087-2095, ISSN 0016-2361

Mahinpey, N., Murugan, P., Mani, T., and Raina, R. (2009). Analysis of Bio-Oil, Biogas, and Biochar from Pressurized Pyrolysis of Wheat Straw using a Tubular Reactor, *Energy & Fuels,* Vol.23, No.5, (21 May 2009), pp. 2736-2742, ISSN 0887-0624

Mohan, D., Pittman, C.U., and Steele, P.H. (2006). Pyrolysis of Wood/Biomass for Bio-Oil: A Critical Review, *Energy & Fuels,* Vol.20, No.3, (May 2006), pp. 848–889, ISSN 0887-0624

Pach, M., Zanzi, R., and Björnbom, E. (2002). Torrefied Biomass a Substitute for Wood and Charcoal, In: *Proceedings of the 6th Asia-Pacific International Symposium on Combustion and Energy Utilization, Kuala Lumpur, Malaysia, May 20–22, 2002.*

Pavia, D.L., Lampman, G.M., Kritz, G.S., and Engel, R.G. (2006). *Introduction to Organic Laboratory Techniques: A Microscale Approach (4th Ed.),* Cengage Learning, Brooks/Cole Publishing Co., St. Paul, MN, (February 2006), pp. 797–817, ISBN 978-049-5016-30-4

Pentananunt R., Mizanur Rahman, A.N.M., and Bhattacharya, S.C. (1990). Upgrading of Biomass by Means of Torrefaction, *Energy*, Vol.15, No.12, (December 1990), pp. 1175–1179, ISSN 0360-5442

Pommer, L., Gerber, L., Olofsson, I., Wiklund-lindstrom, S., and Nordin, A. (2011). Gas composition from biomass torrefaction-preliminary results, *19th European Biomass Conference and Exhibition, Berlin, Germany, 6–10 June 2011,* Available from http://www.conference-biomass.com/fileadmin/DATA/Documents_and_Pictures/PAST_CONFERENCES/Previous_Conferences/18th_EU_BC_E_Proceedings/TOC_Biomass_2010.pdf

Poy, F.S., Visani, F., and Terrosi, F. (1981). Automatic Injection in High Resolution Gas Chromatography: A Programmed Temperature Vaporizer as a General Purpose Injection System, *Journal of Chromatography A*, Vol.217, No.1, (6 November 1981), pp. 81–90, ISSN 0021-9673

Prins, M.J., Ptasinski, K.J., and Janssen, F.J.J.G. (2006). Torrefaction of Wood: Part 1 - Weight Loss Kinetics, *Journal of Analytical and Applied Pyrolysis*, Vol.77, No.1, (August 2006), pp. 28–34, ISSN 0165-2370

Prins, M.J., Ptasinski, K.J., and Janssen, F.J.J.G. (2006a). More Efficient Biomass Gasification via Torrefaction, *Energy*, Vol.31, No.15, (December 2006), pp. 3458–3470, ISSN 0360-5442

Prins, M.J., Ptasinski, K.J., and Janssen, F.J.J.G. (2006b). Torrefaction of Wood: Part 2 - Analysis of Products, *Journal of Analytical and Applied Pyrolysis*, Vol.77, No.1, (August 2006), pp. 35–40, ISSN 0165-2370

Punrattanasi, W and Spad, C. (2011). Environmental Sampling and Monitoring Primer: Gas Chromatography (webpage), Virginia Polytechnic Institute and State University (Virginia Tech) Department of Civil and Environmental Engineering, Available from http://www.cee.vt.edu/ewr/environmental/teach/smprimer/gc/gc.html

Sadaka, S., and Negi, S. (2009). Improvements of Biomass Physical and Thermochemical Characteristics via Torrefaction Process, *Environmental Progress & Sustainable Energy*, Vol.28, No.3, (October 2009), pp. 427–434, ISSN 1944-7442

Scott, R.P.W. (2003). *Chrom-Ed Book Series: Book 4 – Gas Chromatography Detectors* (e-book), LibraryForScience, LLC, Available from http://www.chromatography-online.org/2/contents.html

SHU, Gas Chromatography (webpage). http://teaching.shu.ac.uk/hwb/chemistry/tutorials/chrom/gaschrm.htm

Tumuluru, J.S., Sokhansanj, S., Hess, R.J., Wright, C.T. and Boardman, R.D. (2011). A review on biomass torrefaction process and product properties for energy applications, *Industrial Biotechnology*, Vol. 7, No. 5, (October, 2011), pp.384-401, ISSN: 1550-9087

United Nations (UN). (1998). Kyoto Protocol to the United Nations Framework Convention on Climate Change, Available from http://unfccc.int/resource/docs/convkp/kpeng.pdf

White, R.H., and Dietenberger, M.A. (2001). Wood Products: Thermal Degradation and Fire, *The Encyclopedia of Materials: Science and Technology,* KH Jürgen Buschow, RW Cahn,

MC Flemings, B Ilschner, EJ Kramer, S Mahajan, and P Veyssière (Eds.), Elsevier, Ltd., Amsterdam, the Netherlands, pp. 9712–9716, ISBN 978-008-0431-52-9

Win, D.T. (2005). Furfural-Gold from Garbage, *AU Journal of Technology*, Vol.8, No.4, (Apr. 2005), pp. 185-190, Available from http://www.journal.au.edu/au_techno/2005/apr05/vol8no4_abstract04.pdf

Zanzi, R, Ferro, D.T., Torres, A., Soler, P.B., and Bjornbom, E. (2002). Biomass Torrefaction, In: *Proceedings of the 6th Asia-Pacific International Symposium on Combustion and Energy Utilization, Kuala Lumpur, Malaysia, May 20–22, 2002.*

12

Application of Pyrolysis-Gas Chromatography/Mass Spectrometry to the Analysis of Lacquer Film

Rong Lu, Takayuki Honda and Tetsuo Miyakoshi
Department of Applied Chemistry,
School of Science and Technology, Meiji University,
1-1-1 Higashi-mita, Tama-ku, Kawasaki-shi,
Japan

1. Introduction

1.1 Characteristics of lacquer

Oriental lacquer is a reproducible natural product that has been used for thousands of years in Asia. No organic solvent evaporates during the drying process, only water. Because of the self-drying system, natural lacquer is an eco-friendly product that is expected to be useful in the future as a coating material.

Lacquer trees are members of the family Anacardiaceae, which includes more than 73 genera worldwide and approximately 600 species including mango (*Mangifera indica*) and cashew (*Anacardium occidentale*), but most of them grow in the subtropical region of Southeast Asia [1-3]. In general, only a few kinds of lacquer trees that grow in the evergreen forests of East Asia are able to produce lacquer sap.

Lacquer has been used in Asian countries for thousands of years as a durable and beautiful coating material [4-7]. Cultural treasures coated with lacquer have maintained their beautiful surfaces without loss of their original beauty for more than 9000 years [8-10].

Lacquer sap for the manufacture of lacquerware is collected not only in China and Japan but also in Southeast Asia, as shown in Figure 1-1. In Japan, about 10 years after cultivating a lacquer tree, cuts are made in the tree trunk. A white, milky resin seeps out of the wounds, and this sap is collected in a keg. After the lacquer sap has been collected, the tree is left untouched for 3 or 4 days before cuts are again made in the tree to collect more sap. Laccol (MW=348) is the main component of *Rhus succedanea* trees, which grow in Vietnam and Chinese Taiwan, and thitsiol (MW=348) is the main component of *Melanorrhoea (Gluta) usitata* trees, which grow in Thailand and Myanmar [11-13]. All these saps are used as a surface coating for wood, porcelain, and metal wares in Asia.

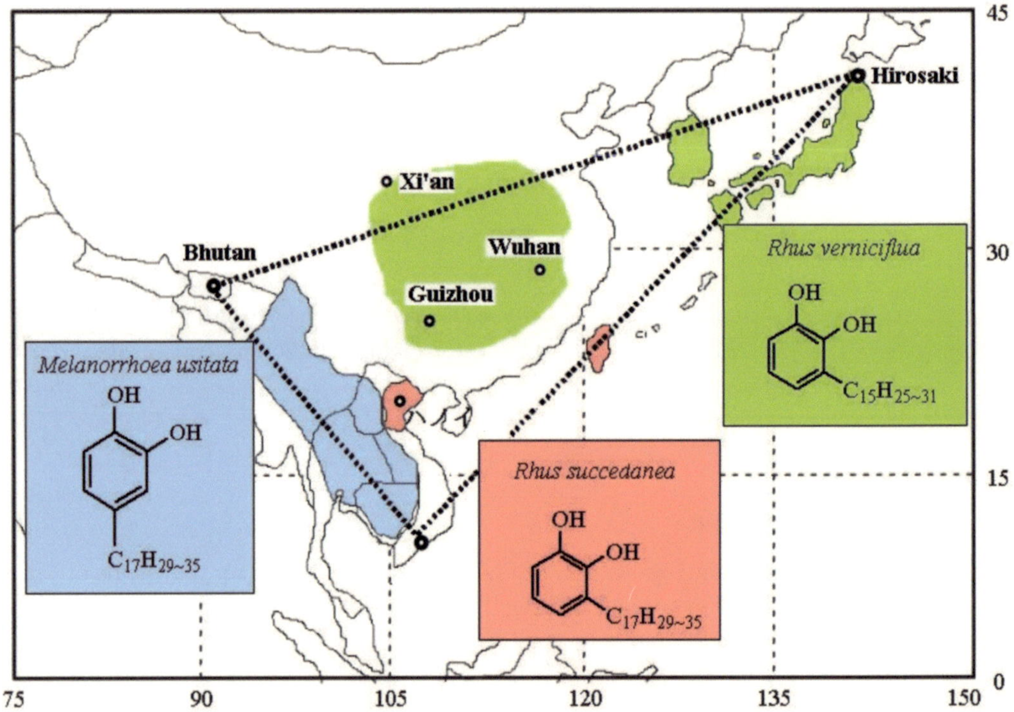

Fig. 1-1. Asian origin and type of lacquer sap

1.2 Components of lacquer

Lacquer is the sap obtained by tapping lac trees, specifically *Rhus vernicifera* (in Japan, China, and Korea), *Rhus succedanea* (in Vietnam and Taiwan), and *Melanorrhoea (Gluta) usitata* (in Thailand and Myanmar). The sap from *Rhus vernicifera* comprises a latex material composed of phenol derivatives (60-65%), water (20-25%), plant gum including saccharides (5-7%), water-insoluble glycoproteins (2-5%), and the laccase enzyme (0.1%) [11-13].

Under the microscope, this milky liquid appears to be a collection of water particles in an emulsion. Small particles of water are dispersed in the oily substance called urushiol. Unlike milk, in which fat is dispersed in water, water is dispersed in the urushiol oil in lacquer. The emulsion droplets contain the dissolved laccase enzyme and polysaccharides. It is believed that nitrogenous substances play an important role in forming a stable emulsion of the aqueous component and urushiol.

The composition of lacquer sap has been investigated using gel permeation chromatography (GPC), high performance liquid chromatography (HPLC), mass spectrometry (MS), and nuclear magnetic resonance (NMR), including two-dimensional NMR spectroscopy. Urushiol is a mixture of 3-substituted catechol derivatives having carbon chains with 15 carbons with 0-3 double bonds. The triene side chain of urushiol makes up 60-70% of urushiol [14]. All of these components have been separated and purified. Figure 1-2 shows photographs of urushiol, polysaccharides, glycoprotein, laccase, and stellacyanin obtained from Chinese lacquer sap isolated according to previously reported methods. Procedures utilized for the separation and purification of lacquer components from *Rhus vernicifera* lacquer sap are shown in Figure 1-3 [15].

Fig. 1-2. Photos of lacquer components

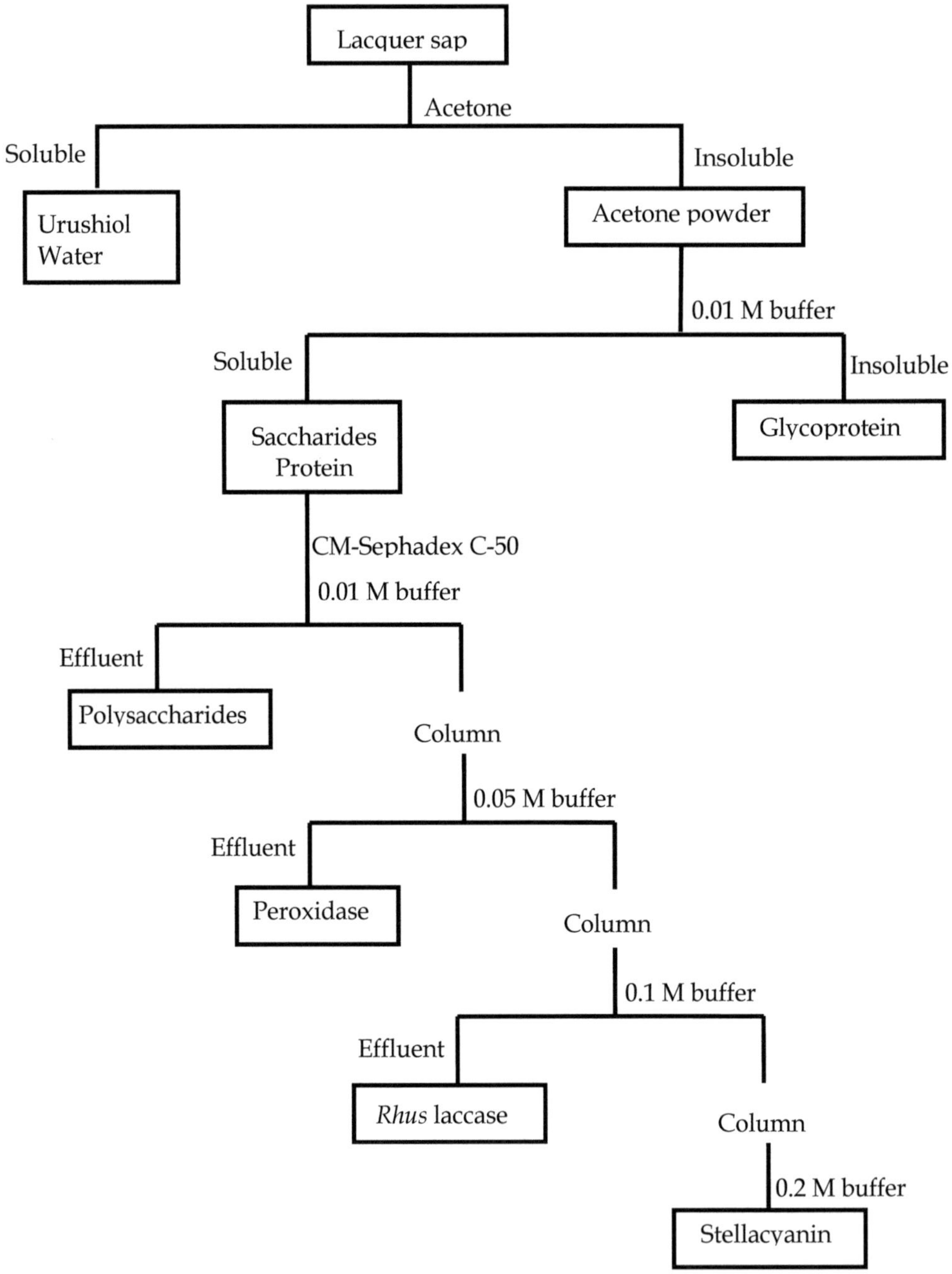

Fig. 1-3. Separation of lacquer sap

1.3 Drying mechanism

The drying mechanism of lacquer sap catalyzed by laccase was first explicated by Kumanotani [16], and the explanation was supplemented and developed by our later studies [17]. The catechol ring of urushiol is first oxidized by laccase to form dimers, trimers, and oligomers, and after the urushiol monomer concentration has decreased to less than 30%, a bridge-construction reaction due to auto-oxidation of the unsaturated side chain occurs. Figure 1-4 shows the laccase-catalyzed oxidation coupling of urushiol [16-17].

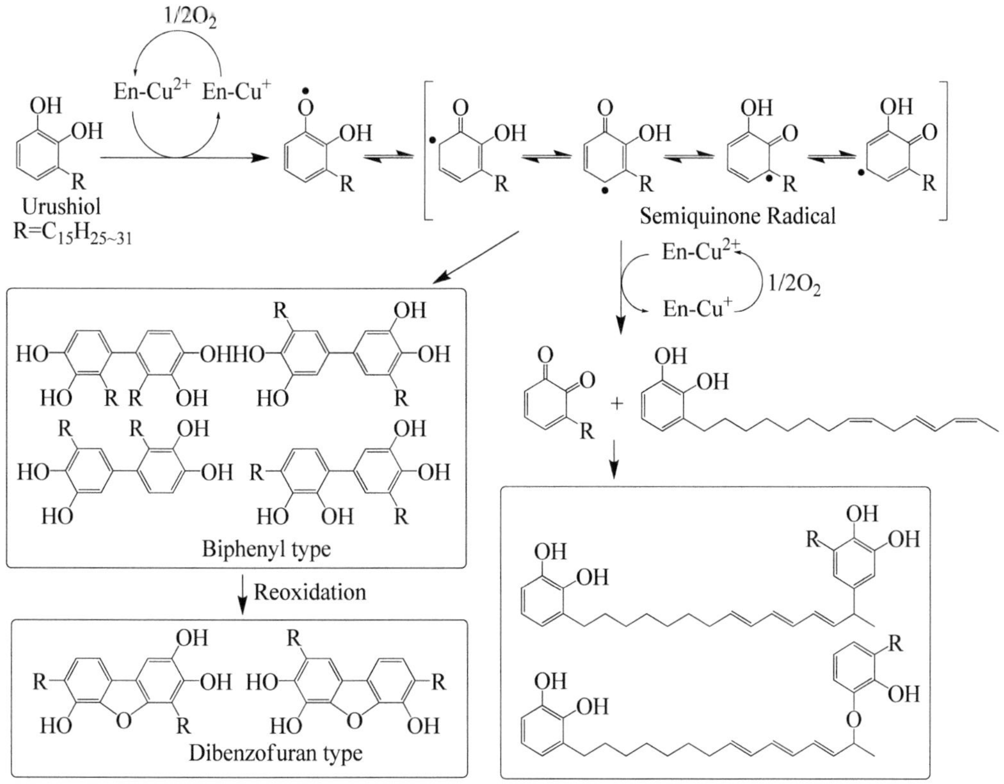

Fig. 1-4. Polymerization mechanism of urushiol catalyzed by laccase

The dimerization of urushiol proceeds through laccase-catalyzed nuclear-nuclear (C-C) coupling. Its detailed mechanism is as follows. During the first step of the dimerization, En-Cu^{2+} oxidizes urushiol to give the semiquinone radical and En-Cu^{+}. During the sap cooking process, the presence of plant gum cause significant foaming, which mixes air and sap during the treatment as well as accelerating the oxidation of Cu^{+} to Cu^{2+} in the laccase. The reformed En-Cu^{2+} takes part in the repeated oxidation of urushiol. The semiquinone radical formed undergoes a C-C coupling reaction to produce biphenyl dimers, and the urushiol quinone formed through the disproportionation reaction undergoes hydrogen abstraction from the triene side chain of urushiol to give a nucleus-side chain C-C coupling dimer. The polymerization of urushiol may proceed through these types of couplings. However, many and various kinds of reactions may occur in the film-making process. Both enzymatic reaction and auto-oxidation (Figure 1-5) are repeated to form a durable network polymer.

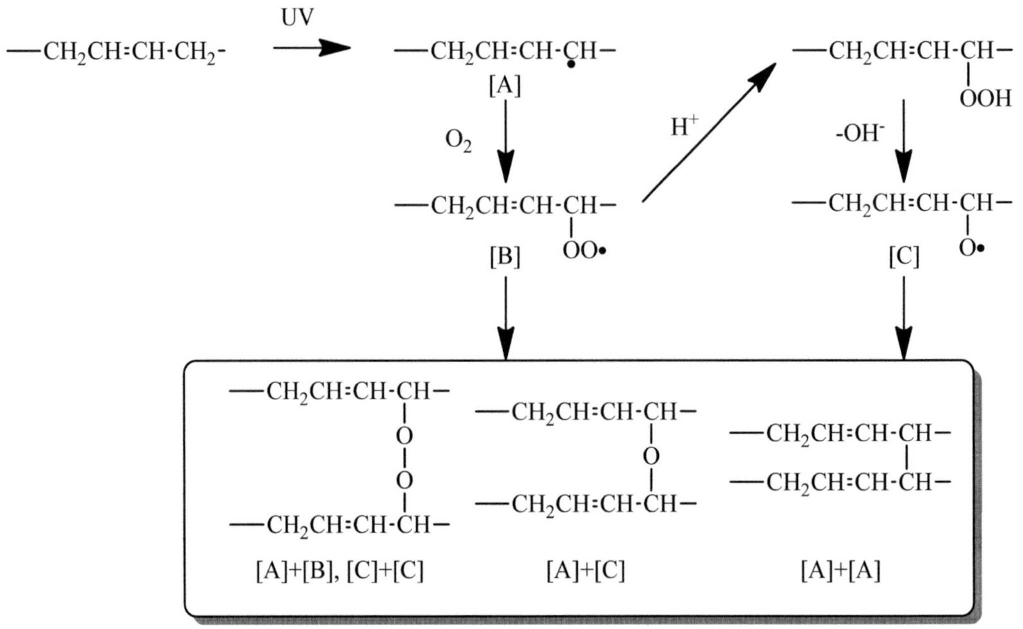

Fig. 1-5. Auto-oxidation of unsaturated lacquer side chain

In order to reveal the enzymatic polymerization mechanism of thitsiol, we reinvestigated the dimer structures produced in the initial stage of enzymatic polymerization by using modern techniques including 1H-1H and 1H-13C correlation two-dimensional NMR measurements [18-19]. The structures of thitsiol dimers from *Melanorrhoea (Gluta) usitata* were characterized by means of high-resolution NMR spectroscopy involving two-dimensional NMR measurements using field gradient DQF-COSY, HMQC, and HMBC. Almost all proton and carbon absorptions were assigned. The results showed that the main products of thitsiol dimers catalyzed by laccase are 1,1',2,2'-tetrahydroxy-3,3'-dialkenyl-5,5'-biphenyl, 1,1',2,2'-tetrahydroxy-3,3'-dialkenyl-6,5'-biphenyl, and 1,1',2,2'-tetrahydroxy-3,4'-dialkenyl-5,5'-biphenyl. The thitsiol dimers are almost all due to nuclear-nuclear (C-C) coupling, which differs from the nuclear-side chain (C-O-C) coupling of urushiol as shown in Figure 1-4 [20].

2. Py-GC/MS

Pyrolysis gas chromatography/mass spectrometry (Py-GC/MS) is a comparatively old analytical method (Figure 2-1). It is one of the techniques developed to analyze polymers. It has problems with reproducibility and long analysis times and it was not used very much, though pyrolysis began to be used to analyze natural fibers and synthetic fibers. The technique for combining this pyrolysis method with gas chromatography and a mass spectrometry was reported by Davison in 1954, and it became appreciated in recent years as a significant technique to analyze the three-dimensional network structure. This technique involves high-temperature treatment of the polymer membrane in order to isolate each element of the polymer using the solvent for gas chromatography, and to perform mass spectrometry of the gas chromatography peak at the same time.

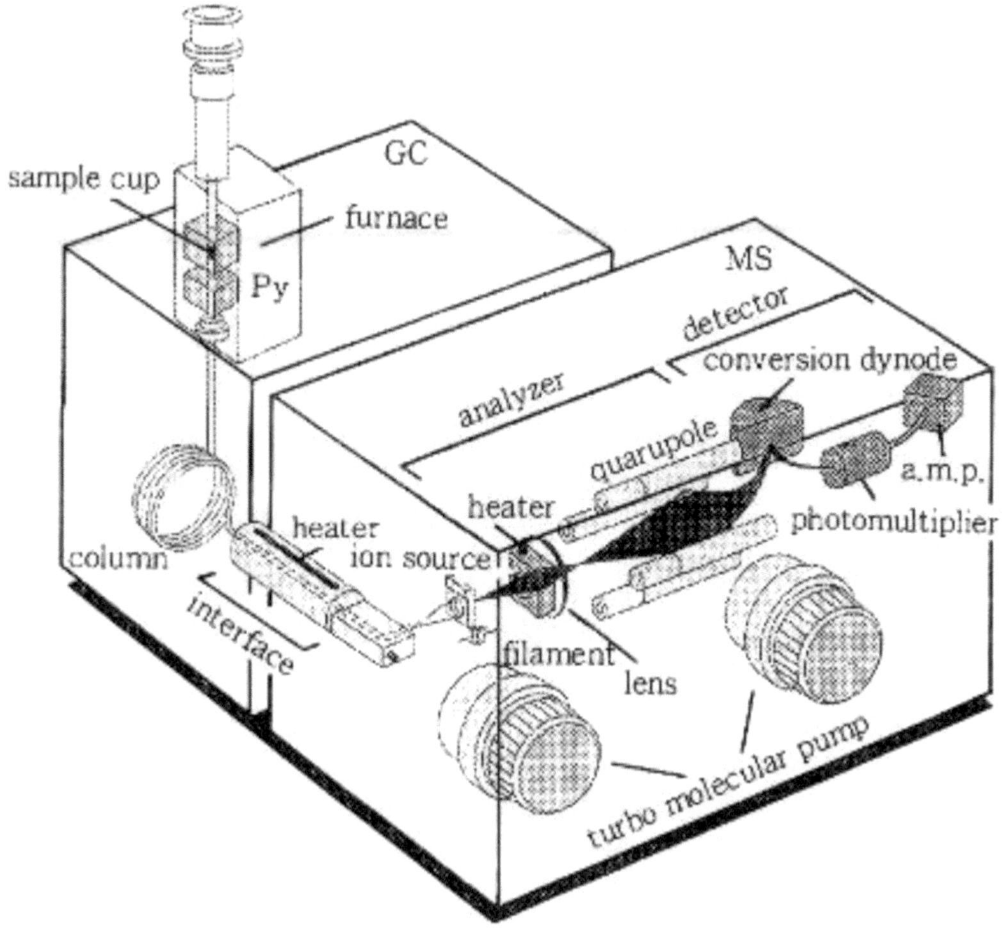

Fig. 2-1. Configuration of Py-GC/MS system

The advantage of this technique is the convenience of the device and pretreatment of the analysis sample. In particular, a polymer can be analyzed using a very small amount of sample (0.01-1mg).

The best feature of pyrolysis gas chromatography/mass spectrometry is pyrolysis. The thermal cracking unit can be heated quickly to a high temperature and is directly related to the reproducibility of the analysis performance. The heating in this method is done in a microfurnace (Figure 2-2).

The method involves putting the sample in the inactivated sample holder of the microfurnace, dropping it into the reactor core from the state set in the device, which is filled with helium as the carrier gas using a switch, and then pyrolyzing it. A feature of this technique is that the measurement result is steady because the small capacity keeps the temperature dispersion is comparatively low [21].

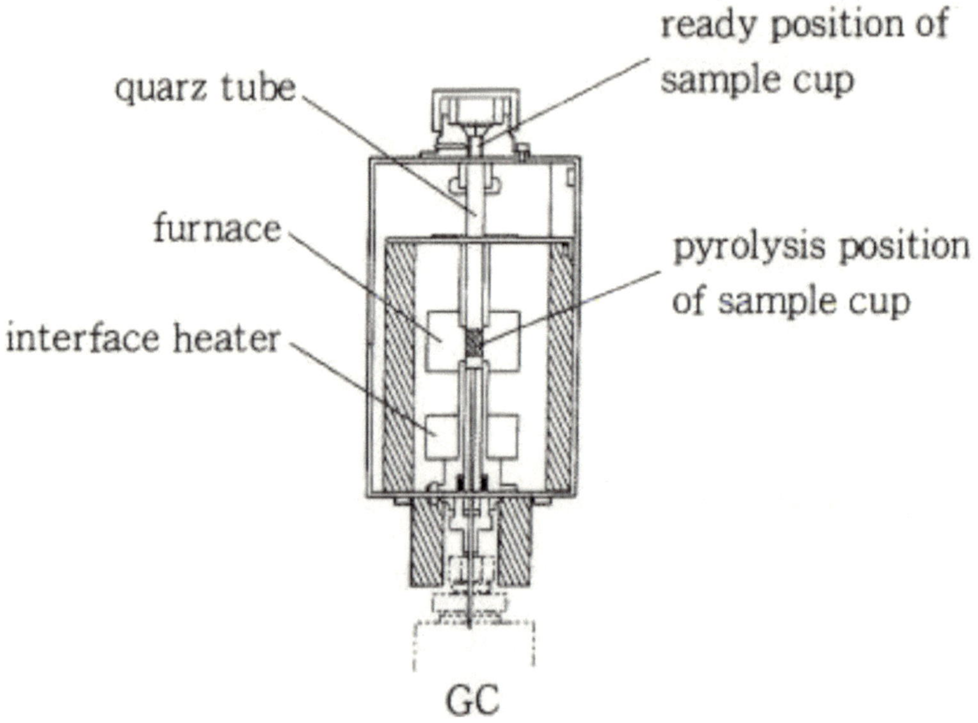

Fig. 2-2. Internal structure of pyrolyzer

2.1 Direct method (single-shot Py-GC/MS)

The easiest Py-GC/MS is a method in which only one heating performs the thermal decomposition, and it is called single shot. The range of temperatures that cause thermal decomposition is 50-1000°C (in the case of the PY-3030). After the sample is thermally decomposed, it was gasified and introduced into a gas chromatography. The gas separated by GC is measured by MS, as shown in Figure 2-3. The advantage of this technique is being able to analyze all components in the original sample by one measurement. This means that the ratio of components in the original solid can be measured.

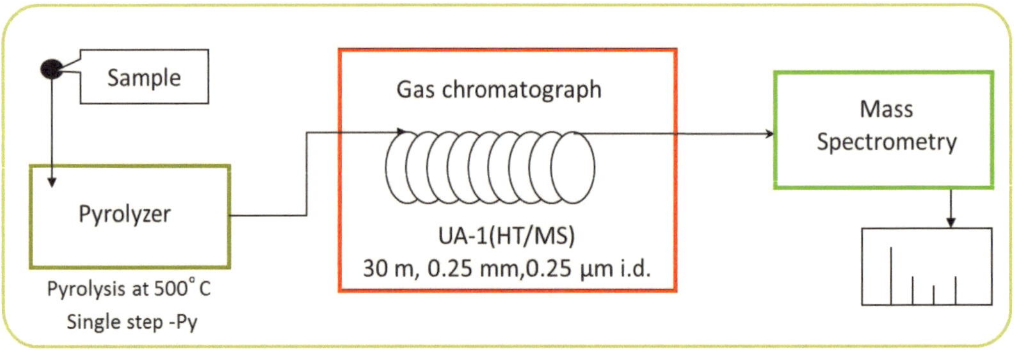

Fig. 2-3. Configuration of Py-GC/MS

The decomposition temperature of lacquer film is 500°C. The GC column is a type of dimethyl polysiloxane or phenyl methylpolysiloxane. Because lacquer film has a catechol configuration, its polarity is very high. Therefore, if a very polar column is used, detection will decrease. Moreover, setting the temperature of the oven high enough and using a heat-resistant column are also important because urushiol is undetectable at temperatures lower than 280°C.

2.2 Fragmentation mechanism of lacquer film

The main component of lacquer is urushiol, which is a catechol derivative, and it is polymerized to form a film by the laccase. When a lacquer film is analyzed by Py-GC/MS, mass fragment m/z=108 from alkyl phenol and mass fragment m/z=123 from alkyl catechol will appear as characteristic fragment ions. The fragment ion m/z=108 is the hydrogen attached to the 3-position carbon of the catechol ring that appears due to McLafferty transference. M/z=123 is a fragment ion which the first CH_2 from a catechol ring enters another catechol ring, and constitutes a seven-member ring, as shown in Figure 2-4.

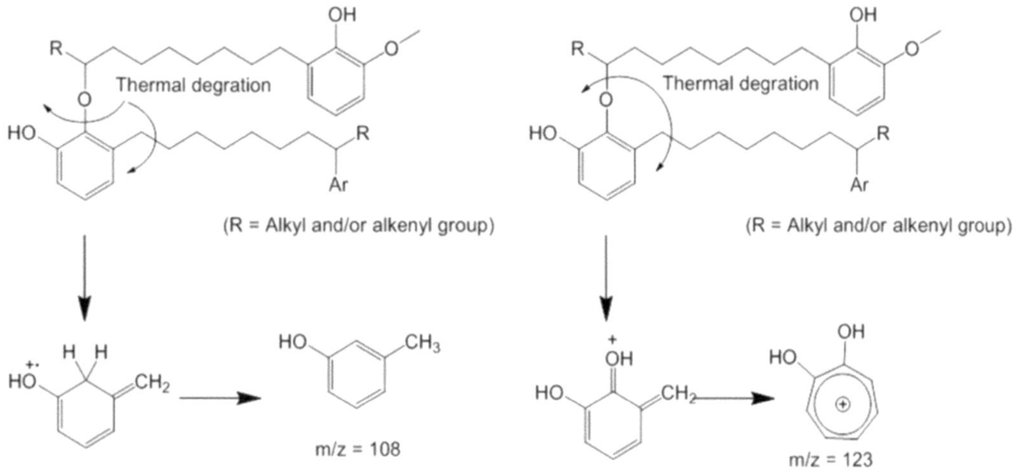

Fig. 2-4. Formation mechanism of fragment m/z = 108 and m/z = 123

The pyrolysis GC/MS Chinese lacquer film is shown in Figure 2-5. Many components are detected in the total ion chromatograph, and this is due to the complex structure of the lacquer film and various components contained in a lacquer sap. The ion chromatographs of (A) m/z=108 and (B) m/z=123 are shown in Figure 2-6. It is a graphic chart that is easily intelligible at a glance. The highest peak is the component that has carbon number seven in the side chain. Because the factor that becomes such is cut next to the 8th carbon in which most double bonds exist in urushiol. In the case of urushiol and thitsiol (Figure 2-7), carbon number seven is the most frequent component, but since the position of the double bond changes in laccol, the alkyl phenol of carbon number nine is detected as the greatest peak (Figure 2-8) [22-24].

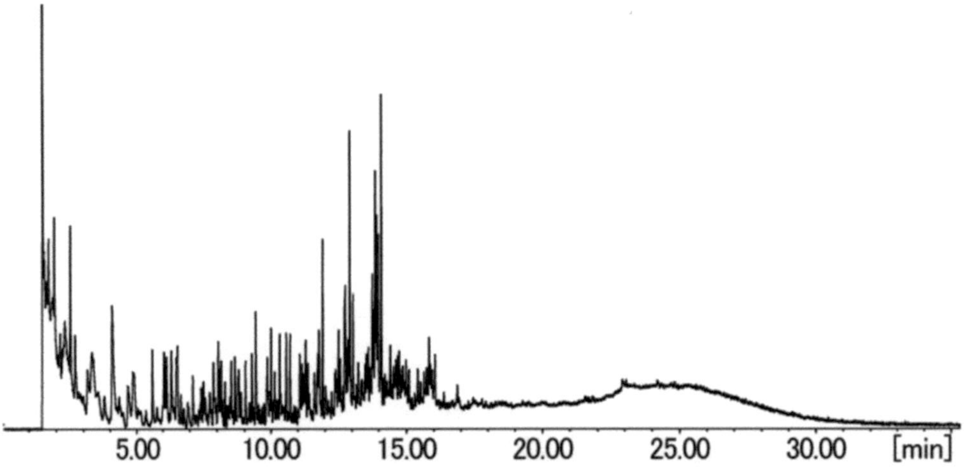

Fig. 2-5. Total ion chromatography (TIC) of lacquer film

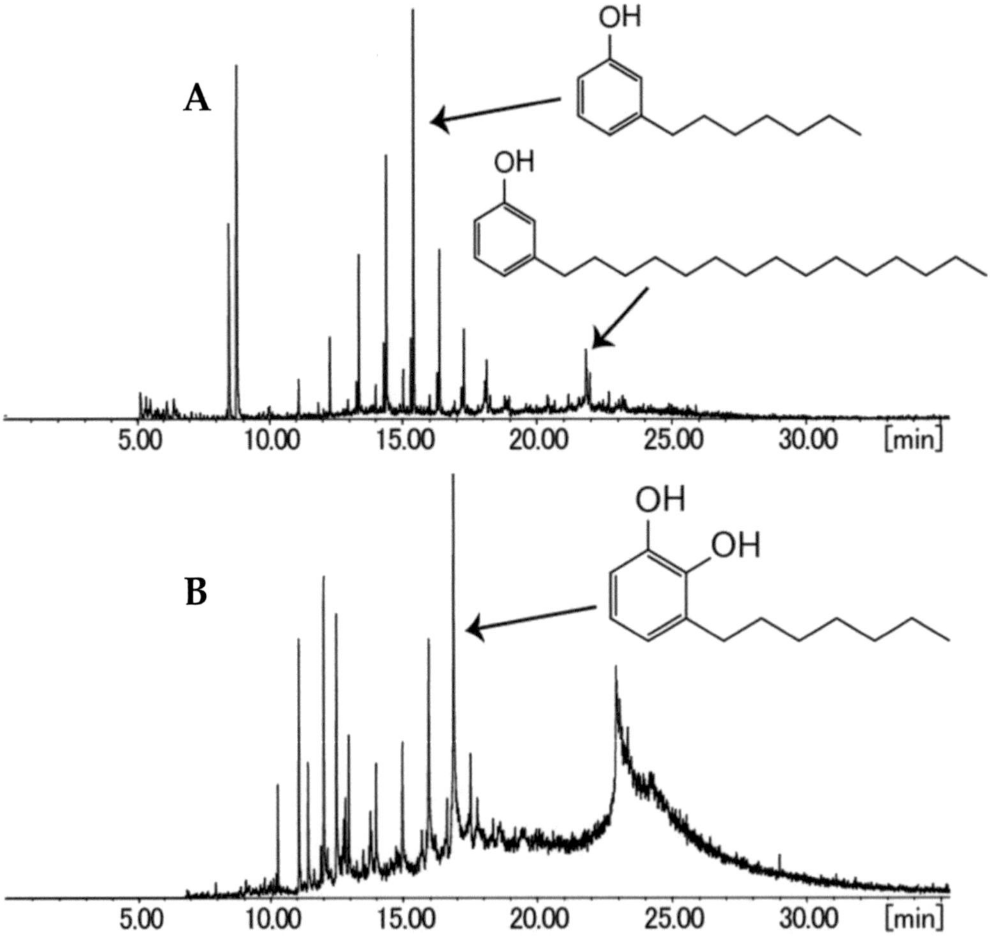

Fig. 2-6. Ion chromatography of Chinese lacquer film: (A) m/z=108 and (B) m/z=123

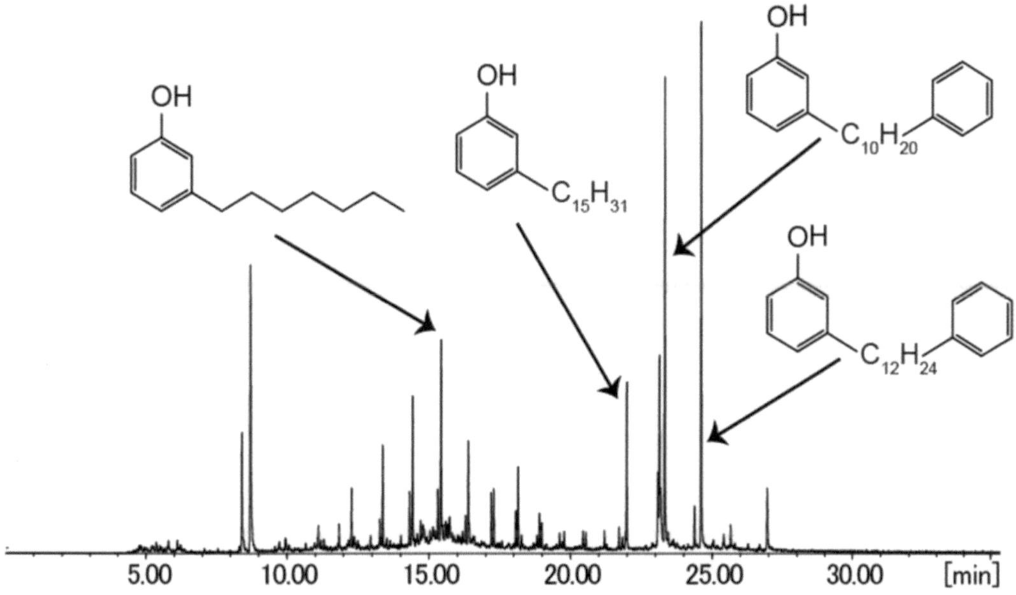

Fig. 2-7. Ion chromatography of Thai lacquer film

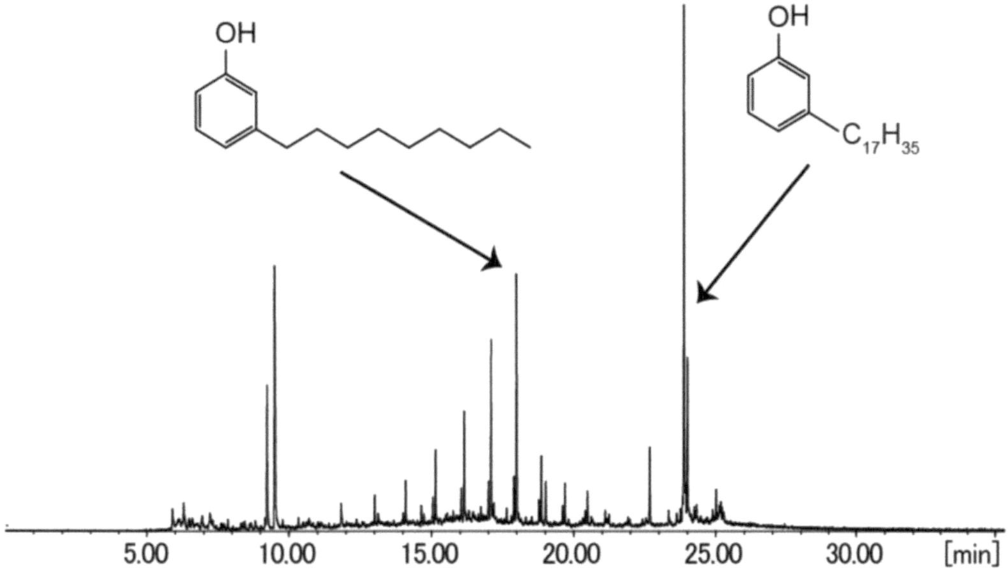

Fig. 2-8. Ion chromatography of Vietnamese lacquer film

2.3 Derivatization method

The principal constituent of Japanese lacquer, urushiol, contains catechol ring. Catechol has two hydroxyl groups of a phenol nature, and because its polarity is high, it tends to adhere to a column. It is said that this leads to a degradation of analytical sensitivity. Adsorption is one reason that the component ratios differ in analysis results. As a method of suppressing alterations of the component ratio by such adsorption, methylation of the hydroxyl group by tetramethylammonium hydroxide (TMAH) was devised. There are reports of the use of TMAH in analysis of a lignin, which generates a component containing the hydroxyl group during thermal decomposition and is an example of analysis of a polysaccharide [25-30]. TMAH can also be used effectively at the time of the thermal decomposition of Japanese lacquer. The lacquer film and TMAH were combined in the sample cup, and the sample cup was put in the pyrolyzer heated to 500°C.

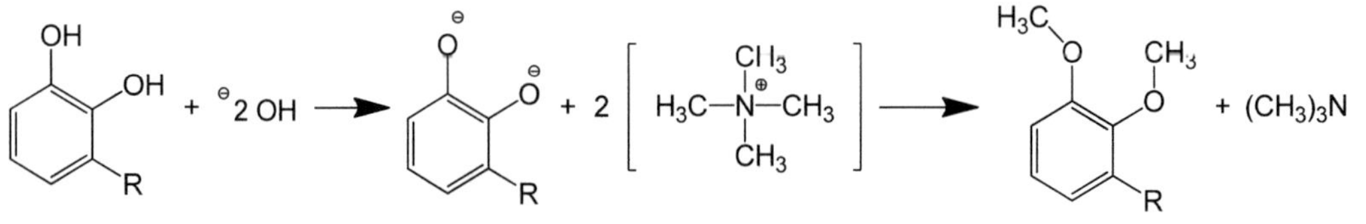

Scheme 2-1. Methylation reaction of hydroxyl group by TMAH

In a lacquer film, a hydroxyl group replaces a methyl group in a thermal decomposition (Scheme 2-1). The example of a Chinese lacquerware analyzed by this method is shown in Figure 2-9. In this examination, the amount of lacquer film used was 0.75 mg.

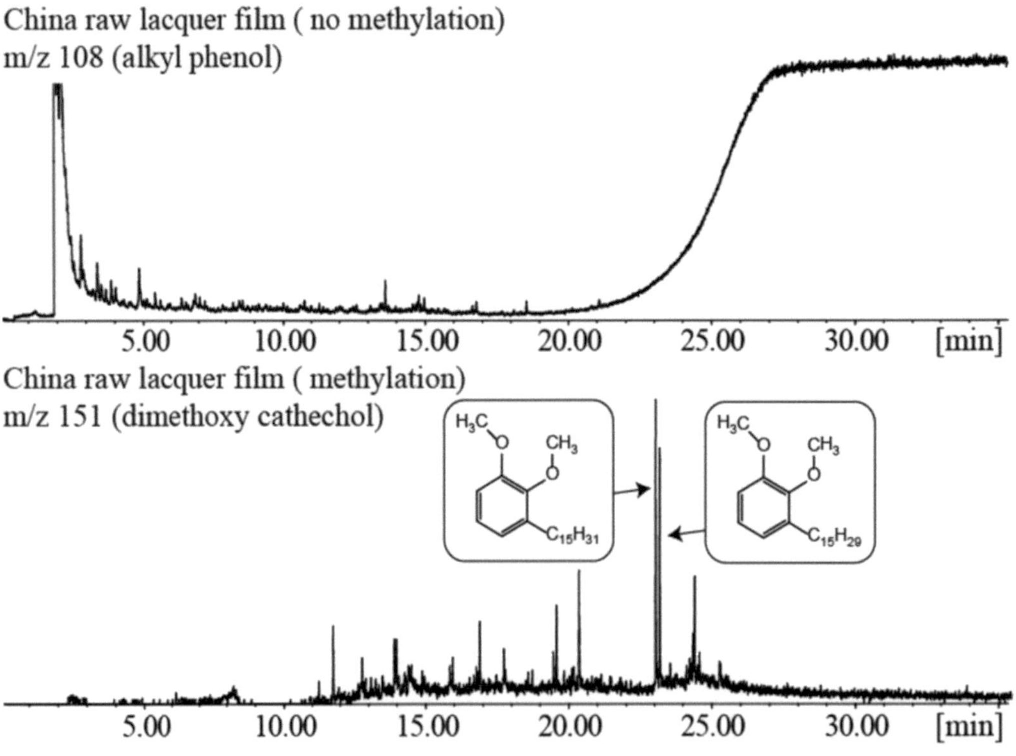

Fig. 2-9. Comparison of sensitivity of methylation

When not methylated, the alkylphenol of the carbon number seven should be detected. On the other hand, when it is methylated, 1,2-dimethoxy-3-pentadecylbenzene, from which urushiol was therefore protected by methylation, was detected. Methylation, even if the amount of lacquer film is small, can reveal urushiol. This method is useful for analysis of cultural properties that can provide only small samples.

2.4 Evolved gas analysis

Evolved gas analysis (EGA) is a method of analyzing the gas emitted during heat is applied to a sample [31-34]. The mechanical configuration characteristically uses a very short column. The column is about 1 m in length, and a fixed zone does not exist. Since many organic macromolecules are decomposed at less than 1000°C, a set of unstring temperatures in the range of 50–1000°C is used. In the case of a lacquer film, the temperature range is set to 50–650°C (Figure 2-10).

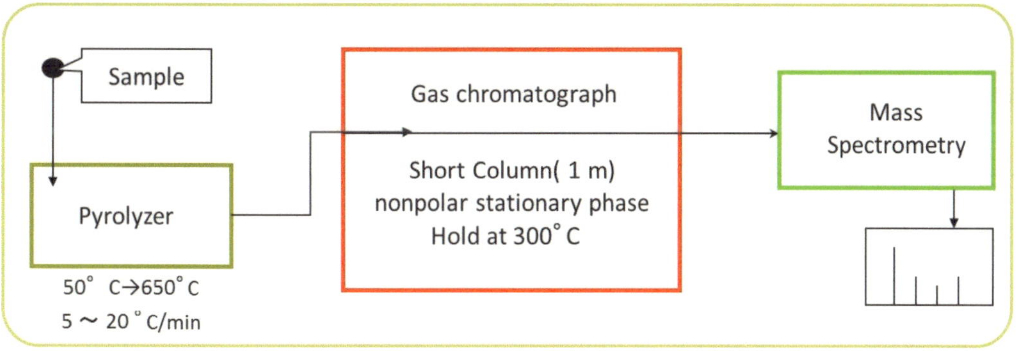

Fig. 2-10. Configuration of EGA system

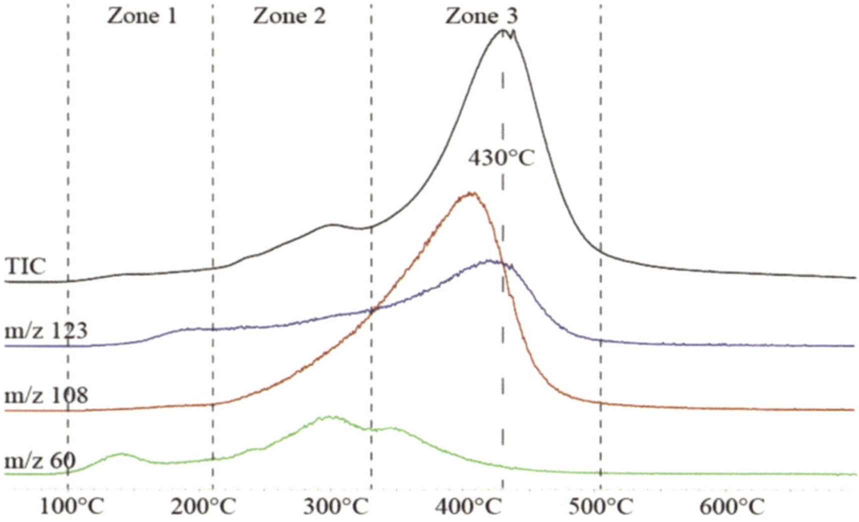

Fig. 2-11. EGA of Chinese lacquer film

EGA analysis decomposes a lacquer film in three steps (Figure 2-11). When the heat decomposition points were divided into three steps, two were the original peaks of m/z=60 and m/z=108, which is a thermal decomposition peak of urushiol, and one was m/z=123. The first peak of m/z=60 at a lower temperature was short carboxylic acid with various carbon chains. The following peak was considered to be due to the decomposition of sugars. The acetone powder (AP) taken from lacquer sap also was analyzed by EGA, and the peak position of acetone powder was mostly in agreement with the second EGA peak of Japanese lacquer film (Figure 2-12).

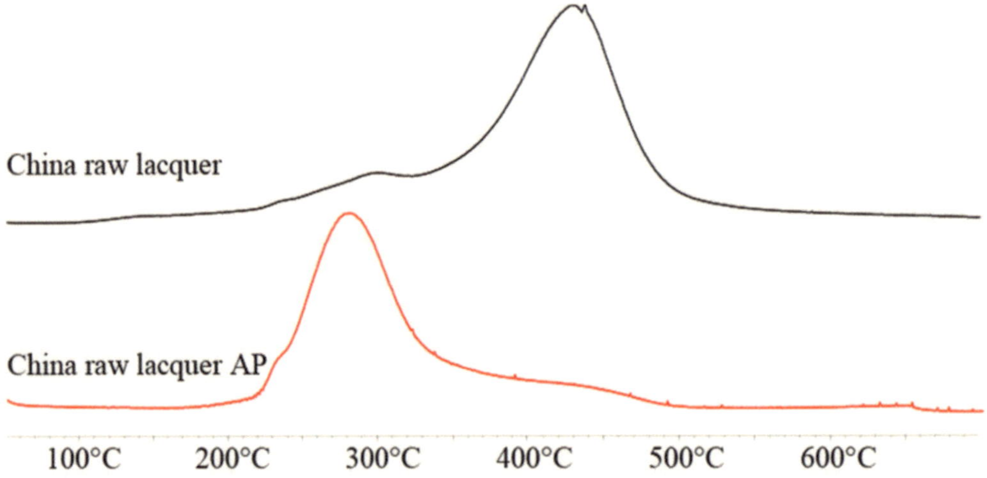

Fig. 2-12. EGA of Chinese lacquer film and AP

2.5 Double-shot pyrolysis

An analysis method that can be performed using the results of EGA is double-shot pyrolysis. Double-shot pyrolysis applies heat to the same sample gradually and analyzes only the gas generated in a certain range (Figure 2-13).

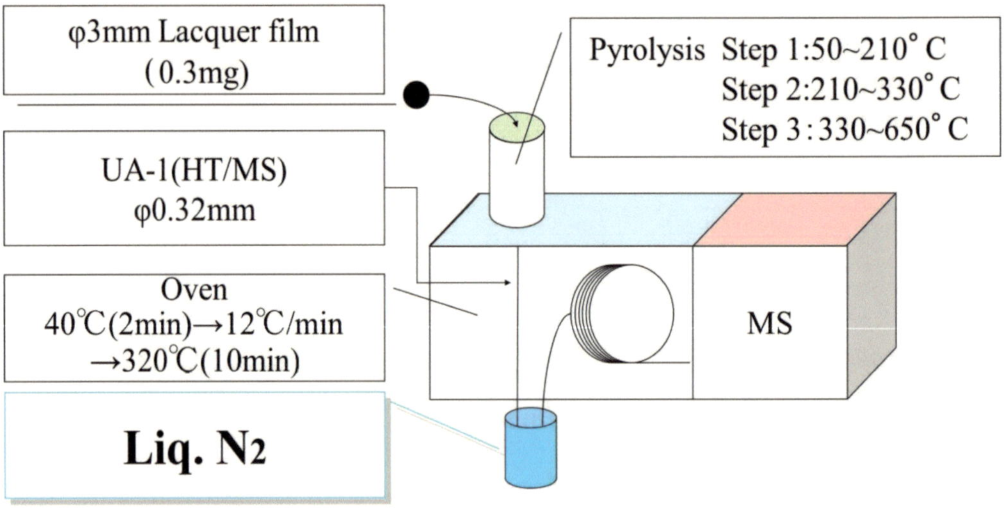

Fig. 2-13. Configuration of double-shot pyrolysis

A sample is placed in a pyrolyzer, and heating is started. The first part of a column is dipped in liquid nitrogen, and the gas chromatograph is not moved. If thermal decomposition occurs at a predetermined temperature, the pyrolyzed sample is removed from the heating block immediately. After raising the sample from the oven, removed the GC column from the liquid nitrogen and then started gas chromatography. Follow this procedure by the number of daylights of a thermal decomposition. In this way, the Py-GC/MS will store a record of each peak of EGA that is attained [35-37].

This technique is convenient to remove only a plasticizer previously added to a sample or to remove a solvent that remains in very small quantity. However, this method cannot be used when a sample reacts with the energy of heating. In the case of a lacquer sap, various additive admixtures (pine drying oil, tar, etc.) may be present. Thus, this technique is very effective when components with clearly different heat decomposition points are present.

3. Application of Py-GC/MS in lacquer analysis

The three kinds of lacquer films analyzed by Py-GC/MS in this section are the lacquer saps from *Rhus vernicifera, Rhus succedanea,* and *Melanorrhoea (Gluta) usitata.* The lacquers were coated on a 70 cm × 40 cm glass plates. They were allowed to polymerize in a humidity-controlled chamber at a relative humidity of 70-90% at 20°C for 12 h, and then removed from the chamber and stored in the open air for 3 years [38].

The pyrolysis-gas chromatography/mass spectrometry measurements were carried out using a vertical microfurnace-type pyrolyzer PY-2010D (Frontier Lab, Japan), an HP 6890 gas chromatograph (Hewlett-Packard), and an HPG 5972A mass spectrometer (Hewlett-Packard). A stainless steel capillary column (0.25 mm i.d. × 30 m) coated with 0.25 µm of Ultra Alloy PY-1 (100% methylsilicone) was used for the separation. The sample (1.0 mg) was placed in the platinum sample cup, and the cup was placed on top of the pyrolyzer at near ambient temperature. The sample cup was introduced into the furnace at 500°C, then the temperature program of the gas chromatograph oven was started. The gas chromatograph oven was programmed to provide a constant temperature increase of 20°C per min from 40° to 280°C, then held for 10 min at 280°C. The flow rate of the helium gas was 18 ml/min. All the pyrolysis products were identified by mass spectrometry. The mass spectrometry ionization energy was 70 eV (EI-mode).

3.1 Pyrolysis of urushiol polymer

The urushiol lacquer film was pyrolyzed at 500°C. The total ion chromatogram (TIC) and mass chromatogram (m/z=320) are shown in Figure 3-1. Although a complex TIC was obtained, the mass chromatogram and mass spectrum revealed that urushiol (MW=320) was the main component of the lacquer.

The pyrolysis products obtained in mass chromatograms and mass spectra of m/z=123 and m/z=108 of urushiol lacquer film are shown in Figure 3-2. 3-Heptylcatechol (C7) and 3-heptylphenol (C7) were detected in the mass chromatogram at m/z=123 and m/z=108, respectively, and had the highest relative intensity.

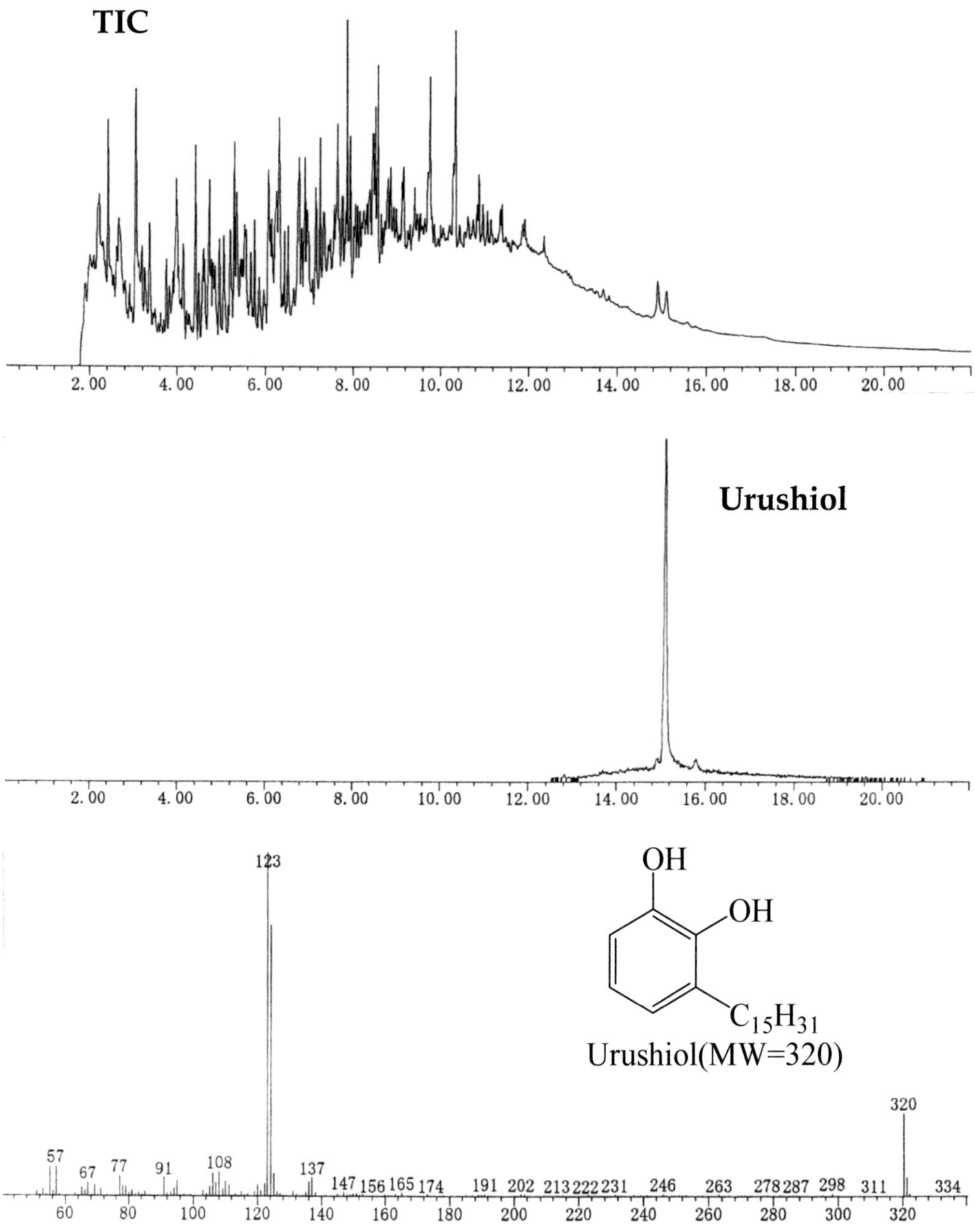

Fig. 3-1. TIC, mass chromatogram (m/z=320), and mass spectrum of urushiol lacquer film

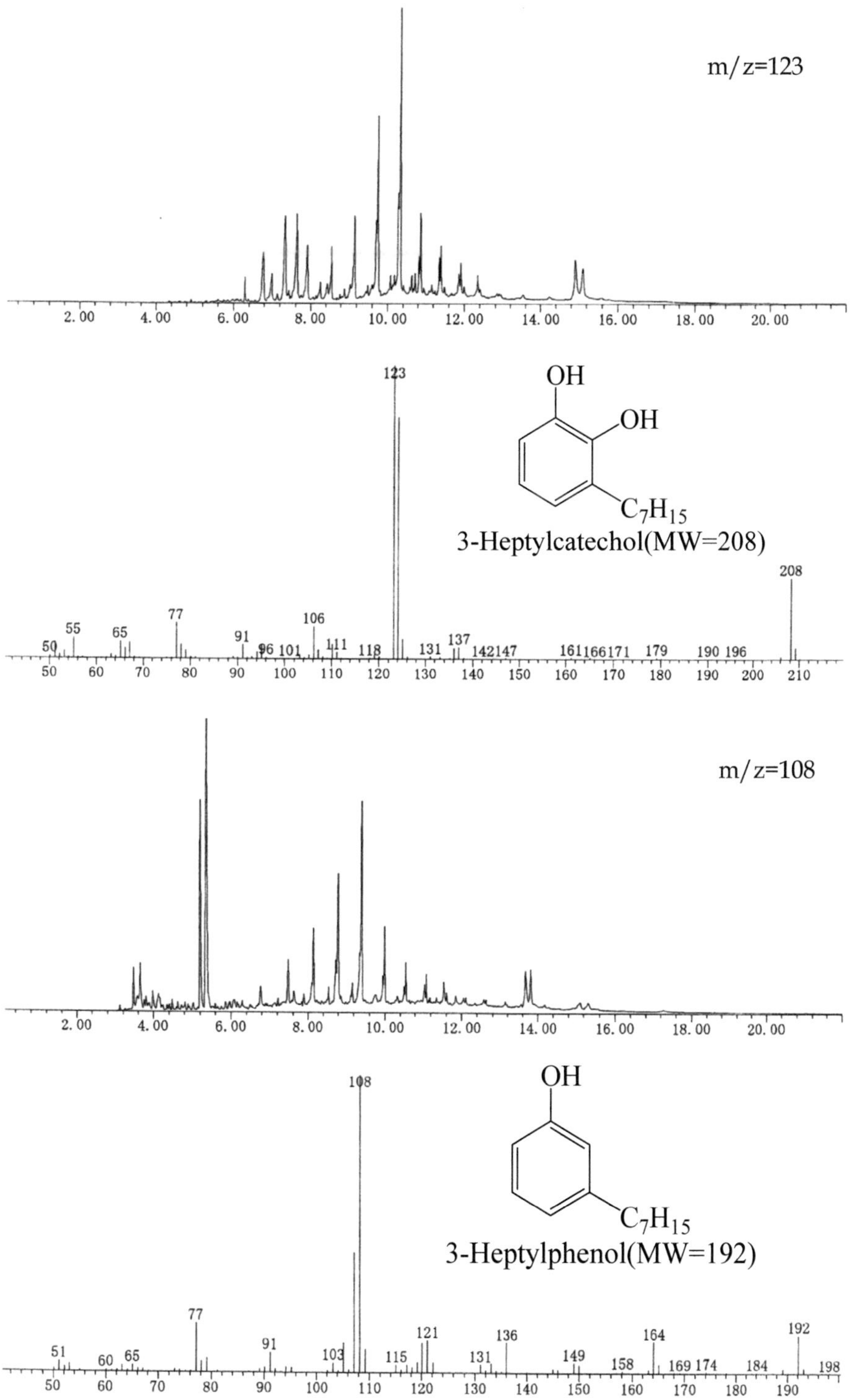

Fig. 3-2. Mass chromatograms (m/z=123 and 108) and mass spectra of urushiol lacquer film

It has been reported that the double bonds of olefins at the α- and β- positions are most susceptible to thermal cleavage [39-40]. Therefore, as shown in Scheme 3-1, the highest yield of 3-heptylcatechol (C7) can be attributed to the preferential cleavage at the α-position of the double bonds of the nucleus-14th chain C-O couplings of the urushiol polymers.

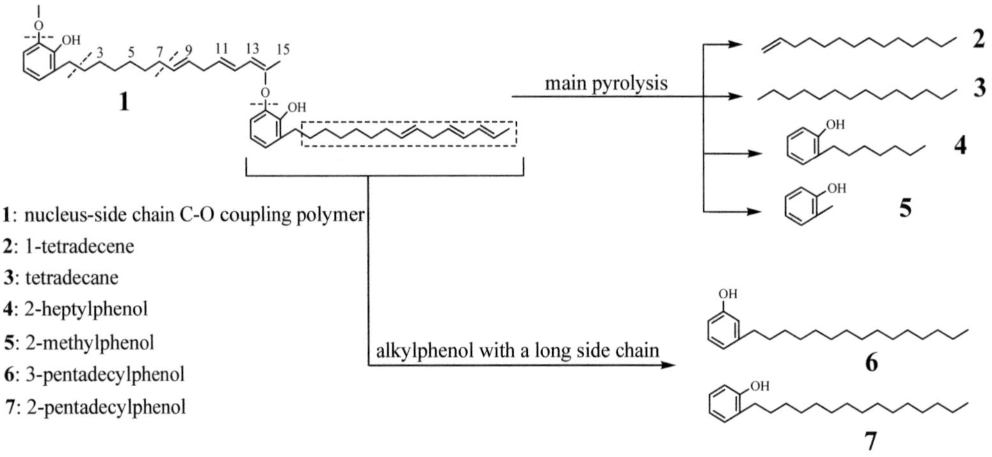

Scheme 3-1. Pyrolysis mechanism of urushiol polymer

3.2 Pyrolysis of laccol polymer

A laccol lacquer film was also pyrolyzed at 500°C at the same time as the urushiol lacquer film. The total ion chromatogram and mass chromatogram (m/z=320) are shown in Figure 3-3. Although a complex TIC was obtained, the mass chromatogram and mass spectrum confirmed that laccol (MW=348) was the main component of the lacquer film.

The pyrolysis products obtained in mass chromatograms and mass spectra of m/z=123 and m/z=108 of laccol lacquer film are shown in Figure 3-4. 3-Nonylcatechol (C9) and 3-nonylphenol (C9) were detected in the mass chromatogram at m/z=123 and m/z=108, respectively, and had the highest relative intensity. The detailed Py-GC/MS results of laccol in our laboratory have been previously reported [23].

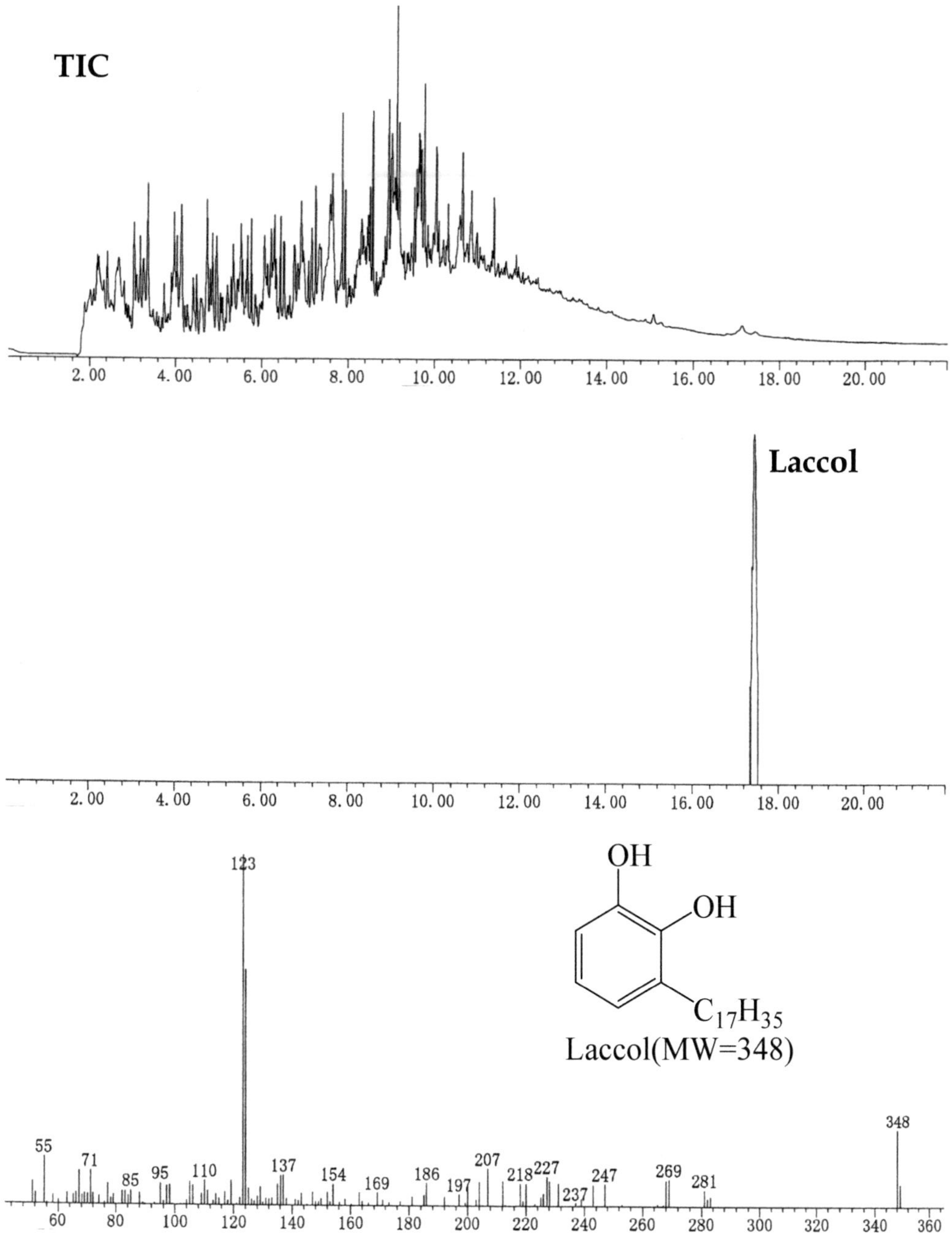

Fig. 3-3. TIC, mass chromatogram (m/z=348), and mass spectrum of laccol lacquer film

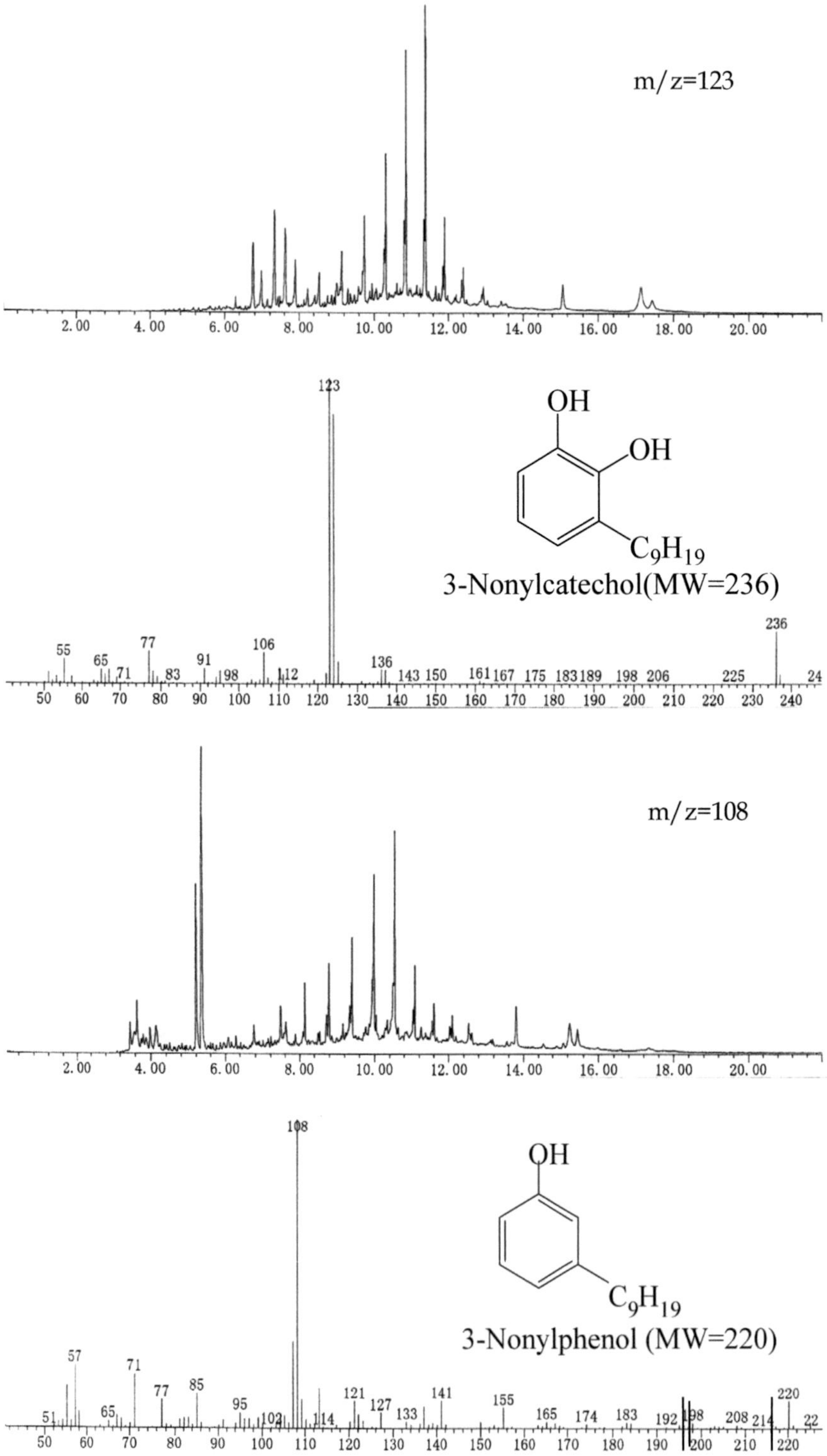

Fig. 3-4. Mass chromatograms (m/z=123 and 108) and mass spectra of laccol lacquer film

3.3 Pyrolysis of thitsiol polymer

Melanorrhoea (Gluta) usitata is a lacquer tree that grows in Myanmar, Thailand, and Cambodia. The main component of *Melanorrhoea (Gluta) usitata* is thitsiol, which contains 3- and 4-heptadecadienylcatechols as well as a series of α- and β- position.

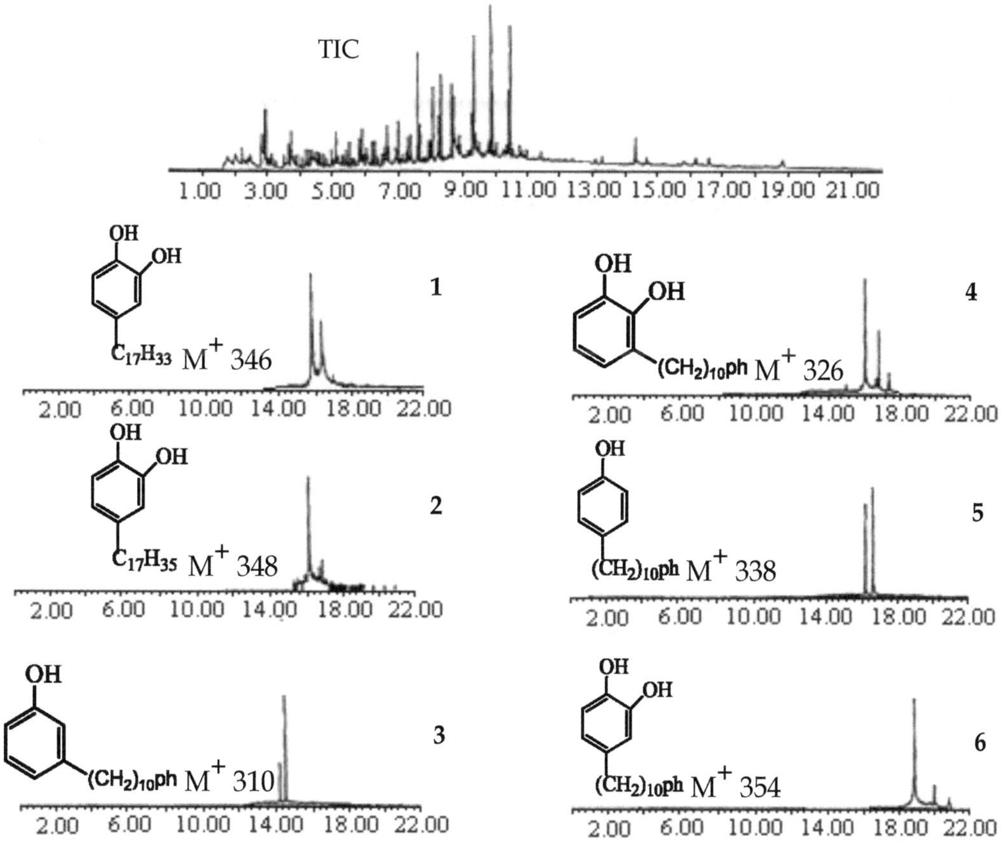

Fig. 3-5. TIC and mass chromatograms (m/z=346, 348, 310, 326, 338, 354) of thitsiol lacquer film

The thitsiol lacquer film was pyrolyzed at 500°C. Figure 3-5 shows the TIC and individual mass chromatograms at m/z=346, 348, 310, 326, 338, and 354. Peaks 1, 2, 3, 4, 5, and 6 of the mass chromatograms were identified as 4-hepatadecenylcatechol (MW=346), 4-hepatadecenylcatechol (MW=348), 3-(10-phenyldecyl)phenol (MW=310), 3-(10-phenyldecyl)catechol (MW=326), 4-(10-phenyldodecyl)phenol (MW=338), and 4-(12-phenyldodecyl)catechol (MW=354), respectively. Because the C-O coupling polymers are terminated with 4-hepatadecenylcatechol, these compounds can be formed from such terminal groups.

In the m/z=123 mass chromatograms of the thitsiol lacquer film, a pair of peaks of the 3- and 4-alkylcatechols were detected (figure not shown). The relative intensities of the 3- and 4-heptylcatechols (C7) were the highest in the pyrolysis products of the thitsiol lacquer. The highest yields of the 3- and 4-heptylcatechols were considered to be mainly due to cleavage at the α-position of the double bonds of the nucleus-8th and 12th chain C=O couplings for the thitsiol polymers. The alkylphenols detected are likely to be the products of pyrolysis of the nucleus-side chain coupling of the thitsiol polymers. Dimerization of the lacquer monomers is considered to proceed through the laccase-catalyzed nucleus-side chain coupling as well as the C-C coupling. The yields of 3-heptylphenol (C7) were the highest, as shown in Figure 3-5. The α- and β-positions of the double bonds of the olefins are susceptible to thermal cleavage so that these highest yields are thought to be produced primarily by cleavage at the α-position of the double bonds of thitsiol, such as the 3- and 4-(8,11-heptadecadienyl)catechols. The detailed Py-GC/MS results of thitsiol in our laboratory have been previously reported [41].

The 3-pentadecylcatechol (MW=320) (urushiol), 3-heptadecylcatechol (MW=348) (laccol), and 4-heptadecylcatechol (MW=348) (thitsiol) are the characteristic pyrolysis products of the three kinds of lacquer films, respectively, as summarized in Table 3-1. Furthermore, the data acquired for the alkylcatechols and alkylphenols in the pyrolysis products also can help to determine the lacquer species.

Species of lacquer film	Features of pyrolysis products		
	Monomer	Alkylcatechol	Alkylphenol
Rhus vernicifera	Urushiol	3-Heptylcatechol	3-Heptylphenol
Rhus succedanea	Laccol	3-Nonylcatechol	3-Nonylphenol
Melanorrhoea usitata	Thitsiol	3-Heptylcatechol, 4-Heptylcatechol	3-Heptylphenol

Table 3-1. Characteristic pyrolysis products of lacquer films

4. Recent finding on lacquer film using Py-GC/MS

4.1 Identification of lacquer species

As described in Section 3, the main pyrolysis products of urushiol are 3-heptylcatechol and 3-heptylphenol, of laccol are 3-nonylcatechol and 3-nonylphenol, and of thitsiol are 3-heptylcatechol, 3-heptylphenol, and 4-heptylcatechol, respectively. In order to confirm these results, we synthesized urushiol [42], laccol [23], and thitsiol [41] lacquer films and analyzed them by Py-GC/MS measurement.

4.1.1 Synthesis of urushiol lacquer film and analysis by Py-GC/MS

Compound **1**, the major component of urushiol, shown in Figure 4-1, was synthesized in good yield via the Witting reaction from a ylide derived from (3E,5Z)-3,5-heptadienyltriphenylphosphonium iodide with 3-(8-oxo-1-octyl) catechol diacetate, using a stepwise procedure based on repeated protection and deprotection of the hydroxyl group of catechol, using the technique we reported previously [43]. The trienyl urushiol compounds **2** and **3** were synthesized in a similar method using ylides derived from (3E,5E)-3,5-heptadienyltriphenylphosphonium iodide and 3E-3,5-heptadienyltriphenylphosphonium iodide, respectively.

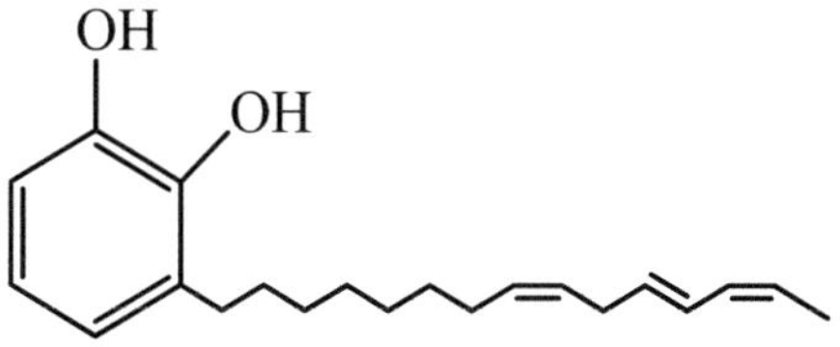

1

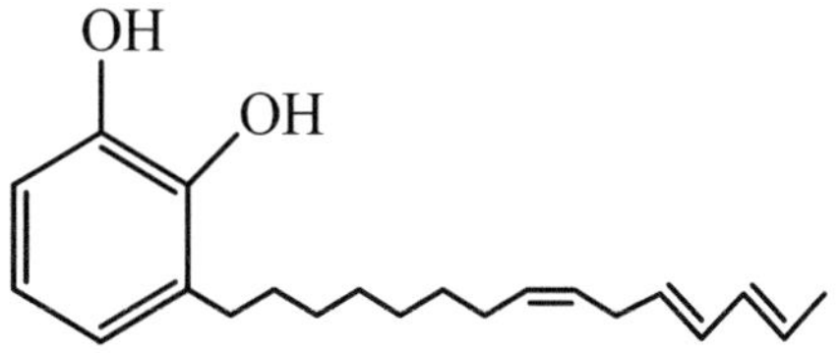

2

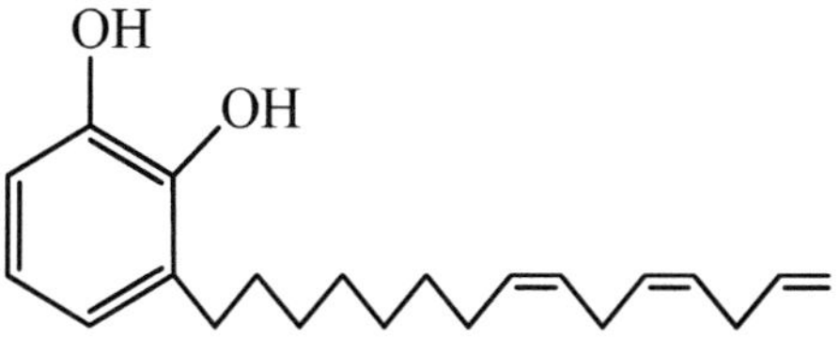

3

Fig. 4-1. Stereo-structure of trienyl urushiols: (1) 3-[(8Z,11E,13Z)-8,11,13-pentadecatrienyl]catechol, (2) 3-[(8Z,11E,13E)-8,11,13-pentadecatrienyl]catechol, (3) 3-[(8Z,11E)-8,11,14-pentadecatrienyl]catechol

Synthesized lacquer films were prepared as follows: 50 mg of compound 1 was added to 10 mg water-isopropyl alcohol (1:1,v/v) containing acetone powder (10 mg), which had been separated as an acetone-insoluble material from *Rhus vernicifera* lacquer sap. The resulting mixture was stirred for about 15 min. The reaction mixture had a viscosity suitable for coating and was coated onto a glass plate and dried in a humidity-controlled chamber with a relative humidity (RH) of 80% at 25–30°C for 10 h. The film was then removed from the chamber and stored in air for 6 months. Films of a single component were prepared. Compounds 2 and 3 were treated in the same manner to obtain the corresponding lacquer films. The films from the synthesized trienyl urushiol components 1, 2, and 3 are called synthesized lacquer films A, B, and C, respectively.

The preparation of natural lacquer was as follows. The sap exuded from a *Rhus vernicifera* lacquer tree in Japan, and composed of the following lipid components: 5% saturated urushiol, 18% monoenyl urushiol, 12% dienyl urushiol, and 65% trienyl urushiol. This lacquer sap was coated onto a glass plate and dried in a humidity-controlled chamber under an RH of 80% at 25–30°C for 10 h. The film was then removed from the chamber and stored in air for 6 months like the synthesized lacquer films.

The three synthesized lacquer films and the natural lacquer film were pyrolyzed at 500°C, and then the resulting pyrolysis products were characterized by GC-MS analysis. The specific ions at m/z 123 and 108 are fragment ions of alkylcatechols and alkylphenols that were produced during the electron ionization process in the mass spectrometer, and the mechanism of the thermal decomposition is shown in Figure 4-2.

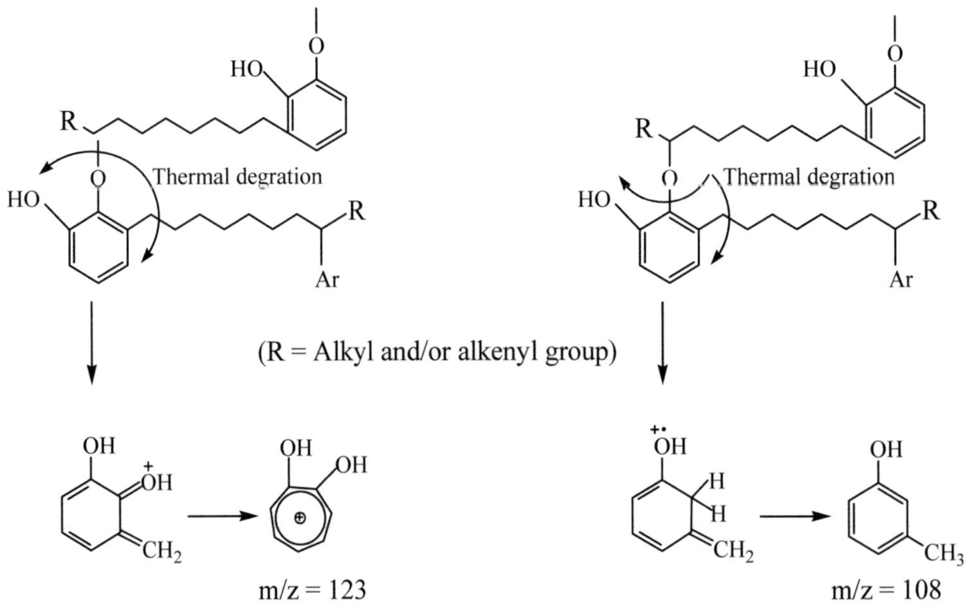

Fig. 4-2. Postulated location of urushiol lacquer film pyrolysis and mechanism of formation of m/z=123 and m/z=108 ion species in Py-GC/MS

The resulting total ion chromatograms (TIC) of the natural and synthesized lacquer films are shown in Figure 4-3. The major components in the TIC peaks were defined as alkanes, alkenes, and alkylbenzenes respectively.

The mass chromatogram of selective scanning of m/z=123 is shown in Figure 4-4. Peaks 1, 2, and 3 were identified as 3-pentylcatechol (M^+=180), 3-hexylcatechol (M^+=194), and 3-heptylcatechol (M^+=208), respectively. Peaks 4 and 5 were observed in only the natural lacquer film, and were identified as the urushiol monomer components 3-pentadecenylcatechol (M^+=318) and 3-pentadecylcatechol (M^+=320), based on analysis of their mass spectra, shown in Figure 4-5.

Figure 4-5 shows the mass spectra of compounds giving rise to peaks 1–5 in the selective plotting of m/z=123 ion species in the spectra from the pyrolysis of synthesized and natural lacquer films shown in Figure 4-4. Thus, it is evident from Figure 4-4 that saturated urushiol (peak 5) and monoenyl urushiol (peak 4) were detected by pyrolysis of the natural lacquer film, but not of the synthesized lacquer films. These urushiols likely produced thermally decomposed fragments from the terminal alkylcatechol and alkenylcatechol side chains of the natural lacquer film. Therefore, it was confirmed that the natural lacquer film has urushiol components with both saturated and monoenyl side chains, unlike the synthesized lacquer films.

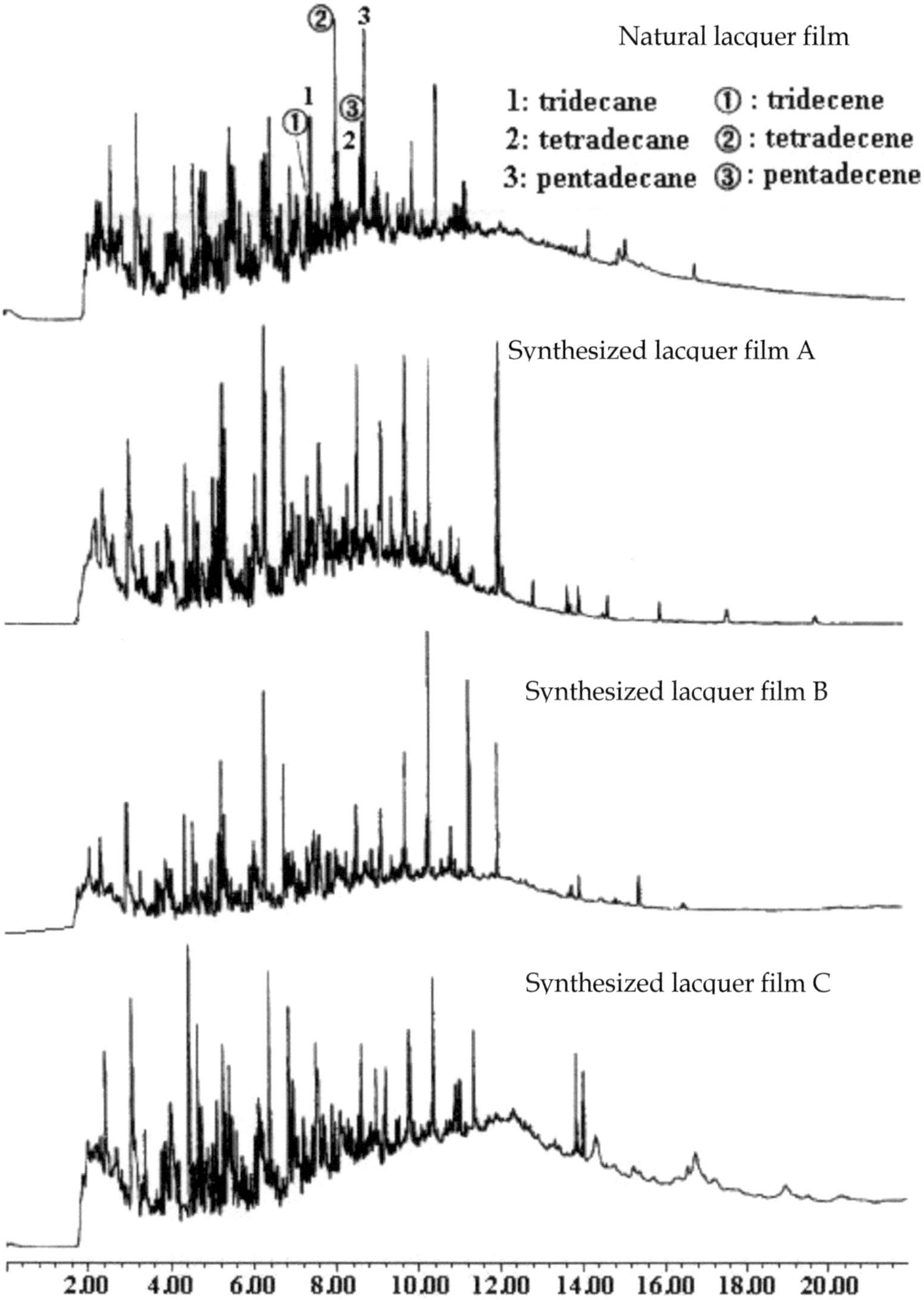

Fig. 4-3. TIC of natural and synthesized lacquer films due to pyrolysis at 500°C

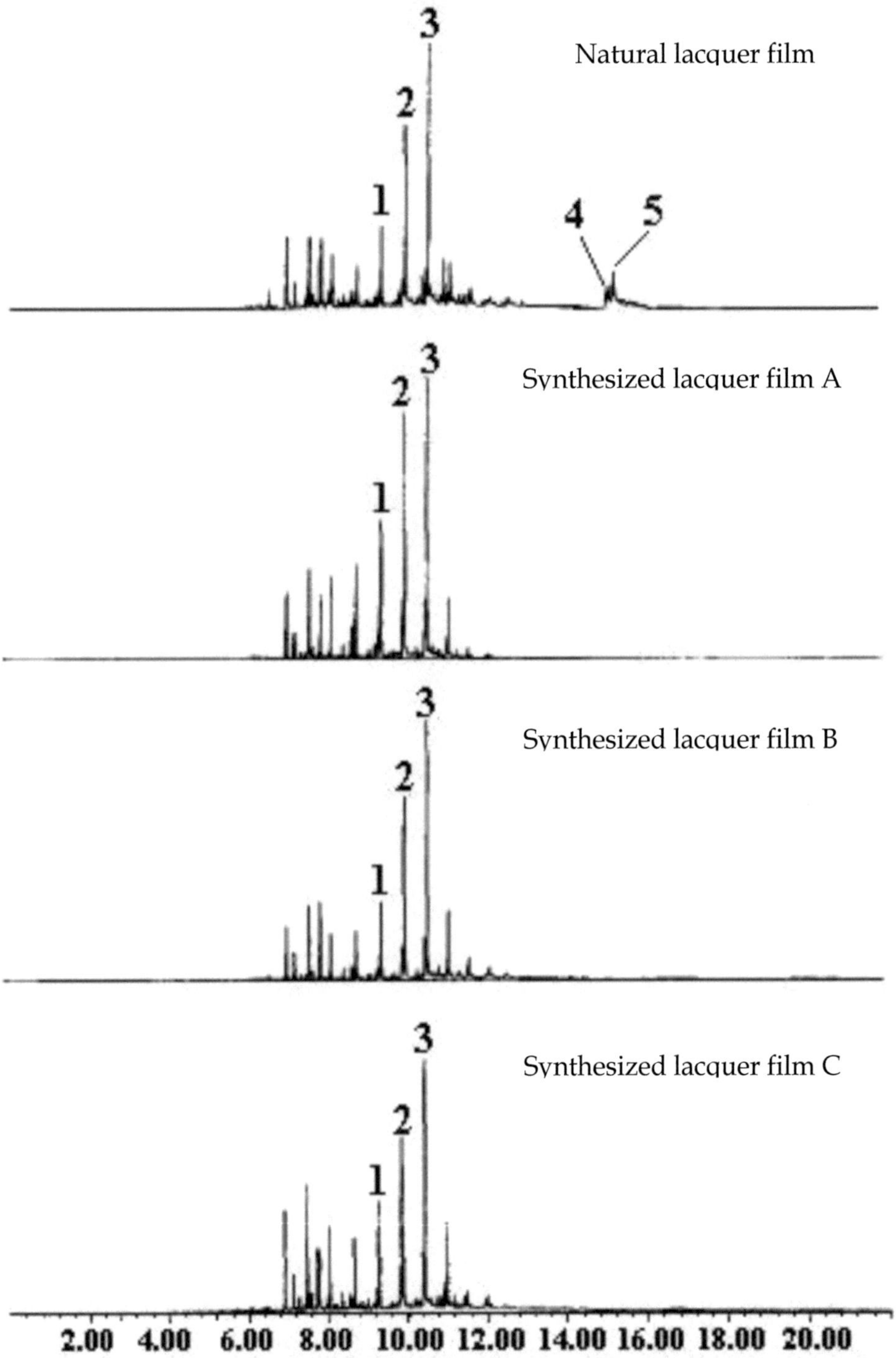

Fig. 4-4. Selective plotting of ion species (m/z=123) in spectra obtained from TIC of one natural and three synthesized lacquer films

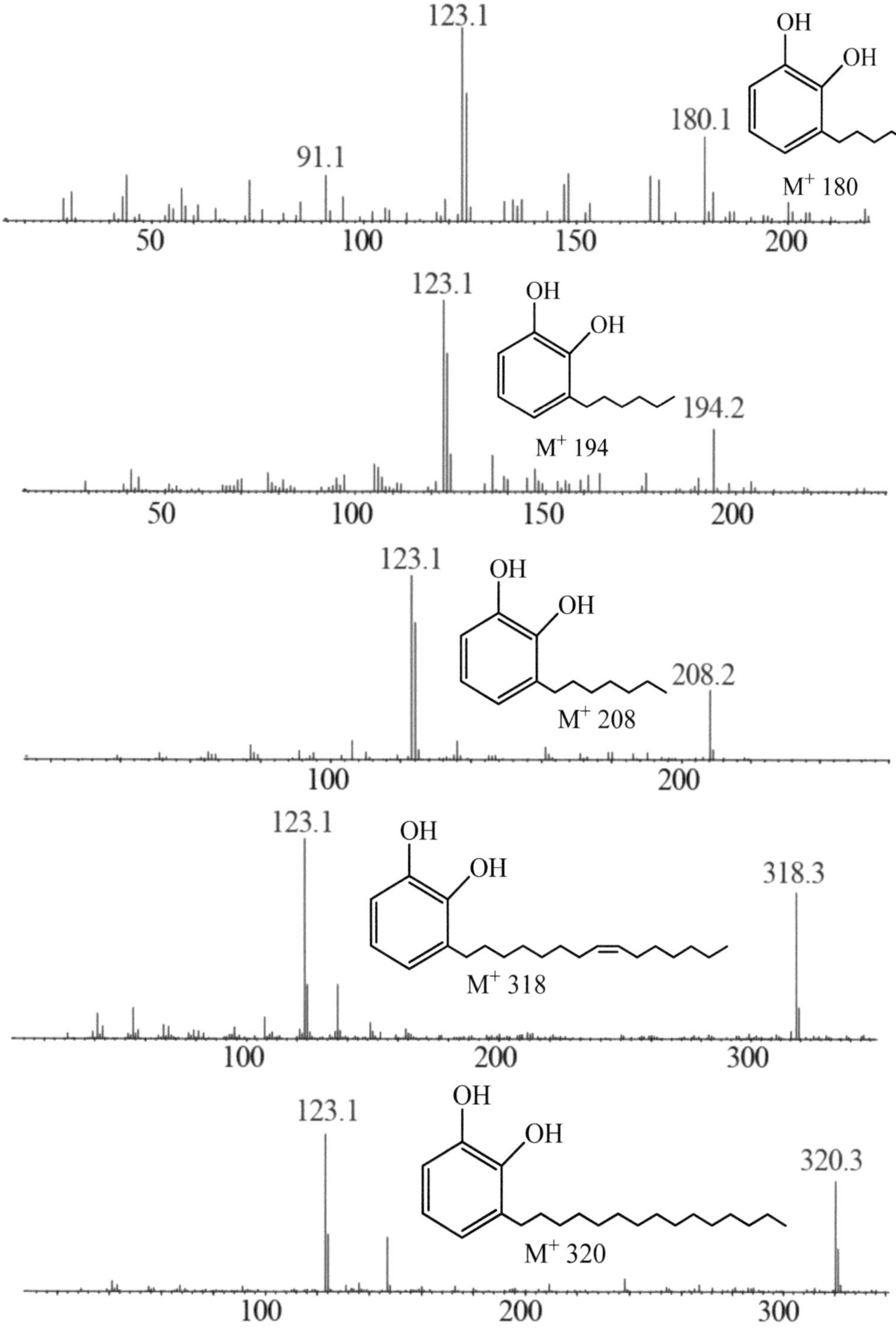

Fig. 4-5. Mass spectra of alkylcatechols from synthesized lacquer film A (peaks 1–3) and natural lacquer film (peaks 4–5) in mass chromatogram (m/z=123)

The highest peak intensity in each mass chromatogram was peak 3, because most urushiol components have a double bond at position 8 of their side chain, and the bond between C7 and C8 is easily thermally decomposed after polymerization as shown in Figure 4-6. The highest peak intensity of *Rhus vernicifera* was 3-heptylcatechol (M^+=208).

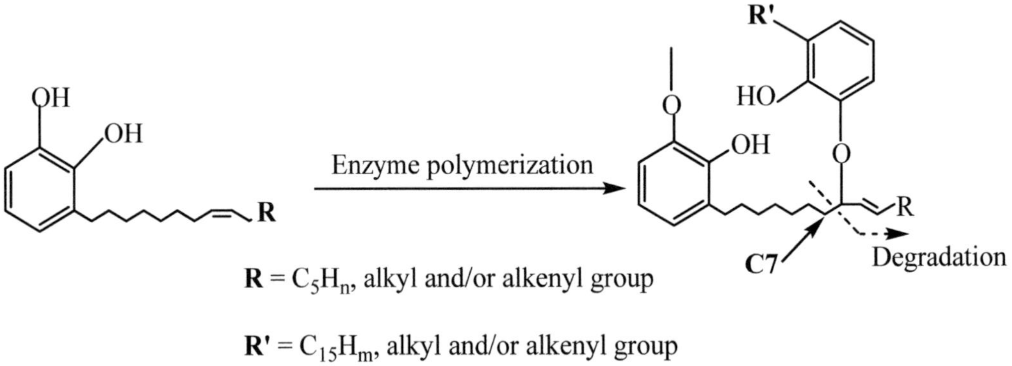

Fig. 4-6. Postulated location of lacquer film in thermal decomposition due to pyrolysis

4.1.2 Synthesis of laccol lacquer film and analysis by Py-GC/MS

The trienyl and dienyl laccols, compounds 4, 5, and 6, shown in Figure 4-7 were synthesized by a method similar to the synthesis of urushiol described in Section 4.1.1. Compound 4, the major component of laccol, was synthesized via the Witting reaction from a ylide derived from (3E,5E)-3,5-heptadienyltriphenylphosphonium iodide with 3-(8-oxo-1-decyl)catechol diacetate. Compounds 5 and 6 were synthesized from 3Z-heptadienyltriphenylphosphonium iodide and 3E-heptadienyltriphenylphosphonium iodide, respectively.

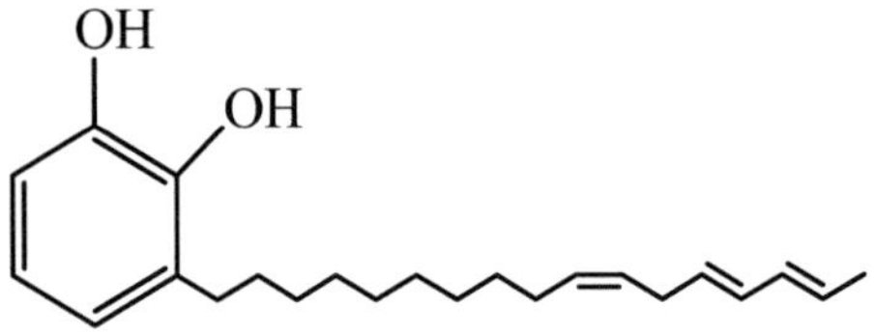

4

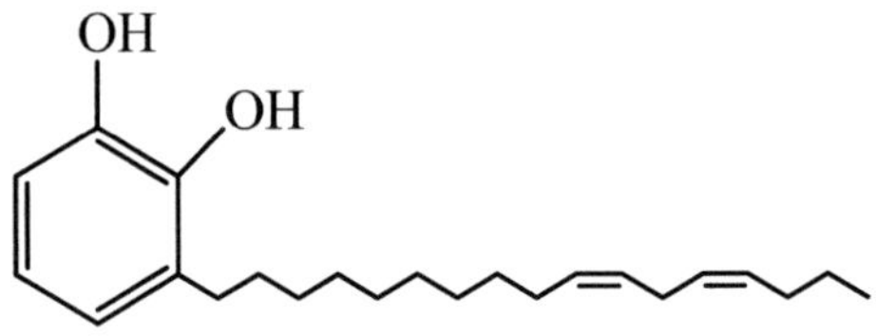

5

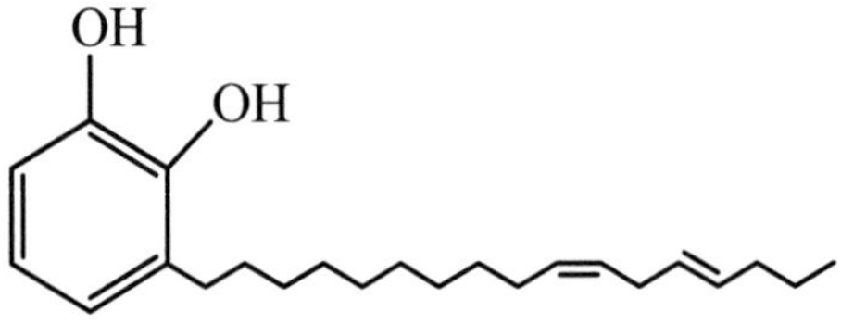

6

Fig. 4-7. Stereo-structure of trienyl and dienyl laccols: (4) 3-[(10Z,13E,15E)-10,13,15-heptadecatrienyl]catechol, (5) 3-[(10Z,13Z)-10,13-heptadecadienyl]catechol, (6) 3-[(10Z,13E)-10,13-heptadecadienyl]catechol

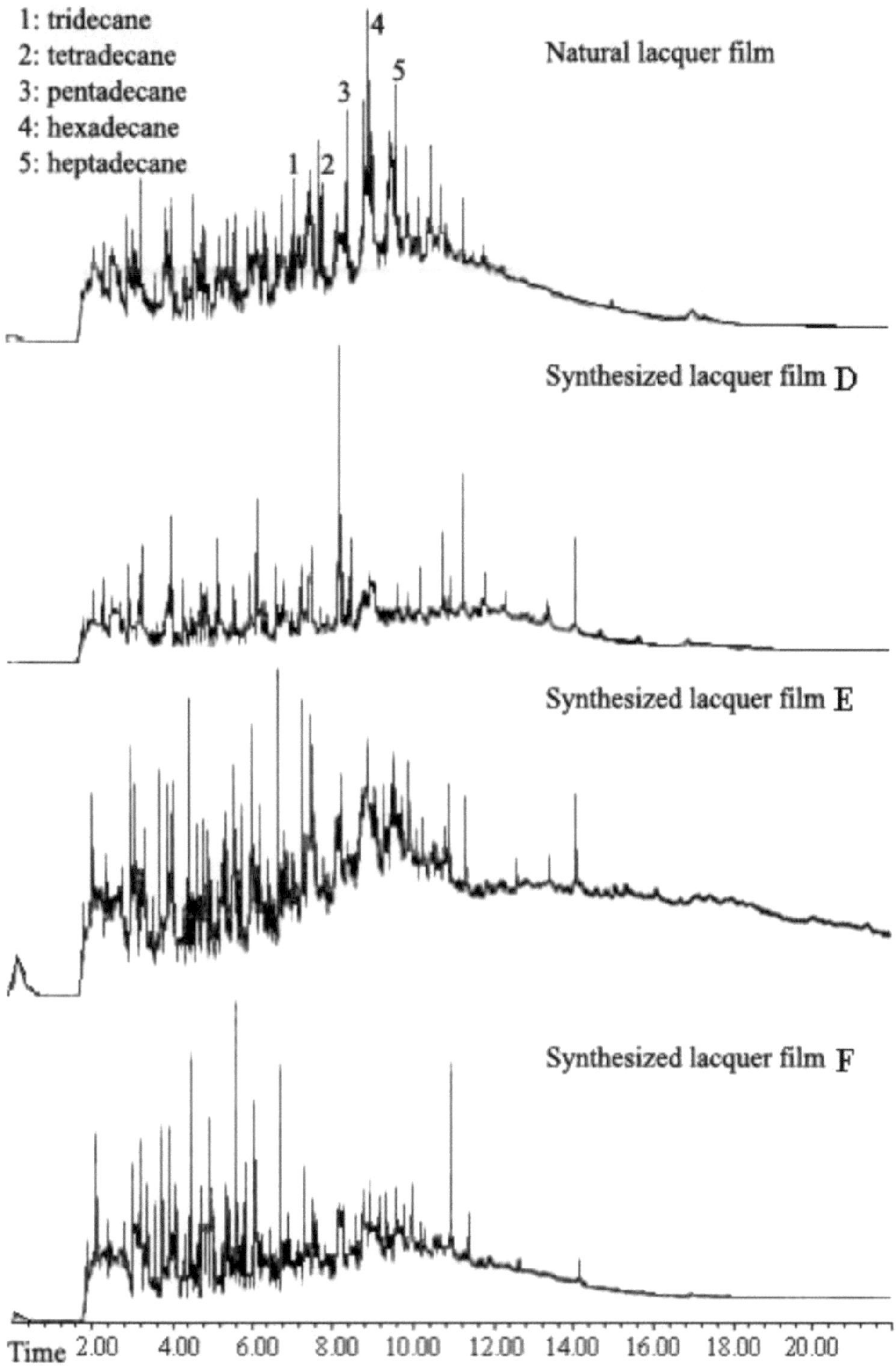

Fig. 4-8. TIC of natural and synthesized laccol lacquer films due to pyrolysis at 500°C

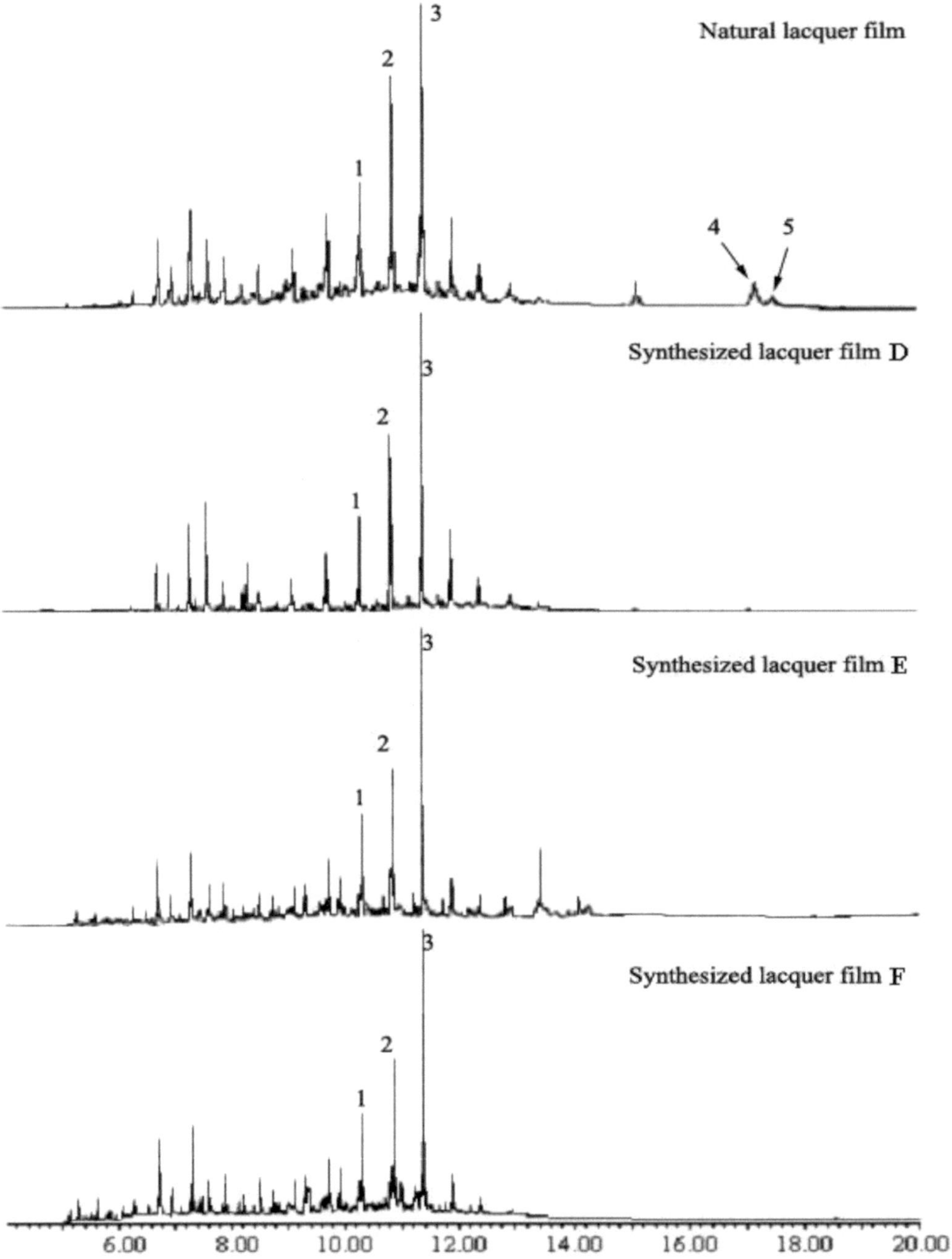

Fig. 4-9. Selective plotting of ion species (m/z=123) in spectra obtained from TIC of one natural and three synthesized lacquer films

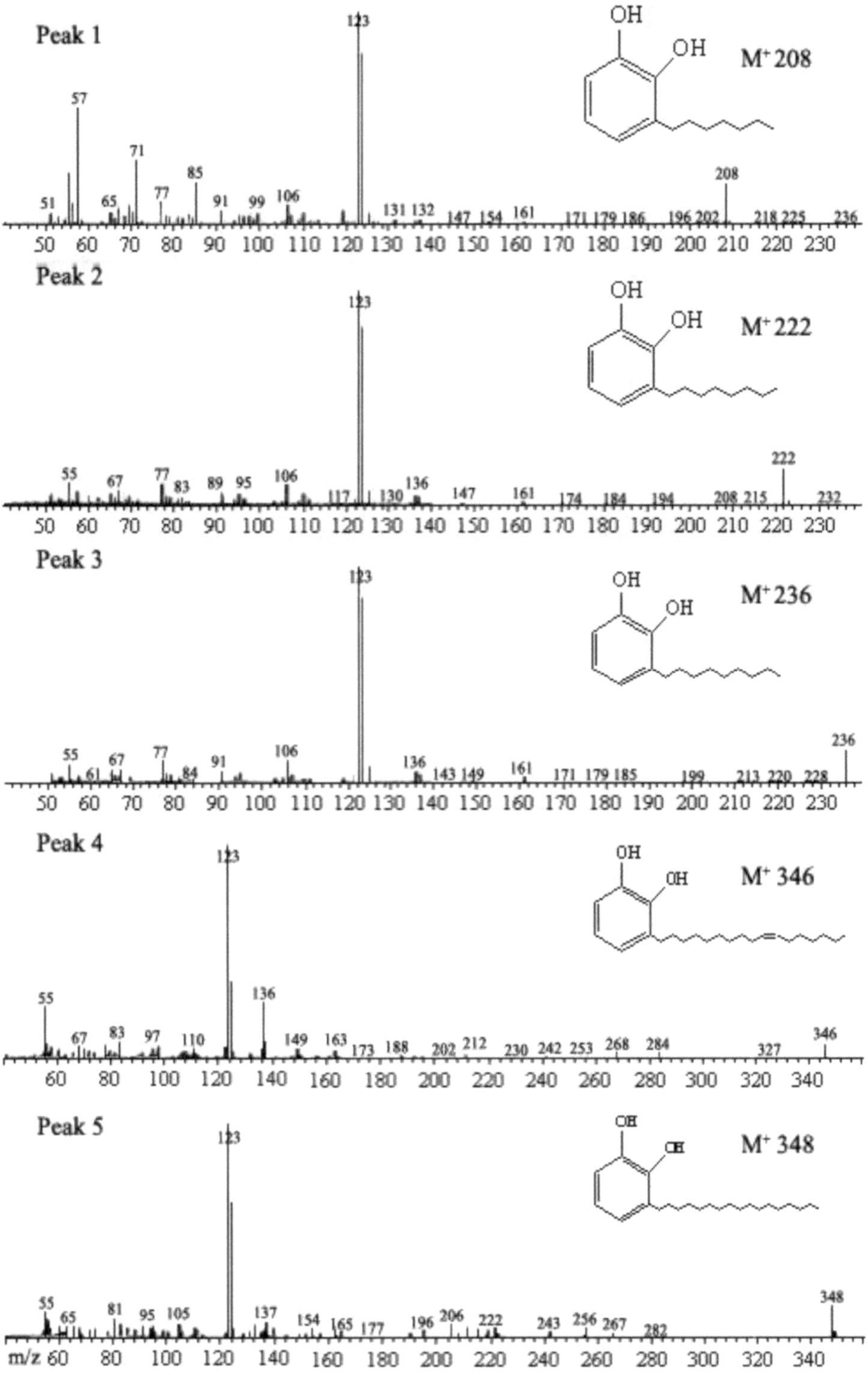

Fig. 4-10. Mass spectra of peaks 1–5 in the mass chromatogram (m/z=123)

Synthesized and natural laccol lacquer films also were prepared by methods similar to those of urushiol lacquer film described in Section 4.1.1. However, because the laccase activity of *Rhus succedanea* is lower than that of *Rhus vernicifera*, the laccol lacquer films were dried in a humidity-controlled chamber with RH of 90% at 30°C for 10 h. Then they were removed from the chamber and stored in air for 6 months. The films from the synthesized laccol components 4, 5, and 6 are called synthesized lacquer films D, E, and F, respectively.

The three synthesized laccol lacquer films and one natural *Rhus succedanea* lacquer film were pyrolyzed at 500°C, and the resulting pyrolysis products were characterized by GC-MS analysis. The specific ions at m/z 123 and 108 are fragment ions of alkylcatechols and alkylphenols that were produced during the electron ionization process in the mass spectrometer, like those of the urushiol lacquer film. The TIC of natural and synthesized laccol lacquer films are shown in Figure 4-8. The major components in the TIC peaks were defined as alkanes, alkenes, and alkylbenzenes, respectively.

The mass chromatogram of the selective scanning of m/z=123 shown in Figure 4-9. Peaks 1, 2, and 3 were 3-heptylcatechol (M^+=208), 3-octylcatechol (M^+=222), and 3-nonylcatechol (M^+=236) respectively. Peaks 4 and 5 were observed in only the natural lacquer film, and were identified as laccol monomer components, 3-heptadecenylcatechol (M^+=346) and 3-heptadecylcatechol (M^+=348), based on their mass spectra (Figure 4-10).

These laccols were likely produced as thermally decomposed fragments from the terminal alkyl and alkenylcatechol of the natural *Rhus succedanea* lacquer film. Therefore, the natural lacquer sap has laccol components with both saturated and monoenyl side chains, but the synthesized laccols did not. This was the same result as that of the natural lacquer film from *Rhus vernicifera*. The highest peak intensity was peak 3 in each mass chromatogram because most laccol components have a double bond at position 10 of their side chain, and the bond between C9 and C10 is easily thermally decomposed after polymerization. This result was different from that of the lacquer film from *Rhus vernicifera*, which has a double bond at position 8 of its side chain. The highest peak intensity of *Rhus vernicifera* was 3-heptylcatechol (M^+=208), of *Rhus succedanea* was 3-nonylcatechol (M^+=236).

4.1.3 Synthesis of thitsiol lacquer film and analysis by Py-GC/MS

The main component of thitsiol, 3-(10-phenyldecyl)catechol, was synthesized by the reaction of dimethyl ether with catechol and 1-phenyl-10-iododecane with n-BuLi, followed by

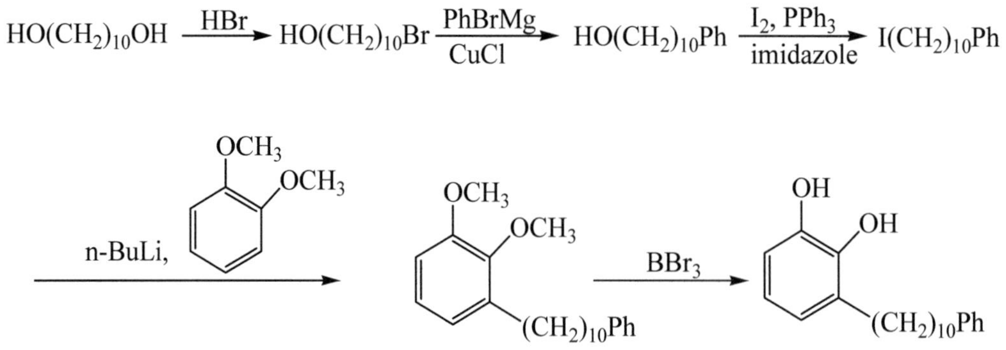

Fig. 4-11. Synthesis of 3-(10-phenyldecyl)catechol

deprotection of the hydroxyl groups of catechol, as shown in Figure 4-11. The detail experimental process can be found in the literature (3 and 4).

The preparation of synthesized thitsiol lacquer films is as follow: 50 mg 3-(10-phenyldecyl)catechol was added to 50 mg of a water-isopropyl alcohol mixture (1:1,v/v) containing 20 mg acetone powder obtained from a natural *Melanorrhoea (Gluta) usitata* lacquer sap. The resulting mixture was stirred for about 15 min, and then applied to a glass sheet and dried in a humidity-controlled chamber with a RH of 90% at 30°C for 10 h. The film was then removed from the chamber and stored in air for 6 months. The natural lacquer film also was prepared by the same method.

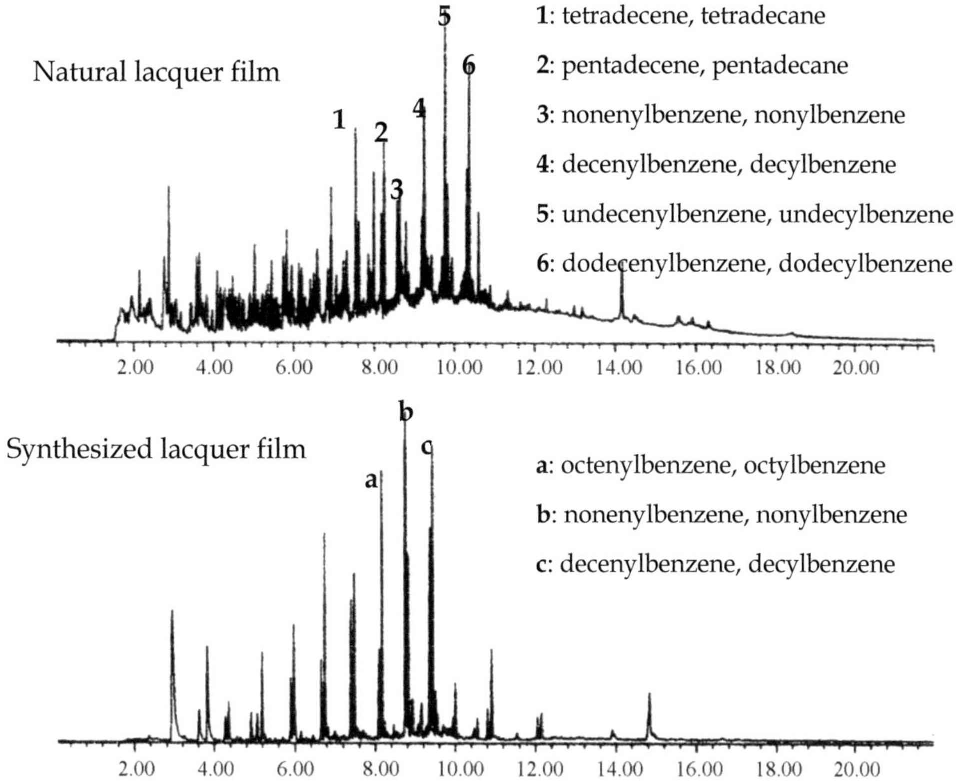

Fig. 4-12. TIC of natural and synthesized thitsiol lacquer films

The synthesized thitsiol lacquer films and a natural *Melanorrhoea (Gluta) usitata* lacquer film were pyrolyzed at 500°C, and then the resulting pyrolysis products were characterized by GC-MS analysis. Figure 4-12 shows the TIC of the natural and synthesized lacquer films. The main pyrolysis products of the natural lacquer film were identified as alkanes, alkenes, alkylbenzenes and alkenylbenzenes by the mass spectra. On the other hand, alkanes and alkenes were not detected in the TIC of the synthesized lacquer film because the alkanes and alkenes were derived from 3- or 4-alkenylcatechols contained in the natural lacquer sap. Each peak in TIC is a pair of peaks of saturated and unsaturated components having one double bond in the termination. It was considered that the alkanes and alkenes were formed by decomposition of the terminal of the lipid components in the polymer.

Figure 4-13 is the mass chromatograms (m/z=123) of natural and synthesized thitsiol lacquer films. Alkylcatechol was detected in the mass chromatogram of the natural lacquer film, and detected as a pair peak of 3-alkylcatechol and 4-alkylcatechol. Due to the analysis of each mass spectrum, the detailed assignment was as follows: ①: 3-methylcatechol, 1: 4-methylcatechol; ②: 3-ethylcatechol, 2: 4-ethylcatechol; ③: 3-propylcatechol, 3: 4-propylcatechol; ④: 3-butylcatechol, 4: 4-butylcatechol; ⑤: 3-pentylcatechol, 5: 4-pentylcatechol; ⑥: 3-hexylcatechol, 6: 4-hexylcatechol; ⑦: 3-heptylcatechol, 7: 4-heptylcatechol; ⑧: 3-octylcatechol, 8: 4-octylcatechol; 9: 3-pentadecylcatechol; 10: 3-(10-phenyldecyl)catechol.

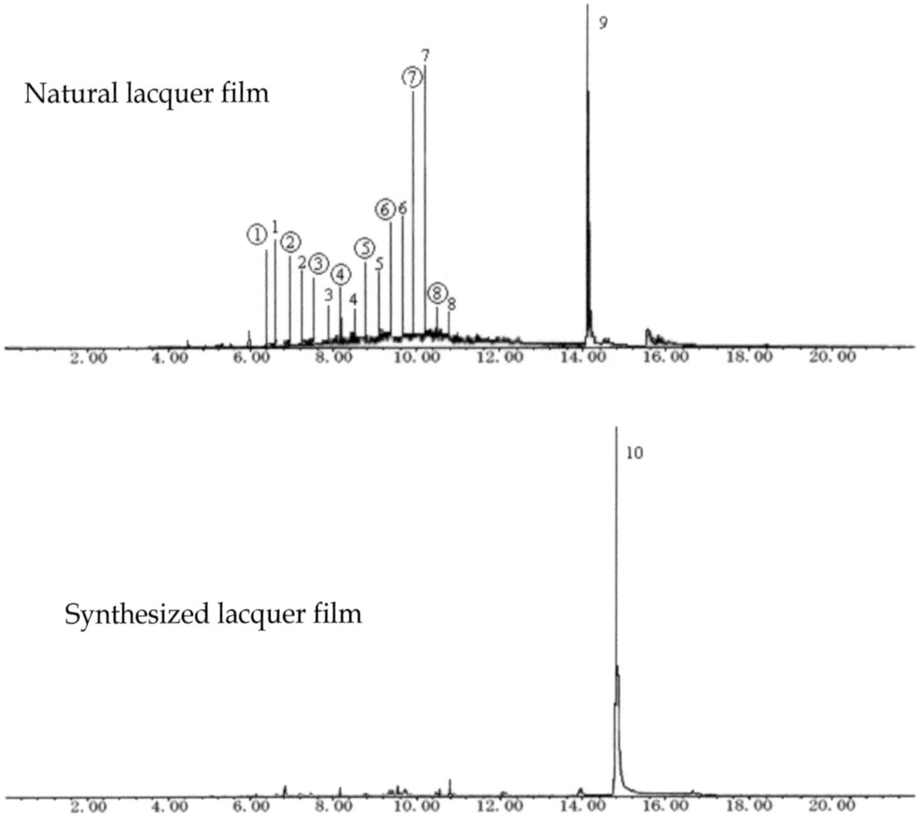

Fig. 4-13. Mass chromatogram (m/z=123) of natural and synthesized films

Figure 4-14 shows the mass spectra of pair of peaks ⑥-6 and ⑦-7. These components were derived from 3-alkylcatechol or 4-alkylcatechol contained as monomeric components in *Melanorrhoea (Gluta) usitata*. On the other hand, alkylcatechol was not detected in the mass chromatogram of the synthesized lacquer film because the synthesized lacquer film included no alkylcatechols as monomeric components. The side chain of the polymerized ω-alkylcatechol was difficult to decompose by heating because ω-alkylcatechol is not polymerized at the side chain by auto-oxidation like a fatty acid. Therefore, alkylcatechol was not detected in the mass chromatogram. However, 3-(10-phenyldecyl)catechol was detected as a monomer. The highest pair-peaks intensity of *Melanorrhoea (Gluta) usitata* was 3-heptylcatechol and 4- heptylcatechol (M^{+}=208, peaks ⑦ and 7 in Figure 4-13).

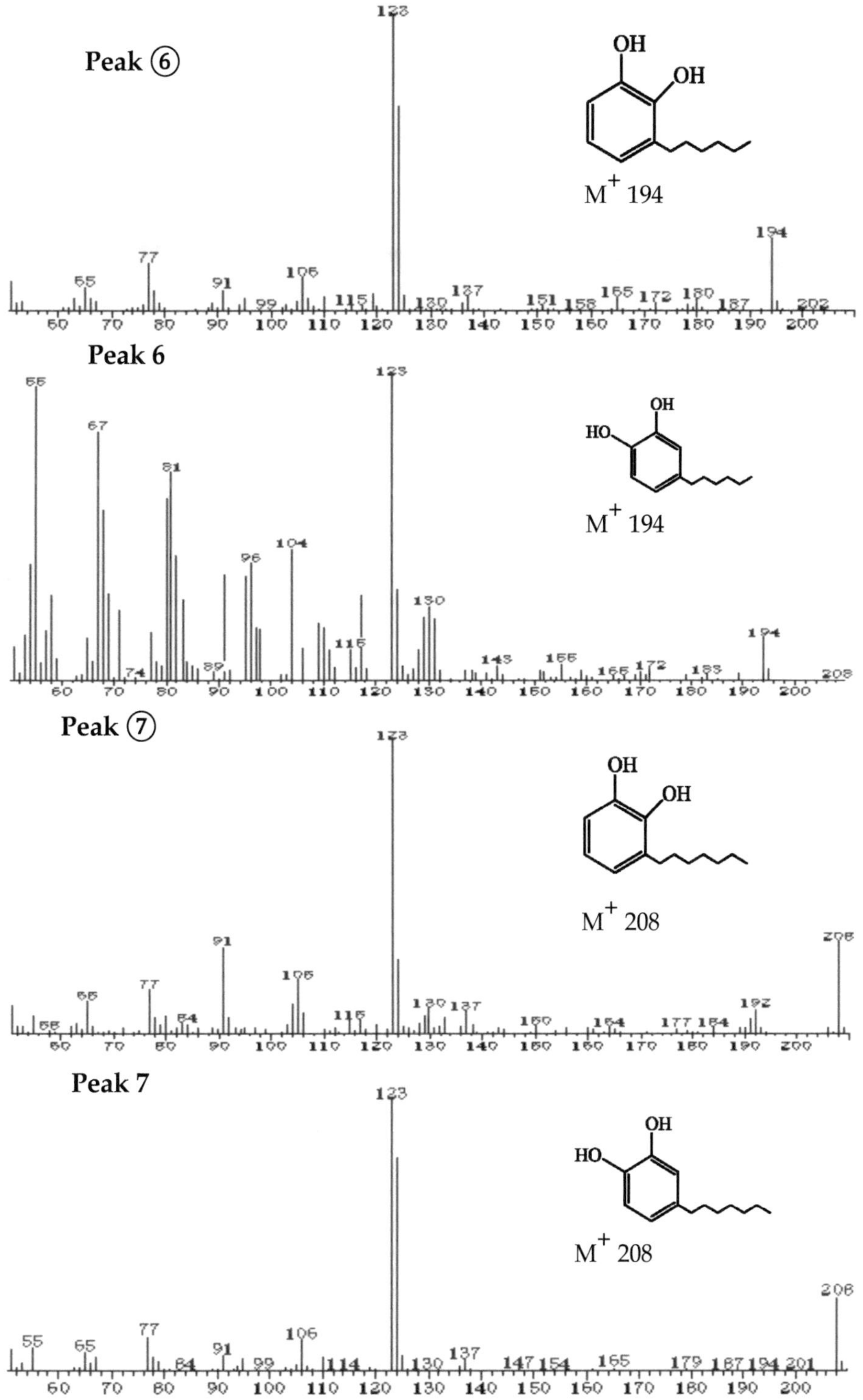

Fig. 4-14. Mass spectra of pair of peaks of ⑥-6 and ⑦-7.

The synthesized lacquer films and corresponding natural lacquer films were characterized using Py-GC/MS. The saturated and monoenyl components were detected in the natural *Rhus vernicifera* and *Rhus succedanea* lacquer films, respectively, but not in the synthesized lacquer films. However, alkylphenols, alkenylphenols, alkanes, and alkenes having longer carbon chains than the side chains of the synthesized lacquers were detected by the Py-GC/MS analysis. Comparing the peak intensity of the mass chromatograms with that of *Rhus succedanea, Rhus vernicifera,* and *Melanorrhoea (Gluta) usitata* lacquer films showed that the highest peak of *Rhus vernicifera* represented the C7 components (heptylcatechol and heptylphenol), and the highest peak of *Rhus succedanea* represented the C9 components (nonylcatechol and nonylphenol) due to a double bond in position 10 of its side chain of laccol. On the other hand, a pair of peaks of 3-alkylcatechol and 4-alkylcatechol was detected in the *Melanorrhoea (Gluta) usitata* lacquer film. The highest pair peaks were 3-heptylcatechol and 4-heptylcatechol due to the side chain attached in position 3 and/or position 4 of catechol ring of thitsiol.

4.2 Identification of ancient lacquerware

Py-GC/MS is a powerful and versatile technique to identify Oriental lacquers as described above. In this section, we describe analysis of several ancient lacquerwares by Py-GC/MS, and compared with the results of natural lacquer films to determine the kind of lacquer.

4.2.1 Ryukyu lacquerware

We have analyzed six kinds of Ryukyu lacquerware by Py-GC/MS to determine the lacquer source [44]. Six pieces of lacquer belonging to Urasoe Art Museum are summarized in Table 4-1, and the photos of each lacquerware object are shown in Photo 4-1. Each sample was removed from the ancient lacquerware objects during restoration. All pieces of Ryukyu lacquerware in Table 4-1 were analyzed by pyrolysis-gas chromatography/mass spectrometry at 500°C.

No.	Collection method	Art object (Museum I.D. number)
1	Naturally flaked off	Black lacquer wood box with *Raden* (43)
2	Naturally flaked off	Black lacquer table box with *Raden* (66)
3	Naturally flaked off	Black lacquer box with *Hakue* (220)
4	Naturally flaked off	Black lacquer box with *Hakue* (416)
5	Collection during restored	Black lacquer box with *Raden* (26)
6	Collection during restored	Red lacquer fabric bowl with *Hakue* (100)

Table 4-1. Samples used for pyrolysis

No. 1

No. 2

No. 3

No. 4

No. 5

No. 6

Photo 4-1. Six lacquer objects belonging to Urasoe Art Museum

Figure 4-15 shows the TIC and mass chromatogram (m/z=320, m/z=123, m/z=108) of Sample 2. The peak at the retention time of 14.326 min (m/z=320) is urushiol, at 10.038 min (m/z=123) is 3-heptylcatechol, and at 9.139 min (m/z=108) is 3-heptylphenol respectively. The mass spectra are shown in Figure 4-16. Samples 1, 4, and 5 showed TIC, mass chromatograms, and mass spectra similar to that of Sample 2, suggested that these four lacquerwares were made from *Rhus vernicifera*.

The TIC and mass chromatogram of Sample 3 are shown in Figure 4-15, and the mass spectrum is shown in Figure 4-16. The results showed that the peak at the retention time of 16.072 min (m/z=348) is laccol, at 11.105 min (m/z=123) is 3-nonylcatechol, and at 10.282 min (m/z=108) is 3-nonylphenol. Sample 6 produced the same pyrolysis results as Sample 3, suggesting that these two lacquerwares were made from *Rhus succedanea*. The pyrolysis products of the six samples are summarized in Table 4-2.

The Py-GC/MS results of Ryukyu ancient lacquerware confirmed that four lacquerware objects belonging to the Urasoe Art Museum were made from lacquer sap of *Rhus vernicifera* and the other two were made from lacquer sap of *Rhus succedanea* lacquer sap. Because Py-GC/MS analysis revealed the answer to the question of what kind of lacquer sap was used to produce Ryukyu lacquerware, the procurement source and production system of the materials of Ryukyu lacquerware could be clarified, and will be very useful in the conservation and restoration of other valuable ancient lacquer ware.

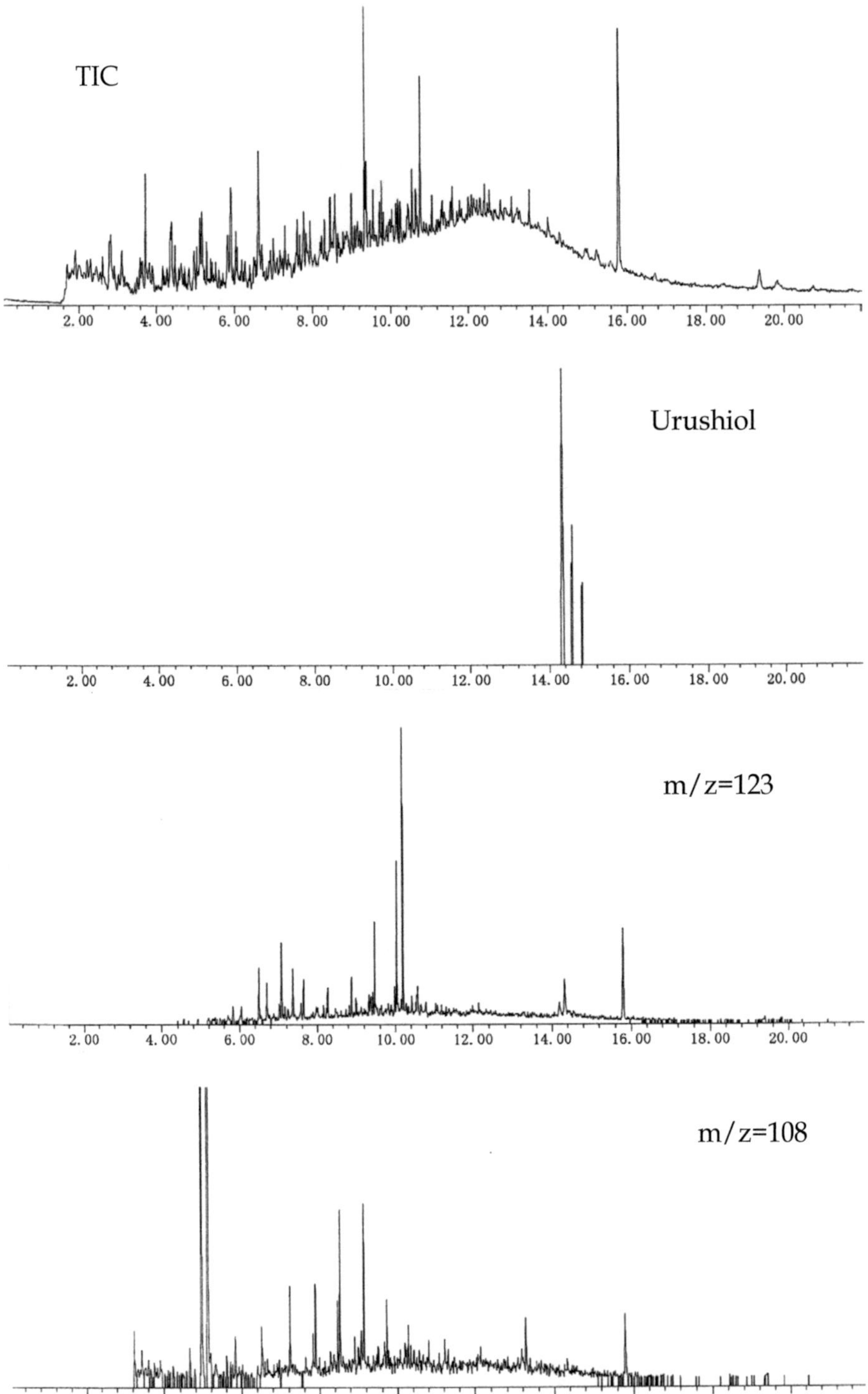

Fig. 4-15. TIC, mass chromatograms (m/z=320, 123, and 108) of sample 2.

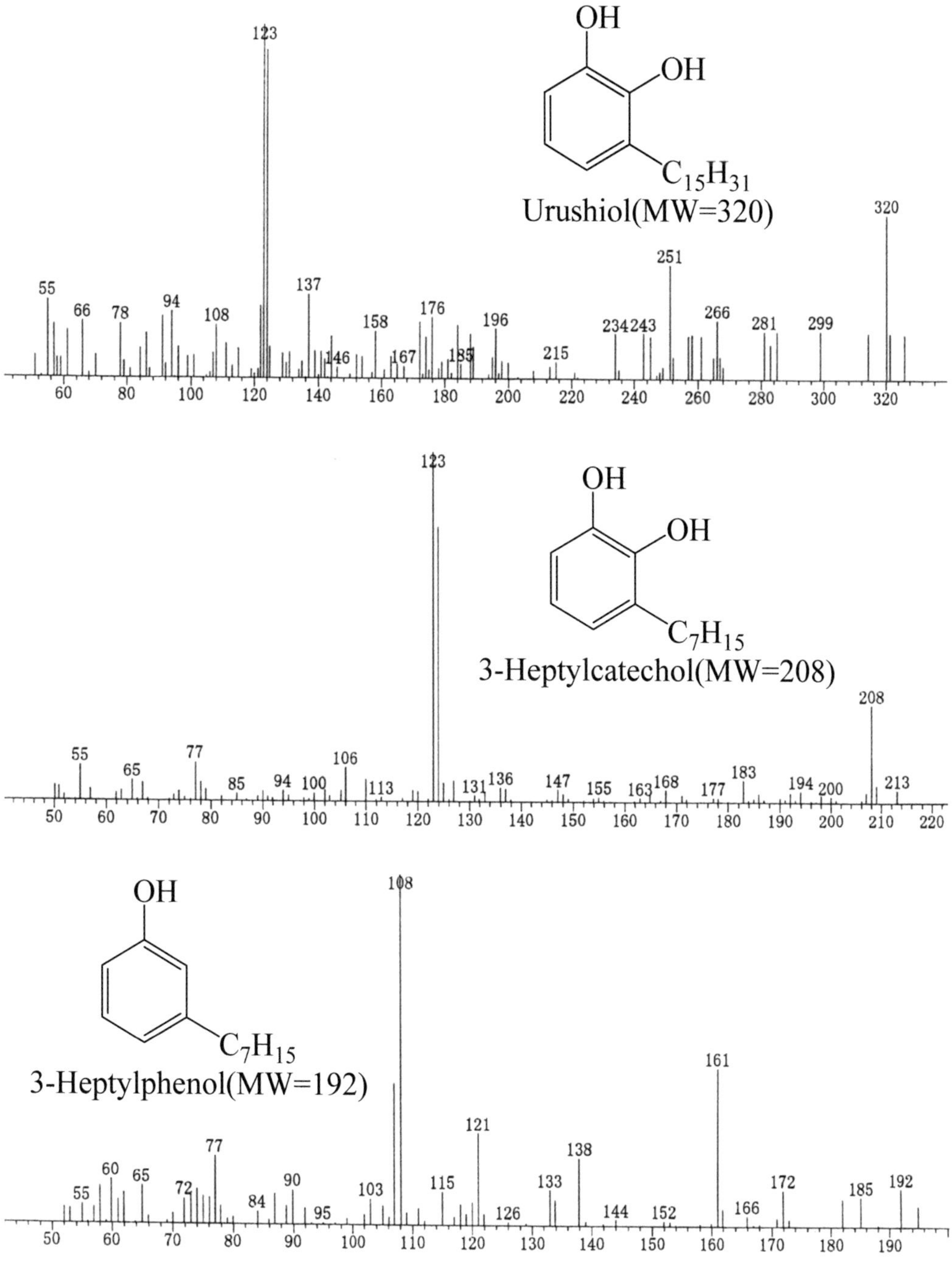

Fig. 4-16. Mass spectra of sample 2

No.	Lacquer species	Pyrolysis products		
		Monomer	Alkylcatechol	Alkylphenol
1	*Rhus vernicifera*	Urushiol	3-heptylcatechol	3-heptylphenol
2	*Rhus vernicifera*	Urushiol	3-heptylcatechol	3-heptylphenol
3	*Rhus succedanea*	Laccol	3-nonylcatechol	3-nonylphenol
4	*Rhus vernicifera*	Urushiol	3-heptylcatechol	3-heptylphenol
5	*Rhus vernicifera*	Urushiol	3-heptylcatechol	3-heptylphenol
6	*Rhus succedanea*	Laccol	3-nonylcatechol	3-nonylphenol

Table 4-2. Pyrolysis products of Ryukyu lacquer films and their species

4.2.2 Other ancient lacquerware

An ancient lacquer film was obtained from the surface of a wooden dish extracted from an excavation site that dates back to the 17th–18th century A.D. at *Kinenkanmae Iseki* on the campus of Meiji University in Tokyo, Japan, called Sample 1. A *Nanban* lacquer film from the 17th century A.D., called Sample 2, and an old lacquer film imported from an Asian country during the 17th–18th century A.D., called Sample 3, were obtained from the surface of wooden crafts objects [38]. Pieces of lacquer taken from a four-eared jar that is a Japanese National Important Cultural Property found in 16th–17th century Kyoto ruins, called Sample 4, was a piece of lacquer obtained from the side of the vessel [22].

The *Baroque* and *Rococo* lacquer films were obtained from the wood surfaces of the Rococo church St. Alto in Altomunster, Munich, Germany, identified as Samples 5 [38]. It is sometimes extremely important to determine whether objects that are claimed to be lacquerware are actually created from lacquer sap or other resins. Whether an object is lacquerware can be precisely determined by the presence of urushiol, laccol, or thitsiol with alkylcatechols and alkylphenols using Py-GC/MS.

The TIC and mass chromatograms of m/z=320 of Sample 1 are shown in Figure 4-17. Urushiol, 3-pentadecylcatechol (MW=320), was identified as the monomer of the lacquer film based on the mass spectrum and retention time. This result was compared to those of the three types of Oriental lacquers. 3-Pentadecylcatechol of MW=320 is the saturated urushiol component, which is the monomer of the *Rhus vernicifera* lacquer. The monomers of the *Rhus succedanea* lacquer are laccol components such as 3-pentadecylcatechol of MW=348. This was not detected in the Sample 1 except for the 3-pentadecylcatechol of MW=320. The monomer of *Melanorrhoea usitata* lacquer is thitsiol, which has saturated and monoenyl side chains, such as 4-hepatadecylcatechol (MW=348) and 3- and 4-(ω-phenylalkyl) phenols and catechols, and these components, except for the 3-pentadecylcatechol of MW=320, were not detected in the Sample 1. It was concluded that the Sample 1 lacquer was made from *Rhus vernicifera* lacquer sap.

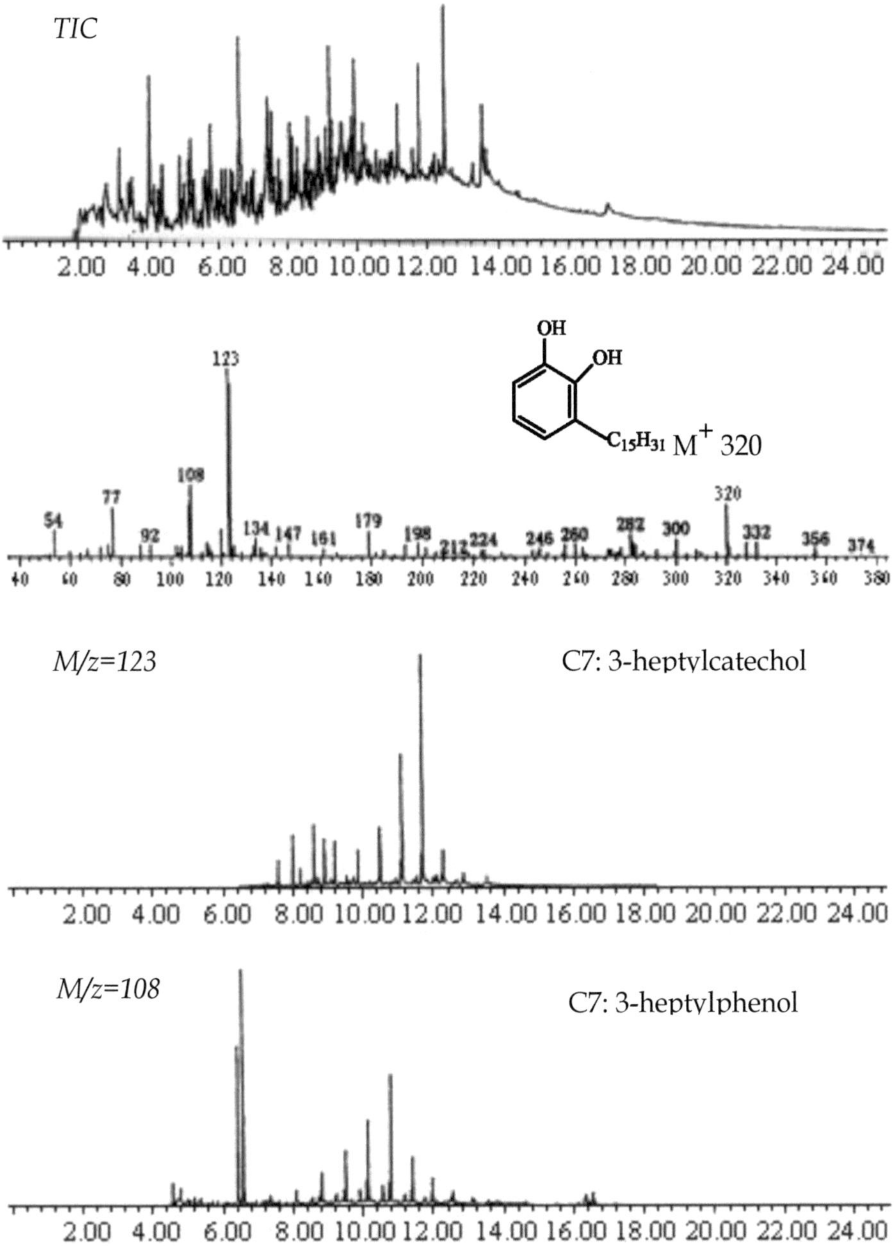

Fig. 4-17. Pyrolysis data of Sample 1 excavated at Meiji University, Tokyo, Japan

The TIC and mass chromatograms of m/z=60, m/z=123, and m/z=108 from the pyrolysis products of the Sample 2 are shown in Figure 18. The mass chromatograms of m/z=60 indicated that the Sample 2 lacquerware included a drying oil [45], which was added to retard the rate of hardening and affected the physical properties of the film. The mass chromatograms of m/z=108 and m/z=123 of the pyrolysis products of the lacquerware are also shown in Figure 4-18. The greatest number of peak were from 3-heptylphenol (C7) and 3-heptylcatechol (C7), as revealed by the mass spectra. It was concluded that the Sample 2 lacquer was made from *Rhus vernicifera* lacquer sap. Urushiol was not detected because the surface of the Sample 2 lacquerware was oxidized by oxygen and light. A type of wax was detected in the mass spectrum of the TIC from the pyrolysis products. It is considered that the wax was used to polish the surface of the lacquerware.

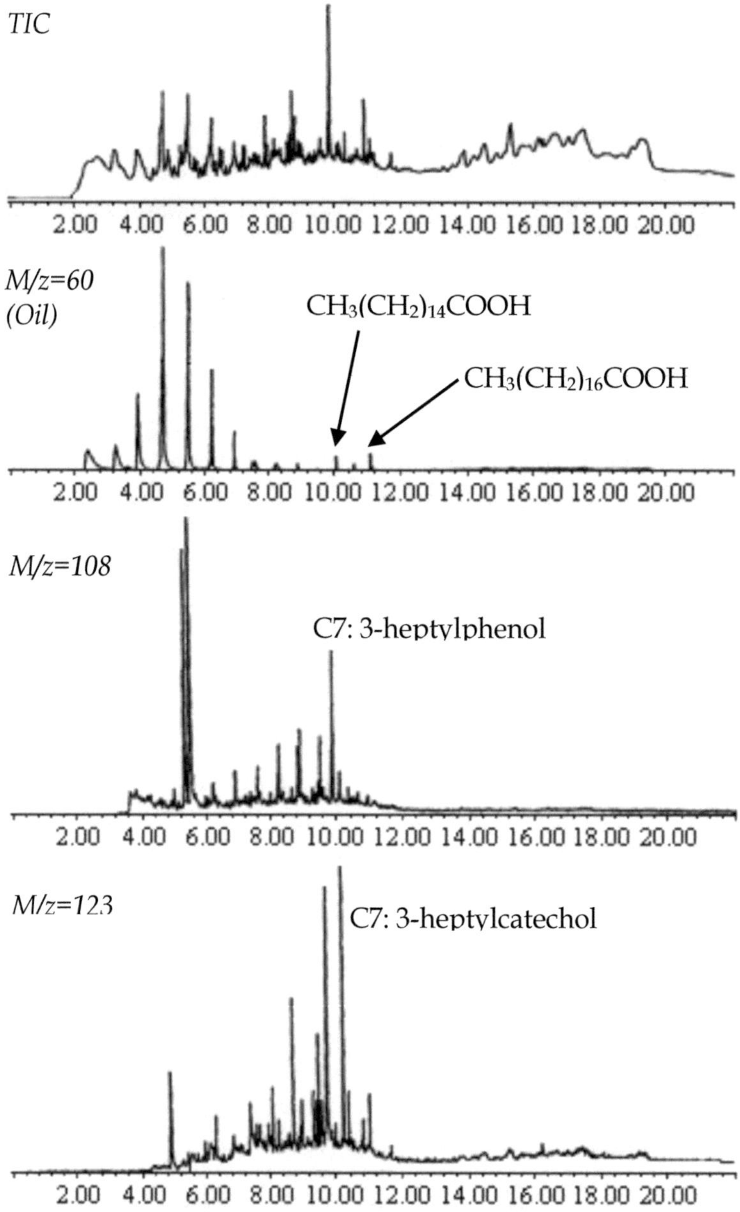

Fig. 4-18. Pyrolysis data of Sample 2 lacquer film

The TIC and mass chromatograms of m/z=123 and m/z=108 for the pyrolysis products of the Sample 3 are shown in Figure 4-19. Alkylcatechols and alkylphenols were detected in the mass chromatograms. The greatest number of peaks was due to the 3- and 4-nonylcatechols (C9) and 3-heptylphenol (C7), as determined from the mass spectra. It was concluded that the Sample 3 lacquerware was produced using *Melanorrhoea (Gluta) usitata* lacquer sap. Thitsiol was not detected in the pyrolysis data because the surface of the lacquerware was oxidized by oxygen and exposure to light. From the TIC and mass spectra of the pyrolysis products, a type of wax was detected. The wax was determined to be a type of beeswax, as revealed by the mass spectra. The beeswax was used to protect and polish the surface of the lacquerware.

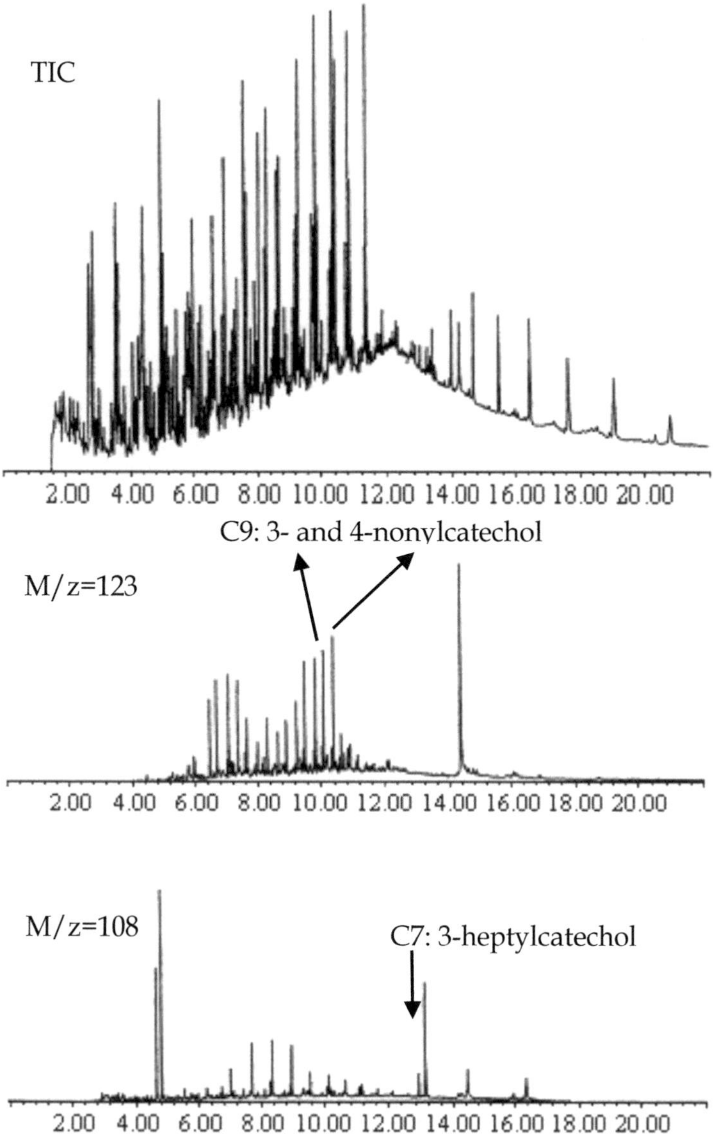

Fig. 4-19. Pyrolysis data of Sample 3

The TIC and mass chromatograms of m/z=123 and m/z=108 for the pyrolysis products of Sample 4 are shown in Figure 4-20. The peak at the retention time of 17.3 min (*m/z* = 123, peak 1) is 3-(10-phenyldecyl)catechol, and at 20.4 min (*m/z* = 123, peak 2) is 3-(10-phenyldodecyl)catechol, according to the results of analysis of the mass chromatogram. It was concluded that the Sample 4 lacquer was made from *Melanorrhoea (Gluta) usitata* lacquer sap.

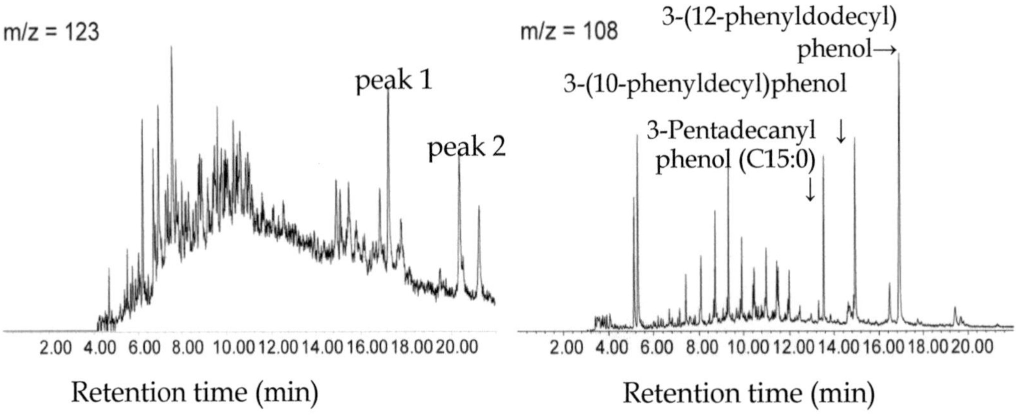

Fig. 4-20. Mass chromatograms of Sample 4

Figure 4-21 shows the Py-GC/MS products of the Sample 5 (Baroque and Rococo) lacquer films. Lacquer components were not detected in the TIC on the mass chromatograms of the alkylcatechols of m/z=123 and alkylphenols of m/z=108 of the pyrolysis products. Monoterpene components and sesquiterpene components were detected in the pyrolysis products of both lacquerwares. It was concluded that the lacquerwares were made from natural resins, but the pyrolysis products of the Baroque and Rococo lacquer and natural resins using this method were not clear.

It was concluded that 3-pentadecylcatechol (MW=320) (urushiol), 3-heptadecylcatechol (MW=348) (laccol), and 4-heptadecylcatechol (MW=348) (thitsiol) are the main products of the pyrolysis of *Rhus vernicifera, Rhus succedanea,* and *Melanorrhoea usitata,* respectively. Compared with the results of the natural lacquer film, the ancient lacquer film (Sample 1) and *Nanban* lacquer film (Sample 2) were assigned to *Rhus vernicifera,* both the old lacquerware objects imported from an Asian country (Sample 3) and the lacquer taken from a four-eared jar that is a Japanese National Important Cultural Property obtained from 16th–17th century Kyoto ruins (Sample 4) were assigned to *Melanorrhoea (Gluta) usitata.* However, although they were also called "lacquer," the Baroque and Rococo (Sample 5) lacquer film were identified as being made from a natural resin. The pyrolysis products clearly showed a good correspondence to the components of the lacquer sap.

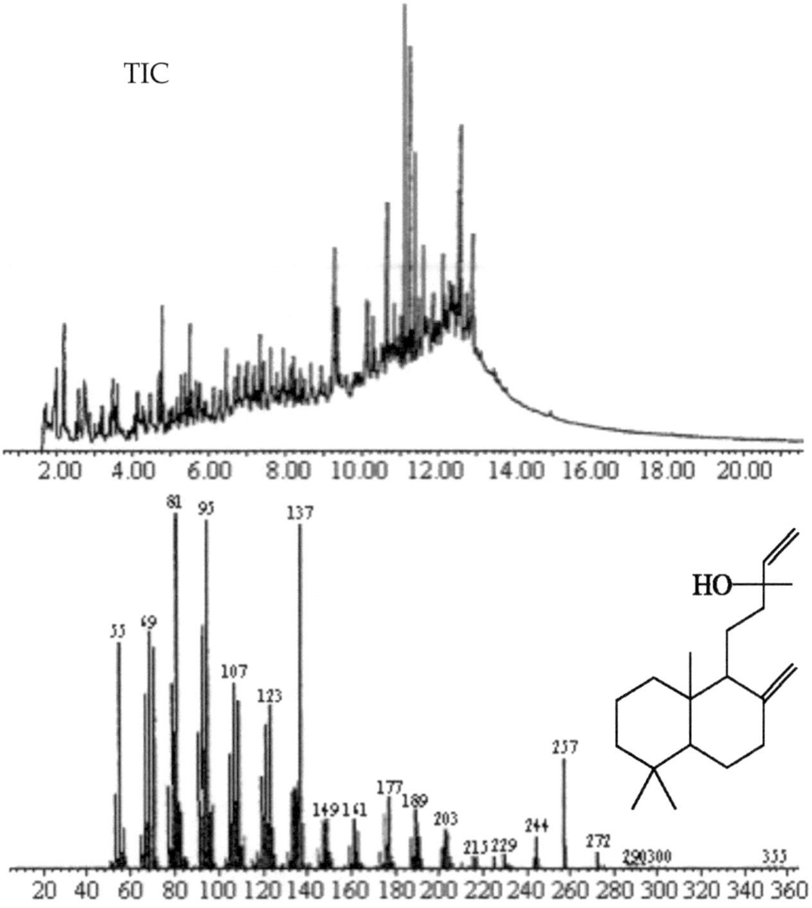

Fig. 4-21. Pyrolysis data of Baroque and Rococo lacquer films

5. Conclusion

This chapter describes the chemical properties of lacquer saps and films that were analyzed by pyrolysis-gas chromatography/mass spectrometry measurement. It was revealed that the advanced Py-GC/MS analytical method is useful for identifying and evaluating the lacquer components and the origin of lacquer species. Py-GC/MS is a well-known method applied to various areas. Due to its ease of control, speed of analysis, and good reproducibility, the Py-GC/MS method not only can be applied to lacquer films, organic coatings, and other materials that cannot be dissolved in solvents, but also to discriminate between lacquer and other resins for the conservation or restoration of lacquerware.

6. References

[1] J. Gan. Lacquer Chemistry, Chinese Academy of Science, 1984.

[2] O. Vogl, J. Bartus, M. Qin, J. Mitchell. Progress in Pacific Polymer Science 3, Spring-Verlag,Berlin, 1994, p. 423.

[3] O. Vogl. Oriental lacquer, poison ivy, and drying oils. *J. Polym. Sci. A: Polym. Chem.* 38, 2000, 4327-4335.

[4] E. J. Kidder. Ancient Peoples and Places, Japan, Thames & Hudson, 1959.

[5] Y. Kuraku. Urushi, N. S. Brommelle and P. Smith (eds.), Getty Conservation Institute, California, 45, 1988.

[6] Y. M. Du. Urushi, N. S. Brommelle and P. Smith (eds.), Getty Conservation Institute, California, 194, 1988.

[7] J. Hu. Conservation and Restoration of Cultural Property: Conservation of Far Eastern Objects, Tokyo National Res. Inst. Of Cultural Property, Tokyo, 89-112, 1980.

[8] A. Huttermann, C. Mai, and A. Kharazipour. Modification of lignin for the production of new compounded materials. *Appl. Microbiol. Biotechnol.*, 55, 2001, 387-394.

[9] http://inoues.net/ruins/torihama.html.

[10] http://www.daily-tohoku.co.jp/kikaku/tyouki-kikaku/jomon/jomon-22.htm.

[11] J. Kumanotani. Urushi (Oriental lacquer) – a natural aesthetic durable and future-promising coating. *Prog. Org. Coat.*, 26, 1995, 163-195.

[12] T. Miyakoshi, K. Nagase, T. Yoshida. Progress of Lacquer Chemistry, IPC Publisher, Tokyo, Japan, 1999, p. 43.

[13] T. Terada, K. Oda, H. Oyabu, T. Asami. Lacquer - the Science and Practice, Rikou Publisher, Tokyo, Japan, 1999, p. 25.

[14] Y. Yamauchi, R. Oshima, J. Kumanotani. Configuration of the olefinic bonds in the heteroolefinic side-chains of Japanese lacquer urushiol: separation and identification of components of dimethylurushiol by means of reductive ozonolysis and high- performance liquid chromatography. *J. Chromatogr.*, 243, 1982, 71-84.

[15] B. Reinhammar, Y. Oda. Spectroscopic and catalytic properties of *Rhus vernicifera* laccase depleted in type 2 copper. *J. Inorg. Biochem.*, 11, 1979, 115-127.

[16] J. Kumanotani. Urushi – cultural relations. *Kagakutokougyo,* 36, 1983, 151-154.

[17] S. Harigaya, T. Honda, R. Lu, T. Miyakoshi, and C. Chen. Enzymatic dehydrogenative polymerization of urushiols in fresh exudates from the lacquer tree, *Rhus vernicifera* DC. *J. Agric. Food Chem.*, 55, 2007, 2201-2208.

[18] R. Lu, S. Harigaya, T. Ishimura, K. Nagase, T. Miyakoshi. Development of a fast drying lacquer based on raw lacquer sap. *Prog Org. Coat.*, 51, 2004, 238-243.

[19] R. Lu, T. Ishimura, K. Tsutida, T. Honda, T. Miyakoshi. Development of a fast drying hybrid lacquer in a low relative-humidity environment based on *Kurome* lacquer sap. *J. Appl. Polym. Sci.*, 98, 2005, 1055-1061.

[20] R. Lu, D. Kanamori, and T. Miyakoshi. Characterization of thitsiol dimer structure from *Melanorrhoea usitata* with laccase catalyst by NMR spectroscopy. *Inter. J. Polym. Anal. Character.*, 16, 2011, 86-94.

[21] M. Herreraa, G. Matuschek. Fast identification of polymer additives by pyrolysis-gas chromatography/mass spectrometry. *J. Anal. Appl. Pyrolysis,* 70, 2003, 35-42.

[22] T. Honda, R. Lu, N. Kitano, Y. Kamiya, T. Miyakoshi. Applied analysis and identification of ancient lacquer based on pyrolysis-gas chromatography/mass spectrometry. *J. Appl. Polym. Sci.*, 118, 2010, 897-901.

[23] R. Lu, Y. Kamiya, Y. Wan, T. Honda, T. Miyakoshi. Synthesis of *Rhus succedanea* lacquer film and analysis by pyrolysis-gas chromatography/mass spectrometry. *J. Anal. Appl. Pyrolysis,* 78, 2007, 117-124.

[24] Y. Wan, R. Lu, Y. Du, T. Honda, T. Miyakoshi. Does Donglan lacquer tree belong to *Rhus vernicifera* species? *Inter. J. Bio. Macromol.*, 41, 2007, 497-503.
[25] K. Urakami, A. Higashi, K. Umemoto, M. Godo, C. Watanabe, K. Hashimoto. Compositional analysis of copoly(DL-lactic/glycolic acid) (PLGA) by pyrolysis-gas chromatography/mass spectrometry combined with one-step thermally assisted hydrolysis and methylation in the presence of tetramethylammonium hydroxide. *Chem. Pharm. Bull.*, 49, 2001, 203-205.
[26] M. Fukushima, M. Yamamoto, T. Komai, K. Yamamoto. Studies of structural alterations of humic acids from conifer bark residue during composting by pyrolysis-gas chromatography/mass spectrometry using tetramethylammonium hydroxide (TMAH-py-GC/MS). *J. Anal. Appl. Pyrolysis*, 86, 2009, 200-206.
[27] X. Chai, T. Shimaoka, Q. Guo, Y. Zhao. Characterization of humic and fulvic acids extracted from landfill by elemental composition, 13C CP/MAS NMR and TMAH-Py-GC/MS. *Waste Management*, 28, 2008, 896-903.
[28] R. Widyorini, T. Higashihara, J. Xu, T. Watanabe, S. Kawai. Self-bonding characteristics of binderless kenaf core composites. *Wood Sci. Tech.*, 39, 2005, 651-662.
[29] R. Martine, G. Nicolas, C. Cecile, R. Christian. New analytical methodology for investigating both volatile and nonvolatile constituents from archaeological ceramic vessels, Abstracts of Papers, 225th ACS National Meeting, New Orleans, LA, United States, March 23-27, 2003, GEOC-107.
[30] H. Araki, N. Tatarazako, K. Kishi, K. Kuroda. Evaluation of bioaccumulation potential of 3,4,5-trichloroguaiacol in a zooplankton (*Daphnia magna*) by pyrolysis-GC/MS in the presence of tetramethylammonium hydroxide (TMAH). *J. Anal. Appl. Pyrolysis*, 55, 2000, 69-80.
[31] T. Arii. Complex thermal analysis: evolved gas analysis-mass spectrometry EGA-MS. *Rigaku Janaru*, 39, 2008, 17-25.
[32] T. Arii. Evolved gas analysis-mass spectrometry (EGA-MS) using skimmer interface system equipped with pressure control function. *J. Mass Spectro. Soc. Jap.*, 53, 2005, 211-216.
[33] T. Tsugoshi, M. Furukawa, M. Ohashi, Y. Iida. Comparison of capillary and skimmer interfaces in evolved gas analysis-mass spectrometry (EGA-MS) with regard to impurities in ceramic raw materials. *J. Therm. Anal. Calorimetry*, 64, 2001, 1127-1132.
[34] G. Matuschek, A. Kettrup, A. Prior. EGA/MS investigations on the thermal degradation of diammoniumhexachloroplatinate. *J. Therm. Anal. Calorimetry*, 56, 1999, 471-477.
[35] J. Lee, G. Chang, J. Kwag, M. Rhee, A. Buglass, G. Lee. Fast analysis of nicotine in tobacco using double-shot pyrolysis-gas chromatography-mass spectrometry. *J. Agric. Food Chem.*, 55, 2007, 1097-1102.
[36] K. Quenea, S. Derenne, F. Gonzalez-Vila, J. Gonzalez-Perez, A. Mariotti, C. Largeau. Double-shot pyrolysis of the non-hydrolysable organic fraction isolated from a sandy forest soil (Landes de Gascogne, South-West France). *J. Anal. Appl. Pyrolysis*, 76, 2006, 271-279.
[37] J. Lee, C. Lee, J. Kwag, A. Buglass, G. Lee. Determination of optimum conditions for the analysis of volatile components in pine needles by double-shot pyrolysis-gas chromatography-mass spectrometry. *J. Chromatogr. A*, 1089, 2005, 227-234.
[38] R. Lu, Y. Kamiya, and T. Miyakoshi. Applied analysis of lacquer films based on pyrolysis-gas chromatography/mass spectrometry. *Talanta*, 70, 2006, 370-376.

[39] N. Niimura, T. Miyakoshi, J. Onodera, and T. Higuchi. Structural studies and polymerization mechanisms of synthesized lacquer films using two-stage pyrolysis-gas chromatography/mass spectrometry. *Inter. J. Poly. Anal. Charact.*, 4, 1998, 309-322.

[40] N. Niimura, T. Miyakoshi, J. Onodera, and T. Higuchi. Identification of ancient lacquer film using two-stage pyrolysis-gas chromatography/mass spectrometry. *Archaeometry*, 41 1999, 137-149.

[41] R. Lu, Y. Kamiya, and T. Miyakoshi. Preparation and characterization of *Melanorrhoea usitata* lacquer film based on pyrolysis-gas chromatography/mass spectrometry. *J. Anal. Appl. Pyrolysis*, 78, 2007, 172-179.

[42] Y. Kamiya, Y. Niimura, and T. Miyakoshi. Evaluation of synthesized lacquer dilms using pyrolysis-gas chromatography/mass spectrometry. *Bull. Chem. Soc. Jpn.*, 73, 2000, 2621-2626.

[43] R. Lu, Y. Kamiya, and T. Miyakoshi. Characterization of lipid components of *Melanorrhoea usitata* lacquer sap. Talanta, 71, 2007, 1536-1540.

[44] R. Lu, X. Ma, Y. Kamiya, T. Honda, Y. Kamiya, A. Okamoto, and T. Miyakoshi. Identification of Ryukyu lacquerware by pyrolysis-gas chromatography/mass spectrometry. *J. Anal. Appl. Pyrolysis*, 80, 2007, 101-110.

[45] N. Niimura, and T. Miyakoshi. Characterization of natural resin films and identification of ancient coating. *J. Mass Spectrom. Soc. Jpn.*, 51, 2003, 439-457.

13

Pyrolysis-Gas Chromatography to Evaluate the Organic Matter Quality of Different Degraded Soil Ecosystems

Cristina Macci, Serena Doni, Eleonora Peruzzi, Brunello Ceccanti and Grazia Masciandaro
Consiglio Nazionale delle Ricerche (CNR), Istituto per lo Studio degli Ecosistemi (ISE), Pisa, Italy

1. Introduction

Soil systems are exposed to a variety of environmental stresses of natural and anthropogenic origin, which can potentially affect soil functioning. For this reason, there is growing recognition for the need to develop sensitive indicators of soil quality that reflect the effects of land management on soil and assist land managers in promoting long-term sustainability of terrestrial ecosystems (Bandick & Dick, 1999). Soil organic matter (SOM) providing energy, substrates, and biological diversity necessary to sustain numerous soil functions, has been considered one of the most important soil properties that contributes to soil quality and fertility (Doran & Parkin, 1994; Reeves, 1997).

SOM comprises a range of humic substances (HS) and non humic substances (NHS). Humic substances make up a significant portion of the total organic carbon (TOC) and nitrogen (TN) in soil. They consist of complex polymeric organic compounds of high molecular weights, which are more resistant to decomposition than the NHS. Moreover, humic substances are widely recognized as an important fraction of soil organic matter because they have several fundamental functions: regulation of nutrient availability, linkage with mineral particles and immobilization of toxic compounds (Ceccanti & Garcia, 1994).

On the other hand, NHS contribute to soil ecosystem functionality by providing metabolically labile organic C and N sources, such as low molecular weight aliphatic and aromatic acids, carbohydrates, aminoacids, and their polymeric derivatives such as polypeptides, proteins, polysaccharides, and waxes (Schnitzer, 1991).

In order to understand the temporal dynamics of SOM in soil ecosystems, it is therefore vital to characterize soil organic carbon quantity and quality. Various analytical methodologies, such as infrared (See et al., 2005), ultraviolet-visible (Alberts et al., 2004), nuclear magnetic resonance spectroscopy (Gondar et al., 2005), potentiometric titration, pyrolysis, electrophoresis, acid/alkali fractionation, have been used to study SOM (Ceccanti et al., 2007). These techniques have been used either alone or in combination with traditional fractionation and purification methods of acid/alkali soluble HS. However, these methodologies may not always reflect native SOM, due to inevitable analytical manipulations and also, sometimes, to a subjective interpretation of the results.

Among these techniques, pyrolysis-gas chromatography (Py-GC) is a technique capable of analysing native SOM, both in bulk soil (without manipulation), and in sodium pyrophosphate extract, considered the favoured solution to extract humic organic matter from soil (García et al., 1992; Masciandaro & Ceccanti, 1999; Clapp & Hayes, 1999). However, some disadvantages exist, since this technique provides information on small molecules but not the overall macromolecular structure (Grandy & Neff, 2008). Nevertheless, Py-GC is a simple and rapid technique which has been used successfully to make a qualitative study of the chemico-structural characteristics of organic matter in soils, sewage sludges (Peruzzi et al., 2011), sediments (Faure et al., 2002), fresh and composted wastes (Garcia et al., 1992; Dignac et al., 2005).

Many papers has been published in the last years on the application of Py-GC to soil analyses (Campo et al., 2011; Sobeih et al., 2008). Nierop et al. (2001) investigated the differences in the chemical composition of SOM within one soil serie from three differently managed fields using a combination of Py-GC/MS and thermally assisted hydrolysis and methylation. The results of this study showed that SOM composition is hardly affected by organic farming compared to conventional management i.e. high tillage intensity and intensive fertilization. Similarly, Marinari et al. (2007a, 2007b) used pyrolytic indices as indicators of SOM quality under organic and conventional management. In these studies has been confirmed that mineral fertilization, showing a relatively large amounts of aliphatic pyrolytic products, contributed to increase the mineralization process.

The Py-GC resulted a promising technique in the assessment of change in soil organic matter composition, even after the application of different types of mulching materials (Ceccanti et al., 2007). In the study of Ceccanti et al. (2007) compost enriched the soil with available organic nutrients, rapidly shown by an increase of pyrolytic aliphatic compounds, while straw application revealed a more stable soil organic matter characterized by a high pyrolytic humification index.

In this chapter the Py-GC technique will be proposed as useful tool to evaluate, on the basis of the chemical structural changes of organic matter, the resistance of degraded soils to different external impacts and rehabilitation practices.

The results obtained from the study of several sites, characterized by local variation of anthropic and natural pressures, will be presented. The investigation has been carried out in the framework of the European project "Indicators and thresholds for desertification, soil quality, and remediation" (INDEX). In this project 12 sites were selected across Europe in order to ensure a broad range of different conditions (natural soil and agro-ecosystems).

The parameters determined by pyrolysis-gas chromatography have been selected as promising indicators of soil degradation and rehabilitation. These parameters were mainly classified as indicator of "State" according to DPSIR (Driving Forces-Pressure-State-Impact-Response) framework of Thematic Strategy for Soil Protection (European Commission, 2006) since making a fingerprint of soil organic matter, they represent the evolution state of organic matter in different climatic and management situation.

2. Methodologies

2.1 Pyrolysis-gas chromatography

Py-GC technique is based on a rapid decomposition of organic matter under a controlled high flash of temperature. A gas chromatograph is used for the separation and quantification of pyrolytic fragments (pyro-chromatogram) originating from organic matter decomposition. 50 micrograms of a representative soil sample, air-dried and ground <100 mesh (bulk soil) or few ml of soil sodium pyrophosphate extract (0.1M) was introduced into pyrolysis quarz microtubes in a CDS Pyroprobe 190.

The textural characteristics of soil can affect the reliability of the Py-GC technique, in particular in degraded soils, for this reason the method can be adapted in order to remove the inert part (for example sand and gross silt) which is not involved in the linkage with the humic substances, in other words, the soil organic matter has to be concentrated. Moreover, for highly organic soils (organic matter > 5%), where a great part of organic matter is linked to quick microbial metabolism, the extraction of humic matter (e.g. sodium pyrophosphate) give more reliable information on native soil organic matter. The Py-GC instrument consisted of a platinum coil probe and a quartz sample holder. Pyrolysis was carried out at 800 °C for 10 s, with a heating rate of 10 °C ms^{-1} (nominal conditions). The probe was coupled directly to a Carlo Erba 6000 gas chromatograph with a flame ionization detector (FID). Chromatographic conditions were as follows: a 3 m x 6 mm, 80/100 mesh, SA 1422 (Supelco inc) POROPAK Q packed column, the temperature program was 60 °C, increasing to 240 °C by 8 °C min^{-1}.

Pyrograms were interpreted by quantification of seven peaks corresponding to the major volatile pyrolytic fragments (Ceccanti et al., 1986): acetic acid (K), acetonitrile (E_1), benzene (B), toluene (E_3), pyrrole (O), furfural (N), and phenol (Y). The retention time (minutes) of the mentioned seven peaks (as standard) are the following: E1, 20.2, K, 23.1, B, 27.8, O, 30.9, E3, 34.7, N, 38.9, Y, 55.6 (Scheme 1). Acetonitrile (E_1) is derived from the pyrolysis of aminoacids, proteins, and microbial cells. Furfural (N), is mostly derived from carbohydates, ligno-cellulosic materials, proteins and other aliphatic organic compounds (Bracewell & Robertson, 1984), indicating the presence of rapidly metabolisable organic substances. Acetic acid (K) is preferentially derived from pyrolysis of lipids, fats, and waxes, cellulose, carbohydrates (Bracewell & Robertson, 1984) and represents relatively less-degraded ligno-cellulose material (Sollins et al., 1996; Buurman et al., 2007). Phenol (Y) is derived from amino acids, tannins and fresh or condensed (humic) lignocellulosic structures (van Bergen et al., 1998; Lobe et al., 2002). Benzene (B) and toluene (E_3) are basically derived from condensed aromatic structures of stable (humified) organic matter, particularly for benzene, since toluene must come from rings with aliphatic chains, albeit short. Pyrrole (O) is derived from nitrogenated compounds such as nucleic acids, proteins, microbial cells and condensed humic structures (Bracewell & Robertson, 1984).

Peak areas were normalized, so that the area under each peak referred to the percentage of the total of the selected seven peaks (relative abundances).

The alphabetic code used was conventional and has already been used in previous papers on natural soils (Ceccanti et al., 1986; Alcaniz et al., 1984). Peak purity of the most important fragments had previously been checked by coupling the same chromatographic system to a mass-detector under the same operating conditions.

Ratios of the relative abundances of the peaks were determined (Ceccanti et al., 1986, 2007) as follows:

N/O: mineralization index of labile soil organic matter. This index expresses the ratio between furfural (N), which is the pyrolytic product arising from polysaccharides, and pyrrole (O), which derives from nitrogenated compounds, humified organic matter, and microbial cells. The higher the ratio, the lower the mineralization of the organic matter, that means a high concentration of polysaccharides may be present.

O/Y: mineralization index of more-stable OM. The higher the ratio, the higher the mineralization of the organic matter, because pyrrole (O) derives from nitrogenated compounds and microbial cells and phenol (Y) mostly derives from polyphenolic compounds.

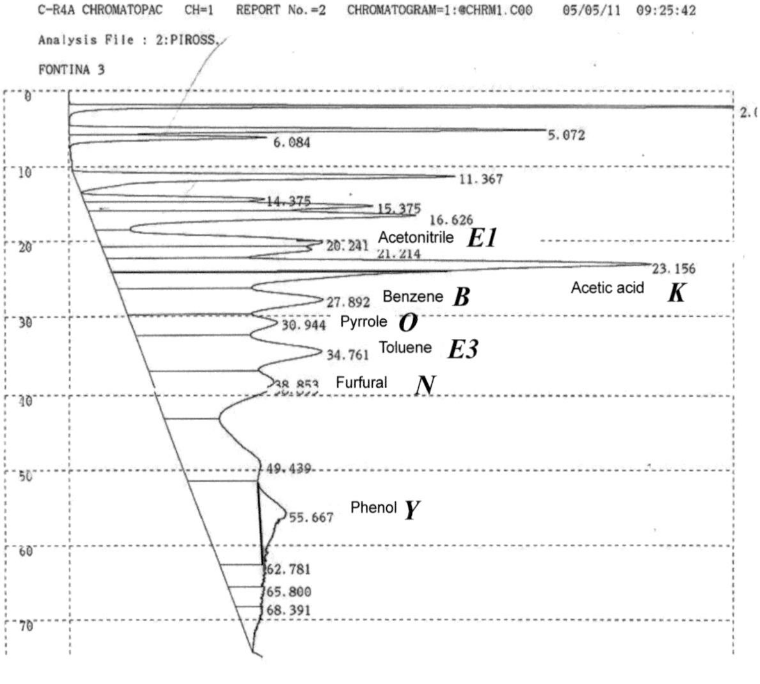

** CALCULATION REPORT **

CH	PKNO	TIME	AREA	HEIGHT	MK	IDNO	CONC	NAME
1	5	2.015	13640908	883455	V		3.9701	
	6	5.072	12345883	366448			3.5932	
	7	6.084	5775108	152592	V		1.6808	
	10	11.367	16618673	293311			4.8367	
	11	14.375	6175923	140019	V		1.7975	
	12	15.375	13101132	220810	V		3.813	
	13	16.626	19371992	251347	V		5.6381	
	14	20.241	15476787	171346	V		4.5044	
	15	21.214	13095291	161156	V		3.8113	
	16	23.156	54121392	417931	V		15.7516	
	17	27.892	25858430	155114	V		7.5259	
	18	30.944	17326828	114021	V		5.0428	
	19	34.761	31244722	139066	V		9.0936	
	20	38.853	24606074	93780	V		7.1614	
	21	49.439	26597246	60914	V		7.7409	
	22	55.667	35184940	66429	V		10.2403	
	23	62.781	5523164	34857	V		1.6075	
	24	65.8	3865950	27964	V		1.1252	
	25	68.391	3661741	20299	V		1.0657	
		TOTAL	343592032	3770857			100	

Scheme 1. Pyrogram of sample with time retention and area of each fragment: Acetonitrile (E1), Acetic acid (K), Benzene (B), Pyrrole (O), Toluene (E3), Furfural (N), Phenol (Y).

B/E_3: humification index. The higher the ratio, the higher the condensation of the organic matter, because benzene (B) derives mostly from the pyrolytic degradation of condensed aromatic structures, while toluene (E_3) comes from aromatic uncondensed rings with aliphatic chains.

AL/AR (Aliphatic/Aromatic compounds): index of "energetic reservoir". It expresses the ratio between the sum of aliphatic products (acetic acid K, furfural N, and acetonitrile E_1) and the sum of aromatic compounds (benzene B, toluene E_3, pyrrole O and phenol Y). This index could indicate an "energetic reservoir" especially in extremely poor soils since it evaluates the labile and stable parts of HS.

In addition, a numeric index of similarity (Sij) between the relative abundances (I) of the homologous peaks (k) in two pyro chromatograms (i and j), was calculated using the following expression: $Sij = (\sum(Ii/Ij)k)/n$ with $Ii < Ij$, and n is the number of peaks (Ceccanti et al., 1986).

The index of similarity Sij compares a pair of pyrograms without discriminating peaks. The index varies in the range of 0–1: the higher the value, the greater the similarity. However, three conventional levels, very high (>0.85), high (0.75-0.85), middle (0.70–0.75), and low (0.60–0.70) have been suggested for the characterization of heterogeneous materials such as SOM (Ceccanti et al., 1986) and compost (Ceccanti & Garcia, 1994; Ceccanti et al., 2007).

2.2 Humic substances extraction

Soil organic matter comprises a range of humified, non-humified, and biologically active compounds, including readily decomposable materials, plant litter and roots, and dead and living organisms. The humified compounds, representing the most microbially recalcitrant and thus stable reservoir of organic carbon in soil may be more appropriate than the labile organic matter compounds in the definition of "soil quality", they can, in fact, be considered a sort of recorder of the soil's history (Masciandaro & Ceccanti, 1999; Piccolo et al., 1996).

Humic substances based on solubility under acidic or alkaline conditions can be fractionated as following: humin, the insoluble fraction of humic substances, humic acid (HA), the fraction soluble under alkaline conditions but not acidic conditions (generally pH <2), and fulvic acid (FA), the fraction soluble under all pH conditions.

Due to the low solubility and complex chemical structure of these compounds, soil extraction and purification are essential steps to study humic substances (Ceccanti et al., 1978). Sodium pyrophosphate is the favoured solution used to extract organic matter from soil, and, neutral conditions permit to extract the greater part of humic carbon, maintaining the functional properties of soil (Masciandaro & Ceccanti,1999).

The extraction of humic substances from the soils were carried out by mechanical shaking of the soil samples with 0.1 M $Na_4P_2O_7$ solution at pH 7.1 for 24 h at 37 °C in a soil/solution ratio of 1/10. The extracts were centrifuged at 24 000 x g for 20 min and filtered on a 0.22 μm Millipore membrane. Identical volumes of these extracts were dialysed in Visking tubes with distilled water. After dialysis, extracts were concentrated by ultrafiltration (AMICON PM10 cut-off membrane) and brought to the volume that extracts had prior to dialysis (Nannipieri et al., 1983; Masciandaro et al., 1997).

3. Case studies

Five case studies have been selected in order to validate the pyrolysis gas-chromatography methodology as useful tool to evaluate chemico-structural changes of soil organic matter: i) two natural soil ecosystems (Catena) with different density of natural plant cover (natural soil), one located in the south of Spain and the other in the south of Italy, ii) two sites under organic (biological) and mineral (conventional) management, one in Cyprus Island and one in the centre of Italy, and iii) a degraded soil undergone to rehabilitation practice with different amount of organic matter (restored soils).

Each soil sample has been collected at 0-15 cm depth and the pyrolysis gas-cromatography has been carried out both in the bulk soil, and in sodium pyrophosphate extract of each sample.

3.1 Natural soil

3.1.1 Spanish catena

A site in Santomera area (Murcia region, Spain) characterized by three different gradual degradation states, related to different natural plant cover establishment and soil slope, has been sampled (sandy loam, USDA classification): i) a natural soil with a 50-60% of vegetation cover of *Pinus halepensis* (high plant density, **Forest**) and a soil slope of 5%, ii) a partially degraded soil with a 20-30% of vegetation cover (autochthonous xerophytic shrub) (low plant density, **Shrub**) and slope of 5%, and iii) a bare soil with only a 5-10% of vegetation cover (scant plant density, **Bare**) and slope of 10%.

The main difference among the soil samples was presented by the percentage of toluene (E_3), the significant increase along the catena ecosystem (Bare > Shrub > Forest) (Table 1), indicated the prevalence of organic matter less-humified in Bare and Shrub with respect to Forest soil, since E_3 comes from aromatic uncondensed rings with aliphatic chains.

In agreement with this result, lower values of benzene (B) and phenol (Y), which derived principally from humified stable organic matter, were found in Bare soil with respect to Forest and Shrub soil, suggesting the reduction of condensed aromatic structure of stable humified organic matter (Masciandaro & Ceccanti, 1999) in consequence of a fast metabolism, involving probably the native humic matter.

bulk soil	E_1%	K%	B%	O%	E_3%	N%	Y%
Forest	11,7	11,7	18,0	22,9	15,4	17,6	2,60
Shrub	11,5	11,3	18,5	20,3	18,5	17,1	2,72
Bare	13,2	10,0	17,4	21,1	19,8	16,8	1,66

Table 1. Relative abundances (%) of main pyrolytic peaks (acetonitrile E_1, acetic acid K, benzene B, pyrrole O, toluene E_3, furfural N, phenol Y) in bulk soils.

Soil differentiation was better shown by the pyrolytic indexes of humification (B/E_3) and mineralization (O/Y and N/O). The humification index B/E_3 which represents the structural condition of a 'condensed aromatic nucleus' of humic substances, increased with the increase of the intensity of vegetation (Forest > Shrub > Bare) (Figure 1), indicating the importance of plant cover in the storage of more stable humus and in the reduction of degradation process (Vancampenhout et al., 2010).

In Bare soil, the scant vegetation seemed unable to protect the soil organic matter from mineralization, as confirmed by the higher value of O/Y (Figure 1). However, the N/O mineralization index, that gave an evidence of the existence of an easily mineralizable fraction of organic matter, did not show many differences among the soils (Figure 1), suggesting that the labile organic matter was less influenced.

Percentages of stable benzene-derived aromatic (AR = B + E_3 + O + Y) and mineralizable aliphatic (AL = E_1 + K + N) structures were very similar in all the three soils (Figure 1). This would indicate that, despite the highest mineralization occurred in Bare soil, the catena ecosystem was self-regulating and still preserved a significative amount of microbial resistant humic substances (Ceccanti & Masciandaro, 2003).

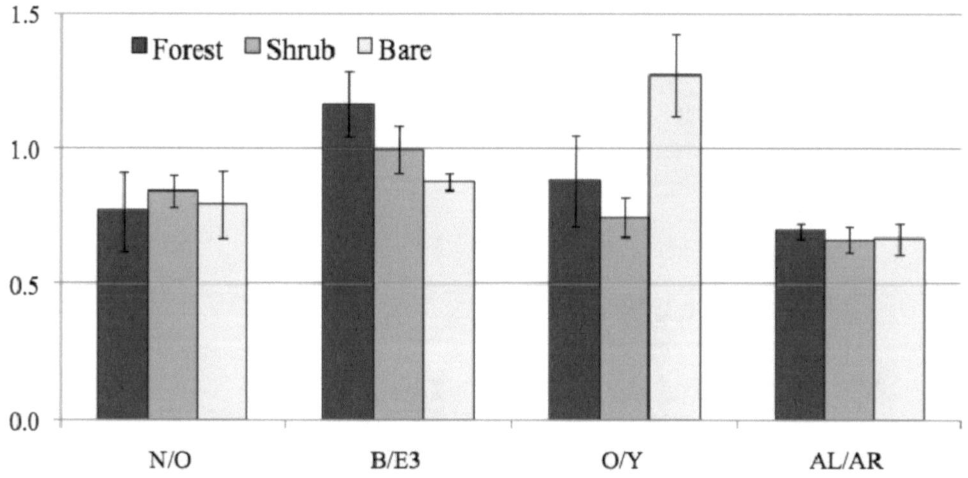

Fig. 1. Indices of mineralization (furfural/pyrrole N/O, pyrrole/phenol O/Y), humification (benzene/toluene B/E_3), and aliphatic/aromatic (AL/AR) ratios (AL=E_1+K+N, AR=B+O+E_3+Y) (mean of three replicates ± standard deviation) in bulk soils The values of O/Y have been divided by 10.

This sort of natural balance in the Catena ecosystem, was confirmed by the high indexes of similarity observed among the three soil samples and, in particular, between Forest and Shrub (Table 2).

bulk soil	Shrub	Bare
Forest	0,938	0,857
Shrub		0,883

Table 2. Pyrolitic index of similarity in the bulk soils.

In the soil extract, the chemical and structural information is amplified by the fact that unlike the bulk soil, the humic fraction is more concentrated and pure and free from interfering substances. In this case no great differences were generally observed among the soils, both in the content of pyrolytic fragments (Table 3) and calculated indices (Figure 2). However, the higher value of N/O found in Bare with respect to Forest soil, suggested that the mineralization of organic matter, hypothesized on the basis of mineralization index O/Y of bulk soil, involved the labile fraction of humic matter. The high similarity indexes observed among the bulk soils (Table 2) were maintained also among the humic extracts (Table 4), underlining a better homogeneity considering the humic part of the organic matter.

Soil extract	E_1%	K%	B%	O%	E_3%	N%	Y%
Forest	8,26	8,53	16,6	19,7	27,2	13,4	6,31
Shrub	8,43	7,48	14,9	18,8	26,3	15,1	9,31
Bare	6,98	10,0	18,2	17,1	27,4	15,4	4,86

Table 3. Relative abundances (%) of main pyrolytic peaks (acetonitrile E_1, acetic acid K, benzene B, pyrrole O, toluene E_3, furfural N, phenol Y) in soil extracts.

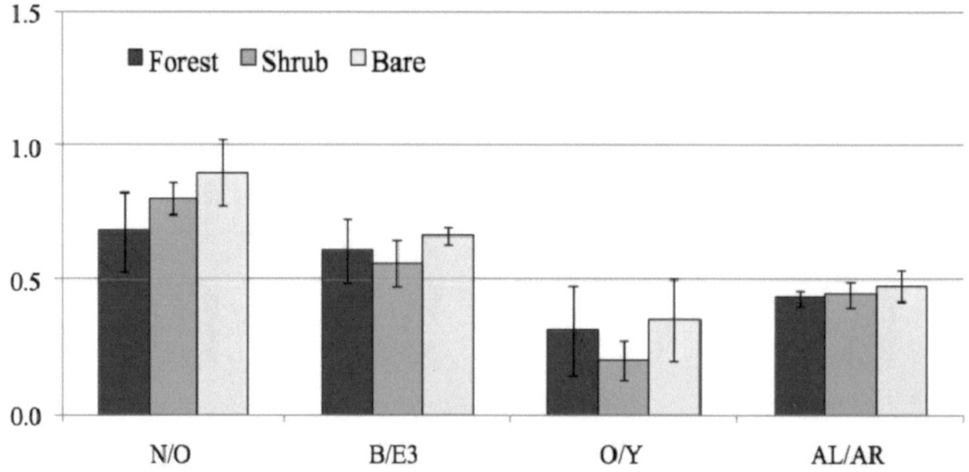

Fig. 2. Indices of mineralization (furfural/pyrrole N/O, pyrrole/phenol O/Y), humification (benzene/toluene B/E_3), and aliphatic/aromatic (AL/AR) ratios (AL=E_1+K+N, AR=B+O+E_3+Y) (mean of three replicates ± standard deviation) in soil extracts. The values of O/Y have been divided by 10.

soil extract	Shrub	Bare
Forest	0,872	0,888
Shrub		0,820

Table 4. Pyrolitic index of similarity in the soil extracts.

3.1.2 Italian catena

A natural catena similar to the Spanish one, were studied in the South of Italy (Basilicata region) (sandy, clay, loam, USDA classification).

Three different gradual degradation states, related to different natural plant cover establishment and soil slope, has been sampled: i) a natural soil with a 60-70% of vegetation cover of *Quercus pubescens* Willd (high plant density, **Forest**) and a soil slope of 15%, ii) a partially degraded soil with a 20-30% of vegetation cover, consisting in autochthonous xerophytic shrub and grass (low plant density, **Shrub**) and slope of 6%, and iii) an eroded soil without plant cover (no plant, **Bare**) and slope of 20%.

In this Italian natural Catena, differently from the previous, the main difference among the three soils concerned the pyrrole (O) and particularly phenol (Y) fragments (Table 5). An increase of phenol in Bare suggested a slower soil metabolism that slightly affected the dinamic of stable humic matter, probably due to the lack of vegetation cover. The simultaneous increase in pyrrole (O) and phenol (Y) fragments in each soil, indicated the same level of mineralization processes of more stable soil organic matter in the three different degraded soils.

However, as already found for Spanish catena, there was a decrease in humification index with the decrease of vegetal cover (Figure 3). This trend, together with the lower value of AL/AR and N/O in Bare soil, suggested that a great mineralization process interesting the labile part of the organic matter affected this environment.

A lower similarity was, in fact, found among the soils and in particular between Forest and Bare (Table 6) soils.

bulk soil	E_1%	K%	B%	O%	E_3%	N%	Y%
Forest	16,0	9,53	16,1	10,6	24,5	10,0	14,2
Shrub	11,9	11,2	14,6	11,2	23,0	11,8	16,2
Bare	8,75	11,2	11,0	14,5	24,6	8,09	21,9

Table 5. Relative abundances (%) of main pyrolytic peaks (acetonitrile E_1, acetic acid K, benzene B, pyrrole O, toluene E_3, furfural N, phenol Y) in bulk soils.

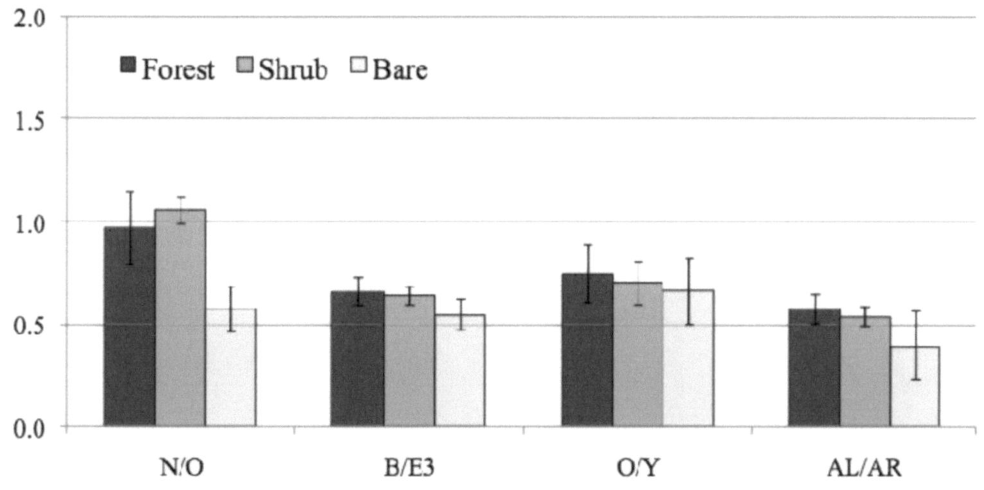

Fig. 3. Indices of mineralization (furfural/pyrrole N/O, pyrrole/phenol O/Y), humification (benzene/toluene B/E_3), and aliphatic/aromatic (AL/AR) ratios ($AL=E_1+K+N$, $AR=B+O+E_3+Y$) (mean of three replicates ± standard deviation) in bulk soils.

bulksoil	Shrub	Bare
Forest	0,845	0,752
Shrub		0,801

Table 6. Pyrolitic index of similarity in the bulk soils

Soil extract	E1%	K%	B%	O%	E3%	N%	Y%
Forest	8,94	11,4	23,4	9,90	20,0	12,3	13,9
Shrub	13,8	13,9	19,2	12,9	14,2	14,5	11,5
Bare	6,39	16,1	25,7	12,0	22,0	9,55	8,32

Table 7. Relative abundances (%) of main pyrolytic peaks (acetonitrile E_1, acetic acid K, benzene B, pyrrole O, toluene E_3, furfural N, phenol Y) in soil extracts.

In the humic extract, the trend of the mineralization indices O/Y and N/O, highlighted the hard mineralization process that affect bare soil, followed by shrub soil (Figure 4, Table 7).

In this Catena ecosystem, differently from the previous, the scant vegetation and probably other environmental factors such as greater soil slope and Mediterranean climate, with dry hot summer and cold winters with strong storm, make the bare soils more susceptible to erosion processes.

However, the high values of B/E_3 detected in each soil, suggested that this ecosystem is able to preserve the more condensed humic matter.

As already observed for the bulk soils, medium-high similarity indices were found among the humic extracts of the different situations (Table 8). Therefore, both Italian and Spanish Catena, were more homogeneous considering the humic part of the organic matter.

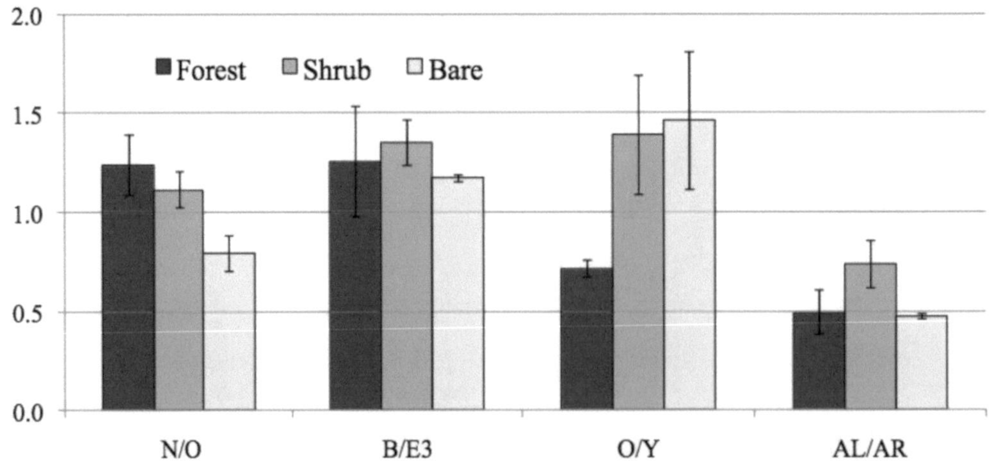

Fig. 4. Indices of mineralization (furfural/pyrrole N/O, pyrrole/phenol O/Y), humification (benzene/toluene B/E_3), and aliphatic/aromatic (AL/AR) ratios ($AL=E_1+K+N$, $AR=B+O+E_3+Y$) (mean of three replicates ± standard deviation) in soil extracts.

Soil extract	Shrub	Bare
Forest	0,778	0,778
Shrub		0,718

Table 8. Pyrolitic index of similarity in the soil extracts

3.2 Organic and mineral management

3.2.1 Site in Cyprus Island

Two adjacent fields in Cyprus Island, where one was managed according to biological, and the other according to conventional farming methods, were studied (sandy clay loam soil, USDA classification). In the biological agriculture, the fields were fertilized with 10 t ha^{-1} y^{-1} of commercial manure (in pellets), while in the conventional treatment mineral fertiliser was added at a total rate of 300 kg ha^{-1}.

A clear difference between biological and conventional management was evident in Py-GC fragments analysis (Table 9): a higher abundance of benzene (B) and phenol (Y) characterized the biological treatment, thus indicating the establishment of a humified stable organic matter nucleus. On the other hand, the greater value of toluene (E_3) and pyrrole (O) found in conventional treatment highlighted the presence of humified organic matter, especially that in less condensed and pseudo-stable form (Table 9). In addition, the high presence of pyrrole (O) in conventional management suggested the triggering of mineralization processes. In fact, the higher value of O/Y and the lower value of N/O in conventional treatment (Figure 5), highlighted the higher level of mineralization of less mineralizable and easily degradable organic matter, respectively, occurred in this soil. Instead, the impressive degree of index of humification B/E_3 found in biological treatment suggested the incorporation of organic matter added to soil addressed the metabolic processes towards the organic matter humification.

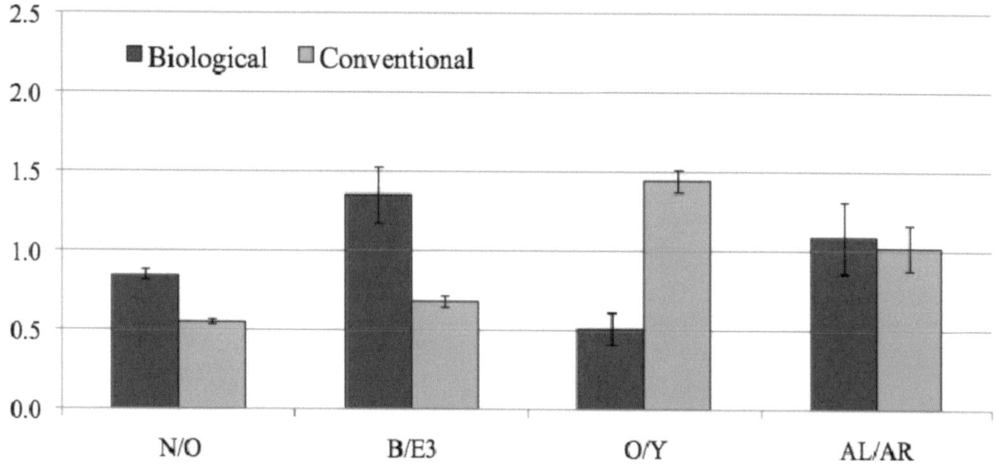

Fig. 5. Indices of mineralization (furfural/pyrrole N/O, pyrrole/phenol O/Y), humification (benzene/toluene B/E_3), and aliphatic/aromatic (AL/AR) ratios (AL=E_1+K+N, AR=B+O+E_3+Y) (mean of three replicates ± standard deviation) in bulk soils.

bulk soil	E_1%	K%	B%	O%	E_3%	N%	Y%
Biological	26,1	20,3	16,8	6,23	12,8	5,29	12,4
Conventional	29,1	15,5	13,0	10,4	19,0	5,75	7,21

Table 9. Relative abundances (%) of main pyrolytic peaks (acetonitrile E_1, acetic acid K, benzene B, pyrrole O, toluene E_3, furfural N, phenol Y) in bulk soils.

The results obtained from the analysis of humic extracts, discriminated more the different influence of the two management practices on soil organic matter quality (Table 10). The effect of compost addition on soil under biological treatments was, in fact, highlighted by the higher level of acetic acid (K) and pyrrole (O) relative abundances with respect to the conventional one, as also found by Ceccanti et al. (2007). The values of these compounds suggested the presence of fresh organic matter and the beginning of a humification process mediated by microbial activity (Peruzzi et al., 2011). The higher relative abundance of toluene (E_3) in conventional site was in agreement with the results of bulk soil pyrolytic fragments, thus showing the presence of less stable humic fraction. The index of mineralization N/O (Figure 6) showed that, in biological treatment, mineralization process of easily degradable organic matter occurred. However, the highest index of humification in biological site, as shown by pyrolytic indices in the bulk soil, suggested that the organic matter added with compost has became part of humic pool of soil (Garcia et al.,1992).

The differences occurred in management practises could be resumed by similarity indices, which clearly expressed the difference between the two treatments both in bulk soil and in pyrophosphate extract (S_{soil}: 0.743, $S_{extract}$: 0.774).

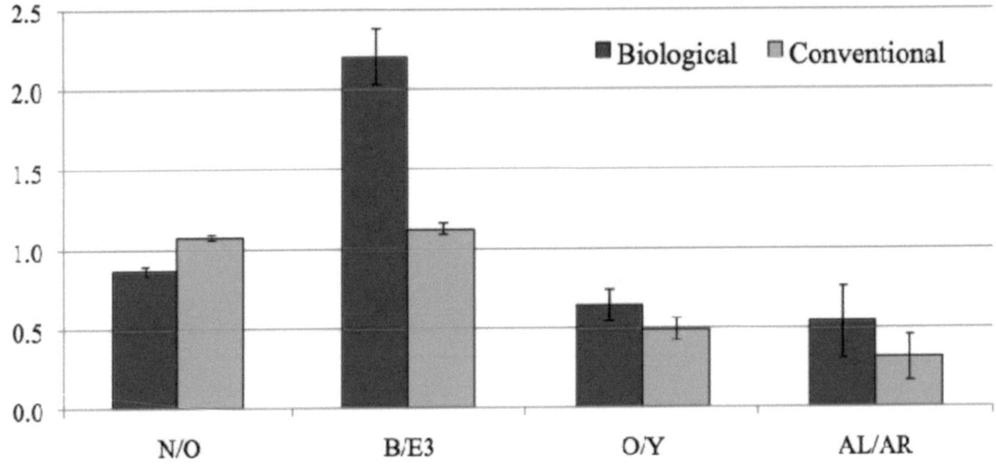

Fig. 6. Indices of mineralization (furfural/pyrrole N/O, pyrrole/phenol O/Y), humification (benzene/toluene B/E_3), and aliphatic/aromatic (AL/AR) ratios (AL=E_1+K+N, AR=B+O+E_3+Y) (mean of three replicates ± standard deviation) in soil extracts.

Soil extract	E_1%	K%	B%	O%	E_3%	N%	Y%
Biological	4,20	17,88	18,9	14,6	8,63	12,6	23,2
Conventional	4,49	6,95	21,5	11,7	19,2	12,4	23,7

Table 10. Relative abundances (%) of main pyrolytic peaks (acetonitrile E_1, acetic acid K, benzene B, pyrrole O, toluene E_3, furfural N, phenol Y) in soil extracts.

3.2.2 Site in Italy

Two adjacent fields in Central Italy (Tuscany region), one managed according to biological, and the other according to conventional farming methods, were also studied. The soil of both fields was classified as sandy clay loam according to the USDA classification. In the biological agricultural soil (biological procedure for five years), green manure (1,5 t ha^{-1} y^{-1}) was applied to the fields, while in the conventional agricultural soil, the fields received mineral fertilizer (0,2 t ha^{-1} y^{-1} ammonium nitrate).

The Italian situation showed fewer differences both in soil and extract, considering the two management practices, with respect to the Cyprus case (Similarity indices S_{soil}: 0.873, $S_{extract}$: 0.870). However, a higher level of benzene (B) and pyrrole (O) characterized biological treatment (Table 11). This result suggested that this kind of soil management promoted: i) the establishment of more humified organic matter, as also suggested by the trend of humification index B/E3, and ii) the stimulation of microbial metabolism driving the mineralization processes of the available organic matter, as suggested by the trend of O/Y and N/O indices (Figure 7). In other words, the biological management stimulated more the process of organic matter turnover, with respect to the conventional one, which on the contrary, started the mineralization of the humified organic matter, as suggested by the highest value of O/Y in the humic extract (Figure 8, Table 12). However, the trend of humification index B/E_3 of the extracts (no difference between conventional and biological treatment), indicating that the conventional treatment did not still affect the stable nucleus of humic fraction.

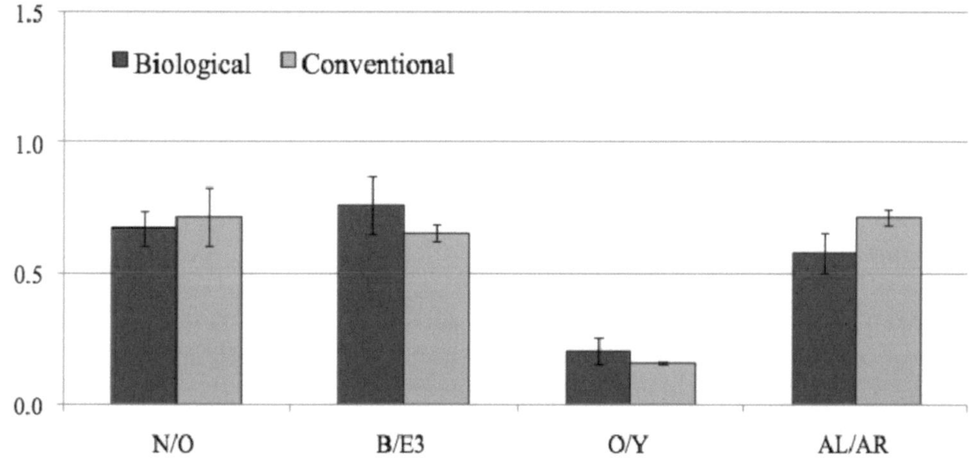

Fig. 7. Indices of mineralization (furfural/pyrrole N/O, pyrrole/phenol O/Y), humification (benzene/toluene B/E_3), and aliphatic/aromatic (AL/AR) ratios (AL=E_1+K+N, AR=B+O+E_3+Y) (mean of three replicates ± standard deviation) in bulk soils. The values of O/Y have been divided by 10.

bulk soil	E_1%	K%	B%	O%	E_3%	N%	Y%
Biological	14,3	11,4	16,8	16,4	22,1	11,0	8,06
Conventional	21,0	11,0	14,3	13,6	21,9	9,70	8,63

Table 11. Relative abundances (%) of main pyrolytic peaks (acetonitrile E_1, acetic acid K, benzene B, pyrrole O, toluene E_3, furfural N, phenol Y) in bulk soils.

Soil extract	E1%	K%	B%	O%	E3%	N%	Y%
Biological	3,38	13,4	21,8	15,6	24,0	13,8	8,02
Conventional	6,00	15,9	22,8	15,8	23,9	11,5	4,10

Table 12. Relative abundances (%) of main pyrolytic peaks (acetonitrile E_1, acetic acid K, benzene B, pyrrole O, toluene E_3, furfural N, phenol Y) in soil extracts.

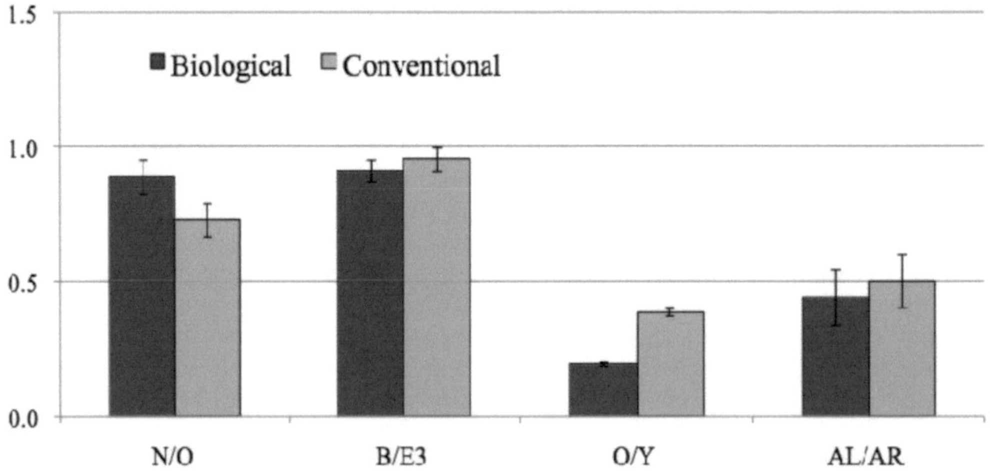

Fig. 8. Indices of mineralization (furfural/pyrrole N/O, pyrrole/phenol O/Y), humification (benzene/toluene B/E_3), and aliphatic/aromatic (AL/AR) ratios (AL=E_1+K+N, AR=B+O+E_3+Y) (mean of three replicates ± standard deviation) in soil extracts. The values of O/Y have been divided by 10.

3.3 Restoration of degraded soils

An experimental field (sandy clay loam soil, USDA classification) in province of Murcia (Spain) was split into plots in which 20 years ago was added a single dose of fresh easily degradable municipal organic waste (MOW) in such dose as to increase the soil organic matter by 0.5 (65t/ha), 1.0 (130 t/ha), 1.5 (200t/ha) and 2.0% (260 t/ha), MOW fraction was incorporated into the top 15 cm of soil using a rotovator. The aim was to restore biochemical and microbial properties and to contrast the erosion through a stimulation of spontaneous establishment of grass cover. One plot without the addition of MOW was used as control (0%). The plots were monitored for three years: following organic amendments an averagely 50-70% plant coverage developed and persisted throughout the experiment until today. The highest plant density were found in 1.5% treatment (80-90%) while only 20-30% plant coverage was developed in the control soil. After 16 years from the treatment the same vegetated plots have been sampled at 0-15 cm depth in order to assess and investigate the changes occurred in consequence of the vegetal cover establishment.

The input of exogenous OM produced qualitative changes of soil OM both in the bulk soil and in the pyrophosphate extracts, which are still evident after 16 years. Considering the different plots, in fact, there were significant differences in the Py-GC fragments analysis: acetonitrile (E_1) and acetic acid (K) increased with the increase of organic matter added (Table 13), at the same time lower values of pyrrole (O) and phenol (Y) were detected in the same plots. The increase of acetonitrile and pyrrole were probably due to the higher cover of vegetation restored in the plots, which bring to higher quantity of cellulose, lipids and easily degradable compounds released into the soil. This trend was also confirmed by AL/AR index, which increased at the increase of the organic matter application (Figure 9).

The plots treated with the higher quantity of organic matter, also showed higher level of humification index (B/E_3), thus suggesting that after 16 years the humification overrides the mineralization process. As suggested by to the humification index, the N/O ratio showed higher values in the plots with the greater application of organic residues, indicating the presence of more evolved (less mineralizable) organic compounds in the soil. The similarity indices confirmed the role of the MOW in affecting the quality of organic matter (Marinari et al., 2007) (Table 14), in that 1%, 1.5% and 2% treatments resulted more similar between them.

bulk soil	E_1%	K%	B%	O%	E_3%	N%	Y%
Organic matter 0.0%	6,39	3,54	11,8	15,1	24,0	14,0	25,2
Organic matter 0.5%	8,81	6,57	12,5	16,9	20,3	16,0	19,5
Organic matter 1.0%	14,4	11,3	15,3	10,6	20,0	10,3	17,4
Organic matter 1.5%	11,9	13,2	14,1	12,7	20,1	14,1	13,9
Organic matter 2.0%	14,6	11,6	14,7	10,7	19,9	12,7	15,7

Table 13. Relative abundances (%) of main pyrolytic peaks (acetonitrile E_1, acetic acid K, benzene B, pyrrole O, toluene E_3, furfural N, phenol Y) in bulk soils.

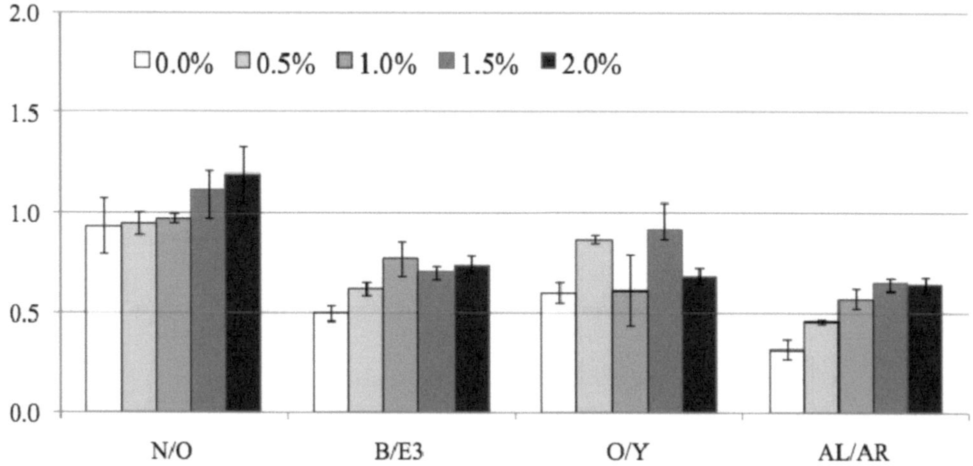

Fig. 9. Indices of mineralization (furfural/pyrrole N/O, pyrrole/phenol O/Y), humification (benzene/toluene, B/E_3), and aliphatic/aromatic (AL/AR) ratios ($AL=E_1+K+N$, $AR=B+O+E_3+Y$) (mean of three replicates ± standard deviation) in bulk soils.

bulk soil	0,5%	1%	1,5%	2%
0%	0,800	0,641	0,695	0,660
0,5%		0,736	0,779	0,747
1.0%			0,853	0,945
1,5%				0,898

Table 14. Pyrolitic index of similarity in the bulk soils.

The results of the humic extract highlighted the great difference among the 1.5% and the others treatments (Table 15,16 and Figure 10). In fact, the higher vegetal cover in 1.5% plots probably promoted the enrichment of soil in labile and less stable fraction of organic matter, which resulted the more sensitive to degradation process, as suggested by the trend of N/O and O/Y. As a consequence, in 1.5% treatment a lower value of B/E_3 was observed, probably due to the masking of humic matter from fresh materials.

Soil extract	E_1%	K%	B%	O%	E_3%	N%	Y%
Organic matter 0.0%	3,66	3,94	16,4	13,2	21,2	18,7	22,9
Organic matter 0.5%	2,62	2,40	17,9	12,5	21,5	19,2	24,0
Organic matter 1.0%	3,46	3,77	18,1	12,0	22,5	17,2	23,2
Organic matter 1.5%	7,88	6,05	12,7	14,9	22,2	17,1	19,0
Organic matter 2.0%	6,07	4,93	15,6	12,2	23,6	15,9	21,7

Table 15. Relative abundances (%) of main pyrolytic peaks (acetonitrile E_1, acetic acid K, benzene B, pyrrole O, toluene E_3, furfural N, phenol Y) in soil extracts.

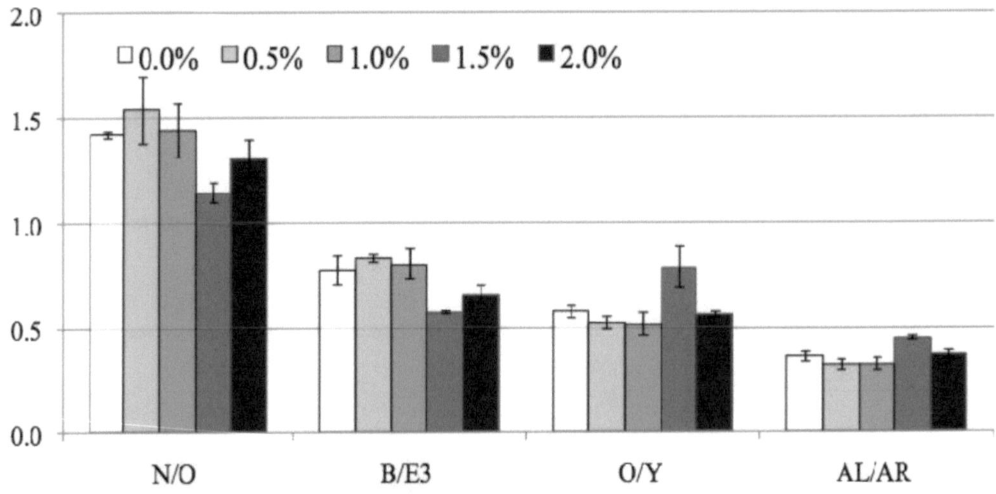

Fig. 10. Indices of mineralization (furfural/pyrrole N/O, pyrrole/phenol O/Y), humification (benzene/toluene B/E_3), and aliphatic/aromatic (AL/AR) ratios ($AL=E_1+K+N$, $AR=B+O+E_3+Y$) (mean of three replicates ± standard deviation) in soil extracts. The values of O/Y have been divided by 10.

Soil extract	0,5%	1%	1,5%	2%
0%	0,872	0,938	0,781	0,888
0,5%		0,88	0,703	0,773
1.0%			0,766	0,855
1,5%				0,852

Table 16. Pyrolitic index of similarity in the soil extracts.

4. Statistical considerations

In order to understand the importance of PY-GC technique in describing the chemico-structural and the functionality of soil organic matter, the principal component analysis (PCA) was used. PCA is a multivariate statistical data analysis technique that reduces a set of raw data into a number of principal components that retain the most variance within the original data to identify possible patterns or clusters between objects and variables.

In order to develop wider conclusions about the state of soil organic matter a data set containing information about different ecosystems was used. 12 sites were selected across Europe in order to ensure a broad range of different conditions (natural soil and agro-ecosystems) and related pressures: Spain (Abanilla, Santomera, El Aguilucho, Tres Caminos, all in Murcia region) Italy, (Castelfalfi and Alberese in Tuscany region, Pantanello in Basilicata region), Germany (Puch, in Bavaria region) and Hungary (Gödöllö). All selected sites were characterized by local variation of anthropogenic and natural pressures additionally to the variation that can be found at the European scale. The details of all sites are listed in Table 17. The selected sites may be grouped onto two different types:

Type one: Catenae. Pressures and soil degradation change along a gradient, for example along a slope with differing steepness or along a landscape with a sequence of plant cover. The selected catenae imply a variation of pressures that is mainly of natural origin. Samples are taken from different locations along the catenae. The sites were situated in Spain (Abanilla, Carcavo, Santomera) and Hungary (Godollo) and are described in Table 17.

Site	Description	Treatment Pression	Experiment
Puch (Germany)	Agricultural field under normal cultivation	Tillage/ Plant Cover	Management variation
	Field kept bare of vegetation with plowing		
	Very limited vegetation maintained		
Abanilla (Spain)	Bare	Erosion	Catena
	Esparto grass		
	High coverage		
	Medium coverage		
	Sludge added 0%	Organic matter	Management variation
	Sludge added 5%		
	Sludge added 10%		
	Sludge added 15%		
	Sludge added 20%		

Site	Description	Treatment Pression	Experiment
Carcavo (Spain)	Abandoned	Erosion/ Exposition	Catena
	Degraded		
	Vegetated with pines		
	North slope		
	South slope		
El Aguilucho (Spain)	Control	Plant Cover/ Erosion/ Organic Matter	Management variation
	Pine		
	Pine + Organic Matter (OM)		
	Terrace + Pine		
	TP+ mycorrize		
	TPm + OM		
	TP+ soil mycorrize		
	TP + OM		
	TPMs + OM		
Santomera (Spain)	Bare	Plant Cover/ Erosion	Management variation/ Catena
	Forest		
	Shrub		
	Removed forest		
	Forest		
Tres Caminos (Spain)	Compost	Plant Cover/ Organic matter	Management variation
	Humus enzymes		
	Control		
	Reforestation		
	Reforestation + mycorrize		
	Seed		
	Sewage Sludge		
Gödöllö (Hungary)	Accumulated	Erosion	Catena
	High erosion		
	Medium erosion		
	No erosion		
Pantanello (Italy)	Organic agriculture	Organic matter/ Agrochemicals	Management variation
	Conventional agriculture		
Castelfalfi (Italy)	Disturbed forest	Animal pressure/ Organic matter/ Erosion	Management variation
	Undisturbed forest		
Alberese (Italy)	Organic agriculture	Organic matter/ Agrochemicals	Management variation
	Conventional agriculture		

Table 17. Site description

Type two: Variation of management. Different types of treatments have been applied to plots at the same site some time ago. These treatments may either represent an anthropogenic pressure, which promotes soil degradation or a remediation action, from which lasting effects may be expected. Type two also contains adjacent sites under different agricultural and forest management, where for instance organic farming may be compared to common farming practice or fenced off forest areas may be compared to the rest of the forest with a high animal population density. The sites were situated in Germany (Puch), Spain (Abanilla, El Aguilucho, Santomera, Tres Caminos) and Italy (Pantanello, Castelfalfi and Alberese) and are described in Table 17.

The PCA of the data set about bulk soils indicated 80% of the data variance as being contained in the first three components (Table 18). PC1, PC2 and PC3 account for 31.0%, 27.9% and 22.0% of the total variance, respectively. PC1 was closely associated with B, O, Y and both indices of mineralization (O/Y and N/O), while PC2 was linked with K, E_3, humification index B/E_3 and AL/AR index. PC3 was associated with E_1 and N. As expected, the indices of mineralization had an opposite behaviour: when O/Y increase, N/O decrease and vice versa, being involved in pseudo-stable and labile organic matter metabolism, respectively. In PC2, the high correlation occurred between humification index B/E_3 and AL/AR index, suggested that both labile and stable organic matter affected soil humification processes.

	Factor 1	Factor 2	Factor 3
Acetonitrile (E_1 soil)	0.027	0.139	*0.899**
Acetic acid (K soil)	0.072	*0.875**	0.190
Benzene (B soil)	*0.730**	-0.014	0.285
Pyrrole (O soil)	*0.804**	0.020	-0.483
Toluene (E_3soil)	0.071	*-0.819**	0.367
Furfural (N soil)	-0.257	0.136	*-0.872**
Phenol (Ysoil)	*-0.786**	-0.283	-0.352
Mineralization index (N/O soil)	*-0.788**	0.031	-0.321
Humification index (B/E_3 soil)	0.290	0.868*	-0.188
Mineralization index (O/Y soil)	*0.895**	0.012	-0.165
Energetic reservoir index (AL/AR soil)	-0.137	*0.871**	0.375
Explained .Variance	3.398	3.068	2.482
Total Proportionality	30.89%	27.89%	22.57%

Table 18. Analysis of principal components on bulk soil pyrolytic fragments and indices.
*Variables with component loading used to read the PC.

The results of PCA of the data set about soil extracts were similar to those about bulk soil, with few important exceptions concerning pyrolytic fragments and index related to the more stable part of humic matter. In the same way of bulk soil, PCA of the extract isolated three principal components (PC) (total variance explained: 70.6%). The 1st PC (29.9 % of the total variance) included E_3 and the same parameters of bulk soil (O, Y, O/Y and N/O) with the exception of B that, in this case, was included in the 3nd PC (17.1 % of the total variance) together with humification index (B/E_3). Therefore, differently from the bulk soil, the humification index resulted independent from the other indices, suggesting the importance of the PY-GC on the pyrophosphate extract to investigate the condensed aromatic nucleus of humic substances. The 2nd PC(23.7% of the total variance) was associated, as for bulk soil, with K and AL/AR index (Table 19).

	Factor 1	Factor 2	Factor 3
Acetonitrile (E_1 extract)	0.074	-0.479	0.203
Acetic acid (K extract)	-0.089	*-0.886**	-0.225
Benzene (B extract)	-0.256	0.358	*-0.804**
Pyrrole (O extract)	*-0.742**	0.260	0.109
Toluene (E_3 extract)	*-0.749**	0.352	0.207
Furfural (N extract)	0.271	0.319	0.494
Phenol (Y extract)	*0.771**	0.485	0.134
Mineralization index (N/O extract)	*0.742**	0.056	0.346
Humification index (B/E_3 extract)	0.355	-0.001	*-0.833**
Mineralization index (O/Y extract)	*-0.859**	-0.102	-0.074
Energetic reservoir index (AL/AR extract)	0.083	*-0.960**	0.067
Explained. Variance	3.279	2.604	1.879
Total Proportionality	29.80%	23.67%	17.08%

Table 19. Analysis of principal components on humic extract pyrolytic fragments and indices. *Variables with component loading used to read the PC.

5. Conclusions

The results obtained by PY-GC carried out on the bulk soil and humic extract of several European sites, characterized by different management practices, permitted to draw these final considerations:

i. Mineralization indices N/O and O/Y resulted the pyrolytic parameters that better discriminate among the different soils, being involved in the dynamics of labile organic matter, linked to microbial metabolism, and pseudo-stable matter, related to the humic fraction.
ii. The pyrolysis on the humic extract often evidenced information hidden by the analysis carried out on the bulk soil. In particular, the pyrolysis on the pyrophosphate extract permitted to obtain clearer information on the condensed aromatic nucleus of humic substances.
iii. Even if the analysis on soil as such (bulk) and on soil humic extracts gave each specific information on the quality of the organic matter; the scientific details obtained from the pyrophosphate extract agreed with those resulted analysing the bulk soil, meaning that this kind of extract does not alter the chemical properties of the soil organic matter.

Finally, it is possible to conclude that the PY-GC can be used as a quick and general reproducible technique to make a qualitative study of the chemico-structural characteristics of soil organic matter turnover related to biological activity (bulk soil), and of the more stable humic fraction (pyrophosphate extract) in soils under different anthropogenic and natural pressures.

Further researches could involve the study of the link between the chemico-structural properties of the humic fraction and their capability to protect enzyme in biologically active form.

6. Acknowledgment

The study was carried out within the framework of the EU project "Indicators and thresholds for desertification, soil quality, and remediation" INDEX (STREP contract n° 505450 2004/2006).

We would like to thank Maurizio Calcinai for his assistance in the pyrolysis-gas chromatography analysis.

7. References

Alberts, J.J.& Takacs, M. (2004). Total luminescence spectra of IHSS standard and reference fulvic acids, humic acids and natural organic matter: comparison of aquatic and terrestrial source terms. *Organic Geochemistry,* Vol. 35, pp. 243–256

Alcaniz, J.M., Seres, A. & Gassiot-Matas, M. (1984). Diferenciacion entre humus mull carbonatado y mull evolucionado por Py-GC. In: *Proceedings of the National Conference Ciencia Del Suelo,* 217-228, Madrid.

Bandick, A.K. & Dick, R.P. (1999). Field management effects on soil enzyme activities. *Soil Biology & Biochemistry,* Vol. 31, pp. 1471-1479

Bracewell, J.M. & Robertson, G.W. (1984). Quantitative comparison of nitrogen containing pyrolytic products and aminoacids composition of soil humic acids. *Journal of Analytical and Applied Pyrolysis,*Vol. 6, pp. 19–29

Buurman, P., Peterse, F. & Almendros Martin, G. (2007). Soil organic matter chemistry in allophanic soils: A pyrolysis-GC/MS study of a costa rican andosol catena. *European Journal of Soil Science,* Vol. 58, pp. 1330-1347

Campo, J., Nierop, K.G.J., Cammeraat, E., Andreu, V. & Rubio, J.L. (2011). Application of pyrolysis-gas chromatography/mass spectrometry to study changes in the organic matter of macro- and microaggregates of a Mediterranean soil upon heating. *Journal of Chromatography A*, Vol. 1218, pp. 4817-482

Ceccanti, B., Nannipieri, P., Cervelli, S. & Sequi, P. (1978). Fractionation of humus-urease complexes. *Soil Biology & Biochemistry*, Vol. 10, pp. 39–45

Ceccanti, B., Alcaniz, J.M., Gispert, M. & Gassiot, M. (1986). Characterization of organic matter from two different soils by pyrolysis–gas chromatography and isoelectrofocusing. *Soil Science*, Vol. 142, pp. 83–90

Ceccanti, B. & Garcia, C. (1994). Coupled chemical and biochemical methodologies to characterize a composting process and the humic substances. In: Senesi, N., Miano, T.M. (Eds.), *Humic Substances in the Global Environment and Implication on Human Health*, 1279–1284, London

Ceccanti, B. & Masciandaro, G. (2003). Stable humus-enzyme nucleus: the last barrier against soil desertification. In: Lobo, M.C., Ibanez, J.J. (Eds.), *Preserving Soil Quality and Soil Biodiversity – The Role of Surrogate Indicators*, 77–82, CSIC-IMIA, Madrid

Ceccanti, B., Masciandaro, G. & Macci C. (2007). Pyrolysis-gas chromatography to evaluate the organic matter quality of a mulched soil. *Soil & Tillage Research*, Vol. 97, pp. 71-78

Clapp, C.E. & Hayes, M.H.B. (1999). Characterization of humic substances isolated from clay- and silt-sized fractions of a corn residue-amended agricultural soil. *Soil Science*, Vol. 164, pp. 899–913

Dignac, M.F., Houot, S., Francou, C. & Derenne, S. (2005). Pyrolytic study of compost and waste organic matter. Organic *Geochemistry*, Vol. 36, pp. 1054-1071

Doran, J.W. & Parkin, T.B. (1994). Defining and assessing soil quality. In: Doran, J.W., Coleman, D.C., Bezdicek, D.F., Stewart, B.A. (Eds.), *Defining Soil Quality for a Sustainable Environment*. Special Publication, Vol. 35, *Soil Science Society of America*, Inc., 3-21, Madison, WI

European Commission (2006). Soil Thematic Strategy. In: *European Commission Environment Soil*. Last access September 2011. Available from: http://ec.europa.eu/environment/soil/index_en.htm

Faure, P., Vilmin, F., Michels, R., Jardé, E., Mansuy, L., Elie, M. & Landais, P. (2002). Application of thermodesorption and pyrolysis-GC–AED to the analysis of river sediments and sewage sludges for environmental purpose. *Journal of Analytical and Applied Pyrolysis*, Vol. 62, pp. 297–318

Garcia, C., Hernandez, T. & Costa, F. (1992). Variation in some chemical parameters and organic matter in soils regenerated by the addition of municipal solid waste. *Environmental Management*, Vol. 16, pp. 763-768

Garcia, C., Hernandez, T., Costa, F., Ceccanti, B. & Calcinai, M. (1992). A chemical-structural study of organic wastes and their humic acids during composting by means of pyrolysis-gas chromatography. *Science of the Total Environment*, Vol. 119, pp. 157-168

Gondar, D., Lopez, R., Fiol, S., Antelo, J.M. & Arce, F. (2005). Characterization and acid–base properties of fulvic and humic acids isolated from two horizons of an ombrotrophic peat bog. *Geoderma*, Vol. 126, pp. 367–374

Grandy, A.S. & Neff, J.C. (2008). Molecular C dynamics downstream: The biochemical decomposition sequence and its impact o soil organic matter structure and function. *Science of the Total Environment*, Vol. 404, pp. 297-307

Lobe, I., Du Preez, C.C. & Amelung, W. (2002). Influence of prolonged arable cropping on lignin compounds in sandy soils of the south african highveld. *European Journal of Soil Science*, Vol. 53, pp. 553-562

Marinari, S., Liburdi, K., Masciandaro, G., Ceccanti, B. & Grego, S. (2007a). Humification-mineralization pyrolytic indices and carbon fractions of soil under organic and conventional management in central Italy. *Soil & Tillage Research*, Vol. 92, pp.10-17

Marinari, S., Masciandaro, G., Ceccanti, B. & Grego, S. (2007b). Evolution of soil organic matter changes using pyrolysis and metabolic indices: A comparison between organic and mineral fertilization *Bioresource Technology*, Vol. 98, pp. 2495-2502

Masciandaro, G., Ceccanti, B. & Garcia, C. (1997). Soil agroecological management: fertirrigation and vermicompost treatments. *Bioresource Technology*, Vol. 59, pp. 199-206

Masciandaro, G. & Ceccanti, B. (1999). Assessing soil quality in different agroecosystems through biochemical and chemico-structural properties of humic substances. *Soil & Tillage Research*, Vol. 51, pp. 129–137

Nannipieri, P., Muccini, L. & Ciardi, C. (1983). Microbial biomass and enzyme activities: production and persistence. *Soil Biology & Biochemistry*, Vol. 15, pp. 679-685

Nierop, K.G.J. (2001). Temporal and vertical organic matter differentiation along a vegetation succession as revealed by pyrolysis and thermally assisted hydrolysis and methylation. *Journal of Analytical and Applied Pyrolysis*, Vol. 61, pp. 111-132

Nierop, K.G.J., Pulleman, M.M. & Marinissen, J.C.Y. (2001). Management-induced organic matter differentiation in grassland arable soil: a study using pyrolysis techniques. *Soil Biology & Biochemistry*, Vol. 33, pp. 755–764

Peruzzi, E., Masciandaro, G., Macci, C., Doni, S., Mora Ravelo, S.G., Peruzzi, P. & Ceccanti, B. (2011). Heavy metal fractionation and organic matter stabilization in sewage sludge treatment wetlands. *Ecological Engineering*, Vol. 37, pp. 771-778

Piccolo, A. (1996). Humus and soil conservation. In: Humic Substances in Terrestrial Ecosystems (ed. A. Piccolo), 225–264, Elsevier, Amsterdam

Reeves, D.W. (1997). The role of soil organic matter in maintaining soil quality in continuous cropping systems. *Soil & Tillage Research*, Vol. 43, pp. 131–167

Schnitzer, M. (1991). Soil organic matter. The next 75 years. *Soil Science*, 151, pp. 41–58

See, J.H. & Bronk, D.A. (2005). Changes in C:N ratios and chemical structures of estuarine humic substances during aging. *Marine Chemistry*, Vol. 97, pp. 334–346

Sobeih, K.L., Baron, M. & Gonzalez-Rodriguez, J. (2008). Recent trends and developments in pyrolysis-gas chromatography. *Journal of Chromatography A*, Vol. 1186, pp. 51-66

Sollins, P., Homann, P. & Caldwell, B.A. (1996). Stabilization and destabilization of soil organic matter: Mechanisms and controls. *Geoderma*, Vol. 74, pp. 65-105

Van Bergen, P.F., Nott, C.J., Bull, I.D., Poulton, P.R. & Evershed, R.P. (1998). Organic geochemical studies of soils from the rothamsted classical experiments - IV. preliminary results from a study of the effect of soil pH on organic matter decay. *Organic Geochemistry*, Vol. 29, pp. 1779-1795

Vancampenhout, K., De Vos, B., Wouters K., Van Calster H., Swennen R., Buurman P. & Deckers J. (2010). Determinants of soil organic matter chemistry in maritime temperate forest ecosystems. *Soil Biology & Biochemistry,* Vol. 42, pp. 220-233

14

Application of Gas Chromatography to Exuded Organic Matter Derived from Macroalgae

Shigeki Wada and Takeo Hama
Shimoda Marine Research Center, University of Tsukuba,
Life and Environmental Sciences, University of Tsukuba,
Japan

1. Introduction

Macroalgae are the most important primary producers in coastal environments, because their productivity per square community area is comparable to that of tropical rain forests (Mann, 1973; Yokohama et al., 1987). Although a half of the products constitute the organic matter of algal body (Hatcher et al., 1977), considerable part of them would be exuded and dissolved into ambient seawater (Khailov & Burlakova, 1969; Sieburth, 1969; Hatcher et al., 1977; Abdullah & Fredriksen, 2004; Wada et al., 2007). The organic materials released into ambient seawater induces some alteration of marine and atmospheric environments. For example, some volatile compounds released from macroalgae would escape into air, and play as ozone-depleting substances (Lovecock, 1975; Laturnus et al., 2010). Phenolic compounds are known as a kind of exudates, which acts as a form of defense to the algal body from herbivores (van Alstyne, 1988). A part of the phenolic compounds are likely to dissolve into seawater, and the light shielding role as a component of colored organic matter was suggested in coastal environments (Wada et al., 2007). In addition to these compounds, carbohydrates such as mucopolysaccharides are important component of released organic matter on the body surface, and a part of them is considered to be released into seawater (Wada et al., 2007). It is known that bacterial community acting on the macroalgae is variable depending on the carbohydrate composition of exudates (Bengtsson et al., 2011), and the bacterial community structure probably changes the carbon flux around algal bed. The characterization of organic compounds derived from macroalgae is necessary to elucidate the biogeochemical role of macroalgae, because the effect to biogeochemical processes is variable depending on each organic compound. Thus, quantitative and qualitative approaches have been carried out so far. Application of gas chromatography (GC) would be effective for some of these exuded compounds. In this review, we focused on the analytical methods using GC for analysis of compounds originated from macroalgae.

2. Volatile halogenated organic matter

2.1 Macroalgal release of volatile halogenated organic matter

Depletion of ozone in stratosphere has been focused as one of the most serious global environmental issues, and reaction of halogenated compounds with ozone has been recognized as the destruction mechanisms of ozone (Farman et al., 1985; Crutzen & Arnold, 1986; Solomon, 1990; Anderson et al., 1991). Such volatile halogenated compounds are not only originated from anthropogenic sources, but also biogenic sources such as macroalgae. Due to such interests, volatile halogenated organic compounds (VHOC) have been intensely studied so far, and development of analytical technique of macroalgal VHOC has been recognized.

2.2 Pre-treatment before injection

When analysis of macroalgal VHOC in seawater samples are carried out, pre-treatment of the sample is important procedure for achievement of high recovery yield. There are several kinds of methods such as closed-loop stripping (CLSP), headspace, liquid-liquid extraction, purge-and-trap (P&T) and solid-phase micro extraction (SPME) methods (Table 1). Within these methods, CLSP had been applied in earlier year (Gschwend et al., 1985; Newman and Gschwend, 1987), but there is a problem that the CLSP method is inappropriate for analyses of extremely volatile compounds (e.g., CH_3Cl and CH_3Br) or relatively involatile compounds (e.g., CBr_4 and CHI_3) (Gschwend et al., 1985). In headspace method, seawater samples were brought into equilibrium with a gas (usually nitrogen) (Lovelock 1975; Manley & Dastoor, 1987; Manley & Dastoor, 1988; Manley et al., 1992). In the case that the concentrations of target compounds were low, cryo-concentration had been also applied (Manley & Dastoor, 1987; Manley & Dastoor, 1988; Manley et al., 1992). Liquid-liquid extraction is that solvent containing internal standard is added into seawater sample, and a part of the solvent phase was injected to GC. This method allows us to make analyses with simple instrumental set-up and the total analysis time is short (Abrahamsson & Klick, 1990; Laturnus et al., 1996; Manley & Barbero, 2001). P&T method have been the most widely-applied method in recent years. Sample seawater is purged with nitrogen or helium gases, and the target compounds in the gas phase were concentrated with cold trap (sometimes with adsorbent; Ekdahl & Abrahamsson, 1997). After degassing, the trapped compounds were transferred into GC instrument by heating (Schall et al., 1994; Nightingale et al., 1995; Laturnus et al., 2004; Weinberger et al., 2007; Laturnus et al., 2010). Recently, SPME technique has been also applied for VHOC originated from macroalgae. SPME fibre is used for trapping of VHOC after purging with pure nitrogen. In case of determination of VHOC in seawater spiked by standards, quantitative quality was confirmed at the concentration level around 100 ng l^{-1} of VHOC (Bravo-Linares et al., 2010).

2.3 GC Instruments

Analyses of extracted VHOC compounds have been mainly carried out by gas chromatography with electron capture detector (GC-ECD) (e.g., Manley & Dastoor, 1987; 1988; Schall et al., 1994; Laturnus et al., 1996; Manley & Barbero, 2001; Laturnus et al., 2004) or gas chromatography/mass spectrometry (GCMS) (e.g., Gschwend et al., 1985; Newman & Gschwend, 1987; Marshall et al., 1999; Bravo-Linares et al., 2010). Although GC-ECD has been commonly applied, it is necessary to confirm the retention time of each compound using authentic standards (Giese et al., 1999) or GCMS (Manley and Dastoor, 1987) to identify each component. On the other hand, application of GCMS has an advantage that it is reliable to characterize the compounds based on the mass spectrum. Capillary column is also an important part of GC for separation of each compound, and various kinds of columns (e.g., SE 54, BP-624, Rtx 502.2) have been applied. Most of them have mid-polarity, which have been commercially recommended for analysis of volatile organic compounds.

Pre-treatment	Target compounds	Sample	Instruments	References
Headspace	CH_3Cl, CH_3Br, CH_3I	Seawater	GC-ECD	Lovelock (1975)
Closed loop stripping	$CHBr_3$, $CHBr_2Cl$, CH_2Br_2	2 Brown algae and 2 Green algae,	GCMS	Gschwend et al. (1985)
Closed loop stripping	$(CH_3)_2CHBr$, $CH_3CH_2CH_2Br$, $CH_3(CH_2)_4Br$, CH_3I, C_2H_5I, C_3H_7I, C_4H_9I, $C_5H_{11}I$, $C_6H_{13}I$, CH_2Br_2, $CHBr_3$, CH_2I_2, $CHBr_2I$, CH_3SCH_3, CH_3SSCH_3	1 Brown algae	GCMS	Newman & Gschwend (1987)
Headspace	CH_3Cl, $CHBr$, CH_3I	1 Brown algae and seawater	GC-ECD	Manley & Dastoor (1987)
Headspace	CH_3I	5 Brown algae	GC-ECD	Manley & Dastoor (1988)
Headspace	$CHBr_3$, CH_2Br_2, CH_3I	6 Brown algae, 3 Red algae, 2 Green algae and seawater	GC-ECD	Manley et al. (1992)
Headspace	$CHBr_3$, CH_2Br_2	1 Brown algae	GC-ECD	Goodwin et al. (1997)
Liquid-liquid	$CHBr_3$, CH_2Br_2, $CHBr_2Cl$, CH_2I_2, CH_2ClI, CCl_4, CH_3CCl_3, $CHClCCl_2$	Seawater	GC-ECD	Klick (1992)
Liquid-liquid	CH_2Br_2, $CHBrCl_2$, CH_2ClI, $CHBr_2Cl$, 1,2-$C_2H_4Br_2$, CH_2I_2, $CHBr_3$	9 Brown algae, 15 Red algae, 2 Green algae and 2 Crysophyta	GC-ECD	Laturnus et al. (1996)
Liquid-liquid	CH_2Br_2, $CHBrCl_2$, CH_2ClI, $CHBr_2Cl$, 1,2-$C_2H_4Br_2$, CH_2I_2, $CHBr_3$	11 Brown algae, 4 Red algae and 6 Green algae	GC-ECD	Laturnus (1996)
Liquid-liquid	CCl_4, $CHCl=CCl_2$, $CCl_2=CCl_2$, CH_2Br_2, $CHBr_3$, CH_2ClI, C_4H_9I, CH_2I_2	Seawater	GC-ECD	Abrahamsson & Ekdahl (1996)
Liquid-liquid	$CHBr_3$	1 Green algae	GC-ECD	Manley & Barbero (2001)

Pre-treatment	Target compounds	Sample	Instruments	References
Purge and trap	CH_3I, $CHCl_3$, CH_3CCl_3, CCl_4, CH_2Br_2, $CHBrCl_2$, CCl_2CCl_2, $CHBr_2Cl$	5 Brown algae, 3 Red algae, 3 Green algae and seawater	GC-ECD	Nightingale et al. (1995)
Purge and trap	CH_3I, CH_2I_2, CH_2ClI, $CH_3CH_2CH_2I$, CH_3CHICH_3, CH_2Br_2, $CHBr_3$, $CHBrCl_2$, $CHBr_2Cl$	3 Brown algae	GC-ECD	Schall et al. (1994)
Purge and trap	C_2H_5Br, 1,2-$C_2H_4Br_2$, C_2H_5I, CH_2ClI, CH_2I_2, CH_3Br, CH_2BrCl, CH_2Br_2, $CHBrCl_2$, $CHBr_2Cl$, $CHBr_3$	4 Brown algae, 2 Red algae and 4 Green algae	GC-ECD	Laturnus (1995)
Purge and trap	C_2HCl_3, C_2Cl_4	5 Brown algae, 17 Red algae and 6 Green algae	GC-ECD	Abrahamsson et al. (1995)
Purge and trap	$CHBr_3$, $CHCl_3$, CH_3-CCl_3, CH_2Br_2, CH_2I_2, CH_3I, $CHBr_2Cl$, $CHCl_2Br$, CH_2ClI, CCl_4, $Cl_2C{=}CCl_2$, $Cl_2C{=}CHCl$, C_4H_9I, sec-C_4H_9I, C_3H_7I	1 Green algae	GC-ECD	Mtolera et al. (1996)
Purge and trap	CH_3I, CH_2I_2, C_4H_9I, CH_2ClI, $CHBr_3$, $CHBr_2Cl$, $CHBrCl_2$, $CHCl_3$, $HCCl{=}CCl_2$, CH_2Br_2	2 Red algae and 1 Green algae	GC-ECD	Pedersen et al. (1996)
Purge and trap	CH_2I_2, CH_2BrCl. $CHBrCl_2$, C_2H_5Br,. 1,2-$C_2H_4Br_2$, C_2H_5I, CH_2ClI	3 Red algae and 1 Green algae	GC-ECD	Laturnus et al. (1998)
Purge and trap	CH_3I, CH_2I_2, CH_2ClI, $CHBr_3$, CH_2Br_2, CH_3CH_2I, $CH_3CH_2CH_2I$, CH_3CHICH_3, $CH_3CH_2CH_2CH_2I$, $CH_3CH(CH_3)CH_2I$, $CH_3CH_2CHICH_3$	2 Brown algae, 19 Red algae and 3 Green algae	GC-ECD	Giese et al. (1999)
Purge and trap	$CHBr_3$, $C_2H_2Br_2$, CH_2Br_2	1 Red algae	GCMS	Marshall et al. (1999)

Pre-treatment	Target compounds	Sample	Instruments	References
Purge and trap	C_2H_5I, 2-C_3H_7I, 1-C_3H_7I, CH_2ClI, 2-n-C_4H_9I, CH_2I_2, $CHBr_3$	1 Red algae	GC-ECD	Laturnus et al. (2000)
Purge and trap	C_2HCl_3, C_2Cl_4	1 Red algae	GCMS	Marshall et al. (2000)
Purge and trap	CH_3Br, CH_3Cl, CH_2Cl_2, $CHCl_3$, CH_3I, CH_2ClI, CH_3CH_2I	2 Brown algae, 1 Red algae and 3 Green algae	GC-ECD	Baker et al. (2001)
Purge and trap	CH_3Cl, CH_3Br, CH_2BrCl, CH_2Br_2, $CHBrCl_2$, $CHBr_2Cl$, $CHBr_3$, 1,2-$C_2H_4Br_2$, CH_3I, CH_2ClI, CH_2I_2, C_2H_5I, 1-C_3H_7I, 2-C_3H_7I, 1,3-C_3H_6ClI, 1-n-C_4H_9I, 20n-C_4H_9I, 1-iso-C_4H_9I	11 Brown algae, 11 Red algae and 8 Green algae	HRGC-ECD/MIP AED	Laturnus (2001)
Purge and trap	$CHCl_3$, C_2HCl_3, C_2Cl_4, $CHBr_3$, CH_2Br_2, $CHClBr_2$, CH_2BrCl, $CHCl_2Br$, CH_2I_2, CH_3I, C_2H_5I, CH_2ClI, C_3H_7I, iso-C_3H_7I, C_4H_9I, sec-C_4H_9I	1 Brown algae, 4 Green algae and 1 Diatom	GC-ECD	Abrahamsson et al. (2003)
Purge and trap	C_2Cl_4, C_2HCl_3, CCl_4, $CHCl_3$, CH_2Cl_2, CH_3CCl_3, CH_3I, CH_2Br_2, $CHBr_3$, $CHBrCl_2$	3 Brown algae, 1 Red algae and 1 Green algae	GC-ECD	Laturnus et al. (2004)
Purge and trap	CH_2ICl, CH_2I_2, CH_2IBr, C_2H_5I, 2-C_3H_7I, CH_3I, CH_2Br_2, $CHBr_3$	seawater	GCMS	Jones et al. (2009)
Purge and trap	CH_3I, CH_2ClI, CH_2I_2, CH_2Cl_2, $CHCl_3$, CH_3CCl_3, CCl_4, C_2HCl_3, C_2Cl_4, CH_2Br_2, $CHBrCl_2$, $CHBr_2Cl$, 1,2-$EtBr_2$, $CHBr_3$	1 Brown algae	Not described	Laturnus et al. (2010)
SPME	CH_3Br, CH_2I_2, $CCl_2{=}CCl_2$, $CHCl_2Br$	4 Brown algae, 2 Red algae and 2 Green algae	GCMS	Bravo-linares et al. (2010)

Table 1. List of studies and methodologies used on VHOC originated from macroalgae

2.4 Estimation of the importance of macroalgal VHOC

In order to evaluate macroalgal release of VHOC, an effective approach is seawater sampling in and out of algal bed in the field or incubation experiment in closed system. Analysis of seawater sample often showed higher concentration of VHOC around algal bed compared with offshore region, strongly suggesting the significant release of VHOC from macroalgae (Lovecock, 1975; Manley & Dastoor, 1987; Klick, 1992; Manley et al., 1992; Nightingale et al., 1995). The incubation experiment can provide production rate of VHOC for each macroalgal species (Gschwend et al., 1985; Manley & Dastoor, 1987). Assuming that the estimated production rate of VHOC is comparable with other macroalgae, some researchers estimated global VHOC production rates (Gschwend et al., 1985; Manley & Dastoor, 1987). Although they multiplied the production rate per algal biomass which is experimentally defined by standing crop in global ocean, the most serious problem is that no detailed investigation on algal biomass in global ocean has been reported yet. Calculation of algal biomass in global ocean was estimated based on coastal length by De Vooys (1979) who also provided an estimation of primary production in global ocean as 0.03 PgC y^{-1}. No estimate of biomass has been published by other researchers, but primary production was revised by Charpy-Roubaud & Sournia (1990). They made estimation of primary production as 2.55 PgC y^{-1} based on algal community area, which is two orders of magnitude higher than the values of De Vooys (1979). Such discrepancy with the previous estimates implies that there is no reliable value of macroalgal parameters such as biomass and productivity in global ocean. In addition to the development of analytical technique, estimates of algal biomass in global ocean will be also required for understanding the macroalgal contribution to ozone-depletion.

3. Phenolic compounds

3.1 Macroalgal phenolic compounds

Macroalgae synthesize phenolic compounds, and a part of them is likely exuded (Paul et al., 2006). Considering that the phenolic compounds accumulated in the outer cortical layer of the thalli (Shibata et al., 2004; Paul et al., 2006), these materials would be actively released. Dissolution of phenolic compounds was also supported by *in situ* field experiment (Wada et al., 2007), in which absorption spectra of macroalgal excretion were relatively similar to those of the materials containing aromatic ring (lower exponential slope of the absorption spectra) (Blough & Del Vecchio, 2002). The UV absorbing property of phenolic compounds (Łabudzińska & Gorczyńska 1995) suggests the attenuation of UV radiation to seawater by phenolic compounds originated from macroalgae. Since biological activity in surface seawater is affected by UV penetration (Blough & Del Vecchio, 2002), analysis of phenolic compounds shows the interaction between macroalgae and other marine organisms.

3.2 Cupric oxide oxidation

Analytical procedure for macroalgal phenolic compounds is cupric oxide (CuO) oxidation method, in which polymeric compounds are degraded to small molecules that can be quantified by GC instrument. Although this technique has been mainly applied to lignin which also contains phenolic structure, some researchers had applied it to macroalgal materials (Goni & Hedges, 1995). Generally, the samples were reacted with CuO at 150-

170°C for 3 h under alkaline conditions, and acidified after the reaction. The oxidized fraction was extracted with ethyl ether, and the solvent was evaporated. The products were dissolved in pyridine, and they are derivatised to trimethylsilyl derivatives for gas chromatographic analysis. Molecules containing aromatic ring were identified using GC or GCMS (Goni & Hedges, 1995), but a possibility that some of them originated from protein, because 3 kinds of amino acids (phenylalanine, tryptophan and tyrosine) have aromatic group in their molecule. Since *m*-hydroxybenzoic acid and 3,5-dihydroxybenzoic acid has been suggested as non-amino acid derived from organic compounds containing phenolic structure such as tannin, these products are likely useful as an indicator of macroalgal materials when we use CuO oxidation method.

3.3 Application of CuO oxidation method for macroalgal exudates

There are just a few studies on analysis of phenolic compounds using CuO oxidation method (Goni & Hedges, 1995), but they examined various species belonging to brown (*Nereocystic luetkeana, Fucus fardneri, Costaria costata, Desmarestia viridis and Sargassum muticum*), green (*Ulva fenestrate and Codium fragile*) and red algae (*Opuntiella californica, Odonthalia floccose*). In their study, there were 11 kinds of aromatic products after CuO oxidation, and m-hydroxybenzoic acid and 3,5-dihydroxybenzoic acid were determined as non-amino acid derived materials as described above. The contents of these two products in body weight of algae were the highest in the brown algae (1.5-2.3 and 4.3-14 times than red and green algae for m-hydroxybenzoic acid and 3.5-dihydroxybenzoic acid), suggesting that production of phenolic compounds by brown algae is larger than other macroalgal groups.

4. Carbohydrates

4.1 Release of carbohydrates

Macroalgal body is covered by sticky mucus due to excretion of materials containing mucopolysaccharides for protection of their body from external stress such as desiccation (Percival & McDowell, 1981; Painter, 1983) and changes in ambient ion condition (Kloareg & Quatrano, 1988). Since the monosaccharide composition of mucopolysaccharides is different among brown, red and green algae, monosaccharide analysis is effective to understand the original source of mucopolysaccharide. Here we described the chemical characteristics of carbohydrate species originated from these algae, and reviewed the methodological aspects of monosaccharide analysis.

4.1.1 Brown algae

The major mucopolysaccharide of brown algae is alginate, which is made up of two kinds of uronic acids (mannuronic acid and guluronic acid). The molar ratio of mannuronic acid to guluronic acid (M/G) ranges from 0.25 to 2.25 (Kloareg & Quatrano, 1988). Another well-known mucopolysaccharide is fucan, which comprises L-fucose and sulphate as major constituents. Actually, fucans are heterogeneous group of polysaccharide, and the fucose content ranges from 55 to 96% of total monosaccharides (Marais & Joseleau, 2001; Bilan et al., 2004; 2006). Considering these monosaccharide composition, quantification of uronic acids and fucose would be available for characterization of mucopolysaccharides of brown algae.

4.1.2 Red algae

Carrageenans are mainly extracted from *Chondrus, Gigartina, Eucheuma* and *Hypnea,* and they are highly sulphated galactans (De Ruiter & Rudolph, 1997). These polysaccharides are categorized into several families based on the position of the sulphate groups. Agars are also a kind of galactan, but this is a low sulphated polymer, which is often extracted from *Gelidium, Gracilaria, Ahnfeltia, Acanthopeltis* and *Pterocladia* (Kloareg & Quatrano, 1988). These polysaccharides are commonly found in red algae, and galactose is the major constituents.

4.1.3 Green algae

Mucopolysaccharides originated from green algae are highly branched sulphated heteropolysaccharides such as xylogalactoarabinans, glucuronoxylorhamnans and rhamnoxylogalacto-galacturonan (Kloareg & Quatrano, 1988).

4.2 Depolymerization and derivatization of carbohydrates

4.2.1 Depolymerization

As mentioned above, monosaccharide composition reflects the original sources of the mucopolysaccharides released into extracellular region. For GC analysis of monosaccharide composition, it is necessary to depolymerize the polysaccharides and derivatize the monosaccharides in volatile forms. Well-known depolymerization and derivatization procedures are 1) hydrolysis-alditol-acetate or -trimethylsilyl and 2) methanolysis-trimethylsilyl methods. In this section, we describe the features of the methods, and introduce the application to macroalgal carbohydrates.

4.2.1.1 Acid hydrolysis

Various acid solutions have been examined for hydrolytic depolymerization of polysaccharides in marine samples, and detail of the results was previously reviewed (Panagiotopoulos & Sempere, 2005). Briefly, acid solutions which are mostly examined are HCl and H_2SO_4, and the recovery has been evaluated. Mopper (1973) compared these two acid solutions at same concentration (2 N), and suggested that HCl efficiently depolymerizes carbohydrate in ancient sediments but it led destruction of those in anoxic sediment (Mopper 1977). Acid strength is also important factor controlling the recovery yields. Two step of hydrolysis reaction, in which pre-treatment was carried out in 72% H_2SO_4 solution at ambient temperature and diluted sample (1-2 N) was heated at 100°C, are often used in order to achieve complete hydrolysis. These two steps reactions would be relatively strong, and sometimes induce loss of recovery of pentoses (Mopper, 1977), but the total yields of aldoses tend to be higher (Skoog & Benner, 1997). Mild hydrolysis was performed in solutions of dilute H_2SO_4, HCl, $CHCl_2COOH$, H_3PO_4, $(COOH)_2$ and trifluoacetic acid. In some cases, their recoveries are comparable with those of strong hydrolysis reaction (Panagiotopoulos & Sempere, 2005). However, they would be inappropriate for refractory species of carbohydrates such as cellulose (Skoog & Benner, 1997).

4.2.1.2 Derivatization and GC analysis

Since carbohydrates are polyhydroxy compounds, it is essential to convert them into the volatile derivatives. Commonly used derivatization methods are trimethylsilylation and alditol-acetate derivatizations. Generally, hexamethyldisilazane and trimethylchlorosilan have been used as derivatizating agents and pyridine as solvent for trimethylsilylation. Analytical procedures of this technique are simple and rapid (Sweeley et al., 1963), and appropriate volatility is obtained. This technique would be also applicable to nonreducing sugars. However, monosaccharide with free carbonyl groups can be present as different tautomers, and each tautomeric form occurs as different peak. Consequently, such derivative method would provide complicated chromatogram in GC analysis. When complex sugar mixture is analyzed, reduction of the monosaccharides should be considered to avoid the overlapping of peaks (Ruiz-Matute et al., 2011). In alditol-acetate method, the carbonyl group of monosaccharide is reduced using reducing reagent (e.g., KBH_4), and hydroxyl group of generated alditol is acetylated. This technique has several advantages, that alditol-acetate derivative produces single peak for each monosaccharide, and that the derivative is stable allowing clean-up for analysis (Knapp 1979). However, this method needs the large number of steps in the experimental procedures, and it is laborious and time consuming (Ruiz-Matute et al., 2011).

4.2.1.3 Methanolysis

Although hydrolysis is commonly used for analysis of neutral aldoses, the method is inappropriate for some kinds of carbohydrates such as uronic acid because of instability of uronic acid in acid hydrolysis reaction (Blake and Richards, 1968). Considering that uronic acid is also an important component of macroalgal mucopolysaccharides as mentioned above, alternative methodology should be examined. To overcome this problem, methanolysis reaction, which provides high recovery yields for uronic acid at 95-100% (Chambers and Clamp, 1971), is available.

Condition of methanolysis reaction commonly accepted is in 0.4-2 N methanolic HCl at 80°C for 5-24 h (Chambers & Clamp, 1971; Doco et al., 2001; Mejanelle et al., 2002). Under this condition, the recoveries of both neutral aldoses and uronic acid are stable (Chambers and Clamp, 1971). Since methanolysis reaction is interfered by water, this reaction should be carried out after drying the samples completely. After depolymerization of polysaccharides by methanolysis reaction, trimethylsilyl derivatization has been usually applied (Dierckxsens et al., 1983; Bleton et al., 1996; Doco et al., 2001; Mejanelle et al., 2002). Since water in the sample interferes in the trimethylsilyl reaction as well as methanolysis reaction, drying of the sample is an essential procedure (Chamber & Clamp, 1971)..

4.2.1.4 GC analysis for methanolysis-trimethylsilyl derivatives

In the methanolysis-trimethylsilyl method, several kinds of isomers are generated from one monosaccharide, and the chromatogram is often complicated due to the presence of a large number of peaks. Quantification of all of the isomers would be ideal, but detection of minor isomer peaks is sometimes difficult. In such a case, quantification is generally achieved by picking up major peaks, because the isomer composition would be constant regardless of initial chemical form of the monosaccharides if the methanolysis reaction was performed under same conditions (Mejanelle et al., 2002). Wada et al. (submitted) actually showed similar isomeric composition using seawater samples from natural environment and authentic standards.

The detail of GC or GCMS detection of methanolysis products has been already reviewed elsewhere (Mejanelle et al., 2004), and here we have simple explanation. When the analysis is carried out for less than 10 neutral and acidic monosaccharides, GC analysis will provide reliable identification of monosaccharide components. However, we should consider the presence of contaminant which is present in natural environment, because ambient seawater contains not only macroalgal exudates but organic constituents existing in seawater. In case that the contaminant peaks overlap with those of target monosaccharides, selective ion monitoring mode of GCMS instrument will be available (Wada et al., submitted). Using electron impact mode, some fragment ions are generated, and base peaks were m/z 73, 204 and 217 in the most cases. Considering that the peaks at m/z 204 and 217 are often found for pyranosides and furanosides, respectively (Mejanelle et al., 2004), the composition of fragment ions will provide useful information about not only quantification, but also identification of the isomers generated by methanolysis reaction.

4.3 Analysis of carbohydrates released from macroalgae

Mucopolysaccharides which are known as extracellular carbohydrates have been analyzed using GC or GCMS after some purification treatments (Lee et al., 2004; Mandal et al., 2007; Karmakar et al., 2010; Rioux et al., 2010; Stephanie et al., 2010). On the other hand, direct analysis for carbohydrates released into ambient seawater is limited (Wada et al., 2007; 2008). In their studies, *in situ* field bag experiment was developed, and the carbohydrates released from a brown alga, *Ecklonia cava*, were obtained by SCUBA diving. By hydrolysis and alditol-acetate derivative method, it was shown that fucose is the major monosaccharide component (36-44% of total carbohydrates), and that values had no significant seasonal variation. Since fucose is the major constituent of fucan derived from brown algae, constant dissolution of extracellular mucopolysaccharides into ambient environments from brown algae was suggested.

5. Conclusion

Macroalgae releases various organic compounds extracellularly and their role and dynamics in aquatic environments partly depend on the chemical composition. Since VHOC has been measured by numerous researchers, the VHOC analysis has been well improved in a few decades. On the other hand, there are possibilities of other unapplied option for phenolic compounds and carbohydrates analysis. For phenolic compounds, pyrolysis-GCMS is also considered as another potential tool, and Van Heemst et al. (1996) had tried to show the macroalgal contribution to marine organic matter pool. However, contamination of proteinaceous phenol under the process of pre-treatment was not as easy (Van Heemst et al., 1999). If this issue can be resolved in future, this instrument would also become a powerful tool for analysis of macroalgal phenolic compounds. Analysis of carbohydrates has been carried out for the extracts from algal body, but there is limited information on direct analysis of carbohydrates released to seawater as described above. Particularly, analysis using methanolysis method for seawater sample has been examined, yet. Thus, examination on the applicability of methanolysis method for seawater sample may be an important issue in the future.

6. Acknowledgement

This report is contribution no. 755 from the Shimoda Marine Research Center, University of Tsukuba.

7. References

Abdullah, M. I. & Fredriksen, S. (2004). Production, respiration and exudation of dissolved organic matter by the kelp *Laminaria hyperborea* along the west coast of Norway. Journal of Marine Biological Association of United Kingdom, Vol. 84, No. 5, pp. 887-894, ISSN 1469-7769

Abrahamsson, K. & Klick, S. (1990). Determination of biogenic and anthropogenic volatile halocarbons in sea water by liquid-liquid extraction and capillary gas chromatography Journal of Chromatography, Vol. 513, pp. 39-45, ISSN 0021-9673

Abrahamsson, K., Ekdahl, A., Collén, J. & Pedersén M. (1995). Marine algae - a source of trichloroethylene and perchloroethylene. Limnology and Oceanography, Vol. 40, No. 7, (November 1995), pp. 1321-1326, ISSN 0024-3590

Abrahamsson, K. & Ekdahl, A. (1996). Volatile halogenated compounds and chlorophenols in the Skagerrak. Journal of Sea Research, Vol. 35, No. 1-3, (February, 1996), pp. 73-79, ISSN 1385-1101

Abrahamsson, K., Choo, K-S., Pedersén, M., Johansson, G. & Snoeijs, P. (2003). Effects of temperature on the production of hydrogen peroxide and volatile halocarbons by brackish-water algae. Phytochemistry, Vol. 64, No. 3, (October 2003), pp. 725-734, ISSN 0031-9422

Anderson, J. G., Toohey, D. W. & Brune, W. H. (1991). Free radicals within the Antarctic vortex: the role of CFCs in Antarctic ozone loss. Science, Vol. 251, No. 4989, (January 1991), pp. 39-46, ISSN 0036-8075

Baker, J. M., Sturges, W. T., Sugier, J. Sunnenberg, G., Lovett, A. A., Reeves, C. E., Nightingale, P. D. & Penkett, S. A. (2001). Emissions of CH_3Br, organochlorines, and organoiodines from temperate macroalgae. Chemosphere - Global Change Science, Vol. 3, No. 1, (January 2001), pp. 93-106, ISSN 1465-9972

Bengtsson, M. M., Sjøtun, K., Storesund, J. E. & Øvreås, L. (2011). Utilization of kelp-derived carbon sources by kelp surface-associated bacteria. Aquatic Microbial Ecology, Vol. 62, No. 2, (January 2011), pp. 191-199, ISSN 0948-3055

Bilan, M. I., Grachev, A. A., Ustuzhanina, N. E., Shashkov, A. S., Nifantiev, N. E. & Usov, A. I. (2004). A highly regular fraction of a fucoidan from the brown seaweed *Fucus distichus* L. Carbohydrate Research, Vol. 339, No. 3, (February 2004), pp. 511-517, ISSN 0008-6215

Bilan, M. I., Grachev, A. A., Shashkov, A. S., Nifantiev, N. E. & Usov, A. I. (2006). Structure of a fucoidan from the brown seaweed *Fucus serratus* L. Carbohydrate Research, Vol. 341, No. 2, (February 2006), pp. 238-245, ISSN 0008-6215

Blake, J. D. & Richards, G. N. (1968). Problems of lactonisation in the analysis of uronic acids. Carbohydrate Research, Vol. 8, No. 3, (November 1968), pp. 275-281, ISSN 0008-6215

Bleton, J., Mejanelle, P., Sansoulet, J., Goursand, S. & Tchapla, A. (1996). Characterization of neutral sugars and uronic acids after methanolysis and trimethylsilylation for recognition of plant gums. Journal of Chromatography A, Vol. 720, No. 1-2, (January 1996), pp. 27-49, ISSN 0021-9673

Blough, N. V. & Del Vecchio, R. (2002). Chromophoric DOM in the coastal environment, In: *Biogeochemistry of marine dissolved organic matter*, Hansell, D. A. & Carlson, C. A., (Eds.), 509-546, Academic Press, ISBN 0-12-323841-2, San Diego, USA

Bravo-Linares, C. M., Mudge, S. M. & Loyola-Sepulveda, R. H. (2010). Production of volatile organic compounds (VOCs) by temperate macroalgae. The use of solid phase microextraction (SPME) coupled to GC-MS as method of analysis. Journal of Chilean Chemical Society, Vol. 55, No. 2, (June 2010), pp. 227-232, ISSN 0717-9707

Chambers, R. E. & Clamp, J. R. (1971). An assessment of methanolysis and other factors used in the analysis of carbohydrate-containing materials. Biochemistry Journal, Vol. 125, No. 4, (December 1971), pp. 1009-1018, ISSN 0264-6021

Charpy-Roubaud, C. & Sournia, A. (1990). The comparative estimation of phytoplanktonic, microphytobenthic and macrophytobenthic primary production in the oceans. Marine Microbial Food Webs, Vol. 4, No. 1, pp. 31-57, ISSN 0297-8148

Crutzen, P. J. & Arnold, F. (1986). Nitric acid cloud formation in the cold Antarctic stratosphere: a major cause for the springtime 'ozone hole'. Nature, Vol. 324, No. 6098, (December 1986), pp. 651-655, ISSN 0028-0836

De Ruiter, G. A. & Rudolph, B. (1997). Carrageenan biotechnology, Trends in Food Science & Technology, Vol. 8, No. 12, (December 1997), pp. 389-395, ISSN 0924-2244

De Vooys, C. G. N. (1979).Primary production in aquatic environments. In: *The Global Carbon Cycle*, Bolin, B., Degens, E. T., Kempe, S. & Ketner, P., (Eds.), 259-292, Wiley, ISBN 978-0471997108, New York, USA

Dierckxsens, G. C., De Meyer, L. & Tonino, G. J. (1983). Simultaneous determination of uronic acids, hexosamines, and galactose of glycosaminoglycans by gas-liquid chromatography. Analytical Biochemistry, Vol. 130, No. 1, (April 1983), pp. 120-127, ISSN 0003-2697

Doco, T., O'Neill, M. A. & Pellerin, P. (2001). Determination of the neutral and acidic glycosyl-residue compositions of plant polysaccharides by GC-EI-MS analysis of the trimethylsilyl methyl glycoside derivatives. Carbohydrate Polymers, Vol. 46, No. 3, (November 2001), pp. 249-259, ISSN 0144-8617

Ekdahl, A. & Abrahamsson, K. (1997). A simple and sensitive method for the determination of volatile halogenated organic compounds in sea water in the amonl l^{-1} to pmol l^{-1} range. Analytica Chimica Acta, Vol. 357, No. 3, (December 1997), pp. 197-209 ISSN 0003-2670

Farman, J. C., Gardiner, B. G. & Shanklin, J. D. (1985). Large losses of total ozone in Antarctica reveal seasonal ClO_x/NO_x interaction. Nature, Vol. 315, No. 6016, (May 1985), pp. 207-210, ISSN 0028-0836

Giese, B., Laturnus, F., Adams, F. C. & Wiencke, C. (1999). Release of volatile iodinated C_1-C_4 hydrocarbons by marine macroalgae from various climate zones. Environmental Science and Technology, Vol. 33, No. 14, (June 1999), pp. 2432-2439, ISSN 0013-936X

Goñi, M. & Hedges J. I. (1995). Sources and reactivities of marine-derived organic matter in coastal sediments as determined by alkaline CuO oxidation. Geochimica et Cosmochimica Acta, Vol. 59, No. 14, (July 1995), pp. 2965-2981, ISSN 0016-7037

Goodwin, K. D., North, W. J. & Lidstrom, M. E. (1997). Production of bromoform and dibromomethane by Giant Kelp: Factors affecting release and comparison to anthropogenic bromine sources. Limnology and Oceanography, Vol. 42, No. 8, (December 1997), pp. 1725-1734, ISSN 0024-3590

Gschwend, P. M., MacFarlane, J. K. & Newman, K. A. (1985). Volatile halogenated organic compounds released to seawater from temperate marine macroalgae. Science, Vol. 227, No. 4690, (March, 1985), pp. 1033-1035, ISSN 0036-8075

Hatcher, B. G., Chapman, A. R. O. & Mann, K. H. (1977). An annual carbon budget for the kelp *Laminaria longicruris*. Marine Biology, Vol. 44, No. 1, (March 1977), pp. 85-96, ISSN 0025-3162

Jones, C. E., Hornsby, K. E., Dunk, R. M., Leigh, R. J. & Carpenter, L. J. (2009). Coastal measurements of short-lived reactive iodocarbons and bromocarbons at Roscoff, Brittany during the RHaMBLe campaign. Atmospheric Chemistry and Physics, Vol. 9, No. 4, pp. 8757-8769, ISSN 1680-7367

Karmakar, P., Pujol, C. A., Damonte, E. B., Ghosh, T. & Ray, B. (2010) Polysaccharides from *Padina tetrastromatica*: structural features, chemical modification and antiviral activity. Carbohydrate Polymers, Vol. 80, No. 2, (December 2009), pp. 513-520, ISSN 0144-8617

Khailov, K. M. & Burlakova, Z. P. (1969). Release of dissolved organic matter by marine seaweeds and distribution of their total organic production to inshore communities. Limnology and Oceanography, Vol. 14, No. 4, (July 1969), pp. 521-527, ISSN 0024-3590

Klick, S. (1992). Seasonal variations of biogenic and anthropogenic halocarbons in seawater from a coastal site. Limnology and Oceanography, Vol. 37, No. 7, (November 1992), pp. 1579-1585, ISSN 0024-3590

Kloareg, B. & Quatrano, R. S. (1988). Structure of the cell walls of marine algae and ecophysiological functions of the matrix polysaccharides. Oceanography and Marine Biology Annual Review, Vol. 26, pp. 259-315, ISSN 0078-3218

Knapp, D. R. (1979). Handbook of analytical derivatization reactions, Vol. 1, Wiley Interscience, ISBN 978-0-471-03469-8, New York, USA

Łabudzińska, A. & Gorczyńska, K. (1995). The UV difference spectra as a characteristic feature of phenols and aromatic amines. Journal of Molecular Structure, Vol. 349, (April 1995), pp. 469-472, ISSN 0022-2860

Laturnus, F. (1995). Release of volatile halogenated organic compounds by unialgal cultures of polar macroalgae. Chemosphere, Vol. 31, pp. 3387-3395.

Laturnus, F. (1996). Volatile halocarbons released from Arctic macroalgae. Marine Chemistry, Vol. 55, pp. 359-366.

Laturnus, F., Wiencke, C. & Klöser, H. (1996). Antarctic macroalgae-Sources of volatile halogenated organic compounds. Marine Environmental Research, Vol. 41, No. 2, pp. 169-181, ISSN 0141-1136

Laturnus, F., Wiencke, C. & Adams, F. C. (1998). Influence of light conditions on the release of volatile halocarbons by Antarctic macroalgae. Marine Environmental Research, Vol. 45, No. 3, (April 1998), pp. 285-294, ISSN 0141-1136

Laturnus, F., Giese, B., Wiencke, C. & Adams, F. C. (2000). Low-molecular-weight organoiodine and organobromine compounds released by polar macroalgae - The influence of abiotic factors. Fresenius Journal of Analytical Chemistry, Vol. 368, No. 2-3, (October 2000), pp. 297-302, ISSN 0937-0633

Laturnus, F. (2001). Marine macroalgae in polar regions as natural sources for volatile organohalogens. Environmental Science and Pollution Research International, Vol. 8, No. 2, pp. 103-108, ISSN 0944-1344

Laturnus, F., Svensson, T., Wiencke, C. & Öberg, G. (2004). Ultraviolet radiation affects emission of ozone-depleting substances by marine macroalgae: results from a laboratory incubation study. Environmental Science and Technology, Vol. 38, No. 24, (December 2004), pp. 6605-6609, ISSN 0013-936X

Laturnus, F., Svensson, T. & Wiencke, C. (2010). Release of reactive organic halogens by the brown macroalga *Saccharina latissima* after exposure to ultraviolet radiation. Polar Research, Vol. 29, No. 3, pp. 379-384, ISSN 1751-8369

Lee, J-B., Hayashi, K., Hashimoto, M., Nakano, T. & Hayashi, T. (2004). Novel antiviral fucoidan from sporophyll of *Undaria pinnatifida* (Mekabu). Chemical Pharmaceutical Bulletin, Vol. 52, No. 9, (September 2004), pp. 1091-1094, ISSN 0009-2363

Lovelock, J. E. (1975). Natural halocarbons in the air and in the sea. Nature, Vol. 256, No. 5514, (July 1975), pp. 193-194, ISSN 0028-0836

Mandal, P., Mateu, C. G., Chattopadhyay, K., Pujol, C. A., Damonte, E. B. & Ray, B. (2007). Structural features and antiviral activity of sulphated fucans from the brown seaweed *Cystoseira indica*. Antiviral Chemistry & Chemotherapy, Vol. 18, No. 3, pp. 153-162, ISSN 0956-3202

Manley, S. L. & Dastoor, M. N. (1987). Methyl halide (CH_3X) production from the giant kelp, *Macrocystis*, and estimates of global CH_3X production by kelp. Limnology and Oceanography, Vol. 32, No. 3, pp. 709-715, ISSN 0024-3590

Manley, S. L. & Dastoor, M. N. (1988). Methyl iodide (CH_3I) production by kelp and associated microbes. Marine Biology, Vol. 98, No. 4, pp. 477-482, ISSN 0025-3162

Manley, S. L., Goodwin, K. & North, W. J. (1992). Laboratory production of bromoform, methylene bromide, and methyl iodide by macroalgae and distribution in nearshore southern California waters. Limnology and Oceanography, Vol. 37, No. 8, (December 1992), pp. 1652-1659, ISSN 0024-3590

Manley, S. L. & Barbero, P. E. (2001). Physiological constraints on bromoform ($CHBr_3$) production by *Ulva lactuca* (Chlorophyta). Limnology and Oceanography, Vol. 46, No. 6, (September 2001), pp. 1392-1399, ISSN 0024-3590

Mann, K. H. (1973). Seaweeds: their productivity and strategy for growth. Science, Vol. 182, No. 4116, pp. 975-981, ISSN 0036-8075

Marais, M-F. & Joseleau, J-P. (2001). A fucoidan fraction from *Ascophyllum nodosum*. Carbohydrate Research, Vol. 336, No. 2, (November 2001), pp. 155-159, ISSN 0008-6215

Marshall, R. A., Harper, D. B., McRoberts, W. C. & Dring, M. J. (1999). Volatile bromocarbons produced by *Falkenbergia* stages of *Asparagopsis* spp. (Rhodophyta). Limnology and Oceanography, Vol. 44, No. 5, (July 1999), pp. 1348-1352, ISSN 0024-3590

Marshall, R. A., Hamilton, J. T. G., Dring, M. J. & Harper, D. B. (2000). The red alga *Asparagopsis taxiformis/Falkenbergia hillebrandii* – a possible source of trichloroethylene and perchloroethylene? Limnology and Oceanography, Vol. 45, No. 2 (March 2000) pp. 516-519, ISSN 0024-3590

Mejanelle, P., Bleton, J., Tchapla, A. & Goursaud, S. (2002). Gas chromatography-mass spectrometric analysis of monosaccharides after methanolysis and trimethylsilylation. Potential for characterization of substances of vegetal origin: application to the study of museum objects. Journal of Chromatography Library, Vol. 66, pp. 845-902, ISSN 0301-4770

Mopper, K. (1973). Aspects of the biogeochemistry of carbohydrates in aquatic environments. Thesis, M. I. T. - Woods Hole Oceanographic Institution Joint Program in Oceanography.

Mopper, K. (1977). Sugars and uronic acids in sediment and water from the Black Sea and North Sea with emphasis on analytical techniques. Marine Chemistry, Vol. 5, No. 4-6, (November 1977), pp. 585-603, ISSN 0304-4203

Mtolera, M. S. P., Collén, J., Pedersén, M., Ekdahl, A., Abrahamsson, K. & Semesi, A. K. (1996). Stress-induced production of volatile halogenated organic compounds in *Eucheuma denticulatum* (Rhodophyta) caused by elevated pH and high light intensities. European Journal of Phycology, Vol. 31, No. 1, pp. 89-95, ISSN 0967-0262

Newman, K. A. & Gschwend, P. M. (1987). A method for quantitative determination of volatile organic compounds in marine macroalgae. Limnology and Oceanography, Vol. 32, No. 3, (May 1987), pp. 702-708, ISSN 0024-3590

Nightingale, P. D., Malin, G. & Liss, P. S. (1995). Production of chloroform and other low-molecular-weight halocarbons by some species of macroalgae. Limnology and Oceanography, Vol. 40, No. 4, (June 1995), pp. 680-689, ISSN 0024-3590

Painter, T. J. (1983). Algal polysaccharides. In: *The polysaccharides*, Aspinall, G. O., (Ed.), Vol. 2, 195-285, Academic Press, ISBN 0-12-065602-7, New York, USA

Panagiotopoulos, C. & Sempéré, R. (2005). Analytical methods for the determination of sugars in marine samples: a historical perspective and future directions. Limnology and Oceanography: Methods, Vol. 3, (January 2005), pp. 419-454, ISSN 1541-5856

Paul, V. J., Puglisi, M. P. & Ritson-Williams, R. (2006). Marine chemical ecology, Natural Product Reports, Vol. 23, (February 2006), pp. 153-180, ISSN 0265-0568

Pedersén, M., Collén, J., Abrahamsson, K. & Ekdahl, A. (1996). Production of halocarbons from seaweeds: an oxidative stress reaction? Scientia Marina, Vol. 60, pp. 257-263, ISSN 0214-8358

Percival, E. & McDowell, R. H. (1981). Algal cell walls-composition and biosynthesis. In: Encyclopedia of plant physiology, Tanner, W. & Loewus, F. A., (Eds.), Vol. 13, 277-316, Springer, ISBN 0-387-11007-0, New York, USA

Rioux, L-E., Turgeon, S. L. & Beaulieu, M. (2010). Structural characterization of laminaran and galactofucan extracted from the brown seaweed *Saccharina longicruris*. Phytochemistry, Vol. 71, No. 13, (September 2010), pp. 1586-1595, ISSN 0031-9422

Ruiz-Matute, A. I., Hernández-Hernández, O., Rodríguez-Sánchez, S., Sanz, M. L. & Martínez-Castro, I. (2011). Derivatization of carbohydrates for GC and GC-MS analyses. Journal of Chromatography B, Vol. 879, No. 17-18, (May 2011), pp. 1226-1240, ISSN 1570-0232

Schall, C., Laturnus, F. & Heumann, K. G. (1994). Biogenic volatile organoiodine and organobromine compounds released from polar macroalgae. Chemosphere, Vol. 28, No. 7, (April 1994), pp. 1315-1324, ISSN 0045-6535

Shibata, T., Hama, Y., Miyasaki, T., Ito, M. & Nakamura, T. (2006). Extracellular secretion of phenolic substances from living brown algae. Journal of Applied Phycology, Vol. 18, No. 6, (December 2006), pp. 787-794, ISSN 0921-8971

Sieburth, J. M. (1969). Studies on algal substances in the sea. III. The production of extracellular organic matter by littoral marine algae. Journal of Experimental Marine Biology and Ecology, Vol. 3, No. 3, pp. 290-309, ISSN 0022-0981

Skoog, A. & Benner, R. (1997). Aldoses in various size fractions of marine organic matter: implications for carbon cycling. Limnology and Oceanography, Vol. 42, No. 8, (December 1997), pp. 1803-1813, ISSN 0024-3590

Solomon, S. (1990). Progress towards a quantitative understanding of Antarctic ozone depletion. Nature, Vol. 347, No. 6291, (September 1900), pp. 347-354, ISSN 0028-0836

Stephanie, B., Eric, D., Sophie, F. M., Christian, B. & Yu, G. (2010) Carrageenan from *Solieria chordalis* (Gigartinales): structural analysis and immunological activities of the low molecular weight fractions. Carbohydrate Polymers, Vol. 81, No. 2, (June 2010), pp. 448-460, ISSN 0144-8617

Sweeley, C. C., Bentley, R., Makita, M & Wells, W. W. (1963). Gas-liquid chromatography of trimethylsilyl derivatives of sugars and related substances. Journal of the American Chemical Society, Vol. 85, No. 16, pp. 2497-2507, ISSN 0002-7863

Van Alstyne, K. L. (1988). Herbivore grazing increases polyphenolic defenses in the intertidal brown alga *Fucus distichus*. Ecology, Vol. 69, No. 3, 655-663, ISSN 0012-9658

Van Heemst, J. D. H., Peulvé, S. & De Leeuw, J. W. (1996). Novel algal polyphenolic biomacromolecules as significant contributors to resistant fractions of marine dissolved and particulate organic matter. Organic Geochemistry, Vol. 24, No. 6-7, (June 1996), pp. 629-640, ISSN 0146-6380

Van Heemst, J. D. H., Van Bergen, P. F., Stankiewicz, B. A. & De Leeuw, J. W. (1999). Multiple sources of alkylphenols produced upon pyrolysis of DOM, POM and recent sediments. Journal of Analytical and Applied Pyrolysis, Vol. 52, No. 2, (November 1999), pp. 239-256, ISSN 0165-2370

Wada, S., Aoki, M. N., Tsuchiya, Y., Sato, T., Shinagawa, H. & Hama, T. (2007). Quantitative and qualitative analyses of dissolved organic matter released from Ecklonia cava Kjellman, in Oura bay, Shimoda, Izu Peninsula, Japan. Journal of Experimental Marine Biology and Ecology, Vol. 349, No. 2, (October 2007), pp. 344-358, ISSN 0022-0981

Wada, S., Aoki, M. N., Mikami, A., Komatsu, T., Tsuchiya, Y., Sato, T., Shinagawa, H. & Hama T. (2008). Bioavailability of macroalgal dissolved organic matter in seawater. Marine Ecology-Progress Series, Vol. 370, (October 2008), pp. 33-44, ISSN 0171-8630

Weinberger, F., Coquempot, B., Forner, S., Morin, P., Kloareg, B. & Potin, P. (2007). Different regulation of haloperoxidation during agar oligosaccharide-activated defense mechanisms in two related red algae, *Gracilaria* sp. and *Gracilaria chilensis*. Journal of Experimental Botany, Vol. 58, No. 15-16, pp. 4365-4372, ISSN 0022-0957

Yokohama, Y., Tanaka, J. & Chihara, M. (1987). Productivity of the *Ecklonia cava* community in a bay of Izu Peninsula on the Pacific coast of Japan. The Botanical Magazine, Tokyo, Vol. 100, No. 2pp. 129-141, ISSN 0918-9440

Part 4

Selected Applications of Gas Chromatography in Industrial Applications

15

Application of Gas Chromatography in Monitoring of Organic and Decontamination Reactions

Pranav Kumar Gutch
Defence Research & Development, Establishment Jhansi Road, Gwalior, India

1. Introduction

Gas chromatography (GC) is a separation technique commonly used for organic compounds which can be converted into gaseous phase without decomposition. The components to be separated are carried through the column by an inert gas (Carrier gas). The separation of the analyte usually takes place in a long capillary tube (the column). The wall of the column is coated with a thin film of a very high boiling liquid (stationary phase), through which an inert gas (mobile phase) flows. The component of a mixture are portioning between the two phases. The stronger the interaction of a component with the stationary phase, the longer will be the retention. The rate of movement of the various components along the column depends upon their tendency to absorb in the stationary phase. Component of mixture are separated at different time interval and it give only a single peak for each separated component of mixture. A plot of the time necessary for a compound to move through the column (retention time) versus its amount (intensity) is called a chromatogram. The time for injection point to peak maxima is called retention time. Retention time depends on various factor viz Flow rate of carrier gas, Temp of column, and nature of stationary phase.

2. Monitoring of organic reactions by GC

GC (Pattision 1973) is generally used for determination of purity, identification of samples and monitoring of organic reactions. Decontamination reaction of chemical warfare agents can also be monitored by GC. *N*-Chloro compounds (Kovari, E. 2007) have been extensively exploited, both for fundamental research and a wide range of industrial applications, (Koval, I. 2002, 2001) owing to their easy handling, commercial availability and high storage stability. As a result, intensive research and studies have been carried out over long period of time on their chlorination, (Ghorbani, Vaghei. 2009) oxidation, water disinfection (Bogoczek, R. 1989)& some other applications (Kowalski, P. 2005) in synthetic organic chemistry. N-chloro compounds (Singh, R. 2011) have also been reported for decontamination of chemical warfare agents.

The polymeric N-Chloro resin proved to be very strong chloridine agent and found to be easily reusable. The additional advantage of polymeric chloramines (Gutch, P. 2011) in higher stability as compared to low chloramines such as chloramines-T and dichloramine – T. Active chlorine content does not decrease with prolonged storage time. Hetrogenised reagent has evinced the attention of researchers because of their handling ease, lower toxicity, nonexplosive, and malodorous nature. Another impotent motive of supported reagent is facile isolation of pure product than their solutions phase equivalents.

N-chloramines (Gutch, P. 2011) in which chlorine is directly attached to nitrogen can generate positively charge chlorine (Cl^+) which is an oxidising species. N-Chloro compound macroporous poly(styrene-co-divinylbenzene) resin having *N,N*-dichlorosulfonamide groups have been used as a polymer supported reagent for chlorination, oxidation for residual sulfides, cyanides, thiocyanates, water disinfection and some application in synthetic organic chemistry. In recent year the use of recyclable reagents has received considerable interest in organic synthesis. This prompted us to explore the possibility of using the stable, non toxic, recyclable, and efficient positive chlorine releasing reagent, N, N-dichloro poly (styrene-co-divinylbenzene) sulfonamide polymeric beads for oxidation, chlorination and decontamination reactions.

2.1 Synthesis of dialkylchlorophosphate from dialkylphosphites at room temperature

Organophosphorus compounds (OPs) (Eto, M. 1974) have attracted the attention of researchers because of their wide applications in industrial, agricultural, biochemical, and medicinal areas. Naturally occurring OPs play an important role in the maintenance of life processes. It is interesting to note that chemical, physical, and biological properties of OPs are governed by the stereochemical disposition of substituents around the phosphorous atom. One such class of compounds is the dialkylchlorophosphates. These chlorophosphates are used for the transformation of various functional groups. Recently, diethylchlorophosphate has been used as an efficient reagent in cyclization reactions and in regioselective ring opening of epoxides.

Methods described in the literature for the preparation of dialkylchlorophosphates, involve chlorination of the corresponding phosphites (dialkylphosphites/trialkylphosphites) with reagents such as elemental chlorine, phosgene, SO_2Cl_2, S_2Cl_2, SCl_2, CCl_4, CCl_3NO_2, $PhSO_2NCl_2$, C_2Cl_6, $ClSCCl_3$, $CuCl_2$, perchlorofulvalene, and *N*-chlorosuccinimide. Among these procedures, only a few can be considered as convenient laboratory methods for the synthesis of dialkylchlorophosphates. Most of these reported methods either use reagents or produce undesired by-products, which are difficult to remove from the sensitive chlorophosphates, while others are time consuming, involve expensive and unstable reagents, and require harsh conditions.

Having established a commercially synthetic procedure for N, N dichloro poly (styrene-co-divinylbenzene)(Gutch, P. 2007) sulfonamide polymeric beads(we investigated its use as an alternative reagent for the conversion of di alkylphosphite to dialkylchlorophosphate (Gupta, Hemendra. 2008).This development enable us to obtained almost all quantitative yields of products in short reaction times. The room temperature (20-25°C) reaction of various dialkylphosphite with N, N dichloro poly (styrene-co-divinylbenzene) sulfonamide polymeric beads afforded the corresponding dialkylchlorophosphates in 3-4 h in excellent yields (Scheme 1).The reaction was monitored by removing an aliquot and analyzing by GC-MS. The progress of a typical reaction as monitored by GC-MS is depicted in fig. 1.

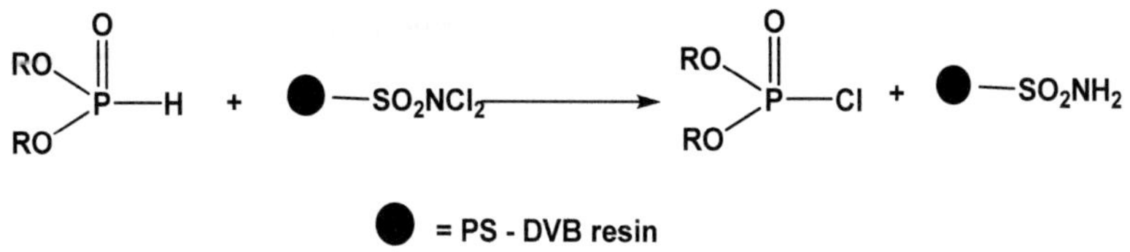

Scheme 1. Dialkylchlorophosphateprepared from dialkylphosphites using polystyrene divinylbenzene (PS-DVB) bound reagent.

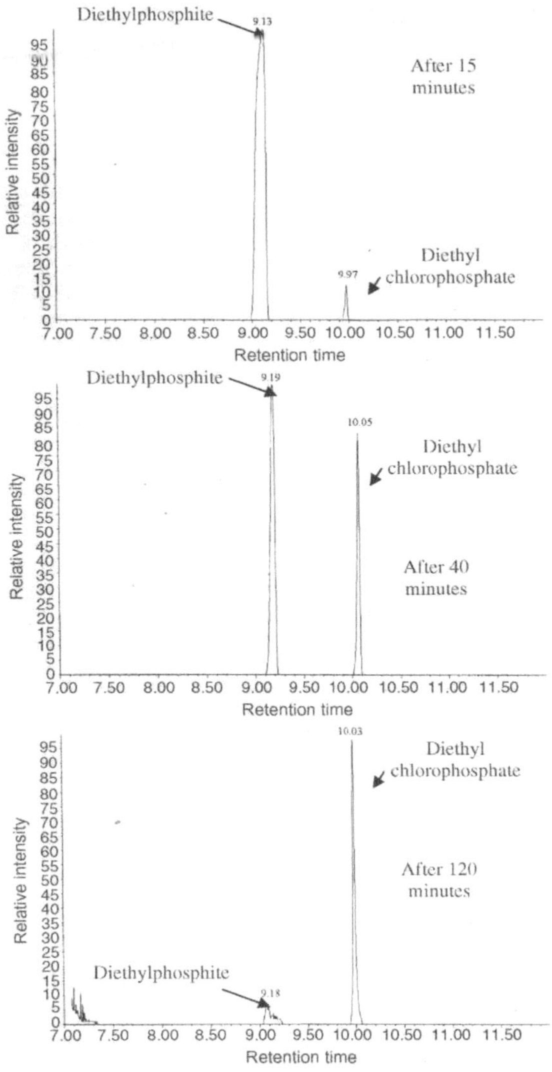

Figure 1. Progress of a typical reaction as monitored by GC–MS analysis at different time intervals.

Fig. 1. Progress of a typical reaction as monitored by GC-MS analysis at different time intervals.

In conclusion, we have described an efficient reagent for the rapid and convenient conversion of dialkylphosphites to dialkylchlorophosphates under mild conditions using *N,N*-dichloro poly(styrene-co-divinyl benzene) sulfonamide polymeric beads, as a stable and non-toxic reagent at room temperature. Here in this work it is proved to be an important and valuable reagent.

2.2 Synthesis of dicyanooxiranes derived from benzylidenemalononitriles

BenzylideneMallononitriles(Rose, S. 1969) have received much attention as cytotoxic agents against tumors and some derivatives have also been used as rodent repellents. 2-Chlorobenzylidene Malononitrile (CS) (Jones, G. 1972) is one of the most potent lachrymator and skin irritants used as riot control agent. Benzylidinemalononitriles (Gutch, PK, Thesis 1997) are highly sensitive to oxygen and form Gem-dicyano epoxides which constitute an important class of synthetic intermediates. These epoxide react with a variety of a reagent such as alcohol, water, hydrazine, and hydroxylamine in presence of halogen acid to give useful products including hydroxamic acid which is an interesting material for the synthesis of intermidiatesaziridinones used in situ as precursor to α- hydroxy and α -amino acids.We have developed an improved methods for the synthesis of substituted phenyl-1,1dicyanooxiranes (Scheme 2) from benzylidenemalononitrile using calcium hypochlorite (Gutch, P. March 2001) in carbon tetrachloride at 25^{0}C. The reaction completion within 90 min and results in 98-99% yields of dicyanooxiranes. The reaction does not require any maintenance of the pH and no side products are formed. The reaction was monitored by GC (fig. 2) using FID as a detector at 140 °C

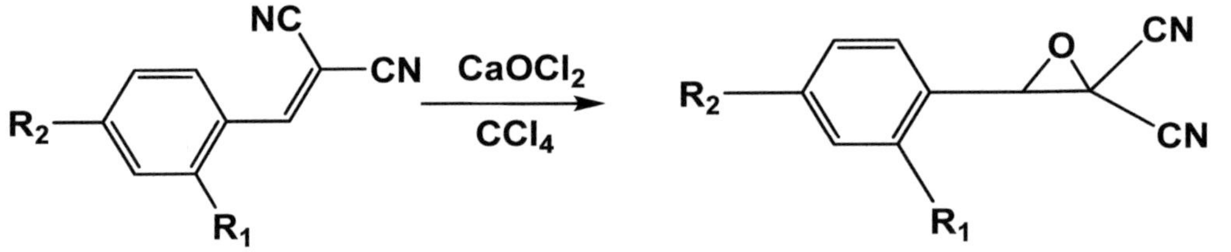

	R_1	R_2
1.	H	H
2.	Cl	H
3.	Br	H
4.	CH_3	H
5.	NO_2	H
6.	H	Cl
7.	H	Br
8.	H	CH_3
9.	H	NO_2

Scheme 2. Synthesis of substituted phenyl-1, 1′-dicyanooxiranes

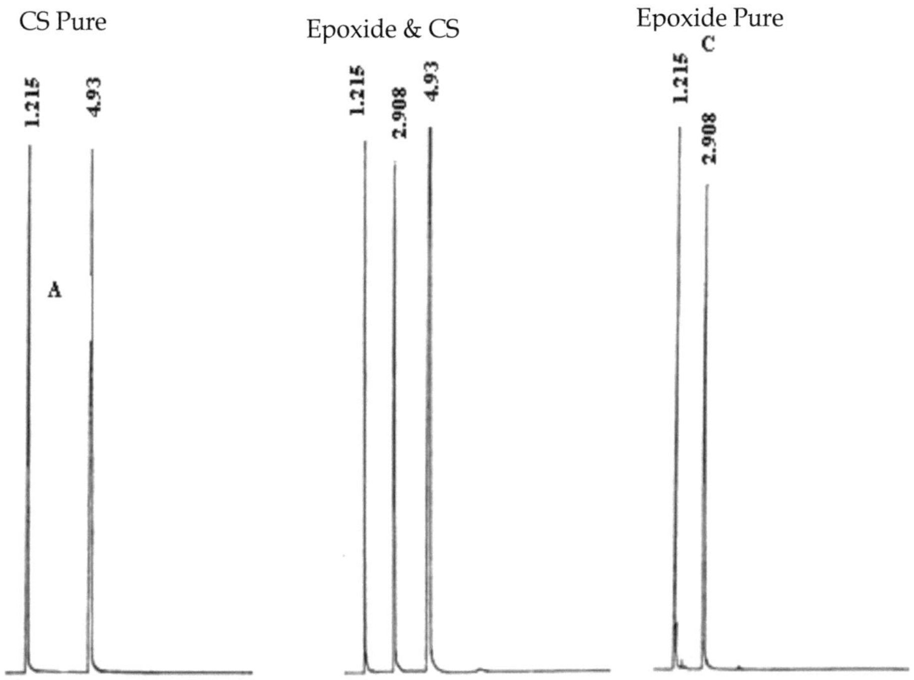

Fig. 2. Reaction of CS with calcium hypochlorite on DB-1701 column at 140 °C: Chromatogram: A= Reaction mixture at zero time showing only CS: B=Reaction Mixture after 20 minutes showing the formation Epoxide: C=Reaction mixture after 45 minutes showing disappearance of CS.

2.3 Synthesis of disulfide from thiols

Disulfides (Pathak, U. 2009)) have found industrial applications as vulcanizing agents and are important synthetic intermediates with many applications in organic synthesis. Disulfides are primarily produced from thiols, which are readily available and facile to prepare. Oxidative coupling of thiols to disulfides under neutral and mild conditions is of practical importance in synthetic chemistry and biochemistry. However, synthesis of disulfides from thiols sometimes can be problematic due to the over oxidation of thiols to sulfoxides and sulfones. On the other hand, many successful reagents such as 1,3-dibromo-5,5-dimethylhydantoin, cerium(IV) salts, permanganates, transition metal oxides, sodium chlorite, peroxides, halogens, solid acid reagents, monochloro poly(styrenehydantoin)(Akin, Akdag. 2006) beads, *N*-Phenyltriazolinedione, poly(*N*-bromomaleimide), (Bahman, Tamani. 2009) have been developed for the synthesis of disulfides from thiols, under a range of experimental conditions.

A novel method for oxidative coupling of thiols (Gutch, P. 2011) to their corresponding disulfides (scheme 3) using *N, N*-dichloro poly (styrene-co-divinyl benzene) sulphonamide (PS-DVS) beads was developed.In this study, we have employed *N, N*- dichloro poly (styrene-co-divinylbenzene) sulfonamide beads in water for the conversion of thiols to symmetric disulfides. These polymeric beads did not over oxidize the thiols to sulfoxides or sulfones and the sole product isolated from these experiments was disulfides. Presumably, the reactions proceed through chlorine transfer from amide *N*-halamine to thiols to yield disulfides and hydrochloric acid.

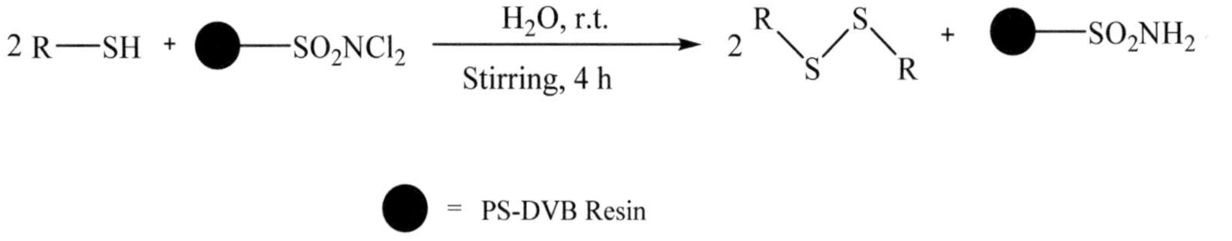

Scheme 3. Synthesis of disulfides from thiols

Thiol(0.05 mol) was added to a suspension of *N, N*-dichloro poly (styrene-co-divinyl benzene) sulphonamide beads (containing 0.05 mol of active chlorine) in water (30mL). The mixture was stirred at room temperature for 4 hrs and reaction was monitored by GC (Fig. 3). PS-DVS beads act as novel and selective oxidative agents which efficiently reduce the reaction time, increase the product yield without producing over oxidized products and performed under air atmosphere.

In conclusion *N, N*-dichloro poly (styrene-co-divinylbenzene) sulfonamide resin is an efficient reagent for fulfill different required objective in organic synthesis. Here in this work it proved to be an important and valuable reagent for the corresponding disulfide. The conversion was very efficient, with excellent yield and polymeric reagent can be recovered activated and reused. The main advantages of the method is that the reaction was very clean and operationally simple. Therefore the method is very attractive for organic chemist.

3. Monitoring of decontamination reaction of chemical reaction for chemical warfare agent

Decontamination (Yu-chu, Yang. 1992) is an important unavoidable part in protection against Chemical warfare agent. The aim of decontamination is to rapidly and effectively render harmless or remove poisonous substances both on personal and equipment. Though a variety of decontaminating agents have been reported over the years, however most of them suffer from drawbacks such as the use of hazardous solvents and prolonged decontamination time. Further, these reagents cannot be used on skin due to the toxicity of solvent and reagents.

Decontamination (Somani, S. (Ed). 1992) may be defined as a method of conversion of toxic chemicals into harmless products which can be handled safely, either by degradation or detoxification using suitable reagents. Decontamination plays a vital role in defence against chemical warfare agents. The toxic materials must be eliminated from the battlefield etc. by application of some efficient decontamination methods as fast as possible for resuming routine activities. Decontamination (Talmage, S. 2007) of CW agents is required on the battlefield, in laboratories, pilot plants and places of chemical agent production, storage, and destruction sites.

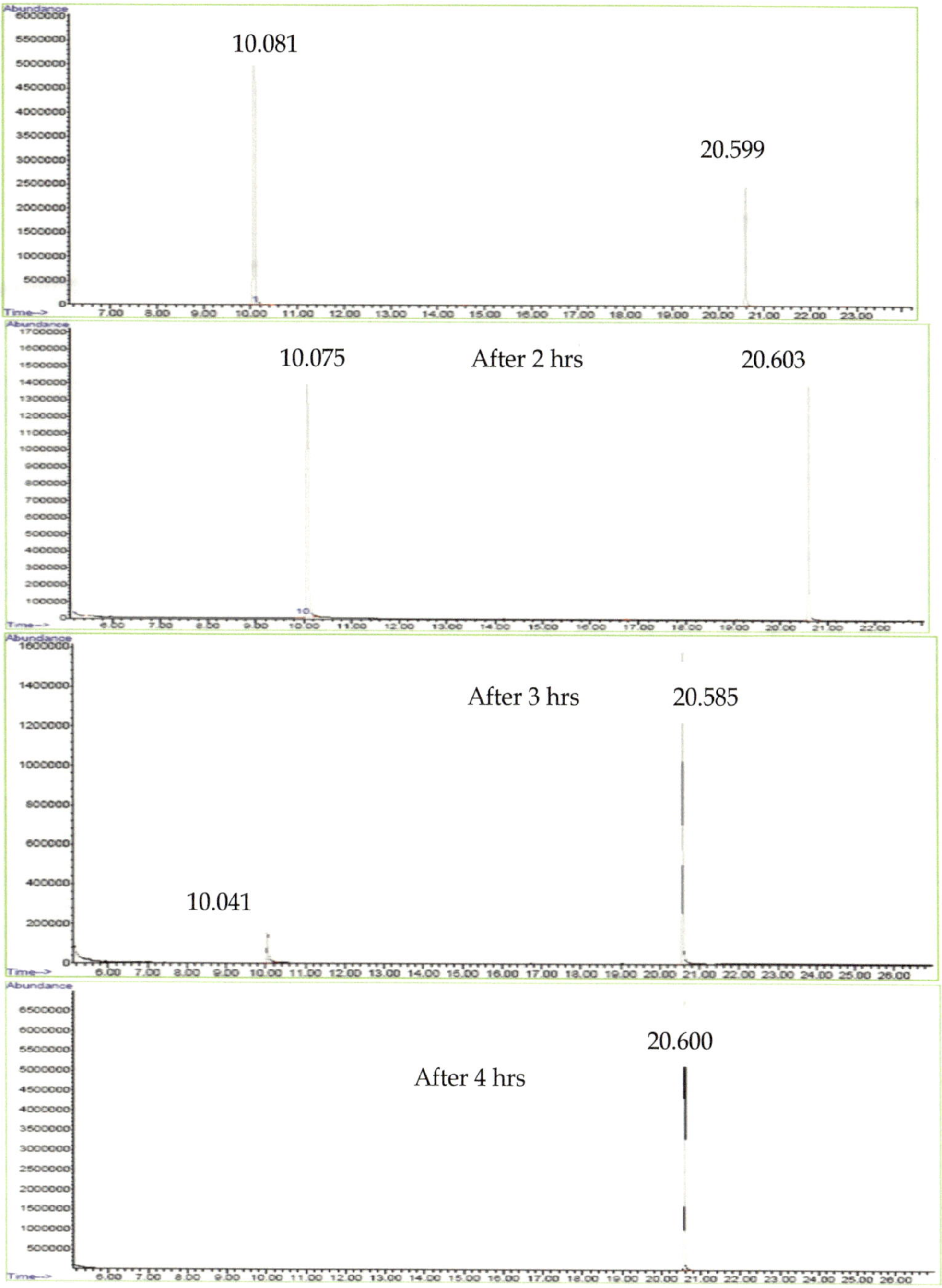

Fig. 3. Progress of disulphide reaction as monitored by GC-MS analysis at different time interval

Toxic substances can also be decontaminated (Yang, Y. 1995) by chemical modification of their toxic structures by using small compounds like hydrogen peroxide. Decontamination (Dubey, D. 1999) through chemical reactions is more reliable and effective as compared to physical methods. Chemical reaction yields less toxic or non-toxic products, while in the physical process they are just removed from the site.

3.1 Simulants

As with all chemical warfare agents, testing with the actual agents which are often very toxic is difficult to synthesize, their handling is also difficult, and they are expensive and are not commercially available. The developmental work on their process is often done on surrogate chemical that possesses many of the important features of the agent. Surrogate, also called a simulant or analog, must be carefully chosen to reasonably substitute for the target agent in the specific reaction pathway or reaction conditions under investigation. In the case of the partial oxidation of sulfur mustard, for example, an appropriate surrogate should clearly have a sulfide bond. Ideally, the surrogate would have similar structural features allowing for the same reaction mechanisms that would detoxify the agent, and would also have similar physical properties like solubility and vapor pressure. Also, one should consider secondary reactivity when possible, allowing for any foreseeable side reaction.

Simulants are those compounds which are less toxic but they have same chemical reactivity like toxicants. For the standardization of the chemical reactions, simulants can be used in place of actual agents. In order to gain a more complete understanding of CW agent chemistry, it is often necessary to study the reactions of a series of agent-analogs under the same conditions. For example, the monofunctional derivatives of mustard, $RSCH_2CH_2Cl$ (R= Me, Et, Ph), $RSCH_2CH_2X$ (X = tosylate, brosylate, Br^-, I^-, or other leaving groups), react via the same mechanism as those of HD, but their reaction products and kinetics rate expressions are much simpler. 2-chloroethylphenylsulfide, 2-chloroethyl ethylsulfide (half mustard) and dibutyl sulfide (DBS) are all simulants of HD. The use of simulants makes it easier to isolate the variable that affects the agent chemistry. VX analogs, (C_2H_5S) (CH_3) P (O)-(C_2H_5O), $(C_2H_5S)(C_6H_5)P(O)$-(C_2H_5O) have also been studied to isolate the effect of the diisopropylamino group on the reaction chemistry of VX. Simulants are those compounds which are less toxic but they have same chemical reactivity like HD & VX. For the standardization of the chemical reactions, simulants can be used in place of actual agents. 2-chloroethyl phenyl sulphide& methyl phosphothioic acid O, S- diethyl ester are simulants of HD & VX respectively.

Our aim in this work is to overcome the limitation and drawbacks of the reported decontaminants (Prasad, G. (2009)). It has been in corrupted as a reactive ingredient in a formulation developed in our laboratory to decontaminated bis (2-chloroethyl) sulfide (SM), a chemical warfare agent.N,N-Dichloro poly(styrene-codivinyl benzene) sulfonamide, a new class of readily available, economical, commercially viable, and recyclable chlorinating reagent was chosen for use as a decontaminating agent in this study.

3.2 Decontamination of Sulfur Mustard (SM)

We have already reported *N, N*-dichloro poly (styrene-co-divinyl benzene) sulphonamide polymeric beads for decontamination of simulant of SM. N,N-dichloro poly(styrene-co-divinylbenzene) sulphonamide reacts with 2-chloro ethyl phenyl sulphide, a simulant of sulfur mustard (SM), at room temperature, yielding corresponding nontoxic sulfone and sulfoxide in aqueous as well as aprotic medium.

Sulfur mustard (SM) is a cytotoxic and persistent blistering agent that was used as a weapon of mass destruction in the World War I and recently in the Iraq-Iran war (1980-1988). Detoxification (Gutch, P. 2008) of the chemical warfare agent (CWA), sulfur mustard is one of the primary goals to get rid of its toxic effects. An efficient and operationally simple method is developed for chemical decontamination of sulfur mustard. A new chlorine bearing reagent *N,N*-dichloro poly (styrene-co-divinyl benzene) sulfonamide was developed to deactivate the sulfur mustard (**Scheme 4**) in aqueous (acetonitrile: water) medium. This decontamination reaction was monitored by GC-FID (Fig.4) and the products were analyzed by GC-MS (Fig.5).

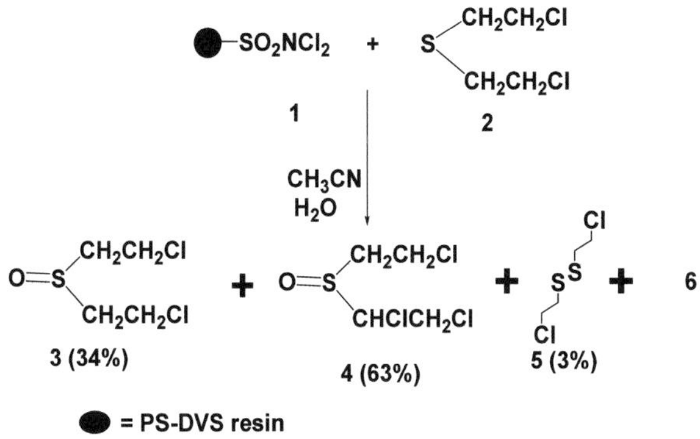

Scheme. 4. Reaction of SM with polymeric Beads in aqueous medium

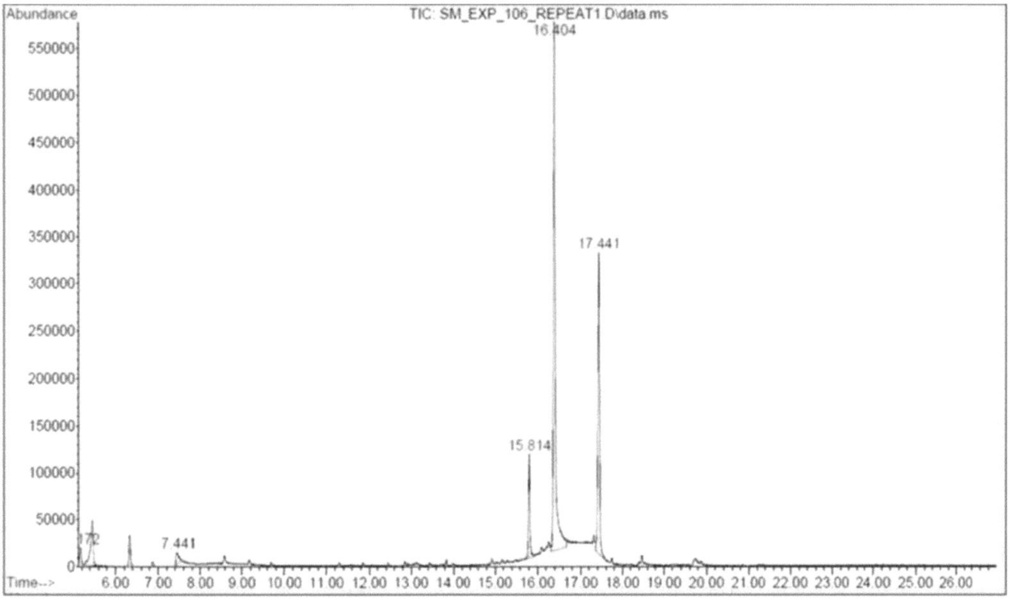

Fig. 4. Gas Chromatogram of SM after decontamination

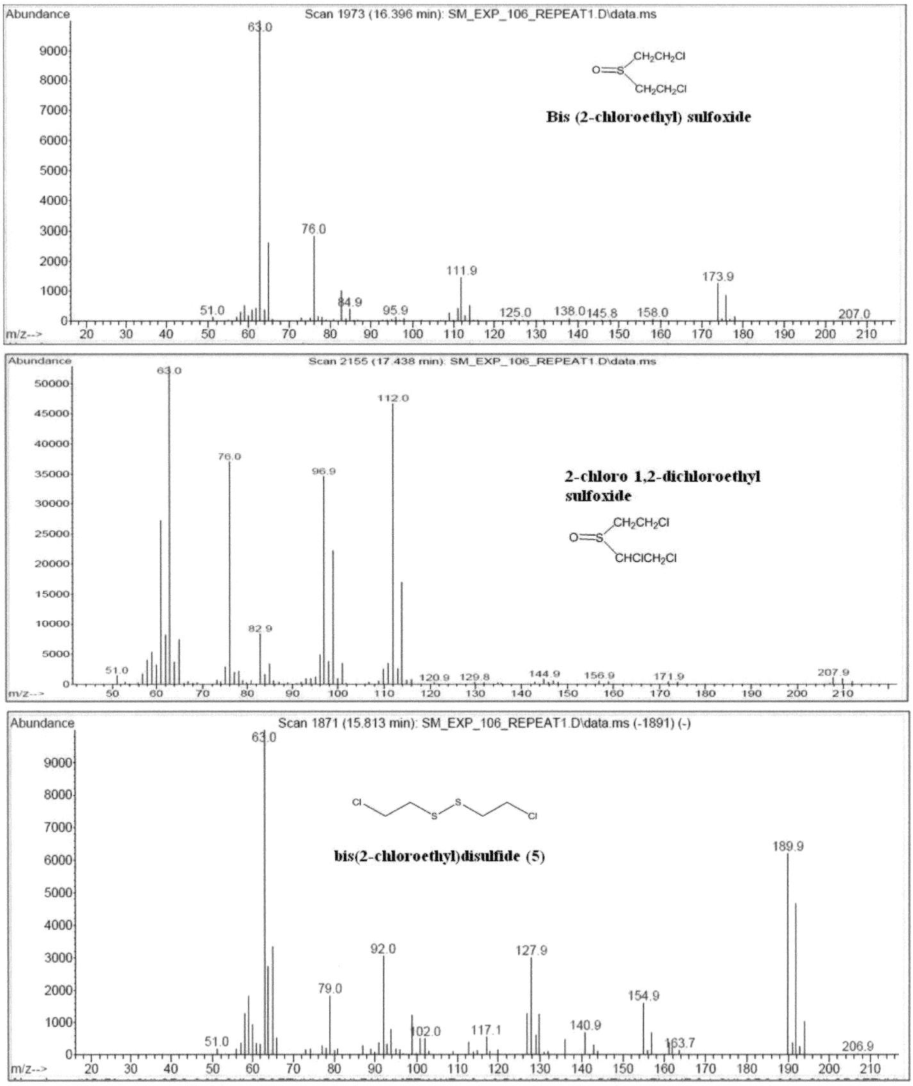

Fig. 5. Mass spectrum of decontaminated products (3-5) of SM in aqueous medium

N, N-dichloro poly (styrene-co-divinyl benzene) sulphonamide (Gutch, P. 2011) polymeric beads (3.0 mmol) was added slowly to a stirred solution of SM (1.0 equiv.) in acetonitrile water mixture (5 ml) at room temperature. It was found that 100 % SM was decontaminated instantaneously in aqueous medium and decontaminated products were identified by Gas Chromatography-Mass Spectrometry. The GC conditions used were as follows; column HP-

5 (30m× 0.250mm i.d., 0.25 μm film thickness) with a temperature programme of 50°C for 2 min followed by a linear gradient to 250°C at 10°Cmin^{-1}, and hold at 250°C for 5 min. The injector temperature was maintained at 250°C.

The decontamination reaction of SM is studied with varying molar concentration of active chlorine available in PS-DVB. At equimolar ratio of SM and active chlorine on PS-DVB, the major product is bis (2-chloroethyl) sulfoxide**(3)**. At 1:2 molar ratio, in addition to **(3)**, 2-chloro 1,2-dichloroethyl sulfoxide(4) is also formed; whereas, at 1:4 molar ratio the major product is 2-chloro 1, 2-dichloroethyl sulfoxide **(4)** with a trace amount of bis(2 chloroethyl)disulfide **(5)**. In all the reactions N, N-dichloro poly (styrene-co-divinyl benzene) sulfonamide (PS-DVB) polymeric beads **(1)** were converted into poly (styrene-co-divinyl benzene) sulfonamide (PS-DVB) polymeric beads **(6)**.

This reagent has advantage over earlier reported reagents in terms of effectiveness, stability, non toxicity, and cost, ease of synthesis, recyclability (collected after filtration, rechlorinated and used for further reaction) and instantaneous decontamination of sulfur mustard at room temperature.

3.3 Decontamination of S-2-(diisopropylamino) ethyl O-ethyl methylphosphonothioate(VX)

Organophosphorus compounds (OPs) have been widely used in the industry, veterinary and human medicine, in the agriculture as pesticides or can be misused for military purpose i.e. as Chemical Warfare Agents (CWA). The chemical warfare nerve agents, commonly known as nerve gases, are in fact not gases but polar organic liquids at ambient conditions. Most of these nerve agents are P (V) organophosphorus esters that are similar in structure to an insecticide.

Highly toxic organophosphorus compounds (OPCs) that are nerve agents such as sarin, soman, tabun and cyclosarin are lethal chemical warfare agents, also known as G-agents and were developed by G. Shrader in Germany during World War-II.It was then followed by the discovery of VX (S-2-(diisopropylamino) ethyl O-ethyl methylphosphonothioate), a nerve agent that was much more potent than all the known G-agents. Currently there are two known V-agent stockpiles; VX (S-2-(diisopropylamino) ethyl O-ethyl methylphosphonothioate), with thousands of tons in the USA, and an analogue and isomer, RVX (Russian-VX, S-2-(diethylamino) ethyl-O-isobutyl methylphosphonothioate) in Russia.

We investigated its use as an alternative reagent for decontamination of simulant of chemical warfare agent VX at room temperature. In order to gain more complete understanding of agent chemistry, a simulant of VX; O, S-diethyl methylphosphonothiolate (OSDEMP) was synthesized initially for decontamination study.

An efficient and operationally simple method is developed for chemical decontamination of simulant of VX. *N, N*-dichloro poly (styrene-co-divinyl benzene) sulfonamide was developed to deactivate the simulant of VX (Scheme 5) in aqueous medium. This decontamination reaction was monitored by gas chromatography.

The decontamination of CWAs from structure, environment, media, and even personal has become an area of particular interest in recent years because of increased homeland security concern. In addition to terrorist attacks scenario such as accidental releases of CWA or from historic, buried, munitions are also subjects from response planning. VX is one of the most difficult CWA to destroy. In general, the decontamination of VX can be achieved either by hydrolysis or by oxidation. However, hydrolysis of VX leads to toxic hydrolyzed products. Therefore, decontamination of VX by oxidation is the preferred method. VX contain bivalent sulfur atoms that can be readily oxidized.

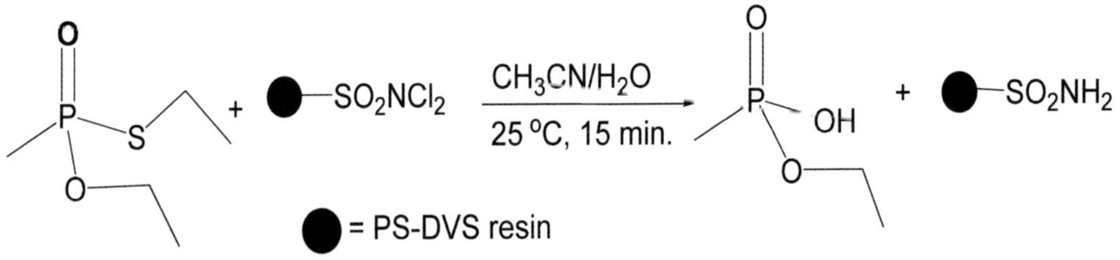

Scheme. 5. Decontamination of OSDEMP with polymeric Beads in aqueous medium

We investigated its use as an alternative reagent for decontamination of simulant of VX at room temperature. To gain more complete understanding of agent chemistry, a simulant of VX, O,S-diethyl methyl phosphonothiolate (OSDEMP) was synthesized initially for decontamination study. The most common and widely used process is the oxidation of VX using N-chloro compounds, which decontaminates it with the formation of nontoxic product.

Decontamination studies of OSDEMP (Gutch, P.2011) were carried out at room temperature in aqueous medium using a mixture of acetonitrile and water (5 : 1). Decontaminated product was separated by GC (Fig 6). OSDEMP was decontaminated 100% with N,Ndichloro poly(styrene-co-divinyl benzene) sulfonamide. The decontaminated product was analyzed as their methyl ester derivatives by reacting with freshly prepared diazomethane in ether. The organic layer did not show any peak corresponding to OSDEMP indicating the reaction of N, N-dichloropoly (styrene-co-divinyl benzene) sulfonamide with OSDEMP. The degradation of OSDEMP with N,N-dichloro poly(styrene-co-divinyl benzene) sulfonamide followed by P-S bond cleavage via oxidation and hydrolysis leading to the formation of nontoxic product ethyl methylphosphonate (EMPA).

In conclusion, the study reveals that N, N-dichloro poly (styrene-co-divinyl benzene) sulfonamide works as an excellent decontaminating agent against OSDEMP, which bears oxidizable bivalent sulfur by its oxidation followed by hydrolysis to nontoxic product EMMP in aqueous medium at room temperature. This reagent has advantage over earlier reported reagent in terms of effectiveness, stable, nontoxic, cheap, easy to synthesize, recyclability (collected after filtration, rechlorinated, and used for further reaction), and decontamination of simulant of VX to give nontoxic product EMPA at room temperature.

N, N-dichloro poly (styrene-co-divinyl benzene) sulphonamide is a efficient reagent to fulfil different required objective to organic synthesis. The main advantage of this reagent is that reactions were clean and operationally simple.

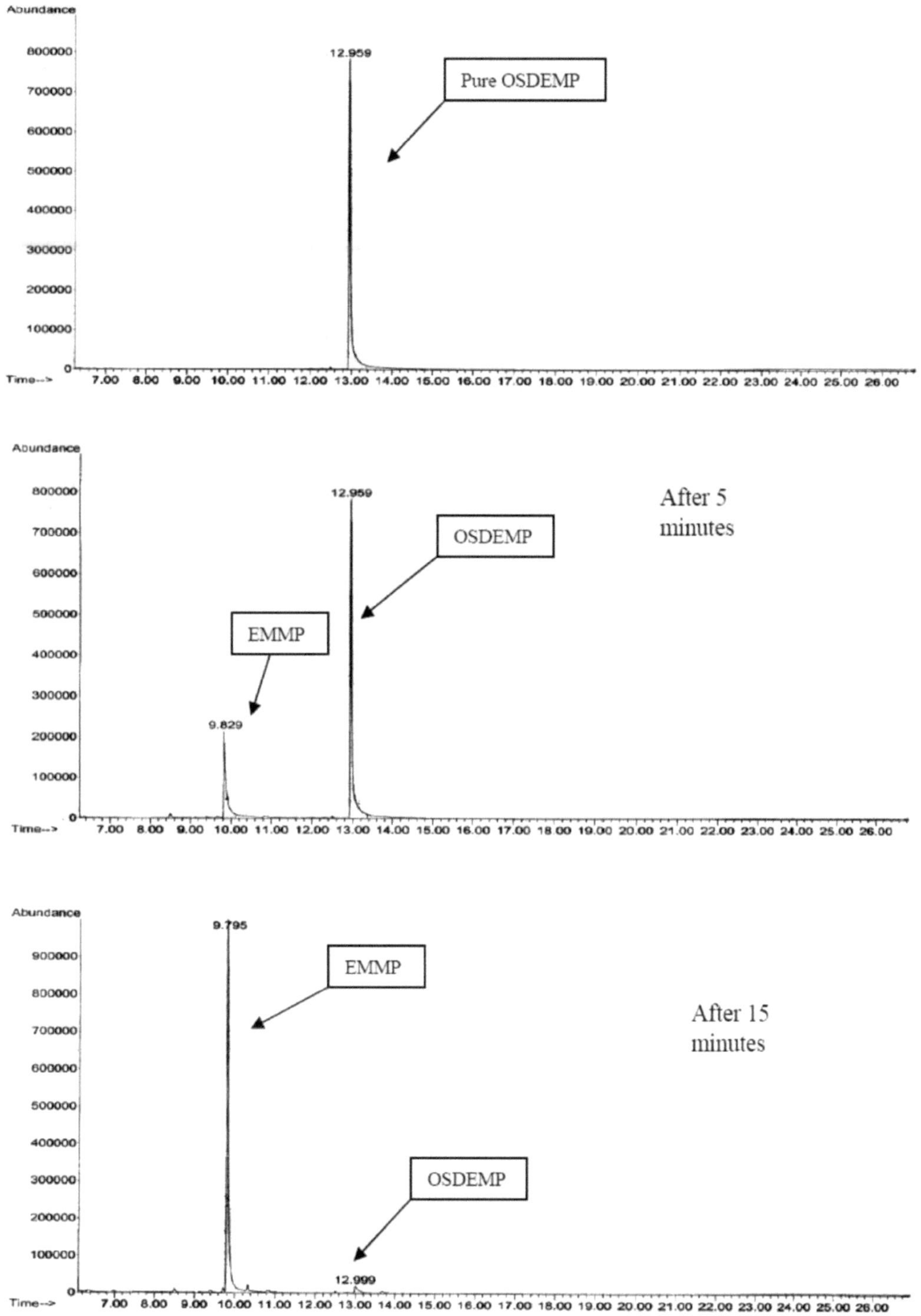

Fig. 6. Progress of a decontamination reaction of OSDEMP as monitored by GC-MS analysis at different time intervals

4. Identification of tear gas compounds in air

Chemical warfare agents can be classified into two general categories, those that exert a lethal effect and those that act in an incapacitating manner. Lethal chemical warfare agents include nerve agents such as Sarin, Soman and Tabun, while incapacitating agents include irritants (tear gases or riot control agents). Acute exposure to irritants causes a number of incapacitating effects including burning or irritation of the skin and eyes, coughing, nausea and vomiting. The incapacitating nature of these chemicals has led to the development of dispersal devices for their use in riot control situations, during military training exercises and to a lesser extent as chemical weapons on the battlefield. The most commonly employed irritants (Malhotra, R. 1987) are o-Chlorobenzylidenemalononitrile, (Gutch, P 2005) often referred to as tear gas, and 2-chloroacetophenone. Dibenz [b,f]-l,4-oxazepin has been used less frequently and 1-methoxycycloheptatriene was evaluated as a possible military training agent.

The text of the 1997 "Convention on the Prohibition of the Development, Production, Stockpiling and Use of Chemical Weapons and their Destruction" states in Article 1 that: "Each state party undertakes not to use riot control agents as a method of warfare". United Nations peacekeeping forces could encounter use of irritants during active duty in regions of the world where there is a threat of chemical warfare agent use. Intelligence gathering, through the collection of contaminated samples, and subsequent analysis of the samples would enable identification of the suspect chemical and confirm use of a controlled chemical for warfare purposes. The results of such analyses would likely contribute to the development of appropriate strategic and political positions. Gas chromatographic (GC) methods, including methods based on GC retention indices, have been used for the detection of irritants in suspect samples.

Methods for detection, identification and quantitative determination of various tear gas compounds such as Dibenz [b,f]-l,4-oxazepin (CR) (Gutch, P (2007), ω-Chloroacetophenone (CN) (Nigam, A. 2010), 2-Chlorobenzylidene malononitrile (CS) and their analogues are required during their production and also for verification of their use for prohibited activities under Organisation of prohibition of chemical weapons (www.opcw.nl)(**OPCW**). The availability of identification data on CS, CR, CN and related compounds would facilities the verification of in case of alleged use of these chemicals. United Nations peacekeeping forces could encounter use of irritants during active duty in regions of the world where there is a threat of chemical warfare agent use. Intelligence gathering, through the collection of contaminated samples, and subsequent analysis of the samples would enable identification of the suspect chemical and confirm use of a controlled chemical for warfare purposes. The results of such analyses would likely contribute to the development of appropriate strategic and political positions. Gas chromatographic (GC) methods, including methods based on GC retention indices, have been used for the detection of irritants in suspect samples. The correlation of retention indices between an unknown and reference compound on two or more columns of different polarities is generally sufficient for identification purposes. For the evolution of retention indices, a homologues series of n-alkanes is commonly used as a reference compound.

4.1 Measurement of retention indices

In order to measure the retention indices generally solutions containing several tear gases in acetone together with n-alkanes standards were injected onto the GLC column, the sample size in each instance was about 0.1 ml. The amounts of individual compounds and n-alkanes present in the range 10-15 mg. The retention times (Fig.7) were recorded with the accuracy of up to 0.01 min with the help of Shimazda CR3A data processor. In order to identify individual components in the mixture, an authentic sample of each tear gas was injected separately and its retention time was compared with that of the component in mixture. Temperature-programmed retention indices for individual compounds was calculated by using equation 1. The measurement of retention indices in two columns with different polarities is shown to be applicable for identification of tear gases compound in air.

4.2 Determination of retention indices under programmed temperature conditions

Programmed temperature GC allowed the analysis of a number of compounds over a wide range of volatilities in a single run. Under programmed temperature conditions, a linear relationship exists between the retention time of n-alkanes and their carbon number. Hence, under these conditions, it is possible to calculate retention index value using retention time only. The retention indices under programmed temperature chromatographic condition (RI_P) were calculated using the van den Dool and Kratz formula. (Equation 1).

$$RI_P = 100\left(\frac{t_c - t_z}{t_{z+1} - t_z}\right) + 100z \quad (1)$$

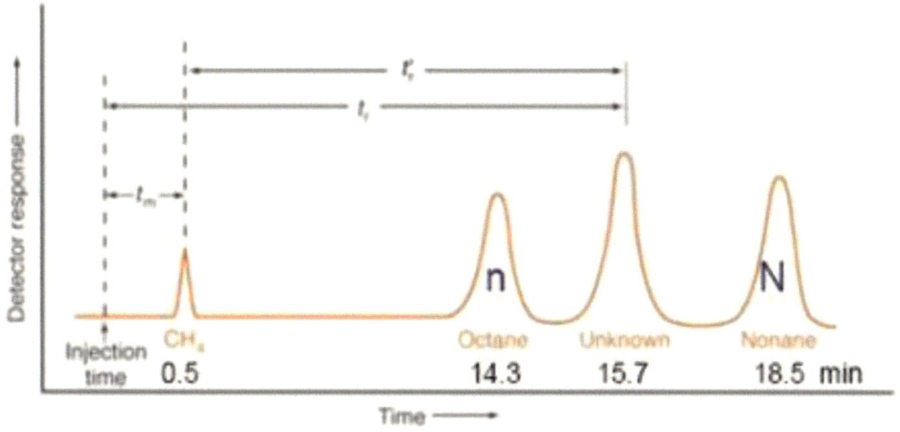

Fig. 7. GC chromatogram of mixture of n-alkane with tear gas.

Here t_c ,$t_z$and t_{z+1} were retention times of the solute, alkane eluted immediately prior to the compound with z number of carbon atoms (lower alkane), and alkane eluted immediately after the compound with z+1 number of carbon atoms (higher alkane), respectively.

4.3 Determination of retention indices under isothermal chromatographic conditions

Under isothermal conditions unlike programmed temperature conditions, there exists a non-linear relationship between retention time and number of carbon atoms of n-alkanes. Therefore, for the calculation of retention indices under these conditions, logarithm of the corrected retention time was taken into account. Retention indices under isothermal experimental conditions (RI_1) were calculated using the Kovats formula (Equation.2)

$$RI_I = 100\left(\frac{log\,t_c^{'} \ - log\,t_z^{'}}{log\,t_{z+1}^{'} \ - log\,t^{'}}\right) + 100z \qquad (2)$$

Here t'_c,t'_z and t'_{z+1} were corrected retention times of the solute, alkane eluted immediately prior to the compound with z number of carbon atoms (lower alkane), and alkane eluted immediately after the compound with z+1 number of carbon atoms (higher alkanes), respectively.

We have reported temperature programmed retention indices for several tear gas compound using Van den Dool's equation. GC of these compounds as a class on a fused silica capillary column and temperature -programmed retention indices for most of them were not previously reported. Further the measurement of retention indices on two columns with different polarities is shown to be applicable for the identification of tear gases in air. We have reported temperature programmed retention indices (Gandhe, B. 1989). on various tear gas compounds such as Methyl ethyl ketone, Chloro acetone, Bromo acetone, Ethyl bromo acetate, Benzoyl chloride, ω-Chloroacetophenone (CN), and 2-Chloro benzylinedinemalononitrile (CS) on polar DB-1 and non polar DB-1701 fused silica capillary column.

Riot control agent2-Chlorobenzylidenemalononitrile (CS) and its analogues have both skin irritating and lacrymating properties. We have also reported retention indices (RI) of CS and its thirteen analogues (Gutch, P 2004) relative to the homologues n-alkanes series. These values are determined on nonpolar BP-1 and polar BP-10 capillary column under programmed temperature and isothermal chromatographic condition. The analogues differ in substitution at ortho or para position of phenyl ring and retention indices are found to vary according to the nature of the substituent.

5. Conclusion

Gas chromatography has been successfully applied to the separation of mixtures of numerous organic and inorganic compounds that have appreciable vapour pressure. Gas chromatography is also used for analysis of pollutants, analysis of alcoholic beverages, analysis of pharmaceutical and drugs, analysis of clinical applications, analysis of essential oils, analysis of Fatty acids, analysis of explosives, and analysis of pesticides. Gas chromatography is very useful analytical technique for identification of known and unknown volatile and thermo labile compounds and monitoring of various organic reactions. Gas chromatography with a fused silica capillary column is the method of choice due to its high resolution and sensitivity.

6. Acknowledgment

We thank Dr. R. Vijayaraghavan, Director, DRDE, Gwalior, for providing necessary facilities and for useful discussion.

7. References

Akin, Akdag. (2006). Oxidation of thiols to disulfides with monochloo poly (styrenehydantoin) beads. *Tetrahedron letter*, Vol. 47, pp. 3509-3510.

Bahman, Tamani. (2009). Synthesis and application of cross-linked Poly (N-bromomaleimide) in oxidation of various organic compounds. *Iranian. Polymer. Journal*, Vol. 18, 12, pp. 957-967.

Bogoczek, R. (1989). Studies on a macromolecular dichloramine: The N, N-dichloro poly (styrene-co-divinyl benzene) sulfonamide. *Angew Makrolmolec Chem*, Vol. 169, pp. 119- .

Dubey, D. (1999). Reaction of bis (2- chloroethyl) sulphide with N,N'-dichlorobis (2,4,6-trichlorophenyl) urea. *J Org Chem*, Vol. 64, pp. 8031-8034.

Eto, M. (1974)*Organoo phosphorous pestisides organic and biological chemistry*. CRC press, USA.

Gandhe, B. (1989). Gas chromatographic retention indices of tear gases on capillary columns. *Journal of Chromatography*, Vol. 479, pp. 165-169.

Ghorbani, Vaghei. (2009). Poly (N, N'-dichloro-N-ethylbengene-1, 3-disulfonamide) and N, N, N', N'-tetra-chlorobengene. *Synhtesis*, Vol. 6, pp. 945-950.

Gupta, Hemendra. (2008). N,N-Dichloro poly (styrene-co-divinyl benzene) sulfonamide polymeric beads: an efficient and recyclable reagent for the synthesis of dialkyl Chlorophosphates from dialkylphosphites at room temperature. *Tetrahedron Letters*, Vol. 49, pp. 6704–6706.

Gutch, P (2004). Chromatographic analysis of BMN analogues of 2-chloro bezylidenemalononitriles. J *Indian Chem.Soc*, Vol. 81, pp. 874-878.

Gutch, P (2005). Structure and biological activity relationship of 2-chloro bezylidenemalononitriles, A Riot Control agent. *Defence Science Journal*, Vol. 55, pp. 447-457.

Gutch, P (2007). A Simple, Convenient And Effective Method For the Synthesis of Dibenz (b,f) 1,4-Oxazepines (CR); A New Generation Riot Control Agent And Its Analogues. *Heterocyclic Communication*, Vol. 13, pp. 339- 396,

Gutch, P. (1997). Studies on the synthesis, structure-activity relationship and degradation of substituted malononitriles. Ph.D Thesis , Jiwaji University Gwalior.

Gutch, P. (2007). N,N-Dichloro Poly(styrene-co-divinyl benzene) Sulfonamide—Synthesis, Characterization and Efficacy against Simulant of Sulfur Mustard. *J. Applied Polymer Science*, Vol. 105, pp. 2203-2207.

Gutch, P. (2008). Polymeric decontaminant 2 (N,N-dichloropolystyrene sulfonamide): Synthesis, characterization, and efficacy against simulant of sulfur mustard. *J. Applied Polymer Science*, Vol. 107, pp. 4109-4115.

Gutch, P. (2011). New Effective Synthetic Method for Preparation of N,N-Dichlorocarbamates. *Synthetic Communications*, Vol. 41, pp. 1554-1557,

Gutch, P. (March 2001). Synthesis, Characterization and Mass Spectrometric fragmentation of gem. Dicyano oxidants derived from Benzilidene malononitriles. *Indian Journal of Chemistry*, Part B, Vol. 40, B, pp. 243-247.

Gutch, P. (2010). N, N-Dichloro poly (styrene-co-divinyl benzene) sulphonamide polymeric beads: An efficient and recyclable decontaminating reagent for Sulfur Mustard. Proceeding in Macro 2010 at IIT Delhi. pp. 15-17.

Gutch, P. (2011). N, N di chloro poly (styrene-co-divinyl benzene) sulphonamide polymeric beads An Efficient recyclable decontaminating reagent for O,S diethyl methyl phosphonothiolate a similuent of VX. *J. Applied Polymer Science*, Vol. 121, pp. 2250-2256.

Gutch, P. (2011). N, N-dichloro poly (styrene-co-divinylbenzene) sulfonamide beads as an efficient, selective and reusable reagent for oxidation of thiols to disulfides. *International J. Polymeric Materials*, Communicated.

Gutch, P. (2011). Polymeric Decontaminant: N, N - Dichloro 2, 6 - Dimethyl 1, 4 - Phenylene Oxide Sulphonamide: Synthesis, Characterization & Efficacy against Simulants of Sulfur Mustard. J. Polymer Materials, Vol. 28, 1, pp. 5-13.

Jones, G. (1972). CS and its chemical relatives. *Nature*, Vol. 235, pp. 257-259.

Kolvari, E. (2007). Application of N-halo reagents in organic synthesis. J. Iran. Chem. Soc, Vol. 4, 2, pp. 126-174.

Koval, I. (2002). N-halo reagent: N-halosuccinimide in organic synthesis and in chemistry, a natural compound. *J Iran Chem Soc*, Vol. 38, pp. 301-337.

Koval, I. (2001). N-Halo Reagents. Synthesis and Reactions of N-Halocarboxamides. *Russ J Org Chem*, Vol. 37, 3, pp. 297-317.

Kowalski, P. (2005). Oxidation of sulphides to sulfoxide, Part 1: Oxidation using halogen derravitives. *Tetrahedron*, Vol. 61, pp. 1933-1953.

Malhotra, R (1987). Chemistry and toxicity of tear gases. *Def. Sci. J*, Vol. 37, pp. 281-296,

Nigam, A. (2010). Thermal decomposition studies of riot control agent Chloro acetophenon (CN) by Pyrolysis-gas chromatography-mass spectrometry. J. of Hazardous Materials. Vol. 184, pp. 506-514.

(1977). *Convention on the Probhition of the development production stockpiling and use of Chemical weapons and on their destruction, technical secretariat of the organization for prohibition of chemical weapons to The Hague*, Available from: < www.opcw.nl.> Accessed 28 May 2008.

Pathak, U. (2009). Efficient and convenient oxidation of thiols to symmetrical disulfide with silica-PCl_5 / $NaNO_2$ in water. *Syn. Comm*,Vol. 39, pp. 2923-2927.

Pattsion, J. (1973). *A Programmed Introduction to gas Chromatography.* Spectrum House Alderton Crescent, London, Great Britain,

Prasad, G. (2009). Modified titania nanotubes for decontamination of sulphur mustard. *J. Hazard. Mater*, Vol. 167, pp. 1192- 1197.

Rose, S. (1969). CS – a case for concern.. *New Sci*, Vol. 43, pp. 468-469.

Singh, R. (2011). Decontamination of Toxic chemicals by N-chlorocompound. Ph.D Thesis , Jiwaji University Gwalior.

Somani, S. (Ed). (1992). *Chemical Warfare Agents*, Academic Press, New York,

Talmage, S. (2007). Chemical warfare agent degradation and decontamination. *Curr. Org. Chem*, Vol. 38, pp. 953-959.

Yang, Y. (1995). Chemical reaction for neutralizing chemical warfare agents. *Chem Ind*, Vol. , pp. 334-336.

Yu-chu, Yang. (1992). Decontamination of chemical warfare agents. *Chem. Rev*, Vol. 92, pp. 1729-1743.

16

Pyrolysis-Gas Chromatography/Mass Spectrometry of Polymeric Materials

Peter Kusch
Bonn-Rhine-Sieg University of Applied Sciences, Department of Applied Natural Sciences, Rheinbach, Germany

1. Introduction

Structural analysis and the study of degradation properties are important in order to understand and improve performance characteristics of synthetic polymers and copolymers in many industrial applications. Polymers/copolymers cannot be analyzed in their normal state by traditional gas chromatography (GC) because of high molecular weight and lack of volatility. However, by heating these macromolecules to temperatures above 500 °C, they are pyrolyzed into many individual fragmentation substances, which can be then separated chromatographically and identified by mass spectrometry.

Pyrolysis technique hyphenated to gas chromatography/mass spectrometry (GC/MS) has extended the range of possible tools for characterization of synthetic polymers/copolymers. Under controlled conditions at elevated temperature (500 - 1400 °C) in the presence of an inert gas, reproducible decomposition products characteristic of the original polymer/copolymer sample are formed. The pyrolysis products are chromatographically separated by using a fused silica capillary column and subsequently identified by interpretation of the obtained mass spectra or by using mass spectra libraries (*e.g.* NIST, Wiley, etc.). Pyrolysis methods eliminate the need for pre-treatment by performing analyses directly on the solid polymer/copolymer sample.

Most of the thermal degradation results from free radical reactions initiated by bond breaking and depends on the relative strengths of the bonds that hold the molecules together. A large molecule will break apart and rearrange in a characteristic way (Moldoveanu, 2005; Wampler, 2007; Sobeih et al., 2008). If the energy transfer to the sample is controlled by temperature, heating- rate and time, the fragmentation pattern is reproducible and characteristic for the original polymer. Another sample of the same composition heated at the same rate to the same temperature for the same period of time will produce the same decomposition products. Therefore the essential requirements of the apparatus in analytical pyrolysis are reproducibility of the final pyrolysis temperature, rapid temperature rise and accurate temperature control. Depending upon the heating mechanism, pyrolysis systems have been classified into two groups: the continuous-mode pyrolyzer (furnace pyrolyzer) and pulse-mode pyrolyzer (flash pyrolyzer) such as the heated filament, Curie-point and laser pyrolyzer. The pyrolysis unit is directly connected to the injector port of a gas chromatograph. A flow of an inert carrier gas, such as helium, flushes the pyrolyzates into the fused silica capillary column. The detection technique of the separated compounds is typically mass spectrometry but other GC detectors have been also used depending on the intentions of the analysis (Sobeih et al., 2008). The currently commercially available pyrolysis equipment was described in detail in previous work of the author (Kusch et al., 2005).

The applications of analytical pyrolysis–gas chromatography/mass spectrometry range from research and development of new materials, quality control, characterization and competitor product evaluation, medicine, biology and biotechnology, geology, airspace, environmental analysis to forensic purposes or conservation and restoration of cultural heritage. These applications cover analysis and identification of polymers/copolymers and additives in components of automobiles, tires, packaging materials, textile fibers, coatings, half-finished products for electronics, paints or varnishes, lacquers, leather, paper or wood products, food, pharmaceuticals, surfactants and fragrances.

In earlier publications of the author (Kusch, 1996; Kusch et al., 2005), the analysis and identification of degradation products of commercially available synthetic polymers and copolymers by using analytical pyrolysis hyphenated to gas chromatography/FID and gas chromatography/mass spectrometry have been presented. In this chapter, examples of application of this analytical technique for identification of different polymeric materials are demonstrated.

2. Experimental

2.1 Instrumentation

Approximately 100 – 200 µg of solid sample were cut out with scalpel and inserted without any further preparation into the bore of the pyrolysis solids-injector and then placed with the plunger on the quartz wool of the quartz tube of the furnace pyrolyzer *Pyrojector II*™ (S.G.E., Melbourne, Australia) (Figs. 1–2). The pyrolyzer was operated at a constant temperature of 550 °C, 600 °C or 700 °C. The pressure of helium carrier gas at the inlet to the furnace was 95 kPa. The pyrolyzer was connected to a *7890A* gas chromatograph with a series *5975C* quadrupole mass spectrometer (Agilent Technologies Inc., Santa Clara, CA, U.S.A.) operated in electron impact ionization (EI) mode. Two fused silica capillary columns (*1*) 60 m long, 0.25 mm I. D. and (2) 59 m long, 0.25 mm I. D. with *DB-5ms* stationary phase, film thickness 0.25 µm were used. Helium, grade 5.0 (Westfalen AG, Münster, Germany) was used as a carrier gas. The gas chromatographic (GC) conditions were as follow:

- (1) programmed temperature of the capillary column from 60 °C (1 min hold) at 7 °C min^{-1} to 280 °C (hold to the end of analysis) and programmed pressure of helium from 122.2 kPa (1 min hold) at 7 kPa/min to 212.9 kPa (hold to the end of analysis),
- (2) programmed temperature of the capillary column from 75 °C (1 min hold) at 7 °C min^{-1} to 280 °C (hold to the end of analysis) and programmed pressure of helium from 122.2 kPa (1 min hold) at 7 kPa/min to 212.9 kPa (hold to the end of analysis),
- (3) programmed temperature of the capillary column from 60 °C (1 min hold) at 7 °C min^{-1} to 280 °C (hold to the end of analysis) and constant helium flow of 1 cm^3 min^{-1} during the whole analysis.

The temperature of the split/splitless injector was 250 °C and the split ratio was 20 : 1. The transfer line temperature was 280 °C. The EI ion source temperature was kept at 230 °C. The ionization occurred with a kinetic energy of the impacting electrons of 70 eV. The quadrupole temperature was 150 °C. Mass spectra and reconstructed chromatograms (total ion current [TIC]) were obtained by automatic scanning in the mass range *m/z* 35 - 750 u. GC/MS data were processed with the *ChemStation* software (Agilent Technologies) and the *NIST 05* mass spectra library (Agilent Technologies).

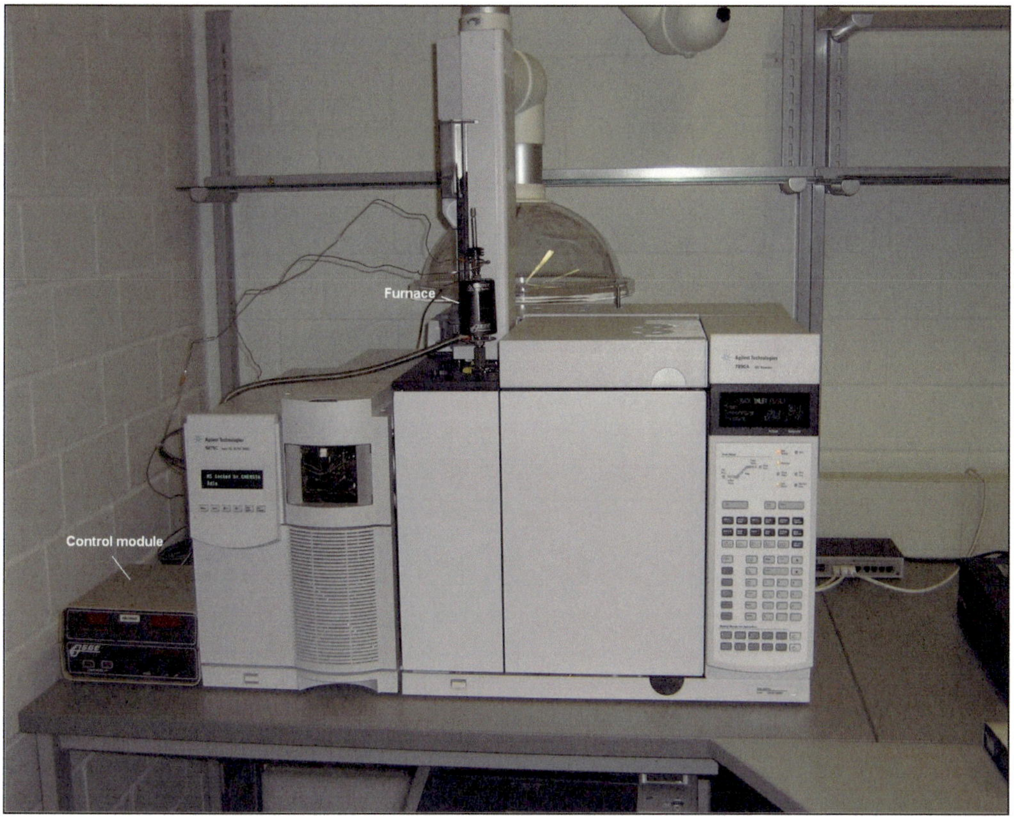

Fig. 1. Pyrolysis-GC/MS system used in this work equipped with a furnace pyrolyzer *Pyrojector II*™ (S.G.E.), control module (S.G.E.), a *7890A* gas chromatograph and a series *5975C* quadrupole mass spectrometer (Agilent Technologies).

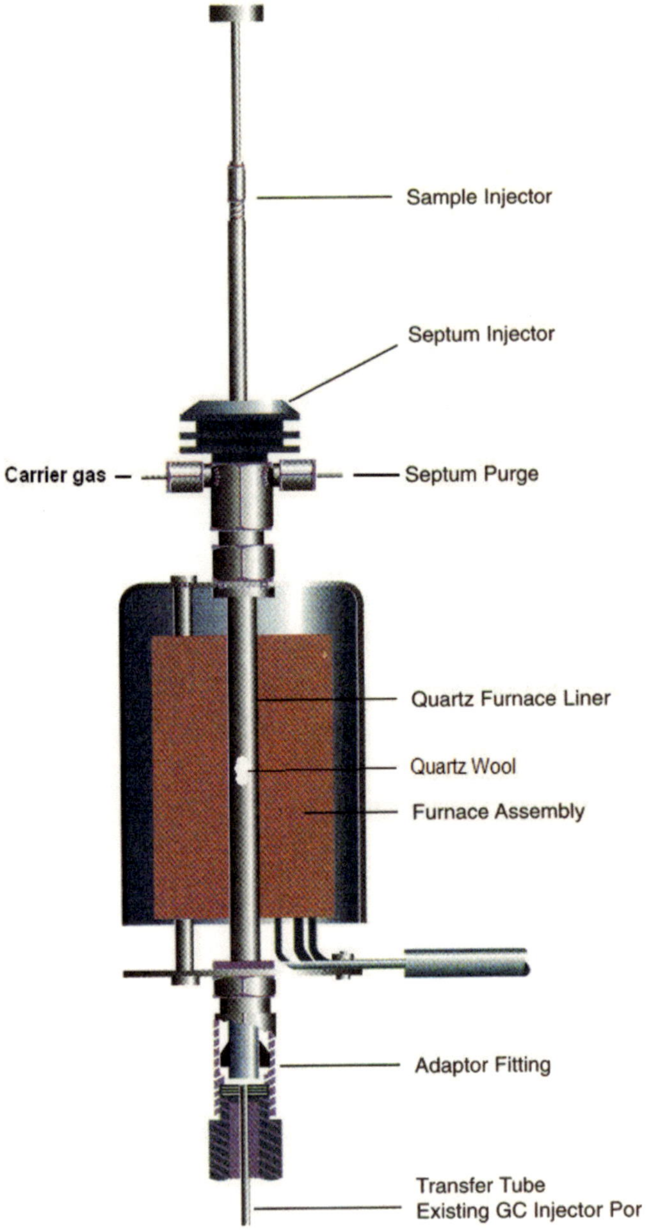

Fig. 2. Schematic view of the furnace pyrolyzer *Pyrojector II*™ (S.G.E., Melbourne, Australia).

2.2 Samples

Samples of an unknown industrial plastic, a valve rubber, a car tire rubber, membranes of hydraulic cylinders from the automotive industry, an O-ring seal, dental filling material and recycled polyethylene were used in the investigation.

3. Results and discussion

3.1 Analytical pyrolysis of synthetic organic polymers/copolymers

Pyrolysis-gas chromatography/mass spectrometry (Py-GC/MS) is used to characterize the structure of synthetic organic polymers and copolymers, polymer blends, biopolymers and natural resins. The traditional physical methods may only be applied to the analysis of technical organic polymers. Several chemical analysis techniques like UV-, FTIR- and NMR-spectroscopy, thermogravimetric analysis (TGA), size-exclusion chromatography (SEC, GPC), headspace-GC (HS-GC) or recently solid-phase microextraction (SPME) with GC/MS (Kusch & Knupp, 2002, 2004; Hakkarainen, 2008) have been established during the last four decades for characterization of macromolecules. These non-destructive methods offer information about functional groups, structural elements, thermal stability, molecular weight, and volatile components.

Pyrolysis-GC/MS is a destructive analytical technique. The most frequent use of this technique is the analysis of polymers and copolymers. Typical fields of interest and application (Hummel & Fischer, 1974; Hummel & Scholl, 1988; Hallensleben & Wurm, 1989; Moldoveanu, 2005; Kusch et al., 2005; Wampler, 2007) are:

- polymer identification by comparison of pyrograms and mass spectra with known references,
- qualitative analysis and structural characterization of copolymers, sequence statistics of copolymers, differentiation between statistical and block polymers,
- determination of the (micro) structure of polymers (degree of branching and cross-linking, compositional analysis of copolymers and blends, co-monomer ratios, sequence distributions, analysis of end-groups),
- determination of the polymers steric structure (stereoregularity, tacticity, steric block length, and chemical inversions),
- investigation of thermal stability, degradation kinetics and oxidative thermal decomposition of polymers and copolymers,
- determination of monomers, volatile organic compounds (VOC), solvents and additives in polymers,
- kinetic studies,
- quality control,
- quantification.

Starting at the University of Cologne (Germany) in the 1960s, Hummel and co-workers studied the process of thermal decomposition of polymers in detail using Py-GC/MS and Py-GC/FTIR (Hummel & Fischer, 1974; Hummel & Scholl, 1988). In general, decomposition proceeds through radical formation, which, due to the high reactivity of radicals, initiates numerous consecutive and parallel reactions. The authors summarized the main pathways of polymer decomposition in four categories:

1. retropolymerization from the end of the polymer-chain, predominantly forming monomers [*e.g.* poly(methyl methacrylate) (PMMA), poly-*a*-methylstyrene],
2. statistical chain scission followed by:

- retropolymerization from radical bearing chain ends (*e.g.* polyisobutylene, polystyrene),
- radical transfer and disproportionation (*e.g.* polyethylene, isotactic polypropylene),

- stabilization of fragments by cyclization (*e.g.* polydimethylsiloxane),
3. splitting side-chain leaving groups (*e.g.* polyvinylchloride, polyacrylonitrile, polyacrylate)
4. intramolecular condensation reactions with loss of smaller molecules (*e.g.* phenol-epoxide resins).

However, this classification is restricted to homogeneous polymers. The situation is more complex in copolymers, depending on the applied monomers and their linking (Hummel & Fischer, 1974; Hummel & Scholl, 1988; Kusch & Knupp, 2007).

In the following, practical application of the analytical pyrolysis connected to GC/MS for the identification of chemical structure and additives of an unknown industrial plastic, rubber products, dental filling material or recycled polyethylene will be presented. The possibility of using this technique in the failure analysis will be also demonstrated.

3.1.1 Identification of an unknown plastic sample

A sample of an industrial plastic was pyrolyzed at 600 °C and 700 °C, respectively in order to identify its composition. The total ion chromatograms (pyrograms) obtained for both pyrolysis temperatures were similar. Figure 3 shows the Py-GC/MS chromatogram of the sample pyrolyzed at 600 °C. Based on the decomposition products (Table 1), the plastic was identified as flexible poly(vinyl chloride) (PVC) with di-(2-ethylhexyl)phthalate (DEHP) plasticizer. The main features of the pyrolysis of PVC are the formation of hydrochloride (HCl) and the formation of unsaturated aliphatic and aromatic hydrocarbons. Benzene is the major pyrolysis product of PVC. Other aromatic components like toluene and the PAHs (polycyclic aromatic hydrocarbons) are also present (Table 1). This is the result of the formation of double bonds by the elimination of HCl from the PVC macromolecules, followed by the breaking of the carbon chain with or without cyclization (Moldoveanu, 2005). The formation of aromatic compounds in PVC pyrolysis is schematically exemplified for benzene in the same monograph (Moldoveanu, 2005).

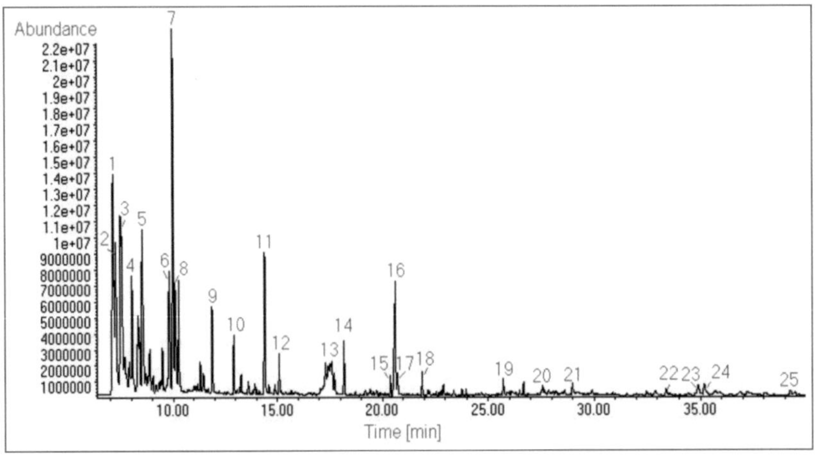

Fig. 3. Pyrolysis-GC/MS chromatogram of a plastic sample pyrolyzed at 600 °C, identified as flexible poly(vinyl chloride) (PVC). For peak identification, see Table 1. GC column 1, GC conditions 2.

The decomposition of the plasticizer di-(2-ethylhexyl)phthalate (DEHP) leads to the formation of 2-ethyl-1-hexene (RT = 9.98 min), 2-ethylhexanal (RT = 12.93 min), 2-ethyl-1-hexanol (RT = 14.40 min) and phthalic anhydride (RT = 20.63 min) (Table 1 and Fig. 3).

The obtained EI mass spectra of the compounds are shown in Fig. 4, while the chemical reaction of the thermal degradation of di-(2-ethylhexyl)phthalate is presented in Fig. 5. The reaction scheme is consistent with the previously published work (Bove & Dalven, 1984).

Peak	Retention time (RT) [min]	Pyrolysis products of polyvinyl chloride (PVC) with DEHP plasticizer at 600 °C
1	7.14	Hydrochloride/Propene
2	7.25	1-Butene
3	7.49	1,4-Pentadiene
4	8.05	2,4-Hexadiene
5	8.53	Benzene
6	9.81	Toluene
7	9.98	2-Ethyl-1-hexene
8	10.09	3-Methyl-3-heptene
9	11.89	Styrene
10	12.93	2-Ethylhexanal
11	14.40	2-Ethyl-1-hexanol
12	15.10	Indene
13	17.57	Benzoic acid
14	18.18	Naphthalene
15	20.38	2-Methylnaphthalene
16	20.63	Phthalic anhydride
17	20.73	1-Methylnaphthalene
18	21.91	Biphenyl
19	25.71	Fluorene
20	27.56	1,2-Diphenylethylene
21	28.97	Anthracene
22	33.45	Fluoranthene
23	34.89	Benzofluorene isomer
24	35.19	Benzofluorene isomer
25	39.35	Benz[a]anthracene

Table 1. Pyrolysis products of flexible polyvinyl chloride (PVC) with DEHP plasticizer at 600 °C. Peak numbers as in Fig. 3. GC column 1, GC conditions 2.

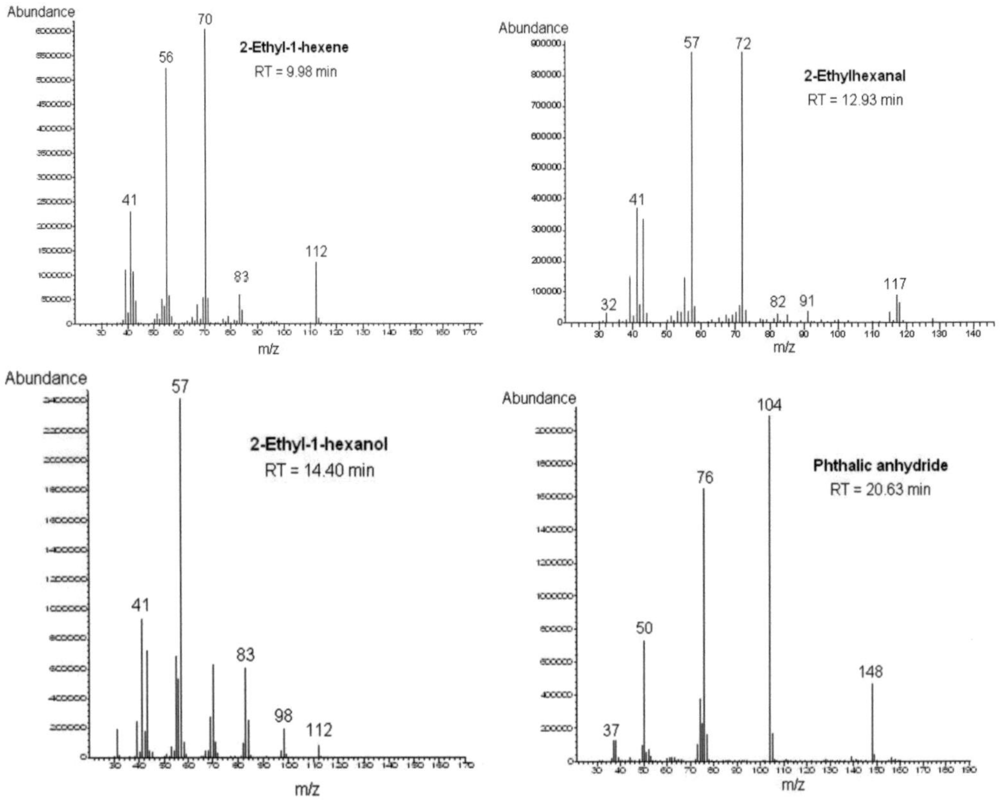

Fig. 4. EI ionization mass spectra at 70 eV of substances formed by the pyrolysis of the plasticizer di-(2-ethylhexyl)phthalate (DEHP) at 600 °C; RT = retention time of substance, as in Fig. 3 and in Table 1.

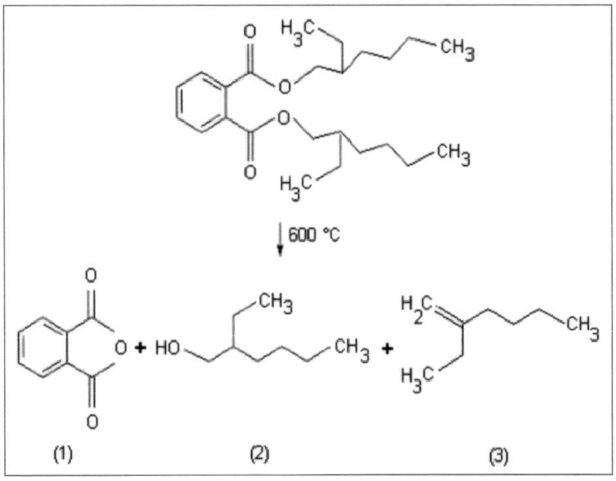

Fig. 5. Chemical reaction of the thermal decomposition of di-(2-ethylhexyl)phthalate (DEHP) at 600 °C leading to obtain (*1*) phthalic anhydride, (*2*) 2-ethyl-1-hexanol and (*3*) 2-ethyl-1-hexene.

3.1.2 Identification of rubber materials

The study of rubbers is the oldest application of analytical pyrolysis-GC (Kusch et al., 2005). Rubbers are frequently filled with opaque materials like carbon black, making them difficult for analysis by spectroscopy. Furthermore, cross-linking makes them insoluble and thus many of the traditional analytical tools for organic analysis are difficult or impossible to apply. Each rubber compound in a tire contains rubber polymers, sometimes one, but often a blend of two or more. The more commonly used polymers are natural rubber (NR, polyisoprene), synthetic polyisoprene (IR), polybutadiene (BR), and styrene-butadiene copolymers (SBR). SBR is widely used for tread compounds of a tire, generally in the tire treads of passenger cars, and NR in the relatively large-sized tires of buses and trucks etc. Inner tubes are usually based on butyl-rubber, a copolymer of isobutylene with a small proportion of isoprene.

Nitrile rubber (NBR) was invented at about the same time as SBR in the German program, in the end of the 1920s, to find substitutes for natural rubber (Graves, 2007). These rubbers are copolymers of acrylonitrile and 1,3-butadiene, containing 15 - 40% acrylonitrile. The major applications for this material are in areas requiring oil and solvent resistance. The largest market for nitrile rubber is in the automotive area because of its solvent and oil resistance. Major end uses are for hoses, fuel lines, O-rings, gaskets and seals. In blends with PVC and ABS, nitrile rubber acts as an impact modifier. Some nitrile rubber is sold in latex form for the production of grease-resistant tapes, gasketing material and abrasive papers. Latex also is used to produce solvent resistant gloves (Graves, 2007).

The criteria for assessing the quality of rubber materials are the polymer/copolymer composition and the additives. Commercial plastics and rubbers always contain a number of additives that are included to give particular physical and/or chemical properties. These additives include plasticizers, extender oils, carbon black, inorganic fillers, antioxidants, heat- and light stabilizers, tackifying resins, processing aids, cross-linking agents, accelerators, retarders, adhesives, pigments, smoke and flame retardants, and others (Hakkarainen, 2008).

The examined rubber materials in our laboratory were a valve rubber, a car tire rubber and an O-ring seal. On the one hand, based on the decomposition products at 700 °C (Fig. 6 and Table 2), the valve rubber was identified as a blend of polyisoprene (NR) and poly(styrene-*co*-butadiene) (SBR). On the other hand, the car tire rubber was determined as the mixture of polybutadiene (BR) and poly(styrene-*co*-butadiene) (SBR) (Kusch & Knupp, 2009).

Compounds, such as additives, are usually applied in low concentrations in plastic and rubbers. They are often recognizable as small irregularities within the characteristic pyrogram of the pyrolyzed material. By using the GC column 1 and the GC conditions 1, it was possible to identify the organic additives in both rubber samples. The peak of benzothiazole (Fig. 6, Peak 21, RT = 20.37 min) corresponds to the product of the thermal decomposition of benzothiazole-2-thiol (2-mercaptobenzothiazole, 2-MBT). In the rubber industry benzothiazole-2-thiol is used as vulcanisation accelerator and as antioxidant.

Furthermore, the antioxidant 2,6-bis-(1,1-dimethylethyl)-4-methylphenol (BHT) was detected in the pyrolyzate of the valve rubber (Fig. 6, Peak 29, RT = 25.50 min). The mass spectra obtained for both additives are shown in Fig. 7.

Peak	Retention time (RT) [min]	Pyrolysis products of the valve and the car tire rubbers at 700 °C
1	8.20	2-Butene
2	8.55	2-Methyl-1,3-butadiene
3	9.34	3-Methyl-2-pentene (isoprene)
4	9.75	5-Methyl-1,3-cyclopentadiene
5	10.08	Benzene
6	11.28	1-Methyl-1,4-cyclohexadiene
7	11.84	Toluene
8	13.51	Ethylbenzene
9	13.69	*p*-Xylene
10	14.08	Styrene
11	15.19	*m*-Ethyltoluene
12	15.52	*a*-Methylstyrene
13	15.81	1,2,4-Trimethylbenzene (pseudocumene)
14	16.28	1,2,3-Trimethylbenzene (hemimelitene)
15	16.78	Indene
16	17.43	*o*-Isopropenyltoluene
17	18.02	1,2,4,5-Tetramethylbenzene (durene)
18	18.75	3-Methylindene
19	18.86	2-Methylindene
20	19.58	Naphthalene
21	20.37	Benzothiazole
22	21.83	2-Methylnaphthalene
23	22.17	1-Methylnaphthalene
24	23.43	Biphenyl
25	24.06	Dimethylnaphthalene isomer
26	24.32	Dimethylnaphthalene isomer
27	24.43	Dimethylnaphthalene isomer
28	25.48	3-Methyl-1,1′-biphenyl
29	25.50	2,6-Bis-(1,1-dimethylethyl)-4-methylphenol (BHT)
30	26.52	1,6,7-Trimethylnaphthalene
31	27.75	Fluorene
32	29.88	1,2-Diphenylethylene (stilbene)
33	31.61	Anthracene
34	34.79	2-Phenylnaphthalene

Table 2. Pyrolysis products of the valve rubber and the car tire rubber materials at 700 °C. Peak numbers as in Fig. 6. GC column 1, GC conditions 1.

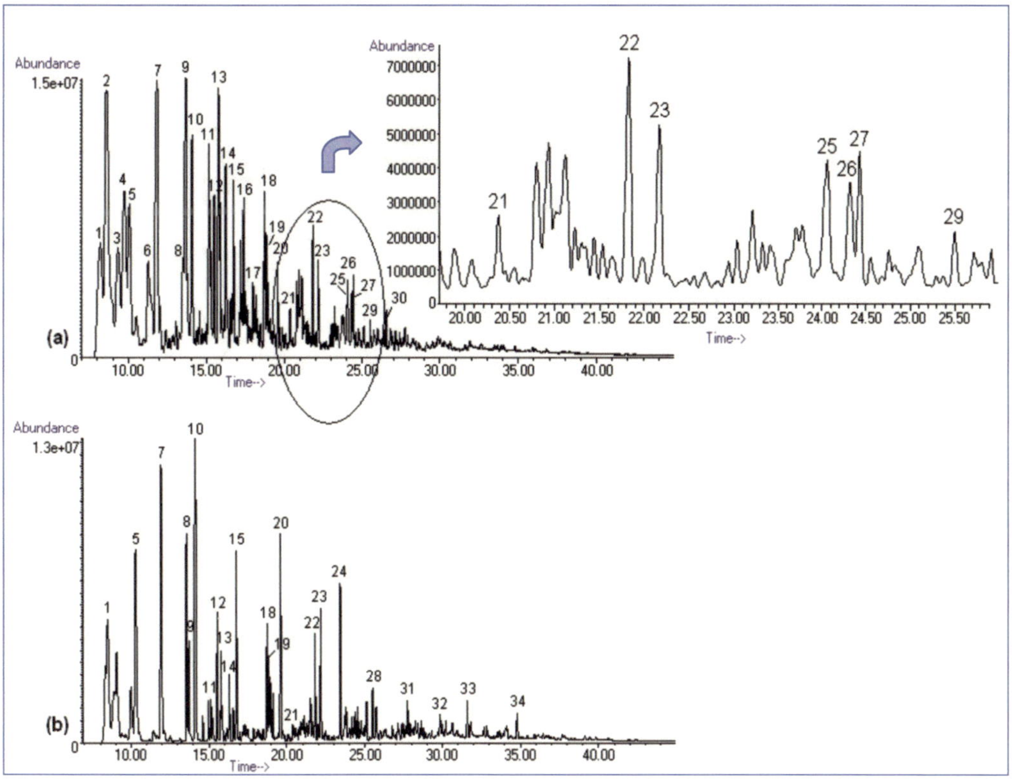

Fig. 6. Pyrolysis-GC/MS chromatograms (TIC) of the rubber materials at 700 °C; (*a*) NR/SBR, (*b*) BR/SBR. Peak 21 = benzothiazole (RT = 20.37 min), peak 29 = BHT (RT = 25.50 min). For other peak identification, see Table 2. GC column 1, GC conditions 1.

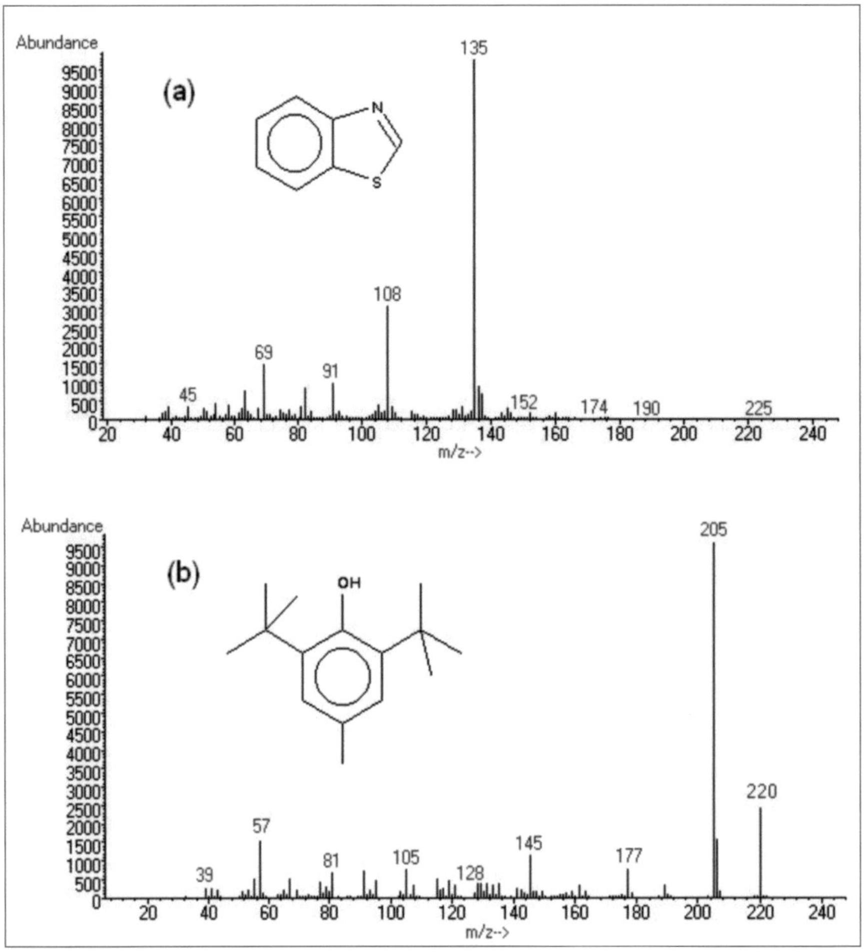

Fig. 7. EI mass spectra at 70 eV of the rubber additives pyrolyzed at 700 °C: (*a*) benzothiazole (thermal degradation product of benzothiazole-2-thiol), (*b*) 2,6-bis-(1,1-dimethylethyl)-4-methylphenol (BHT).

The pyrolysis–GC/MS chromatogram of the O-ring sample at 700 °C is shown in Fig. 8, while the identified degradation products are summarized in Table 3. Based on the decomposition products at 700 °C (Fig. 8 and Table 3), the O-ring rubber was identified as poly(acrylonitrile-*co*-butadiene) (NBR) with high content of the plasticizer di-(2-ethylhexyl)phthalate (DEHP) and with the vulcanization accelerator benzothiazole-2-thiol (2-mercaptobenzothiazole, 2-MBT). The analysis of pyrolyzate of poly(acrylonitrile-*co*-butadiene) indicated in Table 3 shows the presence of compounds generated from the acrylomitrile sequences (acrylonitrile, methacrylonitrile, benzonitrile), from the butadiene sequences (1,3-butadiene) and from the additives of the copolymer (2-ethyl-1-hexene, 2-ethyl-1-hexanol, phthalic anhydride and benzothiazol).

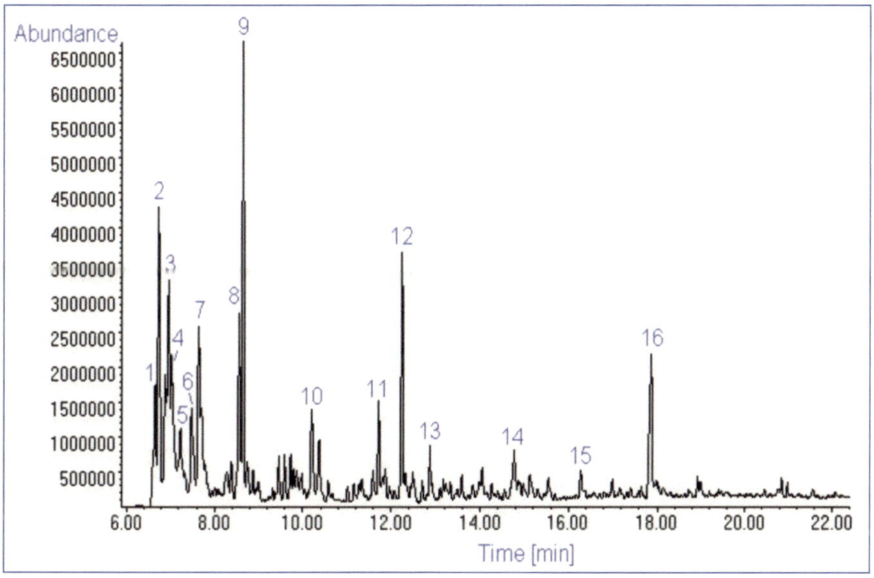

Fig. 8. Pyrolysis–GC/MS chromatogram (TIC) of the O-ring rubber at 700 °C identified as NBR with additives. For peak identification, see Table 3. GC column 2, GC conditions 2.

Peak	Retention time (RT) [min]	Pyrolysis products of the O-ring rubber at 700 °C
1	6.66	Propylene
2	6.74	1,3-Butadiene
3	6.97	Acrylonitrile
4	7.04	1,3-Cyclopentadiene
5	7.24	Methacrylonitrile
6	7.50	1,4-Cyclohexadiene
7	7.65	Benzene
8	8.56	Toluene
9	8.65	2-Ethyl-1-hexene
10	10.20	Styrene
11	11.73	Benzonitrile
12	12.24	2-Ethyl-1-hexanol
13	12.87	Indene
14	14.76	3-Methylindene
15	16.29	Benzothiazole
16	17.87	Phthalic anhydride

Table 3. Pyrolysis products of the O-ring rubber at 700 °C. Peak numbers as in Fig. 8. GC column 2, GC conditions 2.

3.1.3 Application of pyrolysis-GC/MS in the dentistry

A number of dental filling materials are presently available for tooth restorations. The four main groups of these materials, which dentists have used for about 35 years, are the conventional glass-ionomer cements, resin-based composites, resin-modified glass-ionomer cements and polyacid-modified resinous composities (Rogalewicz et al., 2006a). Light-curing glass-ionomer cements contain polyacrylic acid, chemically and/or photo-curing monomers (multifunctional methacrylates, like triethylene glycol dimethacrylate or 2-hydroxyethyl methacrylate), an ion-leaching glass and additives (initiators, inhibitors, stabilizers and others) (Rogalewicz et al., 2006a). Resin-modified glass-ionomer cements are now widely used in dentistry as direct filling materials, liners, bases, luting cements and fissure sealants (Rogalewicz et al., 2006b). These materials mainly consist of polymer matrix and glass-ionomer parts. The polymer matrix is based on a monomer system and different multifunctional methacrylates with additives (Rogalewicz et al., 2006b). Methacrylic monomers, like bisphenol A glycidyl methacrylate (Bis-GMA), urethane dimethacrylate (UDMA), triethylene glycol dimethacrylate (TEGDMA) and 2-hydroxyethyl methacrylate (HEMA) are the main components of resin-based dental filling materials. The presence of additives such as initiators, activators, inhibitors and plasticizers in uncured dental material mixture is necessary (Rogalewicz et al., 2006b).

Figure 9 shows the total ion current Py–GC/MS chromatogram of commercially light-curing dental filling material pyrolyzed at 550 °C. The pyrolysis products identified by using mass spectra library *NIST 05* are summarized in Table 4. The carbon dioxide (RT = 5.85 min) identified in pyrolyzate is formed from polyacrylic acid. The identified substances, like HEMA (RT = 13.65 min), EGDMA (RT = 19.48 min) and TEDMA (RT = 28.72 min) are known as standard composites of dental filling materials. Other compounds in Table 4, such as bisphenol A (RT = 33.10 min) or bisphenol A diglycidylether (RT = 42.42 min) are probably formed by thermal degradation of bisphenol A diglycidyl mono- or dimethacrylates. The presence of the additives, like antioxidant BHT [2,6-bis-(1,1-

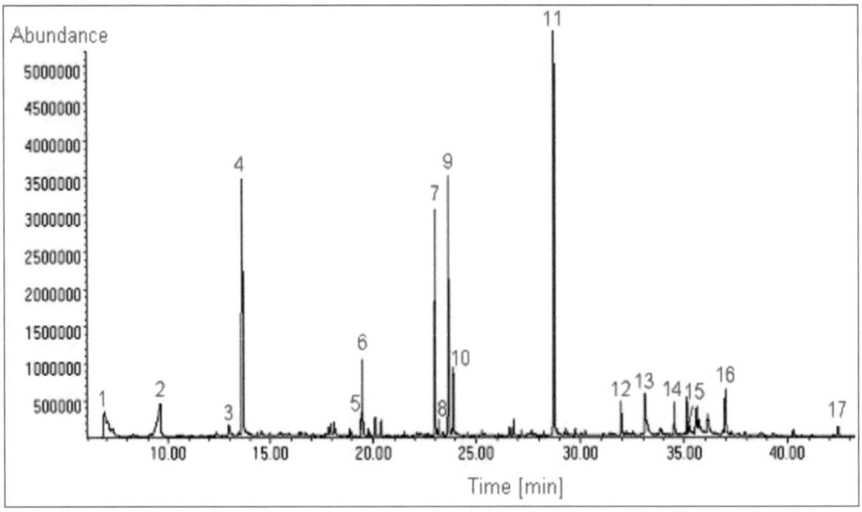

Fig. 9. Pyrolysis-GC/MS chromatogram of commercially light-curing dental filling material pyrolyzed at 550 °C. For peak identification, see Table 4. GC column 2, GC conditions 1.

dimethylethyl)-4-methylphenol, RT = 23.17 min] or the UV-absorber drometrizol (RT = 31.95 min) was also confirmed. The triphenylantimony (RT = 34.55 min) identified in pyrolyzate is used as catalyst in the UV-induced polymerisation.

Peak	Retention time (RT) [min]	Pyrolysis products of the dental filling material at 550 °C
1	5.85	Carbon dioxide
2	9.62	Methacrylic acid
3	12.96	Phenol
4	13.65	2-Hydroxyethyl methacrylate (HEMA)
5	19.40	4-Isopropenylphenol
6	19.48	Ethylene glycol dimethacrylate (EGDMA)
7	23.00	Not identified
8	23.17	2,6-Bis-(1,1-dimethylethyl)-4-methylphenol (BHT)
9	23.65	Not identified
10	23.89	Triethylene glycol (TEG)
11	28.72	Triethylene glycol dimethacrylate (TEDMA)
12	31.95	Drometrizol (Tinuvin-P, UV-absorber)
13	33.10	4,4´-Dihydroxy-2,2-diphenylpropane (Bisphenol A)
14	34.55	Triphenylantimony
15	35.25	Tetraethylene glycol dimethacrylate
16	36.98	4,4´-(1-Methylethylidene)-bis-[2,6-dimethylphenol]
17	42.42	Bisphenol A diglycidylether

Table 4. Pyrolysis products of commercially light-curing dental filling material at 550 °C. Peak numbers as in Fig. 9. GC column 2, GC conditions 1.

3.1.4 Environmental application of pyrolysis-GC/MS

Analytical pyrolysis has been used to study a variety of environmental samples including fossil fuel source rocks, natural resins or aquatic and terrestrial natural organic matter (White et al., 2004; Kusch et al., 2008a). Much work has been published on the analysis of soil (Tienpomt et al., 2001; Evans et al., 2003; Campo et al., 2011). Pyrolysis-GC/MS was also used in plastic recycling for analysis of products generated from the thermal and catalytic degradation of pure and waste polyolefins (Aguado et al., 2007).

Plastic recycling can be divided into mechanical (material) recycling, feedstock (chemical) recycling and energy recovery. In case of material recycling, the plastic waste is washed, ground and used in the plastic processing industry as a raw material. The chemical structure of the preserved material (re-granulate) is maintained. In case of chemical recycling, the residual polymers are converted catalytically into their monomers. The resulting monomers are then used in the manufacture of new plastics. In the energy recovery the plastic waste is used in cement- or in steel industry as an energy carrier.

The following example presents that pyrolysis–GC/MS allows the identification of contaminants in low-density polyethylene (LDPE) re-granulate from mechanical recycling process. Figure 10 shows the obtained Py–GC/MS chromatogram of the LDPE pellets at 700 °C. The pyrogram consists of serial triplet-peaks of straight-chain aliphatic C_3-C_{33} hydrocarbons, corresponding to α,ω-alkadienes, α-alkenes and *n*-alkanes, respectively, in the order of the increasing *n+1* carbon number in the molecule. Such elution pattern is characteristic for the pyrolysis of polyethylene. For example, the triplet-peaks in Fig. 10 of C_{10} hydrocarbons correspond to 1,9-decadiene (RT = 14.01 min), 1-decene (RT = 14.16 min) and *n*-decane (RT = 14.33 min), respectively. The identification of the compounds was carried out by comparison of retention times and mass spectra of standard substances, study of the mass spectra, and comparison with data in the *NIST 05* mass spectra library. The substances identified in pyrolyzate, like 2,4-dimethyl-1-heptene (RT = 11.12 min) and styrene (RT = 12.39 min) (Fig. 10) indicate the contamination of the recycled polyethylene with polypropylene and polystyrene, respectively. The detected palmitic acid (RT = 31.27 min) and oleic acid (RT = 33.99 min) suggest the presence of plant residues in the investigated re-granular material.

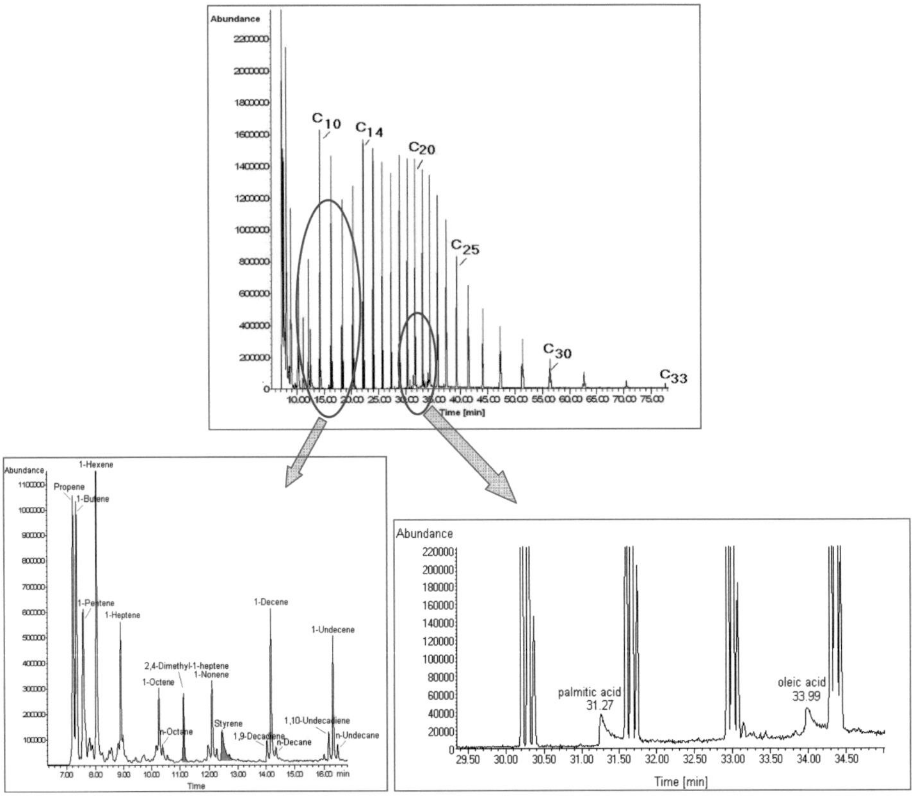

Fig. 10. Full view (top) and time windows between 6.2 – 16.8 min (bottom left) and 29.3 – 35.0 min (bottom right) of the Py–GC/MS chromatogram of recycled HDPE pyrolyzed at 700 °C. GC column 1, GC conditions 3.

3.1.5 Application of pyrolysis–GC/MS in the failure analysis

The increasing use of polymeric materials in the automotive industry requires sensitive and reliable methods for its analysis. For the failure analysis in motor vehicles there is often lack of information about the component itself, such as chemical composition, temperature resistance, possible contaminants or mechanical properties. The damage range is usually limited and not always homogeneous. There are often only small amounts of samples available to clarify the damage, which may be important for recognizing the cause of damage. Traditional analytical techniques used for characterization of polymers/copolymers, such as thermal analysis (TA) and Fourier transform infrared spectroscopy (FTIR) have limitations or are not sufficiently sensitive to demonstrate the change of the structure and the resulting dysfunction of used materials.

Previous work of the author (Kusch et al., 2008b) presents the application examples of Py-GC/MS in the failure analysis of various plastic or metal components and assemblies from the automotive industry. In the following case a glue residue on the surface of a rubber membrane from the hydraulic cylinder from the automotive industry was identified. Figure 11 shows the pyrolysis-GC/MS chromatograms at 700 °C, obtained from the new and operationally stressed rubber membrane coated with the glue residue, respectively.

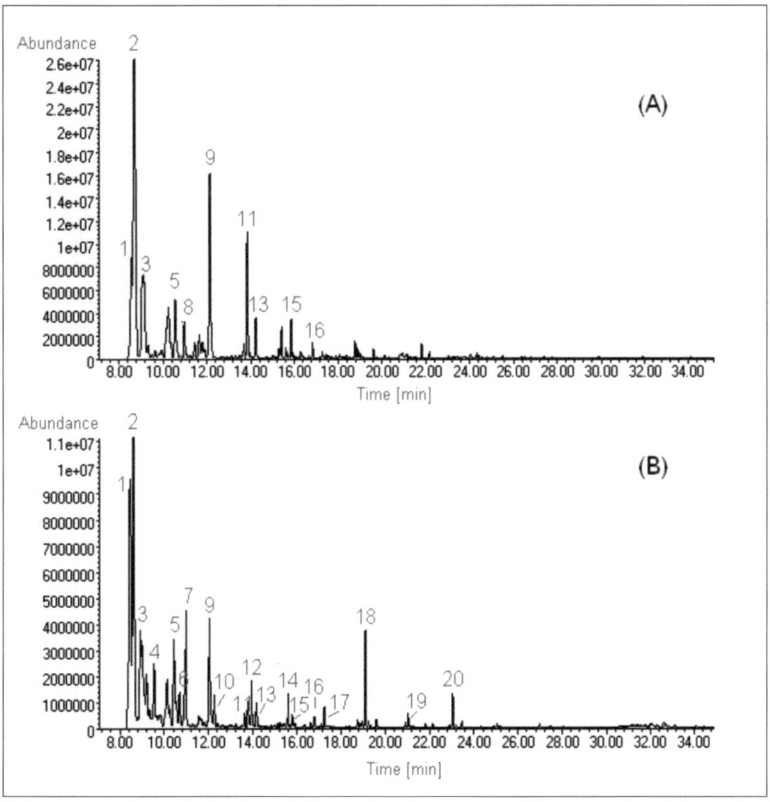

Fig. 11. Pyrolysis-GC/MS chromatograms at 700 °C, obtained from the new rubber membrane (A) and from the operationally stressed rubber membrane coated with glue residue (B). For peak identification, see Table 5. GC column 1, GC conditions 1.

The pyrolysis products identified by using mass spectra library *NIST 05* are summarized in Table 5. On the one hand, the pyrogram from the new rubber membrane obtained in Fig. 11 (A) consists of the characteristic fragments formed by the thermal degradation of poly(styrene-*co*-butadiene) (SBR). On the other hand, as can be seen from Fig. 11 (B) and from Table 5, the chemical structures of the identified substances of the operationally stressed rubber membrane coated with the glue residue build three groups of compounds: (*1*) *n*-alkenes and *n*-alkanes, (*2*) aromatics, and (*3*) ester. The presence of *n*-alkenes and heavy *n*-alkanes in pyrolyzate is characteristic for the pyrolysis of mineral oil (motor oil). The methyl methacrylate monomer (RT = 10.94 min) was produced by the retropolymerization of poly(methyl methacrylate) (PMMA) at 700 °C. Other peaks, like 1,3-butadiene (RT = 8.61 min), benzene (RT = 10.43 min), toluene (RT = 12.04 min), *p*-xylene (RT = 13.84 min), styrene (RT = 14.22 min), trimethylbenzene (RT = 15.87 min) and indene (RT = 16.85 min) are typical pyrolysis products of SBR. Based on the obtained results, it was possible to identify the glue coating from the surface of the rubber membrane from the automotive industry as a mixture of PMMA and mineral oil. The analytical results were then used for troubleshooting and remedial action of the technological process.

Peak	Retention time (RT) [min]	Pyrolysis products of the rubber membrane with glue coating at 700 °C	Pyrolyzed material
1	8.47	Propene	SBR, mineral oil
2	8.61	1-Butene/1,3-Butadiene	SBR, mineral oil
3	8.95	1-Pentene	Mineral oil
4	9.55	1-Hexene	Mineral oil
5	10.43	Benzene	SBR
6	10.72	1-Heptene	Mineral oil
7	10.94	Methyl methacrylate	PMMA
8	11.00	2,4-Dimethyl-1,3-pentadiene	SBR
9	12.04	Toluene	SBR
10	12.26	1-Octene	Mineral oil
11	13.84	*p*-Xylene	SBR
12	13.95	1-Nonene	Mineral oil
13	14.22	Styrene	SBR
14	15.56	1-Decene	Mineral oil
15	15.87	Trimethylbenzene	SBR
16	16.85	Indene	SBR
17	17.25	1-Undecene	Mineral oil
18	19.10	1-Dodecene	Mineral oil
19	21.04	1-Tridecene	Mineral oil
20	23.05	1-Tetradecene	Mineral oil
21	30.0 - 34.0	Heavy *n*-alkanes	Mineral oil

Table 5. Pyrolysis products of the rubber membrane with glue coating at 700 °C. Peak numbers as in Fig. 11. GC column 1, GC conditions 1.

4. Conclusions

Pyrolysis-GC/MS has proved as a valuable technique for the analysis and identification of organic polymeric materials in the plastic and rubber industry, dentistry, environmental protection and in the failure analysis. This technique allows the direct analysis of very small sample amounts without the need of time-consuming sample preparation.

5. Acknowledgements

The author thanks the *Kompetenzplattform "Polymere Materialien"* at the Bonn-Rhine-Sieg University of Applied Sciences in Rheinbach and the firm *Dr. Obst Technische Werkstoffe GmbH* in Rheinbach, Germany for samples and excellent cooperation.

I also thank my daughter Maria Kusch, M.A. for critical reading of the manuscript.

6. References

Aguado, J.; Serrano, D. P. & San Miguel, G. (2007). Analysis of Products Generated from the Thermal and Catalytic Degradation of Pure and Waste Polyolefins using Py-GC/MS. *Journal of Polymers and the Environment,* Vol. 15, No. 2, pp. 107-118

Bove, J. L. & Dalven, P. (1984). Pyrolysis of Phthalic Acid Esters: Their Fate. *The Science of the Total Environment,* Vol. 36, No. 1, pp. 313-318

Campo, J.; Nierop, K. G. J.; Cammeraat, E.; Andrei, V. & Rubio J. L. (2011). Application of Pyrolysis-Gas Chromatography/Mass Spectrometry to Study Changes in the Organic Matter of Macro- and Microaggregates of a Mediterranean Soil Uppon Heating. *Journal of Chromatography A,* Vol. 1218, No. 30, pp. 4817-4827

Evans, C. J.; Evershed, R. P.; Black, H. I. J. & Ineson, P. (2003). Compound-Specific Stable Isotope Analysis of Soil Mesofauna Using Thermally Assisted Hydrolysis and Methylation for Ecological Investigations. *Analytical Chemistry,* Vol. 75, No. 22, pp. 6056-6062

Graves, D. F. (2007). Rubber, In: *Kent and Riegels Handbook of Industrial Chemistry and Biotechnology, Part 1,* J. A. Kent (Ed.), pp. 689-718, Springer Science + Business Media, ISBN 978-0-387-27842-1, e-ISBN 978-0-387-27843-8, New York, U.S.A.

Hakkarainen, M. (2008). Solid-Phase Microextraction for Analysis of Polymer Degradation Products and Additives, In: *Chromatography for Sustainable Polymeric Materials,* A.-C. Albertsson & M. Hakkarainen (Eds.), pp. 23-50, Springer Verlag, ISBN 978-3-540-78762-4, e-ISBN 978-3-540-78763-1, Berlin, Heidelberg, Germany

Hallensleben, M. L. & Wurm, H. (1989). Polymeranalytik. *Nachrichten aus Chemie, Technik und Laboratorium,* Vol. 37, No. 6, pp. M1-M45

Hummel, D. O. & Fischer, H. (1974). *Polymer Spectroscopy,* D. O. Hummel (Ed.), Verlag Chemie, ISBN 3527254110, Weinheim, Germany

Hummel, D. O. & Scholl, F. (1988). *Atlas der Polymer- und Kunststoffanalyse,* D. O Hummel & F. Scholl (Eds.), Hanser Verlag, ISBN 3-446-12586-8, München, Germany

Kusch, P. (1996). Application of the Curie-Point Pyrolysis – High Resolution Gas Chromatography for Analysis of Synthetic Polymers and Copolymers. *Chemia Analityczna (Warsaw),* Vol. 41, No. 2, pp. 241- 252

Kusch, P. & Knupp, G. (2002). Analysis of Residual Styrene Monomer and Other Volatile Organic Compounds in Expanded Polystyrene by Headspace Solid-Phase

Microextraction Followed by Gas Chromatography and Gas Chromatography/Mass Spectrometry. *Journal of Separation Science,* Vol. 25, No. 8, pp. 539-542

Kusch, P. & Knupp, G. (2004). Headspace-SPME-GC-MS Identification of Volatile Organic Compounds Released from Expanded Polystyrene. *Journal of Polymers and the Environment,* Vol. 12, No. 2, pp. 83-87

Kusch, P.; Knupp, G. & Morrisson, A. (2005). Analysis of Synthetic Polymers and Copolymers by Pyrolysis-Gas Chromatography/Mass Spectrometry, In: *Horizons in Polymer Research,* R. K. Bregg (Ed.), pp. 141-191, Nova Science Publishers, ISBN 1-59454-412-3, New York, U.S.A.

Kusch, P. & Knupp, G. (2007). Identifizierung von Verpackungskunststoffen mittels Pyrolyse und Headspace-Festphasenmikroextraktion in Kombination mit der Gaschromatographie/Massenspektrometrie. *LC·GC Ausgabe in deutscher Sprache,* June, pp. 28-34

Kusch, P.; Fink, W.; Schroeder-Obst, D. & Obst, V. (2008a). Identification of Polymeric Residues in Recycled Aluminium by Analytical Pyrolysis-Gas Chromatography-Mass Spectrometry. *ALUMINIUM International Journal,* Vol. 84, No. 4, pp. 76-79

Kusch, P.; Obst, V.; Schroeder-Obst, D.; Knupp, G. & Fink, W. (2008b). Einsatz der Pyrolyse -GC/MS zur Untersuchung von Polymeren Materialien in der Schadenanalyse in der Automobilindustrie. *LC·GC Ausgabe in deutscher Sprache,* July/August, pp. 5-11

Kusch, P. & Knupp, G. (2009). Identifizierung anhand der Zersetzungsprodukte. *Nachrichten aus der Chemie,* Vol. 57, No. 6, pp. 682-685

Moldoveanu, S. C. (2005). *Analytical Pyrolysis of Synthetic Organic Polymers,* Elsevier, ISBN 0-444-51292-6, Amsterdam, the Netherlands

Rogalewicz, R.; Voelkel, A. & Kownacki, I. (2006a). Application of HS-SPME in the Determination of Potentially Toxic Organic Compounds Emitted from Resin-Based Dental Materials. *Journal of Environmental Monitoring,* Vol. 8, No. 3, pp. 377-383

Rogalewicz, R.; Batko, K. & Voelkel A. (2006b). Identification of Organic Extractables from Commercial Resin-Modified Glass-Ionomers using HPLC-MS. *Journal of Environmental Monitoring,* Vol. 8, No. 7, pp. 750-758

Sobeih, K. L.; Baron, M. & Gonzales-Rodrigues, J. (2008). Recent Trends and Developments in Pyrolysis-Gas Chromatography. *Journal of Chromatography A,* Vol. 1186, No. 1-2, pp. 51-66

Tienpont, B; David, F.; Vanwalleghem, F. & Sandra, P. (2001). Pyrolysis-Capillary Gas Chromatography-Mass Spectrometry for the Determination of Polyvinyl Chloride Traces in Solid Environmental Samples. *Journal of Chromatography A,* Vol. 911, No. 2, pp. 235-247

Wampler, T. P. (Ed.) (2007). *Applied Pyrolysis Handbook,* second edition, CRC Press, ISBN 1-57444-641-X, Boca Raton, U.S.A.

White, D. M.; Garland, D. S.; Beyer, L. & Yoshikawa, K. (2004). Pyrolysis-GC/MS Fingerprinting of Environmental Samples. *Journal of Analytical and Applied Pyrolysis,* Vol. 71, No. 1, pp. 107-118

17

Gas Chromatograph Applications in Petroleum Hydrocarbon Fluids

Huang Zeng, Fenglou Zou, Eric Lehne, Julian Y. Zuo* and Dan Zhang
Schlumberger DBR Technology Center, Edmonton, AB, Canada

1. Introduction

1.1 Composition of reservoir hydrocarbon fluids

In the petroleum hydrocarbon fluids, the most commonly found molecules are alkanes (linear or branched paraffins), cycloalkanes (naphthenes), aromatic hydrocarbons, or more complicated compounds like asphaltenes. Under surface pressure and temperature conditions, lighter hydrocarbons such as CH_4, C_2H_6, and inorganic compounds such as N_2, CO_2, and H_2S occur as gases, while pentane and heavier ones are in the form of liquids or solids. However, in petroleum reservoir the proportions of gas, liquid, and solid depend on subsurface conditions and on the phase diagram (envelop) of the petroleum mixture. To obtain compositions of a reservoir fluid, a reservoir sample is flashed into gas and liquid phases at ambient conditions. The volume of the flashed gas, and the mass, molar mass and density of the flashed liquid are measured. Then a gas chromatograph is used to analyze compositions of the gas and liquid phases as described briefly below. The recombined compositions based on the gas and liquid according to the measured gas/oil ratio are those of the reservoir fluid.

Generally speaking, crude oils are made of three major groups:

- Hydrocarbon compounds that are made exclusively from carbon and hydrogen;
- Non-hydrocarbon but still organic compounds that contain, in addition to carbon and hydrogen, heteroatoms including sulfur, nitrogen and oxygen;
- Organometallic compounds: organic compounds, normally molecules of porphyrin type that have a metal atom (Ni, V or Fe) attached to them.

1.2 Hydrocarbons

Hydrocarbons are usually made of few groups:

a. linear (or normal) alkanes (paraffins)
b. branched alkanes (paraffins)
c. cyclic alkanes or cycloparaffins (naphthenes)
d. aromatic alkanes (aromatics)

---

* Corresponding Author

From the GC perspective, the analysis of alkanes is performed using a non-polar column and separation is based on boiling point. Normal alkanes boil few degrees higher than their respective branched ones. In Table 1.1 an example illustrating this point is given.

Compound	Boiling point (°C)
n-Octane	126
2-methylheptane	116
3-methylheptane	118
4-methylheptane	117

Table 1.1. Boiling points of octane isomers

From the data above, the branched alkanes are closer together and the corresponding normal alkanes boil at higher temperature. Thus, branched alkanes elute first, followed by the normal alkanes. For GC analysis, it is recommended to integrate the end of an alkane to the end of the next alkane as a family of one particular alkane, as shown in Figure 1.1.

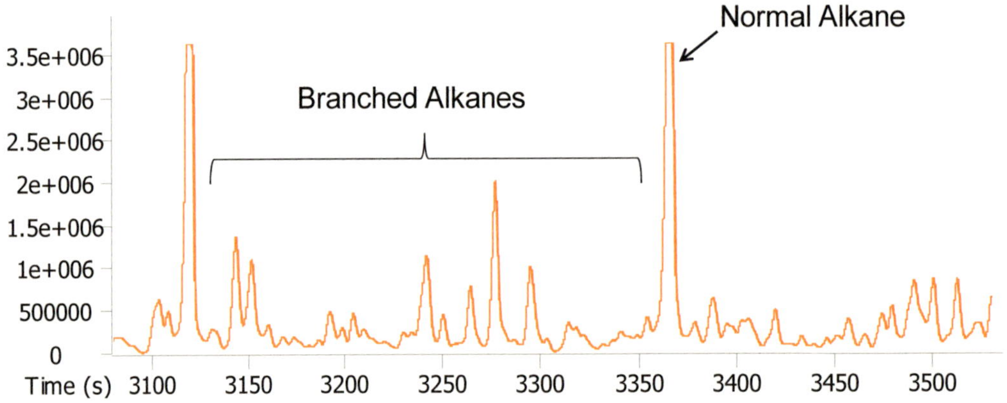

Fig. 1.1. Illustration of integration branched and linear alkanes

In 1873, van der Waals introduced the first cubic equation of state (EOS) by modifying ideal gas law. In 1949, Redlich and Kwong modified the van der Waals EOS which was then modified by Soave (1972). Peng and Robinson (1976) introduced the Peng-Robinson EOS for better liquid volume calculations. Many cubic EOS were developed later. Cubic equations of state such as the Peng-Robinson EOS with volume translation have been widely used for the calculations of fluid phase behaviour for hydrocarbon systems. Based on the recombined composition of the reservoir fluid, the characterization procedure of Zuo and Zhang (2000) can be used to characterize single carbon number (SCN) or true boiling point (TBP) fractions and plus fractions. Then cubic EOS can be employed to calculate phase behaviour of the reservoir fluid. The EOS (compositional model) or simulated fluid properties (black oil model) is used in reservoir simulators such as Eclipse and/or process simulators such as HYSYS. For polar systems, cubic EOS can also be used by coupling complicated mixing rules such as the Huron-Vidal mixing rule and the Wong-Sandler mixing rule. On the other hand, Davarnejad *et al.* (2007, 2008) considered the Regular Solution Equations as a general model for polar and non-polar systems.

1.3 Non-hydrocarbons

The hydrocarbons that contain heteroatoms could vary from very simple one such as thiophene to very complex mixtures such as asphaltenes for which the structure is not well understood, but known to contain sulfur, oxygen and nitrogen at different levels, in addition to carbon and hydrogen (Buenrostro-Gonzalez *et al.*, 2002; Woods *et al.*, 2008).

The most common method to separate petroleum fractions is called SARA, which stands for saturates, aromatics, resins and asphaltenes. It needs to be noted that cyclic compound are included in the fraction of saturates. The light fraction is made mostly of alkanes and aromatics. Light aromatics containing heteroatoms could be distilled off with this fraction. The split between alkanes and aromatics could be performed using supercritical fluid chromatography by changing the solvent strength (Dulaurent *et al.*, 2007).

The heavy fraction is first subjected to asphaltenes precipitation using an excess of normal alkanes such as n-pentane or n-heptane, usually at a oil-to-alkane ratio of 1 to 40. Different methods exist in the literature for asphaltenes separation (Kharrat, 2009). After extraction of asphaltenes, the maltenes are separated into three fractions: saturates, aromatics and resins using solvents with increasing polarity as indicated in Figure 1.2.

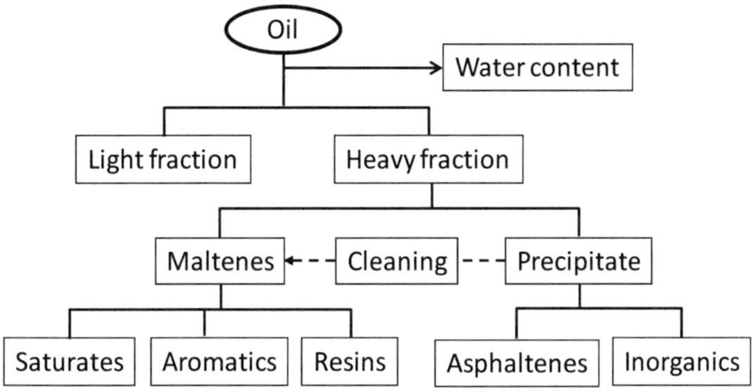

Fig. 1.2. Chart flow of fractionation of a crude oil

The saturate fraction is analyzed by gas chromatography, leading to n-alkanes content. Aromatics are analyzed by Gas Chromatography with Mass Spectrometry (GC-MS). Resins and asphaltenes are the most difficult to analyzed by GC because of their high boiling points. Therefore, the applications of GC on the analysis of heavy oil, which has a high concentration of asphaltene and resin fractions, are limited. In a reported high temperature GC (HTGC) technique, a short 5-m glass capillary column was used to elute compounds in bitumen and bitumen-derived products with boiling points as high as 700°C (equivalent to alkane with carbon number of 90, C_{90}) (Subramanian *et al.*, 1996).

In this chapter, the principles and instrumentations of several GC techniques, and their applications on the analysis of petroleum hydrocarbon fluids are reviewed.

2. High Temperature Gas Chromatography (HTGC)

GC has the advantages of high column efficiency, high sensitivity, fast analysis speed and ease to be combined with other analytical methods (e.g. Mass Spectrometry). Thus, it is widely used to analyze crude oil and its products. Because of the limited thermal stabilities of capillary column and the stationary phase, the maximal column temperature of conventional GC is around 325°C, and the analysis is limited to hydrocarbons with carbon number less than about 35. This fact limits the GC applications on the analysis of alkanes of high carbon numbers (C_{40+}) which are important for some areas including organic geochemistry.

During the past decades, high temperature (325-450°C) GC (HTGC) has been developed rapidly and used for components of high molecular weight in crude oil, *etc.*

2.1 Requirements on the parts of HTGC

The main materials for the HTGC column include stainless steel and fused silica. Steel column has excellent mechanical properties at high temperature, but it has very strong catalytic and adsorptive effects. Therefore, a deactivation inner coating between the steel tubing and the stationary phase is needed. Another drawback of stainless steel column is that it difficult to cut. A protective coating is needed on the outside of the fused silica column to maintain the flexibility of the column. Right now the frequently used coating materials include polyimide and aluminum (Kaal & Janssen, 2008). It has been suggested that the polyimide may be broken down above 360°C and alumina coating can overcome this problem. However, the alumina coating on silica column can become brittle upon repeated heating above 400°C (Application note #59551, Restek Corporation).

The stationary phase also needs to be stable above 400°C upon repeated heating with minimal breakdown. It is mostly based on highly thermostable polysiloxane which can be bonded onto the capillary inner wall via the condensation reaction between the silanol terminal groups of the polysiloxane and the silanol groups on the silica surface during curling process (Mayer *et al.*, 2003; Takayama *et al.*, 1990). The commonly used materials for stationary phase of HTGC include carborane-siloxane polymers (maximum temperature up to 480°C) and silphenylene-modified polysiloxane (maximum temperature up to 430°C), *etc.* (Kaal & Janssen, 2008).

Injection method is very important for HTGC. Cold on-column injection is preferred because of its ability to eliminate discrimination against the most non-volatile compounds (Damasceno *et al.*, 1992). For many compounds with high boiling points, programmed-temperature vapourisation injection (PTV) also gives good results (van Lieshout *et al.*, 1996).

The most frequently used detection method for HTGC is flame ionization detection (FID). Other detection methods have also been used, including mass spectrometery (MS) (Hsieh *et al.*, 2000; Philp, 1994), atomic emission detection (AED) (Asmussen & Stan, 1998) and inductively coupled plasma mass spectrometry (ICP-MS) (Glindemann *et al.*, 2002), *etc.*

2.2 Applications of HTGC

2.2.1 Simulated distillation of crude oil

The crude oil is composed of a large amount of alkanes with different carbon numbers, giving rise to a broad range of boiling point. The understanding of the carbon number distribution of crude oil can help to precisely evaluate the factors affecting the properties of crude oil and the oil products. It is also important for the designing of the distillation, processing equipment and the quality control of the products.

Normal true boiling point (TBP) distillation involves a long procedure and is costly, and the distillation temperature is normally limited even when vacuum is used. GC has been used widely as a fast and reproducible method for simulated distillation (SimDis) to analyze the carbon number distribution of the hydrocarbon in the crude oil. When HTGC is used, SimDis method can reach a boiling point range of 35-750°C, equivalent to n-alkanes with a carbon number distribution of C_5 to C_{120} (Kaal & Janssen, 2008). HTGC SimDis normally uses a short capillary column with a thin film of polydimethylsiloxane stationary phase. Recently, Boczkaj *et al* reported the possibility of using an empty deactivated fused silica column (EC-GC) for HTGC-SimDis (Boczkaj *et al.*, 2011).

Figure 2.1 presents an example of high temperature simulated distilation (HTSD) results for heavy and light oils. The heavy oil starts distillation at a higher temperature (~200°C) than for the lighter oil (~150°C). The residue, which is the fraction that does not distill at 700°C, is much higher for the heavy oil than for the lighter one.

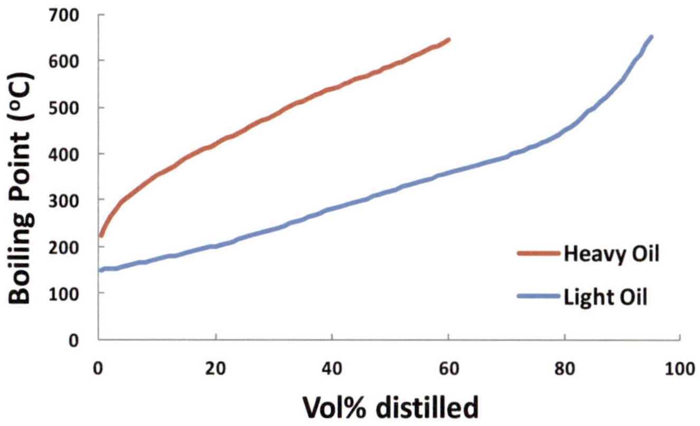

Fig. 2.1. Example of HTSD results for a heavy (red) and light oil (blue).

2.2.2 Wax analysis

Waxes are solids made up of heavy hydrocarbon (C_{18+}) which are mainly normal alkanes (paraffins) (Kelland, 2009, p. 261). The waxy oil can be used to produce wax-based products, and normally has low concentrations of sulphur and metal which are harmful for refinery. But on the other hand, when temperature of the crude oil drops during oil production, transportation or storage, paraffin waxes in the crude oil can precipitate and make serious problems including pipeline blockage and oil gelling, *etc*. Thus, it is important to measure the composition (amount and type) of wax in the crude oil, and to estimate the temperature at which the wax will crystallize (wax appearance temperature, WAT) and the wax precipitation curve (WPC) to understand the potential wax problem and its magnitude.

Compared with conventional GC, HTGC significantly extends the range of detectable hydrocarbon. Therefore, HTGC has become more and more routinely used for wax analysis, and the HTGC results can be correlated to the physical properties of the wax, including melting point, refractive index and kinematic viscosity, *etc* (Gupta & Severin, 1997).

Currently, there are no standard wax-analysis methods, but some methods are developed by the petroleum industry. These methods remain proprietary. Figure 2.2 shows a typical HTGC chromatogram of a waxy crude oil. With a calibrated column, the relative area under the curve for each n-alkane will be converted to the relative abundance of the species. The reported n-alkane composition will be the amount of each n-alkane chain number and relative or absolute abundance (see an example as shown in Figure 2.3).

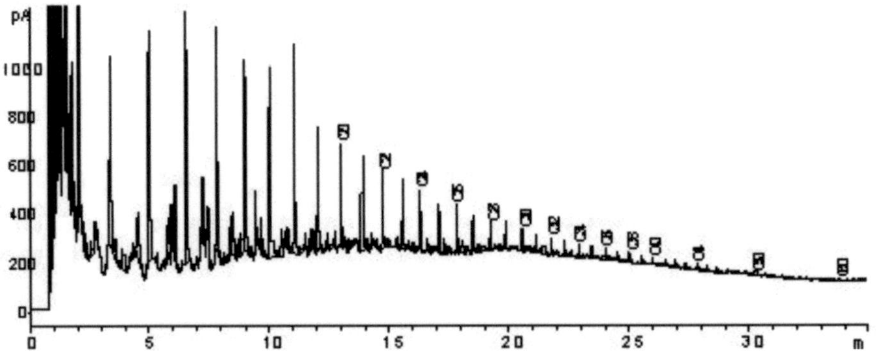

Fig. 2.2. Examples of HTGC traces of a waxy crude oil

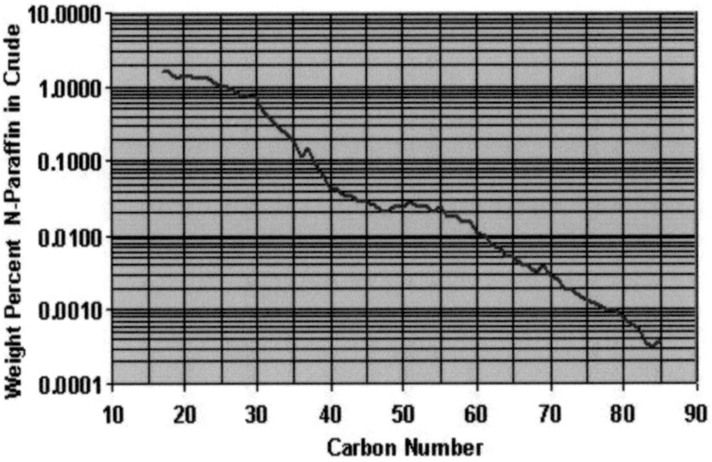

Fig. 2.3. n-Paraffin distribution for a waxy crude oil analyzed by HTGC

An important application of HTGC analysis on waxy crude oils is to measure high molecular weight hydrocarbons (HMWHCs) which provide desirable geochemical information, and some significant findings have been reported.

Del Rio and Philp reported HTGC analysis of some wax samples blocking oil wells and found that the wax deposits were normally composed of hydrocarbons with maximal carbon number around 40 to 50 (del Rio & Philp, 1992). Roehner *et al* used an extended HTGC method to determine the compositions of crude oil solids and waxes up to C_{60} formed in the Trans-Alaska Pipeline. A longer capillary column was used to achieve an improved resolution of high carbon number groups. The n-alkane/non-n-alkane ratio was used to distinguish between certain types of crude oil solids (Roehner *et al.*, 2002).

HTGC has been used to study/monitor the biodegradation of crude oils. For example, using a quantitative HTGC method, Heath *et al* estimated the biodegradation of the aliphatic fraction of a waxy oil. It was found that light hydrocarbon was quickly biodegraded and HMWHCs (C_{40+}) showed resistance to biodegradation (Heath *et al.*, 1997).

HTGC is also an useful tool to study the origin/source of the crude oils. Hsieh and Philp measured HMWHCs from crude oils derived from different sources, including terrigenous, lacustrine, marine source material as well as source rock extracts. Different structural compositions of these HMWHCs (including alkyl-cycloalkanes, methylbranched alkanes, and alkyl-aromatic hydrocarbons) have been revealed by HTGC and it was found that the fraction of the HMWHCs in the whole oil is significantly higher than previously thought (can be up to 8%) (Hsieh & Philp, 2001; Hsieh *et al.*, 2000). The distribution of long chain, branched and alkylcyclohexanes of the HMWHCs analyzed by HTGC has been used as a useful mean to distinguish oils derived from different sources (Huang *et al.*, 2003).

It has been pointed out that the sample extraction/cleaning procedure is crucial to get representative HMWHC samples from crude oil for HTGC measurement (Thanh *et al.*, 1999). More details on the HTGC anslysis of HMWHCs and it applications can be found in an overview given by Philp *et al* (Philp *et al.*, 2004).

WAT and WPC are two important parameters for wax-related flow assurance problems. They can be experimentally measured (Kelland, 2009), and can also be predicted by different thermodynamic models. All these models rely on the experimental data of n-paraffin distribution which is now commonly provided by HTGC. For example, Zuo and Zhang have developed a model to predict the WAT based on the oil composition provided by HTGC (Zuo & Zhang, 2008). The model has been proved to provide prediction results in good agreement with experimental data of synthetic oils and reservoir fluids, and several examples are given in Figures 2.4 to 2.6 (details givein in Zuo & Zhang 2008). Coto *et al* analyzed three parameters to improve the n-paraffin distribution provided by HTGC, including total amount of C_{20+} paraffin, extrapolation of C_{38+} paraffin and molecular weight of the crude oil. The results showed that the distribution of the n-paraffin has great impact on the accuracy of the prediction model (Coto *et al.*, 2011).

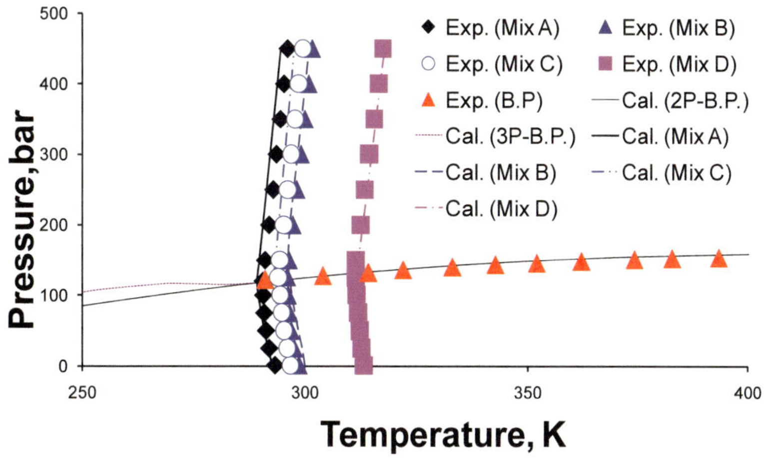

Fig. 2.4. WAT and phase diagram for synthetic oil mixtures.

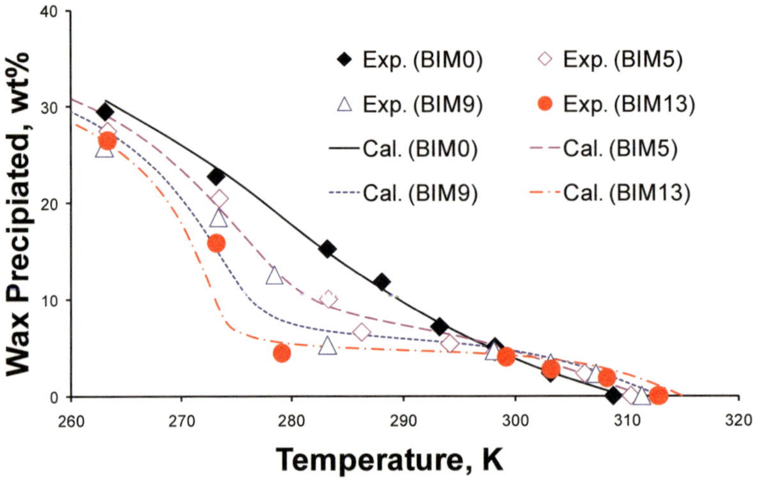

Fig. 2.5. Wax amount for synthetic oil mixtures.

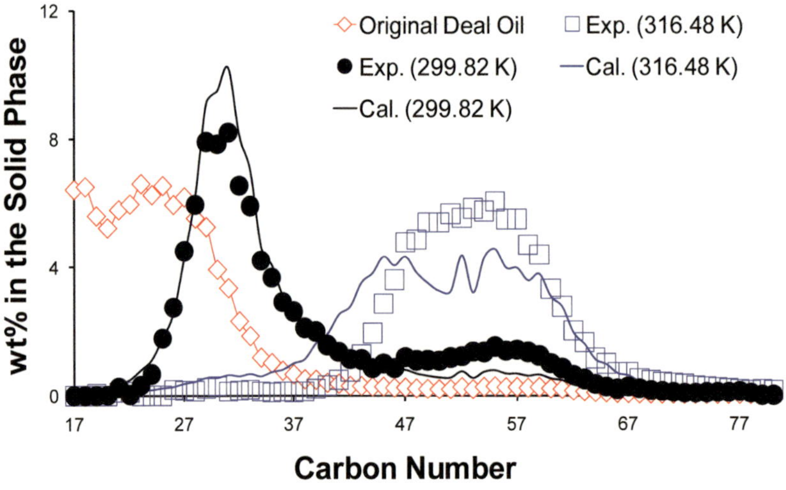

Fig. 2.6. The wax distributions vs. carbon numbers are compared for a crude oil.

3. 2D-GC (GC-GC)

3.1 Concept of comprehensive GC × GC

Since Liu and Phillips depictured comprehensive two dimensional gas chromatography (2D-GC) in 1990s (Liu & Phillips, 1991; Vendeuvre *et al.*, 2007; Zrostlíková *et al.*, 2003), the technology of 2D-GC has developed rapidly and been applied in the areas such as biological/clinical, environmental, food, forensics, petroleum, pharmaceuticals and fragrances in forensic, food and petroleum oil characterization (Wang *et al.*, 2010). 2D-GC employs two capillary GC columns of different selectivity coupled by a modulator (Figure 3.1) which will be introduced here.

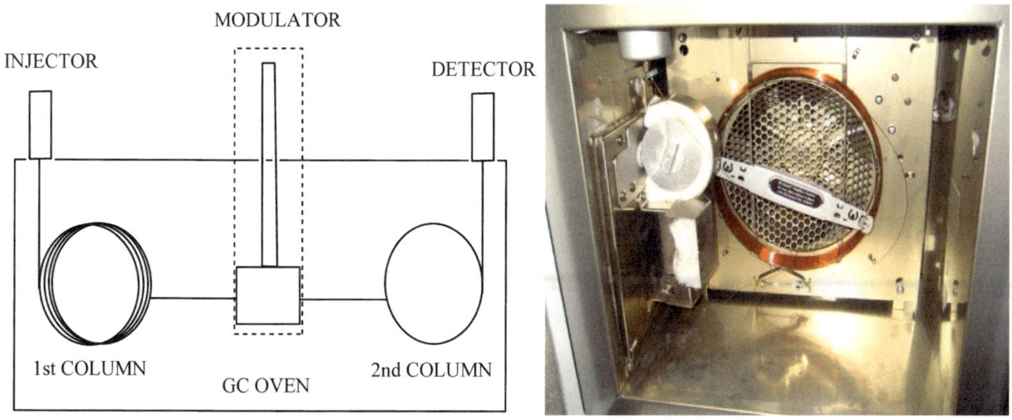

Fig. 3.1. Scheme of comprehensive two dimensional gas chromatography (left). A photo of 2D-GC (right), where the primary column located in the main GC oven, and the second column housed in an independent oven within the main oven.

3.1.1 First column

The first column is typically of conventional length, longer, wider and thicker than the second column, being (15–30) m (long) × (0.25–0.32) mm I.D. (inner diameter) × (0.1–1) µm film (df). The stationary phase of the column can be non-polar or polar. For example, 100% dimethylpolysiloxane is considered as non-polar, and phenyl-substituted methylpolysiloxane is considered as polar column (Scheme 3.1), and the more phenyl contained, the more polar the column is (Betancourt *et al.*, 2009).

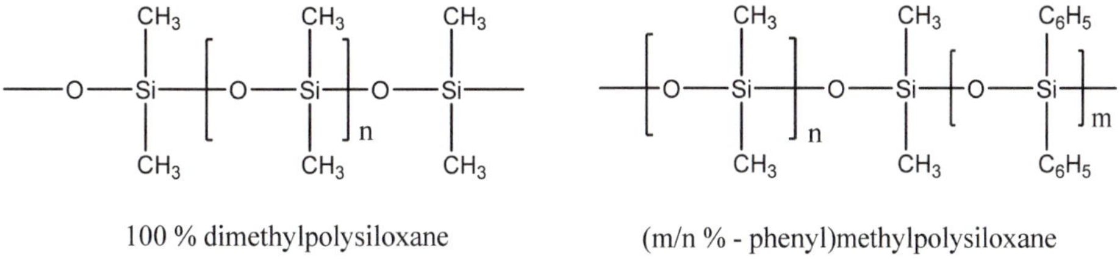

Scheme 3.1. Structures of dimethylpolysiloxane and phenyl-substituted methylpolysiloxane

3.1.2 Second column

The second column is shorter, narrower and thinner, of (0.5–2) m × 0.1mm I.D. × 0.1µm df. It is usually more polar or less polar than the first column. If the second column is more polar than the first column, it is called non-polar × polar configuration; if the second column is less polar than the first one, it is called polar × non-polar configuration. In most time, 2D-GC has non-polar × polar configuration. The two columns can be housed in the same oven or in separate ovens to enable more flexible temperature control (Li *et al.*, 2008).

3.1.3 Modulator

A modulation unit, placed between the two columns, is the most critical component of 2D-GC (Adahchour *et al.*, 2008; Adahchour *et al.*, 2006; Marriott *et al.*, 2003; Pursch *et al.*, 2002;

Wang & Walters, 2007). Modulator periodically samples effluent from the first column and injects it into the second column. After the second-dimension separation is completed the next modulation starts. In this way, all compounds are subjected to two different column separation mechanisms, and are resolved based on two different aspects of chemistry, or selectivity. Each modulation period is so short that each peak from the first column is cut into several smaller slices that go through separation on the second column. Therefore, GC × GC chromatography peaks become much narrower compared with traditional 1D GC peaks. The setup of two columns (the first column is longer and wider and the second column is shorter and narrower) ensures that the total second dimension (2D) separation is completed in the run time of the first-dimension analysis.

It should be noted that comprehensive GC × GC is different from two-dimensional "heart-cut" gas chromatography (Zrostlíková *et al.*, 2003) in which one zone of effluent eluting from the first column is isolated and subsequently separated in a different column.

Modulator is also a reference timing signal at the interface between the two columns (Phillips & Beens, 1999). There are different kinds of modulators such as valve modulator, thermo-modulator and cryogenic modulator, which are very well reviewed (Betancourt *et al.*, 2009; Phillips & Beens, 1999). Nowadays most of modulators are dual-stage cryogenic modulator using CO_2 or liquid nitrogen as cooling agent. Two-stage cryogenic modulator system has two hot points and two cold points; hot point is used to release and reject the slice from the first column, and cold point is used to hold the slice. For the analyses of whole petroleum oil, liquid nitrogen is a better choice because of the existence of very low boiling point compounds in the light ends of the petroleum oils.

3.1.4 Detector

The dimension of the second column is such that eluting peaks have peak widths in the order of 10-100 ms (Adahchour *et al.*, 2008, 2006; Marriott *et al.*, 2003; Phillips & Beens, 1999; Pursch *et al.*, 2002; Wang & Walters, 2007). To properly sample the narrowest peaks, detectors need to have fast response. Therefore, the sampling rate at which the detector signal is sampled should be at least 100 Hz, but a slower data rate of 50 Hz can be used for wider peaks. Flame ionization detectors, FID, which has negligible internal volumes and can acquire data at frequencies of 50–300 Hz, are most widely used. Mass spectrometer (MS) can provide structural information, enable unambiguous identification, and ensure high selectivity throughout the chromatogram, and hence is a good detection method. Quadrupole mass spectrometers operating in full scan mode are too slow to properly sample a GC×GC peak unless that peak is broadened. Fast time-of-flight mass spectrometers (TOFMS) that operate with spectral acquisition rates of 100-200 Hz are well-suited for GC × GC and have been used for numerous studies. On the other hand, element-specific detectors such as sulphur, nitrogen chemiluminescence detections (SCD, NCD), have been used for nitrogen containing and sulphur containing compounds, respectively (Dutriez *et al.*, 2011; Zrostlíková *et al.*, 2003).

3.2 2D-GC results

3.2.1 GC×GC chromatogram

2D-GC chromatograms can be visualized in traditional 1D version, 2D version (contour plot) and 3D image (surface plot), as shown in Figure 3.2. The contour plot of GC×GC chromatogram conventionally demonstrates the advantage of 2D-GC: structured chromatogram. It clearly shows the different group types in certain patch of the 2D plane, for example, tri-aromatics, bi-aromatic, mono-aromatic, paraffins, hopanes, and so on. Within the ring of biaromatics (naphthalene), different methyl-substitute also distribute orderly, from left to right within the ring being: non-methylated biaromatics (naphthalene), mono-methylated naphthalene, bi-methylated naphthalene, tri-methylated naphthalene and tetra-methylated naphthalene. The GC×GC contour plot makes group-type analysis a great advantage for GC×GC analysis, but it should be mentioned that there are some crossing. For instance, nonylbenzene appears in the area of cyclic group, and some bi-cyclic appears in the area of mono-aromatics. The surface plot of a GC×GC chromatogram against a 1D version of the GC×GC chromatogram clearly shows that 2D-GC has better separation: the peaks crowded in 1D are well separated along the second dimension of the 2D plane, and it also demonstrates other advantages of 2D-GC: high sensitivity and bigger peak capacity. Some peaks are invisible in 1D chromatogram but visible in 2D-GC chromatogram.

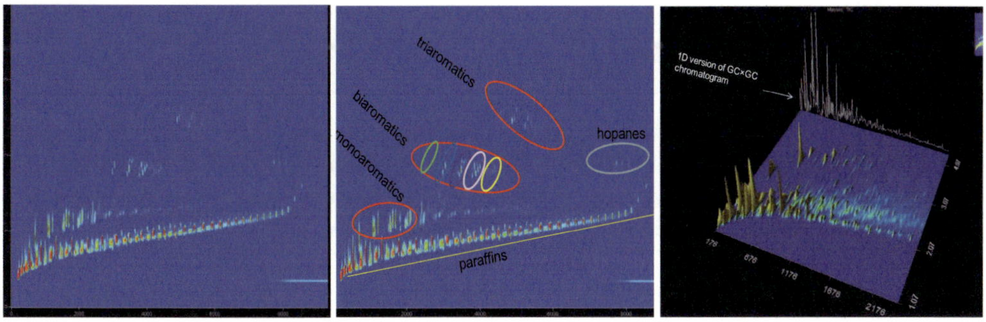

Fig. 3.2. GC×GC chromatograms, first column: VF-1ms 30m × 0.25mm × 0.1μm, second column: BPX50, 1m × 0.1mm × 0.1μm, modulator: 40°C offset to the primary oven, second column: 15°C offset to the primary column. left: Contour plot of a GC×GC chromatogram; middle: contour plot with labelled groups, right: surface plot with the show of 1D version of of GC×GC chromatogram.

3.2.2 Advantages of GC×GC

As demonstrated in the two dimensional contour plot and three dimensional surface plot, it is obvious that GC×GC has the following advantages compared with conventional one demensional GC:

1. Structured chromatograms: the compounds are distributed in the 2D-GC plan according to their group types and each certain group has a certain pattern. Ordered chromatograms have the potential advantage of being much more interpretable than disordered ones. The pattern of peak placement is itself informative and may make it possible in many mixtures to identify most or all of the components or at least to recognize the mixture with good reliability (Adahchour *et al.*, 2008; Li *et al.*, 2008; Phillips & Beens, 1999).
2. Better separation: 2D-GC separates components along the primary dimension and also along the second dimension. Sometimes compounds co-eluted with conventional 1D GC technology can be separated by 2D-GC along the second dimension.

3. Larger capacity: 2D-GC peaks are distributed in the whole plane rather than in one line. 2D-GC has a peak capacity of n1× n2, in which n1 is the capacity along the first dimension and n2 the capacity along the second dimension. Therefore, the capacity of 2D-GC is higher than that of conventional 1D GC.
4. Higher sensitivity: Compared with conventional 1D GC, the sensitivity of 2D-GC is increased by 1.5 - 50 fold (Zrostlíková *et al.*, 2003). Trace amount of analytes can be detected with 2D-GC. The detection limit for 2D-GC is about 2pg (Zrostlíková *et al.*, 2003).

All these characteristics make GC × GC a particularly useful technique for analyzing complex mixture, for example, petroleum oil. The ordered distribution of effluent has been used for quick screening of oil, recognizing the difference between individual oils (Li *et al.*, 2008).

On the other hand, 2D-GC also has some disadvantages. Cooling in cryogenic modulator causes peak tailing along the second column. 2D-GC files are extremely large, and not easy to be applied into different software.

3.2.3 GC × GC column configuration

Normal GC×GC column configuration is from non-polar to polar (the first column is non-polar, and the second column is polar), whereby the sample is separated on the first column based on the boiling point differential of all components, and then further separated on the second column based on the polarity differential. The choice of columns depends on the stationary phase (mainly on polarity), the application temperature of the column, column length and diameter, commercial availability and so on. The normal column configuration is usually good for selecting out high polar components in the samples. Nevertheless, column configuration of polar to non-polar (reversed configuration) is reported to improve the resolution of individual alkanes, cyclokane, branched alkanes, and isoprenoids (Vendeuvre *et al.*, 2005).

Figure 3.3 demonstrates a contour plot of 2D-GC-MS, using a reversed column configuration, from polar (30m×0.25mm×0.25um, DB-17) to non-polar column (1m×0.1 mm×0.1um,

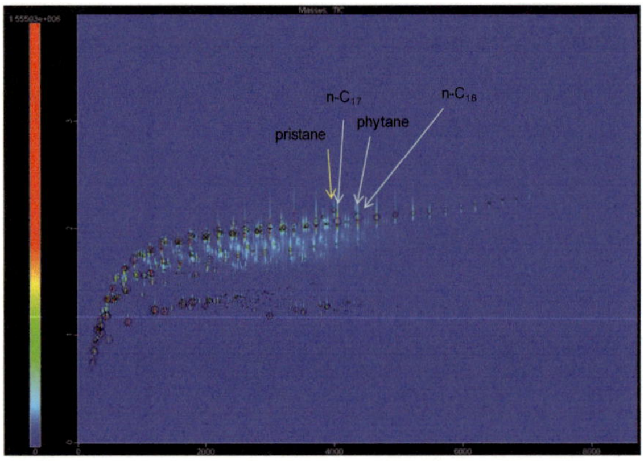

Fig. 3.3. Contour plot of 2D-GC-MS chromatogram using reversed column configuration. First column: polar DB-17 30m×0.25mm×0.25μm; second column: non-polar DB-1MS 1m×0.1 mm×0.1 μm.

DB-1MS). Overall, the contour plot from the reversed configuration looks just like a reversed chromatogram from normal configuration. In the chromatogram of reversed configuration non-polar compounds, like paraffin, have longer second-retention time, and polar compounds have shorter second retention time. In other words, the shorter the second retention time is, the more polar the compounds are. With a reversed column configuration, the compounds are also separated based on boiling point along the first dimensional retention time. For example, with the increase of retention time, the carbon number of normal paraffins increased. However, pristane and phytane come shortly before nC_{17} and nC_{18}, respectively. This is different from the normal column configuration, where pristine and phytane come shortly after nC_{17} and nC_{18}, respectively.

3.2.4 Data processing

GC×GC presents information technology challenges in data handling, visualization, processing, analysis, and reporting due to the quantity and complexity of GC × GC data. Usually, two different softwares are used for data acquisition and data processing, respectively. GC Image (GCImage LLC) is a software system developed at the University of Nebraska-Lincoln that uses advanced information technologies to process and visualized GC × GC data, detect peaks, compare chromatograms, and perform peak deconvolution, pattern recognition and other data mining tasks. ChromaTOF (Leco Corporation) is another software program designed to control Leco's commercially available GC × GC-TOFMS system that has similar functions. Both GCImage and ChromaTOF make effective use of the tremendous amount of data generated when a time-of-flight mass spectrometer is used as a GC × GC detector, including spectral library matching and extracted ion chromatograms. HyperChrom (Thermo Electron Corporation) is a third software program designed to control Thermo Electron's commercially available GC × GC system with flame ionisation detection and employs various data processing and visualization capabilities.

Actually data post process to re-read/re-display the chromatogram in the need of particular application is important, and is proven to be troublesome with the problem of peak alignment, retention time deviation, *etc.* Some particular work has to be done to re-display the chromatogram properly (Aguiar *et al.*, 2011).

3.2.5 Factors affecting 2D-GC retention time and separation

The factors which affect conventional 1D GC retention time and separation all apply to 2D-GC analysis, including:

- inlet temperature
- column temperature and temperature program
- carrier gas and carrier gas flow rates
- the column's stationary phase
- column diameter and length
- sample size and injection technique

Inlet temperature should be set high enough to vaporize injected samples, but not so high that the injected samples can be decomposed (Juyal *et al.*, 2011). Column temperature and temperature program are very important factors affecting GC retention time and separation.

The setup of column temperature and temperature program is to make sure that all of analytes are eluted and separated well. Higher column temperature causes shorter retention times; slower temperature ramp usually leads to better separation. The higher speed of carry-gas results in shorter retention time, which potentially causes peak co-elution. If the amount/concentration of injected sample is too big/high, the chromatogram will look overwhelming crowded with no good separation, especially when the injected samples are crude oil samples. In this case, split injection technique can be employed. When the split ratio is too high (200), the injection accuracy will decrease. Considering the high complexity of crude oil samples, temperature programmed injection technique might be very useful, whereby the sample is introduced in the injector at low temperature and then vaporized by a fast programmed heating process, during which time the split is open all the time and the sample amount entering the column is proportional to the pre-set split ratio (Wang *et al.*, 2010).

Due to the characteristics of 2D-GC, there are several more factors affecting 2D-GC separation, including the modulation period (the second column time), hot pulse time and second column temperature. A short modulation period usually means a short slice from the first column for further separation on the second column, which implies a better separation on the second column. If the modulation period is too long, the separated effluents on the first dimension are accumulated while waiting for injection onto the second dimension, which intends to lose the resolution of the first dimension. If modulation period is too short, on the other hand, peaks wrap-up is unavoidable, resulting in wrong retention times and bad separation.

Currently, the update of 2D-GC operation platform focuses on multi-stage temperature ramp of the second column which will resolve the dilemma between the modulation period and the second column separation time (Betancourt *et al.*, 2009). In some cases, *e.g.* the detection of nitrogen-containing compounds and the analysis on vacuum gas oil (VGO), high modulation period (20s - 30s) is needed. In this case, wide bore first column with bigger diameter are used to make the first dimension peak wider (Dutriez *et al.*, 2009, 2010, 2011).

In 2D-GC analysis with dual stage cryogenic modulator, hot pulse time and cooling period should be adjusted, especially in petroleum oil analysis. The hot pulse time should be long enough to make sure all of samples are re-injected to the second column, and cooling period should be good to make sure all the compounds with low boiling points are refocused on the second column.

3.3 Applications of comprehensive 2D-GC in oil analysis

2D-GC has been used in high temperature analysis of vacuum gas oil (VGO) (HT-2D-GC) (Dutriez *et al.*, 2009, 2010), nitrogen-containing compounds (2D-GC NCD) (Dutriez *et al.*, 2011), sulphur speciation (2D-GC SCD), middle distillates (Vendeuvre *et al.*, 2005), pyrolysis of petroleum source rocks (Py-2D-GC) (Wang & Walters, 2007), biomarkers (Aguiar *et al.*, 2011; Juyal *et al.*, 2011), *etc.*

3.3.1 Group type analysis of two Tarmat oils

Two oils A and B are isolated by a huge tarmat and no connection between the two oils is known. A question arouse regarding to the two oils: are they different? To answer the question, the two oils were subject to independent group type analysis with 2D-GC FID in house and detail analysis with commercial 1D GC-MS. The whole oils were used for group type 2D-GC FID analysis; the two oils were fractionated before detailed 1D GC-MS analysis. The TIC (total ion chromatography) 2D-GC contour plot of oil A is shown in Figure 3.4. There are two steps in the group type analysis: 1) divide each contour plot of the 2D-GC chromatogram of oil A and oil B into 7 areas, tri-aromatic (3-ring-1), bi-aromatic (2-ring-1), mono-aromatic (1-ring-1), hopane (polycyclic-1), sterane (polycyclic-2), before nC_{17} and after nC_{17} at the same retention times (first dimensional retention time, second dimensional retention time), and 2) compare each area of the oil A with the oil B. The analysis results are listed in Table 3.2. The two oils share the same amount of tri-aromatics, bi-aromatics, mono-aromatics, polycyclic-1(hopane compounds), before nC_{17} and after nC_{17}. No sterane compounds (polycyclic-2) were detected in the two whole oil analyses due to too low concentration.

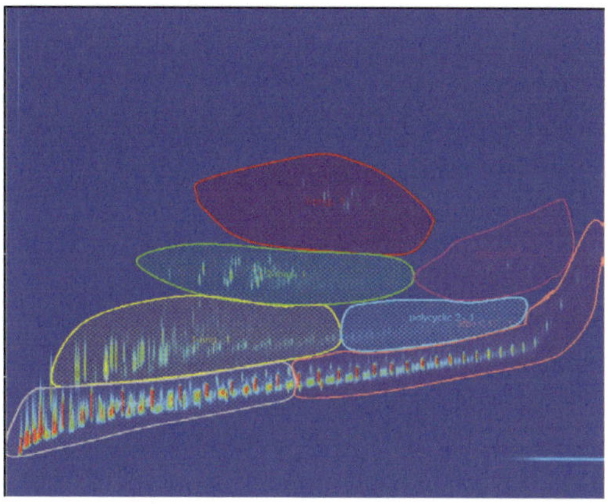

Fig. 3.4. Group type analysis of Tarmat oil A. Dual stage liquid N_2 cryogenic modulator; First column: VF-1MS 30m×0.25mm×0.25µm; second column: BPX50 1m×01.mm×0.1µm. Modulator: 45°C offset to primary oven, second column: 15°C offset to primary column.

The independent 1D GC-MS analyses gave details in hopane, steranes, cheilanthane, adamantanes, naphthalenes, phenanthrenes, benzothiophenes, biphenyls, aromatic steroids and bicyclics. The two oils showed very similar results in each studied items, for example, in biphenyls, as listed in Table 3.3.

So the 2D-GC FID group type analysis of the two oils agrees with the 1D GC-MS detailed analysis on that the two oils are the same.

	Normalized Area%_ Oil A	**Normalized Area %_oil B**
3-rings	0.41	0.43
2-rings	1.55	1.56
1-ring	7.83	7.86
Polycyclic-1	0.05	0.05
Polycyclic-2	0	0
before nC_{17}	69.53	69.32
after nC_{17}	20.63	20.78

Table 3.2. 2D-GC FID group type analysis of Tarmat oil A and B

	BP	2MBP	DPM	3MBP	4MBP	DBF	DBT
Oil A	7.71	2.68	6.49	8.82	4.82	13.45	253.05
Oil B	8.21	2.88	6.69	8.89	4.93	13.60	252.87
	MBP ratio	DPM ratio	DPM/MBP ratio	DBF ratio	BP/DBT ratio		
Oil A	0.42	0.46	0.40	0.64	0.03		
Oil B	0.42	0.45	0.40	0.62	0.03		

BP: biphenyl; 2MBP: 2-methylbiphenyl; DPM:diphenylmethane; 3MBP:3-methylbiphenyl; 4MBP: 4-methylbiphenyl; DBF: dibenzofuran; DBT: dibenzothiophene; MBP ratio: MBP ratio = 4-methylbiphenyl / (2- + 3-methylbiphenyl); DPM ratio = diphenylmethane / (diphenylmethane + biphenyl); DPM / MBP ratio = diphenylmethane / (diphenylmethane + 2- + 3- + 4-methylbiphenyl); DBF ratio = dibenzofuran / (dibenzofuran + biphenyl); BP / DBT ratio = biphenyl / dibenzothiophene.

Table 3.3. Bipheyl compounds in Tarmat oil A and B based on 1D GC-MS analysis

3.3.2 Advanced product back allocation

Production back allocation of commingled oils from different zones or reservoirs is usually done with 1D GC by comparing composition of commingled oils with involved end members, in which process some subtle differences between inter-paraffin peaks (the GC peaks between n-paraffins) are taken into account. Some uncertainties exist in oil product back allocation when related end members are very similar, or when commingled oils contain heavy oil end members, or the inter-paraffin are not well resolved. GC × GC may help to overcome these uncertainties due to enhanced separation in complex oil. Better identification and quantification of single compounds in heavy oils with GC × GC helps to differentiate the heavy oil end members in commingled oils, and therefore improves heavy oil product back allocation.

In a heavy oil blind production back allocation test, artificial commingled heavy oil #1 (C1) was made of 40% heavy oil A and 60% heavy oil B (C1 = 0.4A +0.6B) , artificial commingled heavy oil #2 (C2) was made of 40% heavy oil B and 60% heavy oil C (C2=0.4B+0.6C), and commingled heavy oil #3 (C3) made of 35% A, 35% B and 30%C (C3=0.35A+0.35B+0.30C). 2D-GC FID analysis was used for allocating all the three samples, of which results was compared with 1D GC FID method. Figure 3.5 shows the 2D-GC FID contour plot of commingled heavy oil C1, where much more peaks are well separated along the second dimension compared with the 1D GC FID. There are two big peaks of CS_2 and toluene, respectively, separated from the other main paraffins in the contour plot, and the peaks are so big that they look like contaminants in the oils. So in the product back allocation, the peaks were omitted to avoid any kind of contamination. The production back allocation results based on 2D-GC FID were compared with that based on 1D GC FID, as shown in Figure 3.6. It shows that overall 2D-GC FID and 1D GC FID show very similar accuracy in the production back allocation of two-end-member commingled samples, but 2D-GC FID shows much high accuracy (error % = 2) than 1D GC FID (error % = 10) in production back allocation of three-end-member commingled sample.

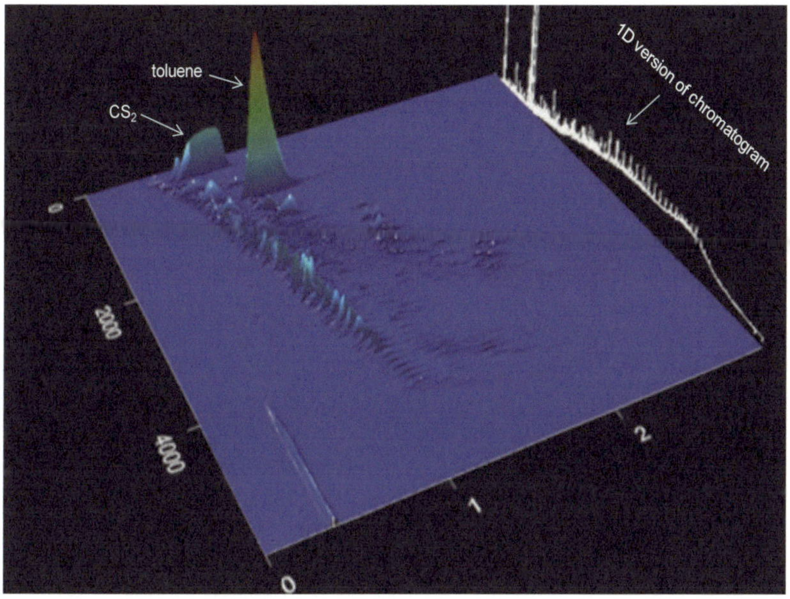

Fig. 3.5. 2D-GC contour plot of commingled heavy oil C1

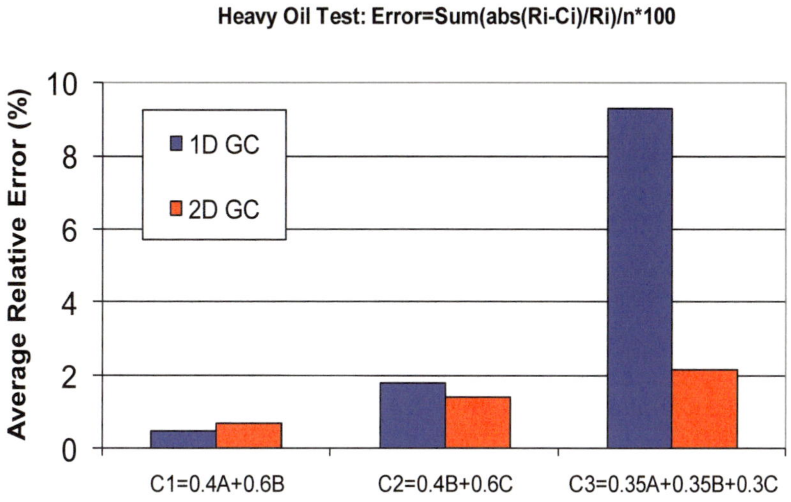

Fig. 3.6. Comparison of heavy oil production back allocation based on 2D-GC FID with 1D GC FID. Ri= real percentage of end members; Ci= calculated percentage of end members. n= number of end members.

4. Gas chromatography fingerprinting

Gas chromatography (GC) analysis is used in the petroleum industry to provide information related to fluid composition, which is needed in petroleum engineering and petroleum geochemistry. Engineering and geochemical approaches, however, require different evaluation techniques for GC chromatograms. Most common application in petroleum engineering is the equation-of-state (EOS) for PVT characterization. Geochemical evaluations include using GC chromatograms for determination of reservoir connectivity, or back allocate commingled production. All these geochemical approaches are based on comparing GC chromatograms on their similarity to each other rather than obtaining compound specific quantitations. This comparison of GC chromatograms is commonly called "geochemical fingerprinting" in the petroleum industry. Fingerprints obtained by gas chromatography (GC) of crude oils is one of the less expensive, less risky and less time consuming methods to study different oils in terms of their similarity to each other. However, fingerprinting using GC chromatograms requires a defined and highly accurate analytical workflow to ensure high precision for assessing the exact similarity index. The analysis and fingerprinting evaluation of reservoir fluids also is affected by uncertainties. Crude oils are complex mixtures containing thousands of different hydrocarbons having huge differences in polarity and molecule size. Consequently, even for a same crude oil sample GC chromatograms can significantly differ when analyzed at different labs or when analyzed using different GC operating conditions. Therefore, it is necessary to keep analytical procedures, the column and the GC operational conditions the same for all GC runs when geochemical fingerprinting is needed. In addition, regular checks on GC performance need to be done to ensure comparability of GC runs. However, a slight shift in retention time is commonly still present when crude oils are analyzed consecutively. The problem mainly results from a slight deterioration in the columns separation performance between consecutive GC runs.

In order to make GC chromatograms directly comparable, it is required to eliminate this retention time shift, which is called warping. Different warping algorithms for retention time alignment of GC chromatograms have been published in the last decade. The most common techniques are dynamic time warping (DTW) and correlation optimized warping (COW) (Nielsen *et al.*, 1998; Tomasi *et al.*, 2004). The comparison of chromatogram similarity index for GC fingerprinting can be done after the preprocessing warping method. The fingerprinting technique relies on comparison of the chemical composition of several chromatograms acquired with the same chromatographic conditions and is based on the differences between peak height ratios of the different crude oil samples. The most advanced and newest technique to determine the similarity index between different chromatograms is the Malcom distribution analysis, recently described in Nouvelle and Coutrot (2010). Malcom distribution analysis uses a statistical method, based on consistent quantification of the uncertainty from chromatography peak height measurements, which provides absolute distances between fingerprints on a universal scale. The method is able to discriminate samples even if the amplitude of the compositional differences is about the same as the error in peak height measurements, and distances between samples are independent of the number of peak ratios available, and the uncertainty in the peak height measurements. The distribution analysis method uses the matrix of all neighboring non-alkane peak height ratios. This method provides a "chemical distance" between each couple of analyses on the basis of the statistical analyses of their respective peak ratios.

In a first step, the distribution histogram of all peak ratio differences available between pairs of chromatograms is built. Then, the inter-quartile range (IQR), as a measurement of the spread of the distributions, is used to characterize the statistical distance between the fingerprints. Since the GC column becomes progressively degraded during analysis, the method requires the peak ratios of two chromatograms from the same sample to calibrate the uncertainty in the measurements each time a new batch of analyses is performed. The uncertainty is used to fix a threshold which distinguishes significant from insignificant distances ("distance threshold"). The distance threshold is about the same as the error for replicate analysis. Hereby, the method builds the distribution histogram of the differences between two chromatograms from the same sample. The shape of the histogram is used to precisely determine the uncertainty in the measurements. This uncertainty is used to obtain the expected profile of the distribution that the differences between two chromatograms should have if they were from the same sample (Nouvelle & Coutrot, 2010).

For GC fingerprinting purposes, the peak quality depends on two main quantities:

- Retention time,
- Peak height ratios

Peak height ratios are generally used for GC fingerprinting. In contrast to peak heights, the peak high ratios avoid the uncertainties dealing with discrepancies between the lightest and the heaviest compounds in the samples. The origin of such discrepancies can be linked to sampling procedure issues, evaporation of the lightest hydrocarbons or also blocking of the heaviest hydrocarbons into the column or the injection device. For all GC runs the retention times or the peak height ratios need to be checked on the tolerance deviations.

The peak height ratios are calculated from the indexation. In practice, several hundreds of peak height ratios are used. A maximal retention time difference between two peaks implied in a ratio is settled to 25 Kovats indices. A Kováts indice is a retention time measurement relative to two consecutive *n*-paraffins (Kováts, 1958):

$$KovatsID(i) = 100 \cdot \left(nC_{i-1} + \frac{\log_{10}(t_i) - \log_{10}(t_{C_{i-1}})}{\log_{10}(t_{C_i}) - \log_{10}(t_{C_{i-1}})} \right) \tag{1}$$

where:

- KovátsID(i): Kovats index of the compound,
- nC_{i-1}: number of carbon on the n-paraffin located just before the compound,
- t_i: retention time of the compound,
- $t_{C_{i-1}}$: retention time of the n-paraffin located just before the compound,
- t_{C_i} : retention time of the n-paraffin located just after the compound.

The aim of the indexation tool is to compare several chromatograms with a reference chromatogram. The peaks of the reference chromatogram are searched in the other chromatograms on the basis of a topological analysis. Figure 4.1 shows as example the chromatograms of two similar oils in the range of nC_{11} and nC_{12}. Even though both chromatograms are similar, they differ in certain peak height ratios, which clearly differentiate both chromatograms belonging to different oil samples.

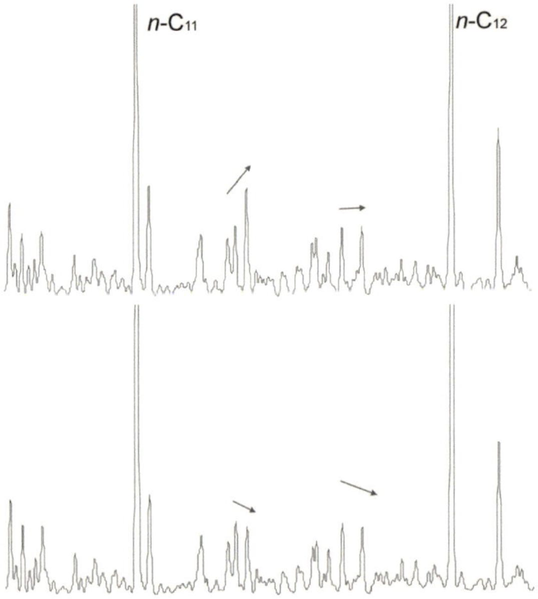

Fig. 4.1. Chromatographic comparison of 2 oil samples based on nC_{11}-nC_{12} inter-paraffin peak height ratios. Arrows indicate differences in neighboring peak height ratios, which indicates compositional differences between both crude oil samples.

The final outcome for a set of oil samples in terms of their similarity is commonly presented graphically as star diagram or cluster analysis. As example, Figure 4.2 shows the star diagram that compares 11 oils from a single well based on 37 n-C_7 — n-C_{20} range peak height ratios.

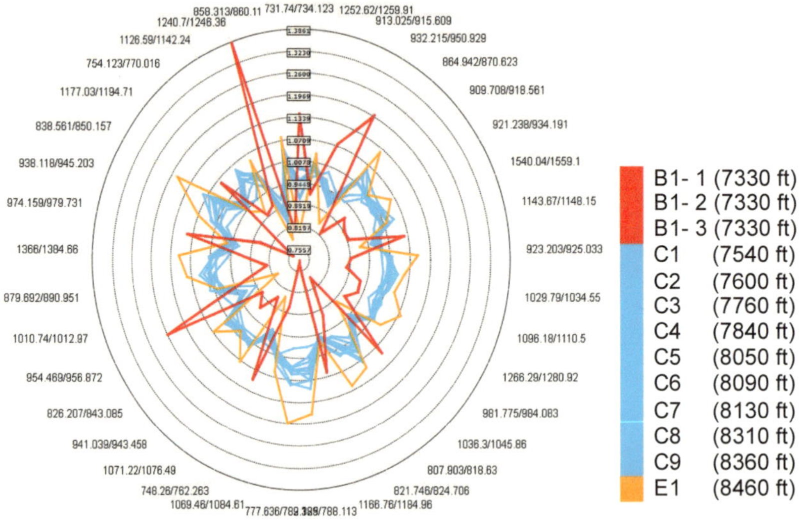

Fig. 4.2. Star diagram comparison of 11 oil samples from a single well based on 37 nC_7-nC_{20} range peak ratios. Each axis on the star shows the values for a different ratio of a pair of GC peaks. Peak labels are in Kovats Indices where peaks eluting from the GC between C_7 and C_8 are labelled in the 700's, those eluting between C_8 and C_9 are in the 800's.

Reduction of the number of variables was performed by using the Reduction by Inertia Constraint (RIC) method, with optimization by Inter/Intra-class maximization. The quality ω of each ratio is proportional to the standard deviation between the pre-groups, and inversely proportional to the standard deviation for analysis of samples within the same pre-group (Nouvelle, 2010). Thus, for a given ratio, the quality ω is the best when:

- The pre-groups are well separated.
- The analyses belonging to the same groups are close.

$$\omega_j = \frac{\sigma_j^{INTER}}{\sigma_j^{INTRA}} \tag{2}$$

with σ_j^{INTER} : standard deviation calculated for inter-class maximization

σ_j^{IINTRA} : standard deviation calculated for intra-class maximization

Triplicate analyses of the oil sample B are included to determine the distance threshold for the dataset. The replicate analysis of sample B is then grouped to determine the quality of the different variables attached to the dataset. Comparison of the star patterns indicates that the oil sample B (red; three replicate analyses) is significantly different than the remaining oils marked as oils C (blue) and oil E (yellow).

Another graphical evaluation method for the similarity of GC chromatograms is the hierarchical cluster analysis. Hierarchical clustering is general mathematical approaches, in

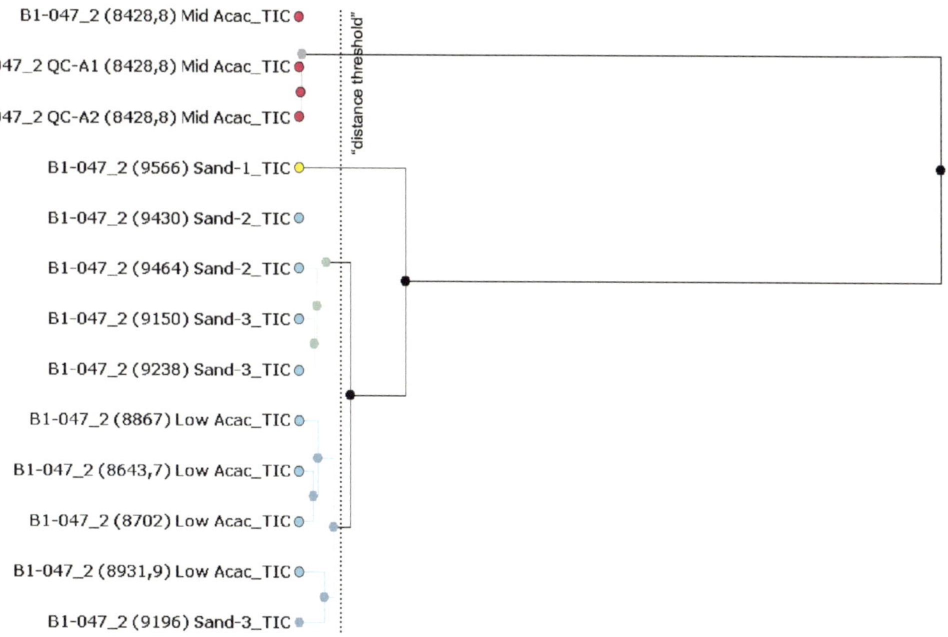

Fig. 4.3. Cluster analysis of the 11 oils from a single well based on 37 nC_7-nC_{20} range peak height ratios.

which the oils are grouped together that are closer to one another based on the distribution analysis using GC chromatograms. A key component of the analysis is repeated calculation of distance measures between data, and between clusters once oils begin to be grouped into clusters. The outcome is represented graphically as a dendrogram (Figure 4.3). For the same samples set of 11 oils in Figure 4.2, the dendogram is shown in Figure 4.3. The colouring of the groups on this diagram allows one to more easily discern the groups visually. Instrumental error is low as indicated by the tie line connection between the three replicate analyses of oil sample B.

5. Conclusions

Since its introduction, gas chromatography (GC) has been widely used as an imnportant method in the analysis of petroleum hydrocarbon which have complex compositions. New techniques have extended the applications of GC in the petroleum composition analysis. In this chapter, several such techniques have been briefly reviewed, including high temperature GC (HTGC), two-dimensional GC (2D-GC, or GC×GC) and GC fingerprinting. Although some of these techniques, such as 2D-GC and GC fingerprinting, are still very young, it is expected that, with the advance of the research work, they will used more routinely to give more precise composition in a broader range, and to give more important geochemical information of the petroleum fluids in the near future.

6. References

Adahchour, M., Beens, J & Brinkman, U. A. T. (2008). Recent developments in the application of comprehensive two-dimensional gas chromatography. *Journal of Chromatography. A, 1186*(1-2), pp. 67-108, ISSN 0021-9673

Adahchour, M., Beens, J., Vreuls, R. J. J. & Brinkman, U. A. T. (2006). Recent developments in comprehensive two-dimensional gas chromatography (GC × GC). II. Modulation and detection. *Trends in Analytical Chemistry, 25*(6), pp. 540-553, ISSN 0165-9936

Aguiar, A., Aguiar, H. G. M., Azevedo, D. A. & Aquino Neto, F. R. (2011). Identification of Methylhopane and Methylmoretane Series in Ceara Basin oils, Brazil, Using Comprehensive Two-Dimensional Gas Chromatography Coupled to Time-of-Flight Mass Spectrometry. *Energy and Fuels, 25*(3), pp. 1060-1065, ISSN 1520-5029

Asmussen, C. & Stan, H. (1998). Determination of non-ionic surfactants of the alcohol polyethoxylate type by means of high temperature gas chromatography and atomic emission detection. *Journal of High Resolution Chromatography, 21*(11), pp. 597-604, ISSN 1521-4168

Betancourt, S. S., Ventura, G. T., Pomerantz, A. E., Viloria, O., Dubost, F. X., Zuo, J., Monson, G., Bustamante, D., Purcell, J. M., Nelson, R.K., Rodgers, R. P., Reddy, C. M., Marshall, A. & Mullins O. C.(2009). Nanoaggregates of asphaltenes in a reservoir crude oil and reservoir connectivity. *Energy and Fuels, 23*(3), pp. 1178-1188, ISSN 1520-5029

Boczkaj, G., Przyjazny, A. & Kamiński, M. (2011). A new procedure for the determination of distillation temperature distribution of high-boiling petroleum products and fractions. *Analytical and Bioanalytical Chemistry, 399*(9), pp. 3253-3260, ISSN 1618-2650

Buenrostro-Gonzalez, E., Andersen, S. I., Garcia-Martinez, J. A. & Lira-Galeana, C. (2002). Solubility/molecular structure relationships of asphaltenes in polar and nonpolar media. *Energy and Fuels*, *16*(3), pp. 732-741, ISSN 1520-5029

Coto, B., Coutinho, J. A. P., Martos, C., Robustillo, M. D., Espada, J. J. & Peña, J. L. (2011). Assessment and improvement of n-Paraffin distribution obtained by HTGC to predict accurately crude oil cold properties. *Energy and Fuels*, *25*(3), pp. 1153-1160, ISSN 1520-5029

Damasceno, L. M. P., Cardoso, J. N. & Coelho, R. B. (1992). High temperature gas chromatography on narrow bore capillary columns. *Journal of High Resolution Chromatography*, *15*(4), pp. 256-259, ISSN 1521-4168

Davarnejad, R., Kassim, K. M., Zainal, A. & Suhairi. A. S. (2007). Phase equilibrium study in supercritical fluid extraction of ethanol to octane mixture using CO_2. *Asean Journal of Chemical Engineering*, 7(2), pp. 127-136, 1655-4418

Davarnejad, R., Kassim, K. M., Zainal, A. & Suhairi. A. S. (2008). Mutual solubility study for 94.2:5.8 of ethanol to octane with supercritical carbon dioxide solvent. *Journal of Chinese Institute of Chemical Engineers*, 39(4), pp. 343-352, ISSN 0368-1653

Dulaurent, A., Dahan, L., Thiébaut, D., Bertoncini, F. & Espinat, D. (2007). Extended simulated distillation by capillary supercritical fluid chromatography. *Oil and Gas Science and Technology*, *62*(1), pp. 33-42, ISSN 1294-4475

Dutriez, T., Borras, J., Courtiade, M., Thiébaut, Didier, Dulot, H., Bertoncini, F. & Hennion, M. C. (2011). Challenge in the speciation of nitrogen-containing compounds in heavy petroleum fractions by high temperature comprehensive two-dimensional gas chromatography. *Journal of Chromatography. A*, *1218*(21), pp. 3190-3199, ISSN 0021-9673

Dutriez, T., Courtiade, M., Thiébaut, Didier, Dulot, H., Bertoncini, Fabrice, Vial, J. & Hennion, M. C. (2009). High-temperature two-dimensional gas chromatography of hydrocarbons up to nC60 for analysis of vacuum gas oils. *Journal of Chromatography A*, *1216*(14), pp. 2905-2912, ISSN 0021-9673

Dutriez, T., Courtiade, M., Thiébaut, D., Dulot, H., Borras, J., Bertoncini, F. & Hennion, M. C. (2010). Advances in quantitative analysis of heavy petroleum fractions by liquid chromatography–high-temperature comprehensive two-dimensional gas chromatography: breakthrough for conversion processes. *Energy and Fuels*, *24*(8), pp. 4430-4438, ISSN 1520-5029

Glindemann, D., Ilgen, G., Herrmann, R. & Gollan, T. (2002). Advanced GC/ICP-MS design for high-boiling analyte speciation and large volume solvent injection. *Journal of Analytical Atomic Spectrometry*, *17*(10), pp. 1386-1389, ISSN 1364-5544

Gupta, A. K. & Severin, D. (1997). Characterization of petroleum waxes by high temperature gas chromatography - correlation with physical properties. *Petroleum Science and Technology*, *15*(9), pp. 943-957, ISSN 1532-2459

Heath, D. J., Lewis, C. A. & Rowland, S. J. (1997). The use of high temperature gas chromatography to study the biodegradation of high molecular weight hydrocarbons. *Organic Geochemistry*, *26*(11-12), pp. 769-785, ISSN 0146-6380

Hsieh, M. & Philp, R. P. (2001). Ubiquitous occurrence of high molecular weight hydrocarbons in crude oils. *Organic Geochemistry*, *32*(8), pp. 955-966, ISSN 0146-6380

Hsieh, M., Philp, R. P. & del Rio, J. C. (2000). Characterization of high molecular weight biomarkers in crude oils. *Organic Geochemistry, 31*(12), pp. 1581-1588, ISSN 0146-6380

Huang, H., Larter, S. R. & Love, G. D. (2003). Analysis of wax hydrocarbons in petroleum source rocks from the Damintun depression, eastern China, using high temperature gas chromatography. *Organic Geochemistry, 34*(12), pp. 1673-1687, ISSN 0146-6380

Juyal, P., McKenna, A. M., Yen, A., Rodgers, R. P., Reddy, C. M., Nelson, R. K., Andrews, A. B., Atolia, E., Allenson, S. J., Mullins, O. C. & Marshall, A. G (2011). Analysis and identification of biomarkers and origin of color in a bright blue crude oil. *Energy and Fuels, 25*(1), pp. 172-182, ISSN 1520-5029

Kaal, E. & Janssen, H. G. (2008). Extending the molecular application range of gas chromatography. *Journal of Chromatography A, 1184*(1-2), pp. 43-60, ISSN 0021-9673

Kelland, M. A. (2009). *Production Chemicals for the Oil and Gas Industry*. CRC Press, ISBN 978-142-0092-90-5

Kharrat, A. M. (2009). Characterization of Canadian heavy oils using sequential extraction approach. *Energy and Fuels, 23*(2), pp. 828-834, ISSN 1520-5029

Kovát, E. (1958). Gas-chromatographische Charakterisierung organischer verbindungen. Teil 1: Retentionsindices aliphatischer Halogenide, Alkohole, Aldehyde und Ketone. *Helvetica Chimica Acta*, 41(7), pp. 1915–1932, ISSN 1522-2675

Li, M., Zhang, S., Jiang, C., Zhu, G., Fowler, M., Achal, S., Milovic, M., Robinson, R. & Larter, S. (2008). Two-dimensional gas chromatograms as fingerprints of sour gas-associated oils. *Organic Geochemistry, 39*(8), pp. 1144-1149, ISSN 0146-6380

van Lieshout, M., Janssen, H., Cramers, C. A., Hetem, M. J. & Schalk, H. (1996). Characterization of polymers by multi-step thermal desorption/programmed pyrolysis gas chromatography using a high temperature PTV injector. *Journal of High Resolution Chromatography, 19*(4), pp. 193-199, ISSN 1521-4168

Liu, Z. & Phillips, J. B. (1991). Comprehensive two-dimensional gas chromatography using an on-column thermal modulator interface. *Journal of Chromatographic Science, 29*(6), pp. 227-231, ISSN 0021-9665

Marriott, P., Dunn, M., Shellie, R. & Morrison, P. (2003). Targeted multidimensional gas chromatography using microswitching and cryogenic modulation. *Analytical Chemistry, 75*(20), pp. 5532-5538, ISSN 0003-2700

Mayer, B. X., Rauter, W., Kählig, H. & Zöllner, P. (2003). A trifluoropropyl-containing silphenylene–siloxane terpolymer for high temperature gas chromatography. *Journal of Separation Science, 26*(15-16), pp. 1436-1442, ISSN 1615-9314

Nielsen, N.-P. V., Carstensen, J. M. & Smedsgaard, J. (1998). Aligning of single and multiple wavelength chromatographic profiles for chemometric data analysis using correlation optimised warping. *Journal of Chromatography A, 805*(1-2), pp. 17-35, ISSN 0021-9673

Nouvelle, X. & Coutrot, D. (2010). The Malcom distribution analysis method: A consistent guideline for assessing reservoir compartmentalisation from GC fingerprinting. *Organic Geochemistry, 41*(9), pp. 981-985, ISSN 0146-6380

Phillips, J. B. & Beens, J. (1999). Comprehensive two-dimensional gas chromatography: a hyphenated method with strong coupling between the two dimensions. *Journal of Chromatography A, 856*(1-2), pp. 331-347, ISSN 0021-9673

Philp, R. P. (1994). High temperature gas chromatography for the analysis of fossil fuels: A review. *Journal of High Resolution Chromatography, 17*(6), pp. 398-406, ISSN 1521-4168

Philp, R. P., Hsieh, M. & Tahira, F. (2004). An overview of developments related to the characterization and significance of high molecular weight paraffins/hydrocarbons (>C40) in crude oils. *Geological Society, London, Special Publications, 237*(1), pp. 37 -51, ISSN 0305-8719

Pursch, M., Sun, K., Winniford, B., Cortes, H., Weber, A., McCabe, T. & Luong, J. (2002). Modulation techniques and applications in comprehensive two-dimensional gas chromatography (GC×GC). *Analytical and Bioanalytical Chemistry, 373*(6), pp. 356-367, ISSN 1618-2650

del Rio, J. C. & Philp, R. P. (1992). High-molecular-weight hydrocarbons: A new frontier in organic geochemistry. *Trends in Analytical Chemistry, 11*(5), pp. 194-199, ISSN 0165-9936

Roehner, R. M., Fletcher, J. V., Hanson, F. V. & Dahdah, N. F. (2002). Comparative compositional study of crude oil solids from the Trans Alaska Pipeline System using high-temperature gas chromatography. *Energy and Fuels, 16*(1), pp. 211-217, ISSN 1520-5029

Subramanian, M., Deo, M. D. & Hanson, F. V. (1996). Compositional analysis of bitumen and bitumen-derived products. *Journal of Chromatographic Science, 34*(1), pp. 20-26, ISSN 0021-9665

Takayama, Y., Takeichi, T., Kawai, S. & Morikawa, M. (1990). Behaviours of siloxane polymers containing phenyl or silarylene as stationary phases for high-temperature gas chromatography. *Journal of Chromatography A, 514*, pp. 259-272, ISSN 0021-9673

Thanh, N. X., Hsieh, M. & Philp, R. P. (1999). Waxes and asphaltenes in crude oils. *Organic Geochemistry, 30*(2-3), pp. 119-132, ISSN 0146-6380

Tomasi, G., van den Berg, F. & Andersson, C. (2004). Correlation optimized warping and dynamic time warping as preprocessing methods for chromatographic data. *Journal of Chemometrics, 18*(5), pp. 231-241, ISSN 0886-9383

Vendeuvre, C., Ruiz-Guerrero, R., Bertoncini, F., Duval, L. & Thiébaut, D. (2007). Two-dimensional gas chromatography for detailed characterisation of petroleum products. *Oil & Gas Science and Technology - Revue de l'IFP, 62*(1), pp. 43-55, ISSN 1294-4475

Vendeuvre, C., Ruiz-Guerrero, R., Bertoncini, F., Duval, L., Thiébaut, D. & Hennion, M. C. (2005). Characterisation of middle-distillates by comprehensive two-dimensional gas chromatography (GC × GC): A powerful alternative for performing various standard analysis of middle-distillates. *Journal of Chromatography A, 1086*(1-2), pp. 21-28, ISSN 0021-9673

Wang, F. C.-Y., & Walters, C. C. (2007). Pyrolysis comprehensive two-dimensional gas chromatography study of petroleum source rock. *Analytical Chemistry, 79*(15), pp. 5642-5650, ISSN 0165-9936

Wang, Y. W., Chen, Q., Norwood, D. L. & McCaffrey, J. (2010). Recent development in the application of comprehensive two-dimensional gas chromatograph. *Journal of Liquid Chromatography & Related Technologies, 33*(9-12), pp. 1082-1115, ISSN 1082-6076

Woods, J. R., Kung, J., Kingston, D., Kotlyar, L., Sparks, B. & McCracken, T. (2008). Canadian crudes: A comparative study of SARA fractions from a modified HPLC separation technique. *Oil and Gas Science and Technology, 63*(1 SPEC. ISS.), pp. 151-163, ISSN 1294-4475

Zrostlíková, J., Hajslová, J. & Cajka, T. (2003). Evaluation of two-dimensional gas chromatography-time-of-flight mass spectrometry for the determination of multiple pesticide residues in fruit. *Journal of Chromatography A, 1019*(1-2), pp. 173-186, ISSN 0021-9673

Zuo, J. Y. & Zhang, D. (2000) Plus fraction characterization and PVT data regression for reservoir fluids near critical conditions. Paper SPE 64520 presented at the SPE Asia Pacific Oil and Gas Conference and Exhibition, Brisbane, Australia, 16–18, October, 2000.

Zuo, J. Y. & Zhang, D. (2008). Wax formation from synthetic oil systems and reservoir fluids. *Energy and Fuels, 22*(4), pp. 2390-2395, ISSN 1520-5029

18

Determination of Organometallic Compounds Using Species Specific Isotope Dilution and GC-ICP-MS

Solomon Tesfalidet
Department of Chemistry, Umeå University
Sweden

1. Introduction

Isotope dilution is used in speciation analysis to determine the concentration of specific chemical forms of an element, offering the possibility for correcting eventual loses or alteration of the species that can occur during the analytical process. Gas Chromatography (GC) is known as the strongest separation method used for separation of volatile and thermo stable organic compounds. Among the detection methods that are used, inductively coupled plasma - mass spectrometry (ICP-MS) is marked as the detector of choice for organometallic compounds. The coupling of GC with ICP-MS is extended by making use of the capability of ICP-MS to detect several isotopes simultaneously. Isotope dilution (ID) which was originally used in organic mass spectrometry is now extensively used in speciation analysis of inorganic compounds. This requires that the analyte of interest has more than one stable isotope. Chemical forms of mono-isotopic elements such as Arsenic (m/z 75) can therefore not be determined using isotope dilution. Other elements such as lead (4 isotopes), tin (10 isotopes), mercury (7 isotopes) and chromium (4 isotopes) have all been determined using isotope dilution [1]. According to IUPAC (International Union of Pure and Allied Chemistry) definition speciation of an element is "*the distribution of that element amongst defined chemical species in a system*" and speciation analysis is "*analytical activities of identifying and/or measuring the quantities of one or more individual chemical species in a sample*". Information about the total concentration of an element is usually not enough for a satisfactory characterization of a sample. Inorganic tin is for example harmless compared to organotin compounds (OTCs). The total concentration of tin, determined in a sample contaminated with the more toxic OTCs, does therefore not give meaningful data for risk assessment, making speciation analysis of the organotin compounds very essential.

In ID a known amount of an isotopically enriched element is added to a sample material. A simple question that one may raise is: what does this addition of an isotopically enriched element lead to? To answer this question Mikael Berglund referred to a work done by an enthymologist named C.H.N Jackson back in 1933[2]. Jackson used fly dilution to study the density of tsetse flies in a region in the village Ujiji at Tanganyka lake in Africa [3]. Jackson released a 'synthetic'' Tsetse fly population, which consisted of marked Tsetse flies, into a natural Tsetse fly population. After complete mixing he caught a representative sample of Tsetse flies and counted the marked and unmarked flies. Knowing the total number of marked and released flies in the sample he could calculate the number of unmarked flies in the region. In a similar fashion we can imagine of counting the number of blue balls (?) in container A in figure 1. To this we add a known amount of identical, red colored, balls from container B. After a thorough mixing we take a portion from the mixture M, and count how many of each ablue nd red balls we have in this portion {P}.

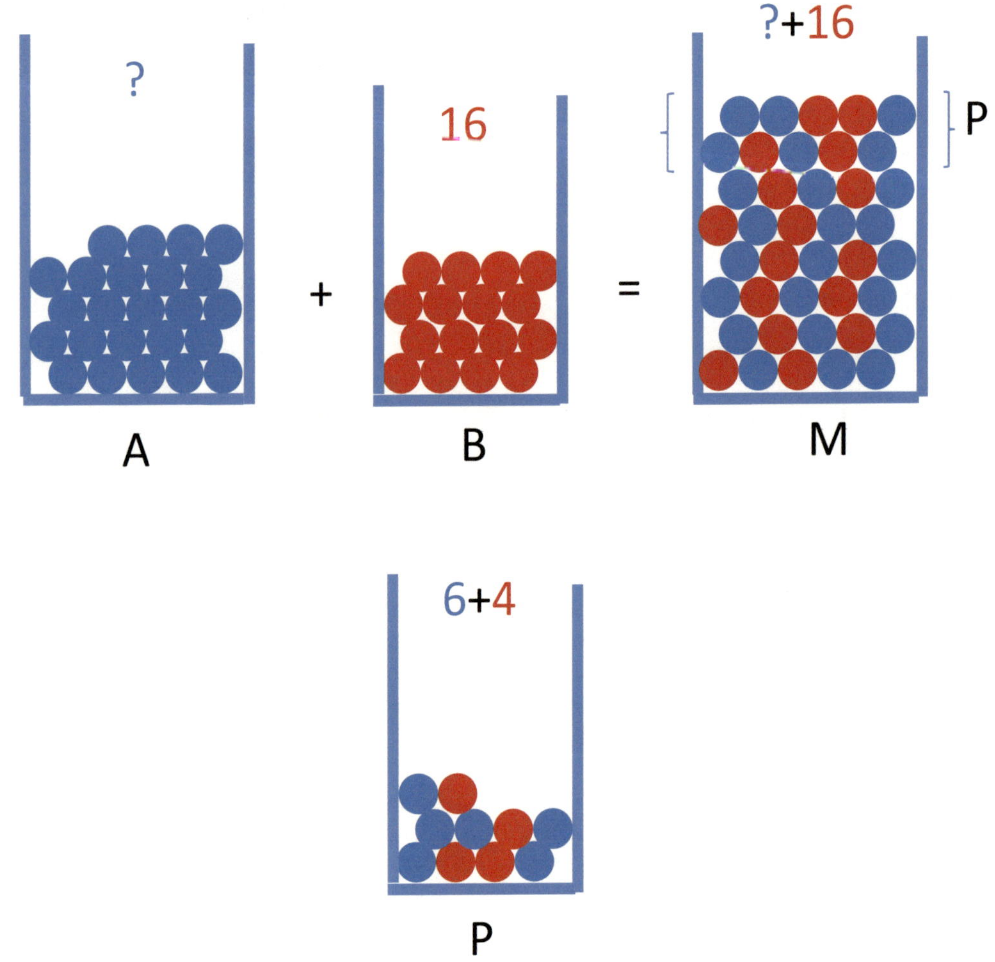

Fig. 1. A: container containing unknown number of blue balls; B: container containing known number of red balls; M: mixture of blue balls in A and red balls from B; P a portion of balls taken from the mixture M.

Using analogous equation to that used by Jackson back in 1933 we can calculate the number of blue bolls in M which also is equal to that we have in container A.

$$\frac{Number\ of\ red\ bolls\ in\ P}{Number\ of\ blue\ bolls\ in\ p} = \frac{Total\ number\ of\ red\ bolls\ in\ C}{Total\ number\ of\ blue\ bolls\ in\ C} \quad (1)$$

$$\frac{4}{6} = \frac{16}{Total\ number\ of\ blue\ bolls\ in\ C}$$

Total number of blue bolls in M = 24 = number of blue bolls in container A.

This quantification principle, exemplified by the ball addition, is in fact the same as that used in isotope dilution (ID) where isotopically marked analyte (called spike) is added to a sample, containing a natural analyte.

In ID a known amount of enriched isotope (called the "spike") is added to a sample. After equilibration of the spike isotope with the natural isotope in the sample the isotopic ratio is measured. Normally a mass spectrometer is used to measure the altered isotopic ratio(s). The measured isotope ratio of isotope A to isotope B, R_m, can be calculated using equation 1.

$$R_m = \frac{A_x C_x + A_s C_s W_s}{B_x C_x W_x + B_s C_s W_s} \quad (2)$$

where A_x and B_x are the atom fractions of isotopes A and B in the sample; A_s and B_s are the atom fractions of isotopes A and B in the spike; C_x and C_s, are the concentrations of the analyte in the sample and spike, respectively; and W_x and W_s are the weights of the sample and spike, respectively. The concentration of the analyte in the sample can then be calculated using equation 2.

$$C_x = \left(\frac{C_s W_s}{W_x}\right)\left(\frac{A_s - R_m B_s}{R_m B_x - A_x}\right) \quad (3)$$

It should however be noted that depending on the analytical method at hand, the stability of the species under investigation, and the need for addressing the resulting uncertainty in the analytical result the mathematical expression used for calculation of species concentration can look different and get more complicated [2, 4, 5]. In the forthcoming sections speciation analysis of organtin compounds will be discussed. Organotin compounds are one of the most investigated organometallic compounds, using isotope dilution GC-ICPMS, and will therefore fit perfect in this chapter for elaborating the analytical methodology and explain why isotope dilution is needed for speciation analysis.

2. Speciation of organotin compounds (OTCs)

Tin in its inorganic form is generally accepted as being non-toxic, but the toxicological pattern of organotin compounds is very complex. The biological effects of the substances depend on both the nature and the number of the organic groups bound to the Sn cation. Figure 2 shows the ionic form of the most extensively studied types of organotin compounds.

The ecotoxicological effects of organotin compounds (OTC), mainly tributyltin (TBT) and triphenyltin (TPhT) but also their di- and monosubstituted degradation products are well documented. Nowadays, the release of TBT from antifouling paints is recognized worldwide as being one of the main contamination problems for the marine environment, and the use of TBT-based antifouling paints is almost everywhere restricted by law. In order to evaluate the environmental distribution and fate of these compounds, and to control the effectiveness of these legal provisions, many analytical methods have been developed among which gas chromatography coupled to inductively coupled plasma mass spectrometry (GC-ICP-MS) is the most powerful.

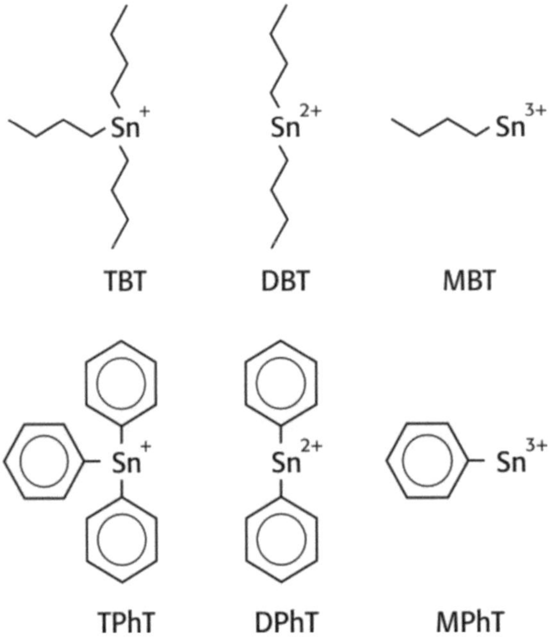

Fig. 2. The ionic form of butyltins and phenyltins of varying substitution.

TBT-based paints are used on vessels hulls to prevent growth of aquatic organisms that create roughness giving rise to increased drag, resulting in reduced vessel speed per unit energy consumption. The antifouling paint consists of a film-forming material with a biocidal ingredient and a pigment. It works by releasing small amounts of the biocide from the painted hull into the water, forming a thin envelope of highly concentrated TBT around the boat. The toxic concentration repels the settling stages of fouling organisms, like barnacles, seaweeds, or tubeworms on the boat's water-immersed surfaces [6].

However, the constant release of OTC from anti-fouling paints has led to toxic effects for nontarget aquatic species in the aquatic environment, where they cause deleterious effects, such as shell anomalies in oysters and imposex in gastropods, even at concentrations as low as nanograms per liter. Despite the restrictions organotin compounds (mainly TBT) are still used in paint formulation for large vessels, and about 69 % of all large ships are reported to use them [7]. Once released into the aquatic environment, organotin compounds may undergo a variety of degradation reactions until they are finally adsorbed onto suspended solids and sediments. Sediments are considered to be the ultimate sinks for organotin compounds [8].

In water, TBT decomposes into less toxic DBT and MBT species. The problem is that this favourable decomposition takes place far more slowly in sediments, creating an ecotoxicological risk long after its release into a given area, making sediments a secondary source of pollution. TBT and its degradation products DBT and MBT have been detected in different environmental compartments, both marine (waters, sediments, and biota) and terrestrial (waters and soils). The occurrences of the less toxic MBT and DBT compounds in the environment have so far been related to the degradation of TBT caused by microbial activity and/or photochemical reactions, but recently evidence for direct input of MBT and DBT was found. In fact the major application of organotin compounds (about 70 %) is the use of mono and dialkyltin derivatives as heat and light stabilizer additives in PVC processing [6].

3. Speciaition analysis of OTCs

Several techniques, based on species specific analytical methods, have been developed for the determination of butyltins in environmental matrices. Hyphenated systems, based on on-line coupling of gas chromatography (GC), liquid chromatography (LC), or supercritical fluid chromatography (SFC), to mass spectrometry (MS), inductively coupled plasma mass spectrometry (ICP-MS), atomic absorption spectrometry (AAS), and microwave induced plasma atomic emission spectrometry (MIP-AES) are in current use [7]. Among the different techniques, the coupling of GC to ICP-MS appears to be one of the most popular techniques, due to high sensitivity and multi-elemental and multi-isotopic capabilities [9]. Some of the most frequently used sample pretreatment methods for the determination of OTC, in various samples, are summarized in Table 1.

The coupling of chromatography with ICP-MS and application of calibration methods that are based on ID permits detection of several compounds. ID is a technique based on isotope ratio measurement whereby the natural isotopic abundance ratio of an analyte is altered by spiking with a standard that has a different isotopic abundance ratio. The prerequisite for the technique is that the analyte of interest should have more than one stable isotope [1]. In the case of tin (10 isotopes) the isotope of highest abundance is ^{120}Sn, usually referred as the reference isotope, and the spike isotope is generally one of the less abundant natural isotopes. For the purposes of speciation analysis, where OTCs are to be determined, there is a requirement for the isotopically enriched element-species to be synthesized. If two interference-free isotopes of a given element can be found, isotope dilution ICP-MS (ID-ICP-MS) can be performed, which generally provides superior accuracy and precision over other calibration strategies, including external calibration and the method of standard addition, because a ratio rather than an absolute intensity measurement is used in the quantification of the analyte concentration. Once equilibration is achieved between the analyte in the sample and the added spike, ID-ICP-MS is theoretically capable of compensating for any subsequent loss of analyte during sample manipulation, suppression of ion sensitivities by concomitant elements present in the sample matrix, and instrument drift. ID-ICP-MS may therefore be considered as a primary method of analysis and can play a crucial role for quality assurance in trace element chemical speciation of environmental and biological samples. ID-ICP-MS coupled with gas chromatography (GC) ID-GC-ICP-MS [9] and to less extent liquid chromatography (LC) ID-LC-ICP-MS [1] are the techniques that are used for speciation of organotin compounds. The derivatisation step is not needed in the case of LC (omitting one source of error), but the coupling with GC is superior when it comes to analysing samples of very low concentrations (pg/g) of OTC.

The presently available techniques for the determination of OTCs involve several analytical steps such as extraction, pre-concentration, cleanup, derivatisation (when gas chromatography is used), separation, and finally detection by element- or molecule specific techniques such as ICP-MS. The multitude of analytical steps is causing errors at various levels, making speciation analysis a difficult task.

4. Derivatization of OTCs for gas chromatographic separation

In order to enable separation by GC, the ionic organotin compounds need to be converted into volatile species such as the hydrides, or their fully alkylated form (derivatisation). For the reduction to hydrides, sodium tetrahydroborate ($NaBH_4$) is commonly used, whereas derivatisation through alkylation can be carried out with Grignard reagents, sodium tetraethylborate ($NaBEt_4$), or sodium tetra (n-propyl) borate ($NaBPr_4$). Derivatisation with $NaBEt_4$ makes the sample preparation faster and easier because it combines an *in situ* derivatisation with extraction of the ethylated organotin compounds into an organic phase [1].

Sample type	Sample pretreatment	Compound studied (amount found (ng))	Separation and detection technique	Ref
Marine sediment from contaminated harbour area in Mar piccolo (Italy)	Add 3 mL HCl and 6 mL methanol to 1 g of sample. Shake and sonicate for 15 min in ultrasonic bath. Add 3 mL of acetate buffer (pH 5.3) and centrifuge the leached sediment at 4000 rpm for 15 min. Pipette 1 mL supernatant into a 15 mL glass vial and add 6 mL of HOAc/NaOAc buffer (pH 5.3). Close the vial with a septum and add 1 mL of $NaBEt_4$ solution with a syringe. Sonicate the reaction mixture and pierce the SPME needle into the septum and expose the fiber into the headspace.	MBT = 8 DBT = 10 TBT = 1	Headspace micro-extraction GC/MS	28
Sediment samples Samples 1 and 2 from the harbour of Ostend, sample 3 from a dry dock in the harbour of Antwerp, and sample 4 from a leisure craft maintenance place located in the province of Limburg, all in Belgium	Add 2 ml of HCl (32 %) and 8 ml H_2O to 1 g sample in a centrifugation vessel. Add 25 ml of hexane-ethyl acetate mixture (1:1) containning 0.05 % tropolone. Sonicate the mixture for 1 h, followed by centrifugation at 3000 rpm for 5 min. Transfer the organic phase into an extraction vessel and evaporate to dryness using rotary evaporation. Add 0.5 zml of hexane containing Pe_3SnEt as an internal standard and derivatize by adding 1 ml of $NaBEt_4$ solution together with 50 ml of acetate buffer solution. Shake the mixture manually for 5 min	**Sample 1** MBT= 0.14 ± 0.02 DBT= 0.44 ± 0.03 TBT= 0.14 ± 0.02 **Sample 2** MBT= 0.36 ± 0.02 DBT= 1.11 ± 0.12 TBT= 2.33 ± 0.14 **Sample 3** MBT= 8.13 ± 0.33 DBT= 10.0 ± 0.6 TBT= 26.4 ± 1.8 **Sample 4**	GC interfaced with AAS and AES	14

Sample type	Sample pretreatment	Compound studied (amount found (ng))	Separation and detection technique	Ref
	and seperate the hexane phase. Introduce the extract into a clean-up column (a pasteur pipette filled with alumina to form a plug of 5 cm). Add an additional volume of 1 ml diethyl ether and evaporate the added diethyl ether using a gentle stream of nitrogen.	MBT= 1.55 ± 0.09 DBT= 1.67 ± 0.20 TBT= 6.60 ± 0.18		
Sediment samples Two samples (1&2) from the harbour of Ostend (Belgium).	Add 4 ml H_2O, 1 ml acetic acid (96 %), 1 ml DDTC in pentane, and 25 ml hexane into a 100 ml Erlenmeyer flask containing 1 g sample. Sonicate the mixture for 30 min and decant the organic phase into a 100 ml beaker. Repeat the extraction with 25 ml of hexane and stir magnetically for 30 min. Centrifuge the mixture for 5 min at 3000 rpm. Dry over Na_2SO_4 and evaporate to dryness on a rotary evaporator. Add 250 µl of n-octane, containing Pr_3SnPe and pentylate with 1 ml of 1 M n-PeMgBr. Destroy excess Grignard reagent by adding 10 ml of 0.5 M H_2SO_4. Introduce the octane layer into a clean-up column (a pasteur pipette filled with alumina to form a plug of 5 cm). Add an additional volume of 1 ml diethyl ether and evaporate the added diethyl ether using a gentle stream of nitrogen.	**Sample 1** DBT= 0.43 ± 0.02 TBT= 0.31 ± 0.03 **Sample 2** DBT= 1.39 ± 0.06 TBT= 2.67 ± 0.08	GC interfaced with AAS and AES	14
Water 1. Sea water from Sahrm el Sheikh harbour in South Sinai (Egypt) 2. Harbour water from Wädenswil, Lake Zurich (Switzerland)	Add 0.5 mL of acetic acid/acetate buffer solution (pH 5) and 1.45 g of sodium chloride to 50 mL water sample. After shaking spike with 100 µl of deuterated standard solution mixture (12.5 ng/L of each species in MeOH). Shake and add 150 µl of 1.5 % (w/v) $NaBEt_4$ aqueous solution. Add 1 ml of hexane and shake in the dark for	**1. Sahrm el Sheikh** MBT = 3.4 ± 7.6 DBT = 2.1 ± 24 TBT = 2.6 ± 17 MPhT= 1.5 ± 6.7 DPhT = 0.5 ± 76 TPhT = 4.8 ± 2.1 **2. Wädenswil** MBT = nd	LLE, Large volume injection GC/MS	29

Sample type	Sample pretreatment	Compound studied (amount found (ng))	Separation and detection technique	Ref
	12 h at 25 °C. Transfer 180 µl of the hexane extract to a 1 mL auto-sampler vial and spike with 10 µl of 0.2 ng/µl TeBT in hexane. Inject 50 µl in to the GC.	DBT = 3 ± 17 TBT = nd MPhT = 37 ± 3 DPhT = 23 ± 6 TPhT = 353 ± 3		
Sediment Certified reference material PACS-2 (0.98 ± 0.13 TBT ng/L)	Put 0.5 g of PACS-2 in a Prolabo microwave digester and spike with 0.04 mL of ^{117}Sn-enriched TBT solution. Add 5 mL of acetic acid and heat at 60 % power for 3 min. Centrifuge at 2000 rpm for 5 min. Transfer 1 mL volume of the supernatant to a reaction vial and add 1 mL of deionized water. Adjust to pH 5-6 with 1.2 mL ammonium hydroxide. Buffer the content with 0.8 mL of ammonium citrate (2 mol L^{-1}) and dilute to 10 mL with methanol.	TBT = 1.018 ± 0.0315	Microwave extraction, ID-HPLC-ICP-MS	30
Sediment/Sludge 1. Sediment pore water from (1-20 cm) from Stansstaad harbour, Lake Lucerne (Switzerland) 2. Sewage sludge from four wastewater treat-ment plants in the Zurich canton (Switzerland)	Weigh 2.5 g freeze dried sediment or sludge in a beaker. Spike homogeneously with 500 µl of deuterated standard solution mixture (12.5 ng/L of each species in MeOH). Mix with 9 g of quartz sand and transfer the mixture to 11 mL extraction cells. After two hours fill the extraction cells with a mixture of 1 M sodium acetate and 1 M acetic acid in MeOH, using ASE. Extract with three to five static cycles of 5 min. Renew 4 mL of solvent between each static extraction. Rinse the cells with 4 mL of solvent and purge with nitrogen. Transfer the combined extracts to 250 mL volumetric flasks containing 7.3 g of NaCl. Add water and adjust the pH to 5 with 1 M NaOH. Add 1 mL of aqueous solution of 5 % (w/v) $NaBEt_4$ and fill the bottles to 250 mL with water. Add 2	1) Sediment pore water MBT = 11.0± 1.9 DBT = 4.5 ± 4.7 TBT = 9.6 ± 4.4 MPhT = 12.3 ± 4 DPhT = 2.5 ± 14 TPhT = 4.1± 3 2) Sewage sludge MBT = 300 ± 4 DBT = 253 ± 5 TBT = 45 ± 5 MPhT = 7 ± 21 DPhT = nd TPhT = 7 ± 38	ASE, Large volume injection GC/MS	29

Sample type	Sample pretreatment	Compound studied (amount found (ng))	Separation and detection technique	Ref
	mL of hexane and shake for 12 h. Transfer 500 µl of the hexane extract to 2 mL GC vials and spike with 10 µl of TeBT (5 ng/µl). For sewage sludge transfer the hexane extract to 10 mL centrifuge tubes containing 0.9 g of deactivated silica gel and 2 mL water. Shake vigorously and centrifuge.			

Table 1. Some of the most commonly used methods for the determination of OTCs in various samples along with the analysis results obtained.

5. Species Specific Isotope Dilution (SSID)

In the last few years different procedures for the speciation of organotin compounds have been proposed, among which species specific isotope dilution (SSID) is one [10-11]. The use of isotope enriched spike solutions has not only the potential for accurate, precise and simultaneous determination of OTCs but also the evaluation of different extraction and derivatization protocols for the analysis of sediments and biological sample in which species interconversion/decomposition can take place during sample preparation. Organotin compounds synthesized from isotopically enriched tin metals are used for preparing spikes that can be used for the determination of concentrations using species specific isotope dilution (SSID). Different types of calibration strategies based on SSID have been used for the determination of organotin compounds:

i. Single isotope - species specific isotope dilution (SI-SSID): where a mixture containing organotin species, all with the same isotope, is used as a spike. For example: a mixture containing ^{116}Sn-enriched MBT, ^{116}Sn-enriched DBT, and ^{116}Sn-enriched TBT is used for the determination of butyltins.
ii. Multiple isotope - species specific isotope dilution (MI-SSID): where a mixture containing the organotin species, each with different isotope, is used as a spike. For example: a mixture containing ^{119}Sn-enriched MBT, ^{118}Sn-enriched DBT, and ^{117}Sn-enriched TBT is used for the determination of butyltins.

Single isotope species specific isotope dilution (SI-SSID) has been used during the last fifteen years for speciation analysis of bromine [12], chromium [13], iodine [14], selenium [15,16], mercury [17] and tin [18-20]. Recently multi-isotope species specific isotope dilution (MI-SSID) has been used to monitor and correct for the degradation/transformation that takes place during sample preparation [21-23]. The use of multi isotope spike can also be extended to compare different extraction procedures and study the inherent procedural parameters that govern the process of degradation/transformation. If these problems are fully understood and addressed, appropriate strategies can be set to facilitate accurate speciation analysis of the compounds of interest in various types of samples.

One of the drawbacks with SI-SSID is the difficulty to assess the degradation/redistribution of the OTs that can take place during sample workup. Another problem is the inability to match the concentrations of the organotin compounds in the spike with those present in the sample, thus affecting the accuracy and precision of the results [18, 24-26]. The latter problem can however be circumvented by using MI-SSID [5, 23].

Owing to the problems with degradation of phenyltin species during the sample pretreatment steps, poor precisions within and large spreads between laboratory results were obtained when certifying organotin compounds in the BCR CRM-477. As a consequence, only indicative values were given for phenyltin species [27]. Figure 3 shows deconvoluted chromatograms obtained for the multi-isotope standard where no peaks, corresponding to redistribution products of the enriched phenyltin standard, are visible. This indicates that the integrity of the phenyltin species is preserved during the ethylation and the detection procedure.

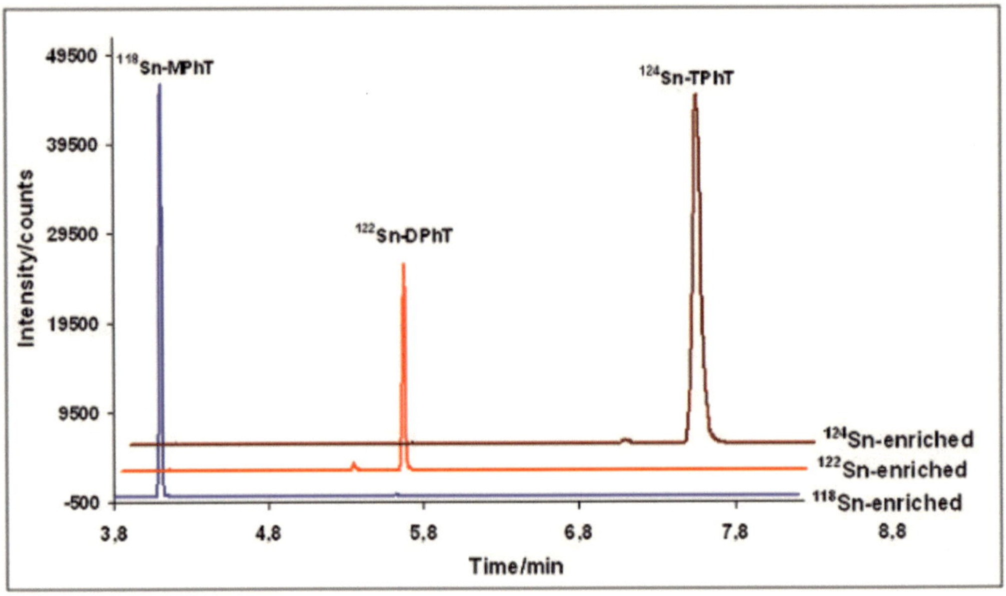

Fig. 3. Deconvoluted GC-ICPMS sub-chromatograms obtained for a mixture of ^{118}Sn-enriched MPhT (106.5 ng g^{-1}), ^{122}Sn-enriched DPhT (57.4 ng g^{-1}) and ^{124}Sn-enriched TPhT (239.0 ng g^{-1}). The chromatograms are shifted from each other by 0.1 minute and 3000 counts for clarity. [*From Van D.N., Muppala S.R.K., Frech W., Tesfalidet S., Preparation, preservation and application of pure isotope-enriched phenyltin species Anal. Bioanal. Chem., 2006, 386, 1505*]

The presently available techniques for the determination of OTCs involve several analytical steps such as extraction, pre-concentration, cleanup, derivatisation (when gas chromatography is used), separation, and finally detection by element- or molecule specific techniques such as ICP-MS. The multitude of analytical steps is causing errors at various levels, making speciation analysis a difficult task.

In the determination of PhTs, species transformation/interconversion is one of the most serious obstacles which affects the reliability of the analytical results. Using MI-SSID, one can assess the degradation/transformation processes, enabling calculation of degradation-corrected concentrations for the species of interest. Possible degradation/transformation pathways of PhTs are depicted in Fig. 4

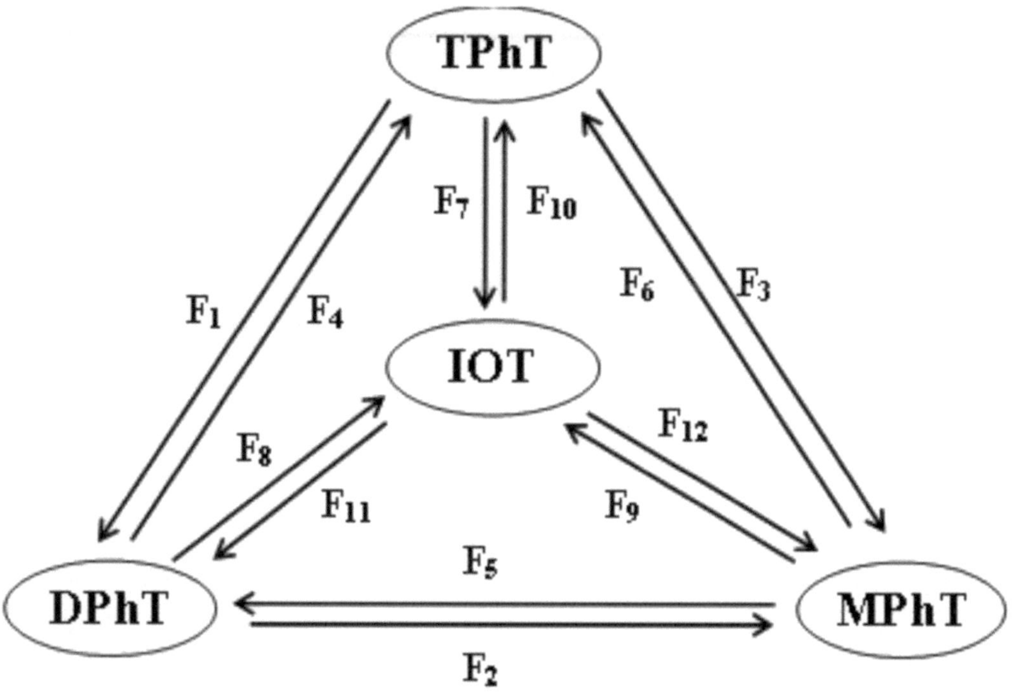

Fig. 4. Possible interconversion pathways of inorganic tin (IOT) and PhTs. F_i is interconversion factor corresponding to the interconversion reaction i. TPhT triphenyltin, DPhT diphenyltin, MPhT monophenyltin. [*From Van D.N., Bui T.T.X., Tesfalidet S. "The transformation of phenyltin species during sample preparation of biological tissues using multi-isotope spike SSID-GC-ICPMS, Anal. Bioanal. Chem 2008, 392, 737*]

6. Instrumentation and operating conditions for the GC and ICP-MS

A Varian 3300 gas chromatograph (Palo Alto, CA, USA) fitted with an on-column injector liner and a methyl silicone capillary column (30 m by 0.53 mm i.d., 1.5 μ film thickness; SPB-1, Supelco, Bellafonte, PA) was used for separation of Sn species and an Agilent 7500 ICP-MS (Foster City, CA) was used for detection. The GC was coupled to the ICP-MS *via* a custom made interface. The operating parameters of the ICP-MS were selected by optimizing the sensitivity for ^{129}Xe, by adding Xe gas at 0.5 ml/min to the Ar carrier gas flow. Oxygen was added to the plasma to prevent carbon deposits on the sampler/skimmer Pt cones. The operating parameters of the GC and ICP-MS are given in Table 2.

GC parameters	
Injection volume	1 μL
Carrier helium gas flow	22 mL min^{-1}
Injector temperature	180 °C
Oven temperature	130 °C 🕐 40 °C min^{-1} 🕐 210 °C, hold 0.5 minutes 210 °C 🕐 7 °C min^{-1} 🕐 225 °C, hold 0.5 minutes 225 °C 🕐 40 °C min^{-1} 🕐 280 °C, hold 2.0 minutes
Transfer line temperature	200 °C
ICP-MS parameters	
ICP RF power	1200 W
Plasma argon gas flow	15 L min^{-1}
Nebulizer argon gas flow	1.0 L min^{-1}
Auxiliary argon gas flow	0.9 L min^{-1}
Auxiliary oxygen gas flow	3 mL min^{-1}
Sampler/Skimmer cones	Platinum
Dwell time	100 ms for $^{116}Sn^+$, $^{117}Sn^+$, $^{118}Sn^+$, $^{119}Sn^+$, $^{120}Sn^+$ and $^{124}Sn^+$

Table 2. Operating conditions for the GC and ICP-MS.

7. Determination of OTCs in biological samples

Various extraction procedures have been used for the analysis of organotin compounds in biological samples. Pellegrino et al. [31] compared twelve selected extraction methods to evaluate the extraction efficiencies obtained for a certified reference material (CRM). The organic solvent, the nature and concentration of the acid used for leaching, and the presence/absence of tropolone used as complexing agent, were all found to influence the grade of transformation that takes place during extraction [32]. Although the performance of the extraction procedures can be validated by recovery tests using certified reference materials (CRMs) or fortified samples, the validity of these procedures on real samples cannot be guaranteed [18,33]. This is because the adsorption/binding forces of the species of interest to the solid are strongly dependent on the matrix [10]. A mild extraction technique could, for example, give rise to incomplete extraction of organotin species from the solid sample while a harsher extraction technique facilitates the degradation/rearrangement of the species of interest [32,34,35]. In both cases, the results of speciation analysis will fail to reflect the real speciation in the solid sample. As a result, it has yet not been able to certify the concentration of phenyltins in the reference material BCR CRM-477 and only indicative values are available [36].

8. Analysis of PhTs in mussel tissue

When analyzing PhTs in certified reference material, mussel tissue, BCR CRM-477, 8.7±0.9% of TPhT was converted to DPhT and 9.0±0.2% of DPhT to MPhT. Other dephenylation/phenylation products were not observed. Peaks indicating the degradation of ^{124}Sn-enriched TPhT and ^{122}Sn-enriched DPhT are shown in Fig. 5 by signals for ^{124}Sn-enriched DPhT and ^{122}Sn-enriched MPhT [5].

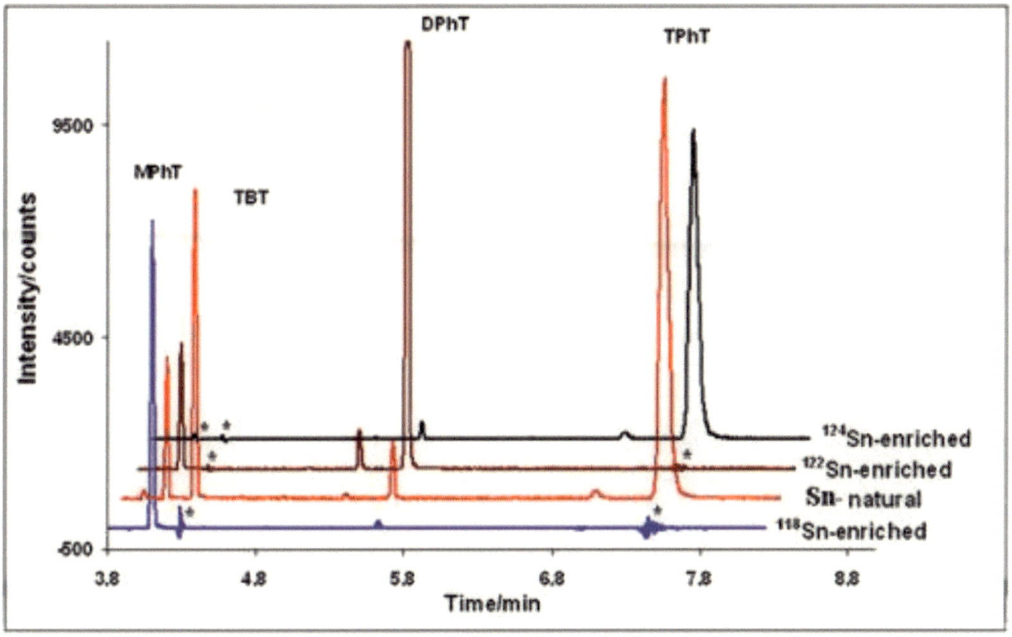

Fig. 5. Deconvoluted GC-ICPMS sub-chromatograms obtained for 0.2g BCR CRM-477 spiked with 0.14g ^{118}Sn-enriched MPhT (90.2 ng g^{-1}), ^{122}Sn-enriched DPhT (367 ng g^{-1}) and ^{124}Sn-enriched TPhT (315 ng g^{-1}). (*) is the fluctuation of the baseline due to deconvolution. The chromatograms are shifted from each other by 0.1 minute and 700 counts for clarity. [*From Van D.N., Muppala S.R.K., Frech W., Tesfalidet S., Preparation, preservation and application of pure isotope-enriched phenyltin species Anal. Bioanal. Chem., 2006, 386, 1505*]

8.1 Equations for calculating the inter-conversion factors and concentrations of PhT species

Various equations have been proposed for the calculation of concentrations in isotope dilution [17, 37-38]. The system of equations used for calculating the inter-conversion factors and concentrations of PhT species presented here are similar to those presented by Rodriguez-Gonzalez et.al. [37] and Nguyen D.N. et.al [5].

A deconvoluted chromatogram of the type shown in figure 5 visualizes how each added isotope-enriched species is distributed between the PhT species. Deconvolution is performed for each species by solving the system of four linear equations (Eq. 4) in an Excel spreadsheet:

$$\begin{aligned}
\sum S_{118} &= R_{118/118}.S_{118} + R_{118/120}.S_{120} + R_{118/122}.S_{122} + R_{118/124}.S_{124} \\
\sum S_{120} &= R_{120/118}.S_{118} + R_{120/120}.S_{120} + R_{120/122}.S_{122} + R_{120/124}.S_{124} \\
\sum S_{122} &= R_{122/118}.S_{118} + R_{122/120}.S_{120} + R_{122/122}.S_{122} + R_{122/124}.S_{124} \\
\sum S_{124} &= R_{124/118}.S_{118} + R_{124/120}.S_{120} + R_{124/122}.S_{122} + R_{124/124}.S_{124}
\end{aligned} \tag{4}$$

To determine the degree of inter-conversion between PhTs, the isotope ratios 120/118, 120/122 and 120/124 for each PhTs were calculated. The following inter-conversion factors were considered: F_1; TPhT degradation to DPhT, F_2; DPhT to MPhT and F_3; TPhT to MPhT F_4; DPhT phenylation to TPhT, F_5; MPhT to DPhT and F_6; MPhT to TPhT. The calculations of two inter-conversion factors related to the formation of TPhT (F_4 and F_6) are exemplified in equations 5&6. Once all the inter-conversion factors are calculated, mass balance equations for each species can be established and used for computing concentrations of each species in the sample. For example, the third row in the matrix (equation 8) is obtained by rearranging Equation 4, the mass balance of TPhT.

The symbols used in the equations represent the following: N_s^{MPhT} , N_s^{DPhT} and N_s^{TPhT} are the number of moles of each species present in the sample; N_{sp}^{X} is the number of moles of the PhTs X (X = MPhT, DPhT or TPhT) spiked into the sample; At_s^i is the abundance of tin isotope ***i*** (***i*** = 118, 120, 122 or 124) in the sample; $At_{i,sp}^{MPhT}$, $At_{i,sp}^{DPhT}$ and $At_{i,sp}^{TPhT}$ are the abundance of tin isotope ***i*** in the spiked ^{118}Sn-enriched MPhT, ^{122}Sn-enriched DPhT and ^{124}Sn-enriched TPhT, respectively. $N_{124,m}^{X}$, $N_{122,m}^{X}$, $N_{120,m}^{X}$ and $N_{118,m}^{X}$ are the number of moles of the corresponding isotopic PhTs X in the blend; $R_{j/k,m}^{X}$ is the tin isotope ratio of the reference isotope ***j*** (***j*** = 120) to the spiked isotope ***k*** (***k*** = 118, 122, 124) measured by GC-ICPMS for the phenyltin **X** in the blend after mass bias correction.

$$\begin{aligned} & N_{sp}^{TPhT}\left(\frac{At_{120,sp}^{TPhT}-R_{120/124,m}^{TPhT}At_{124,sp}^{TPhT}}{R_{120/124,m}^{TPhT}At_{124,s}-At_{120,s}}-\frac{At_{120,sp}^{TPhT}-R_{120/122,m}^{TPhT}At_{122,sp}^{TPhT}}{R_{120/122,m}^{TPhT}At_{122,s}-At_{120,s}}\right) \\ & +F_4N_{sp}^{DPhT}\left(\frac{At_{120,sp}^{DPhT}-R_{120/124,m}^{TPhT}At_{124,sp}^{DPhT}}{R_{120/124,m}^{TPhT}At_{124,s}-At_{120,s}}-\frac{At_{120,sp}^{DPhT}-R_{120/122,m}^{TPhT}At_{122,sp}^{DPhT}}{R_{120/122,m}^{TPhT}At_{122,s}-At_{120,s}}\right) \\ & +F_6N_{sp}^{MPhT}\left(\frac{At_{120,sp}^{MPhT}-R_{120/124,m}^{TPhT}At_{124,sp}^{MPhT}}{R_{120/124,m}^{TPhT}At_{124,s}-At_{120,s}}-\frac{At_{120,sp}^{MPhT}-R_{120/122,m}^{TPhT}At_{122,sp}^{MPhT}}{R_{120/122,m}^{TPhT}At_{122,s}-At_{120,s}}\right)=0 \end{aligned} \tag{5}$$

$$\begin{aligned} & N_{sp}^{TPhT}\left(\frac{At_{120,sp}^{TPhT}-R_{120/124,m}^{TPhT}At_{124,sp}^{TPhT}}{R_{120/124,m}^{TPhT}At_{124,s}-At_{120,s}}-\frac{At_{120,sp}^{TPhT}-R_{120/118,m}^{TPhT}At_{118,sp}^{TPhT}}{R_{120/118,m}^{TPhT}At_{118,s}-At_{120,s}}\right) \\ & +F_4N_{sp}^{DPhT}\left(\frac{At_{120,sp}^{DPhT}-R_{120/124,m}^{TPhT}At_{124,sp}^{DPhT}}{R_{120/124,m}^{TPhT}At_{124,s}-At_{120,s}}-\frac{At_{122,sp}^{DPhT}-R_{120/118,m}^{TPhT}At_{118,sp}^{DPhT}}{R_{120/118,m}^{TPhT}At_{118,s}-At_{120,s}}\right) \\ & +F_6N_{sp}^{MPhT}\left(\frac{At_{120,sp}^{MPhT}-R_{120/124,m}^{TPhT}At_{124,sp}^{MPhT}}{R_{120/124,m}^{TPhT}At_{124,s}-At_{120,s}}-\frac{At_{120,sp}^{MPhT}-R_{120/118,m}^{TPhT}At_{118,sp}^{MPhT}}{R_{120/118,m}^{TPhT}At_{118,s}-At_{120,s}}\right)=0 \end{aligned} \tag{6}$$

$$\begin{aligned} & F_6N_s^{MPhT}+F_4N_s^{DPhT}+N_s^{TPhT}=N_{sp}^{TPhT}\frac{At_{120,sp}^{TPhT}-R_{120/124,m}^{TPhT}At_{124,sp}^{TPhT}}{R_{120/124,m}^{TPhT}At_{124,s}-At_{120,s}} \\ & +F_4N_{sp}^{DPhT}\frac{At_{120,sp}^{DPhT}-R_{120/124,m}^{TPhT}At_{124,sp}^{DPhT}}{R_{120/124,m}^{TPhT}At_{124,s}-At_{120,s}}+F_6N_{sp}^{MPhT}\frac{At_{120,sp}^{MPhT}-R_{120/124,m}^{TPhT}At_{124,sp}^{MPhT}}{R_{120/124,m}^{TPhT}At_{124,s}-At_{120,s}} \end{aligned} \tag{7}$$

$$\begin{pmatrix} 1 & F_2 & F_3 \\ F_5 & 1 & F_1 \\ F_6 & F_4 & 1 \end{pmatrix} \begin{pmatrix} N_s^{MPhT} \\ N_s^{DPhT} \\ N_s^{TPhT} \end{pmatrix} = \begin{pmatrix} X_{MPhT} \\ Y_{DPhT} \\ Z_{TPhT} \end{pmatrix} \quad (8)$$

Solving Matrix 8 by setting the F values (both on the left side and even included in X_{MPhT}, Y_{DPhT}, and Z_{TPhT}) equal to zero will correspond to degradation-uncorrected concentrations of the phenyltins.

9. References

[1] Hill S.J., Pitts L.J., Fisher A.S. *Trends Anal. Chem.*, 2000, 19, 120.

[2] Berglund M., 'Introduction to Isotope Dilution Mass Spectrometry (IDMS)', *Handbook of Stable Isotope Analytical Techniques, ed. P. De Groot, Elsevier, Amsterdam*, 2004, vol. 1, 820–834.

[3] Jackson C.H.N. *J. Anim. Ecol.*, 1933, 2, 204.

[4] Vogl J. *J. Anal. At. Spectrom.*, 2007, 22, 475.

[5] Nguyen V.D., Bui T.T., Tesfalidet S., *Anal. Bioanal. Chem.*, 2008, 392, 737.

[6] Hoch M., *Applied Geochemistry*, 2001, 16, 719.

[7] Tao H., Rajendran R. B., Quetel C. R., Nakazato T., Tominaga M., Miyazaki A., *Anal. Chem.*, 1999, 71, 4208.

[8] Schubert P., Rosenberg E., Grasserbauer M. *Intern. J. Environ. Anal. Chem.*, 2000, 78, 185.

[9] Encinar J. R., Alonso J. I. G., Sanz-Medel A., *J. Anal. Atom. Spectrom.*, 2000, 15, 1233.

[10] Berg M., Arnold C. G., Muller S. R., Muhlemann J., Schwarzenbach R. P., *Environ. Sci. Technol.*, 2001, 35, 3151.

[11] Arnold C. G., Berg M., Muller S. R., Dommann U., Schwarzenbach R. P., *Anal. Chem.*, 1998, 70, 3094.

[12] Diemer J, Heumann KG., *Fresenius J. Anal. Chem.*, 1997, 357, 74–79.

[13] Tirez K, Brusten W, Cluyts A, Patyn J, De Brucker N., *J. Anal. Atom. Spectrom.*, 2003, 18, 922–932.

[14] Reifenhäuser C, Heumann K.G., *Fresenius J. Anal. Chem.*, 1990, 336, 559.

[15] Minami H., Cai W.T., Kusumoto T., Nishikawa K., Zhang Q., Inoue S., Atsuya I., *Anal. Sci.*, 2003, 19, 1359.

[16] Hinojosa R. L., Marchante G.J.M., Alonso J.I.G., Sanz-Medel A., *J. Anal. At. Spectrom.*, 2004,19, 1230.

[17] Hintelmann H, Evans R.D., *Fresenius J. Anal. Chem.*, 1997, 358:375.

[18] Kumar S.J., Tesfalidet S., Snell J.P., Van D.N., Frech W., *J. Anal. At. Spectrom.*, 2004,19, 368.

[19] Kumar S.J., Tesfalidet S., Snell J., Frech W., J., *Anal. At. Spectrom.*, 2003, 18, 714.

[20] Encinar J.R., Villar M.I.M, Santamarıa V.G., Alonso J.I.G., Sanz-Medel A., *Anal. Chem.*, 2001, 73, 3174.

[21] Qvarnström J., Frech W., *J. Anal. At. Spectrom.*, 2002, 17, 1486.

[22] Encinar J.R, Rodríguez-Gonzalez P, Alonso J.I.G., Sanz-Medel A., *Anal. Chem.*, 2002, 74, 270.

[23] Van D.N., Muppala S.R.K., Frech W., Tesfalidet S., *Anal. Bioanal. Chem.*, 2006, 386, 1505.

[24] Dauchy X., Cottier R., Batel A., Jeannot R., Borsier M., *J. Chromatogr. Sci.*, 1993, 31, 416.

[25] Kumar S.J., Tesfalidet S., Snell J., Frech W. *J. Anal. At. Spectrom.*, 2003, 18, 714.

[26] Tesfalidet S. Screening of organotin compounds in the Swedish environment, Final report; SNV contract: 219 0102, 2004. http://www.ivl.se/miljo/projekt/dvss/pdf/organotenn.pdf

[27] Morabito R, Soldati P., de la Calle M.B., Quevauviller Ph., *Appl. Organomet. Chem.*, 1998, 12, 621.

[28] Cardellicchio N., Giandomenico S., Decataldo A., Di Leo A., *Fresenius J. Anal. Chem.*, 2001, *369*, 510.

[29] Ceulemans M., Adams F. C., *Anal. Chim. Acta.*, 1995, *317*, 161.

[30] Yang L., Mester Z., Sturgeon R. E., *Anal. Chem.*, 2002, *74*, 2968.

[31] Pellegrino C., Massanisso P. , Morabito R., *Trends Anal. Chem.*, 2000, 19, 97.

[32] Abalos M. , Bayona J.M. , Quevauviller Ph., *App., Organomet. Chem*, 1998, 12, 541.

[33] Kumar S.J., Tesfalidet S., Snell J., Frech W., *J. Anal. At. Spectrom.* 2003, 18, 714.

[34] Ariese F., Cofino W., Gomez-Ariza J.L., Kramer G.N., Quevauviller Ph., *J. Environ. Monit.* 1999, 2, 191.

[35] Alonso I.J.G., Encinar J., González R.P., Sanz-Medel A., *Anal. Bioanal. Chem.* 2002, 373, 432.

[36] Morabito R., Soldati P., de la Calle Gutinas M.B., Quevauviller Ph., *Appl. Organomet. Chem.* 1998, 12, 621.

[37] Gonzalez P.R., Alonso J.G., Sanz-Medel A., *J. Anal. At. Spectrom.* 2005, 20, 1076.

[38] Meija J., Yang L., Caruso J.A., Mester Z., *J. Anal. At. Spectrom.* 2006, 21, 1294.

Part 5

New Techniques in Gas Chromatography

19

Multidimensional Gas Chromatography – Time of Flight Mass Spectrometry of PAH in Smog Chamber Studies and in Smog Samples

Douglas Lane[1] and Ji Yi Lee[2]
[1]Environment Canada,
[2]Chosun University,
[1]Canada
[2]Republic of Korea

1. Introduction

Hans Falk (Falk et al., 1956; 1960) published the first papers demonstrating that polycyclic aromatic hydrocarbons underwent a transformation when exposed to ultraviolet light and oxidants. His experiments, in which concentrated spots of individual PAH were irradiated under UV light and exposed to ozone, suggested that PAH were oxidized to unknown products at a very slow rate of about 1 to 5% per hour. However, in 1977, it was shown (Lane & Katz, 1977) that PAH were, indeed, very sensitive to ozone by exposing monolayer distributions of PAH adsorbed on glass plates to ozone and irradiation which closely matched solar irradiation in the 2500A to 5000A region of the spectrum from a quartz lamp. Under lamp irradiation and 0.19 ppm O_3, the half lives of Benzo[a]pyrene, Benzo[k]fluoranthene and Benzo[b]fluoranthene were 0.58 h, 3.9 h, and 4.2 h respectively (Lane, 1975). They also were the first to describe how multi-layering of PAH on particles could influence the apparent reaction rate of the PAH in Falk's experiment, and demonstrated that a rapid surface oxidation reaction formed a "crust" of oxidized products and was then followed by a slower penetration reaction involving the ozone accessing the subsurface material by penetrating the surface oxidized material - much like passing through a crust of bread. In the 1980s the group at the University of California, lead by Roger Atkinson began smog chamber studies to determine the rate of decomposition of a wide variety of PAH and PAH-related compounds when exposed to various oxidants. Their work clearly indicated that the OH radical was the dominant oxidant in the atmosphere (Atkinson et al., 1984; Atkinson, 1988). Most subsequent smog chamber studies to investigate the decomposition of PAH have utilized the OH radical as the oxidant of choice. The first report of a variety of oxidized products of the reaction of Naphthalene with the OH radical in a smog chamber was published in 1994 (Lane & Tang, 1994). Numerous publications have followed, describing the products of the OH radical reactions with naphthalene (Atkinson & Arey, 1994; Bunce et al., 1997; Sasaki et al., 1997; Mihele et al., 2002; Wang et al., 2007; Lee & Lane, 2009; Nishino & Arey, 2009); with acenaphene (Sauret-Szczepanski & Lane, 2004); with Alkylated naphthalenes (Banceu et al., 2001; Wang et al., 2007) and with phenanthrene (Helmig & Harger, 1994; Esteve et al., 2003; Lee & Lane, 2010).

Gas Chromatographic methods, because of their resolution and separation power over other analytical methods have been used to investigate the myriad of products found. With the advent of multidimensional gas chromatography, (GCxGC), the resolution and separation capabilities of gas chromatographic methods took a quantum leap forward. The combination of GCxGC with Time of Flight Mass Spectrometry, (GCxGC-TOFMS) has resulted in the most powerful instrument for the determination of oxidation products in chemical reactions. Numerous papers demonstrating the power of GCxGC-TOFMS for the determination of the products of the OH radical reactions with various PAH in smog chamber studies have been published. (Lee & Lane, 2009; 2010; Lane & Lee, 2010). Lee and Lane then searched for, and found, many of the oxidation products found the smog chamber studies in smog samples collected in Seoul, Republic of Korea (Lane & Lee, 2010).

An important aspect of looking for secondary organic compounds produced in smog chamber studies, is to determine how the products partition themselves between the gaseous and particulate phases. This has been facilitated by the development of Annular Diffusion Denuders for organic compounds. Johnson and co-workers (Johnson et al., 1985; Lane et al., 1988; 1992) utilized 8-channel, multi-annular denuders, coated with a combination of SE-54 silicone oil and crushed Tenax®. Although these denuders were primarily designed for the collection of ambient air samples for the analysis of chlorinated aromatics, they were also shown to be suitable for the determination of the gas/particle partitioning of PAH. In 1995 Gundel (Gundel et al., 1995) introduced a single-channel denuder coated with finely ground XAD-4 resin for the analysis of PAH in tobacco smoke. In 1993 a collaboration between Gundel and Lane resulted in the first Integrated Organic Gas And Particle Sampler (IOGAPS) (Gundel & Lane 1995) which was manufactured by URG in Chapel Hill NC. The denuder in this instrument was an 8-channel, 30cm long denuder operating at a flow rate of 16.7 L/min. For application to smog chamber studies, a smaller 5-channel 20-cm long denuder was utilized (Mihele et al., 2002). Coupled with a filter pack comprising a quartz or Teflon coated glass fiber filter backed up with 2 sorbent impregnated filters (SIFs) (Gundel & Herring, 1998; Galarneau et al., 2006), the denuder-filterpack proved extremely efficient for the determination of the gas particle partitioning of the products during the course of the smog chamber reaction studies.

In this chapter, the application of GCxGC-TOFMS to the determination of the phase-partitioned products of the OH radical reactions of PAH will be discussed. The utility of thermal desorption of samples directly into the GCxGC-TOFMS for the analysis of smog samples from the Republic of Korea will also be described.

2. Experimental

2.1 Smog chamber

The smog chamber in our laboratory is a 10 m^3 cylindrical chamber, essentially a cylindrical Teflon bag which is surrounded by a bank of fluorescent black lights. These may be turned on in 12 stages giving 12 levels of UV intensity. For the reactions described in this paper, full illumination was used. A fan, the blades of which were also coated with Teflon, inside the chamber caused efficient air mixing in the chamber. Air is supplied to the chamber by an AADCO model 737 pure air generator. The air in the chamber was monitored using a Columbia Science Chemiluminescence monitor for the appearance of ozone, and, when ozone started to appear, it indicated that the NO, added to prevent the presence of ozone, had been used up. The reaction study was then terminated. NO was monitored by a Thermo Environmental Instruments INC. Model 42C NO-NO_2-NO_x instrument.

The OH radical was generated in the chamber by the photochemical reaction of isopropyl nitrite in the presence of NO (Mihele et al., 2002). When the fluorescent lamps were turned on, the OH radical was produced to a steady state in just a few milliseconds (Bunce et al, 1997). When the lamps were turned off, the OH radical disappeared almost as quickly, effectively stopping the reactions and freezing the product mix. Samples could then be withdrawn from the chamber for analysis of the products. After the samples had been collected, the lamps were turned back on for a specified period of time after which the lamps were turned off and another sample was collected. This process continued until the NO was used up and ozone began to appear.

2.2 Annular diffusion denuder sample collection and analysis

A 5-channel, 30 cm long annular diffusion denuder coated with finely ground (to an average particle diameter of 0.7 μm) XAD-4 resin and filterpack were used to collect samples from the smog chamber to evaluate the gas/particle partitioning of the products produced during the reactions. During collection of the air samples, the gas phase products were adsorbed on the denuder while the particle phase material passed through the denuder and were trapped on the filter in the filter pack. The filter pack consisted of a Teflon coated glass fiber filter and two XAD-4 sorbent impregnated filters (Gundel & Herring, 1998; Galerneau et al., 2006) to trap any products which might volatilize from the particles collected on the filter and pass through the filter. The denuder and the filters were solvent extracted with hexane (Lane & Gundel, 1996; Gundel & Lane, 1999). The solutions were reduced in volume, internal standards were added and the samples were injected into the GC × GC-TOFMS.

2.3 Multidimensional Gas Chromatography-Time of Flight Mass Spectrometry (GCxGC-TOFMS)

A Pegasus IV GCxGC-TOFMS (LECO Instruments Inc, Dearborne, MI) was used for the chemical separation and analysis of the products. The first dimension column (for separation based on vapor pressure), was a DB5-MS capillary column (30 m × 0.25 mm i.d. × 0.25 μm film thickness) and the second dimension column (for separation in the polar dimension) was a DB17-MS column (1.1 m × 0.18 mm i.d. × 0.18 μm film thickness). The secondary oven temperature was offset +10 °C relative to the main oven and the modulator temperature was offset +20 °C relative to the main oven. The temperature modulator was a liquid nitrogen cooled dual jet configuration with a cool time of 1.90 s between heating and cooling cycles. The main oven was commenced at 60 °C for 3.0 minutes then programmed at 3 C° /min to 300 °C and then held isothermally for 5 minutes. The interface between the Gas Chromatograph and the mass spectrometer was held at 320 °C and the ion source was maintained at 225 °C. Mass scans were taken from 35 to 600 daltons at 200 full scans.s^{-1}. LECO's ChromaTOF software v 3.32 was used for the control of the system and for the collection of and processing of the data. National Institute of Standards and Technology (NIST), Willey and in-house PAH mass spectral libraries were used for the identification of the analytes. When the mass spectrum of a compound in the sample agreed with a library reference with a match of greater than 800 out of 1000, a positive identification was indicated.

2.4 Thermal desorption Multidimensional Gas Chromatography–Time of Flight Mass Spectrometry (TD-GC × GC-TOFMS)

Being provided with only a portion of a 47 mm diameter glass fiber filter that had collected particulate matter over a 24 hour period of time (less than 24 m^3 of sampled air) sensitivity concerns arose. It was realized that to extract and reduce the sample to one mL would result in injecting less than 0.1 percent of the total sample whereas, if two 4 mm punches of the filter were thermally desorbed, about 6.5% of the entire sample or 65 times more than a single liquid injection could be introduced into the GC column in a single injection. We coupled a Gerstel Thermal desorption (TDS-G) system to the front end of the Pegasus and thermally desorbed the filter punches. The TDS was interfaced directly to the GC of the Pegasus by a transfer line that was maintained at 320°C. To desorb the filter punches, the thermal desorption tube was ramped from 20°C to 300°C at 25°C/min and held for 3 min under a flow of He (BIP grade, Linde Gas, Canada). The extracted analytes were trapped at -60 °C using liquid N_2 as coolant in a programmable temperature vaporizing inlet system (PTV-CIS, Gerstel Inc., Baltimore, MD). The inlet was operated in the solvent venting mode. The thermal desorption system was then warmed to 20 °C. While the CIS was heated to 320 °C at 12 C° min^{-1}, the analytes were transferred to the first dimension GC capillary column. During the transfer of analytes the TD transfer line was fixed at 320 °C.

3. Results

3.1 Chromatographic results

Analyses of the smog chamber samples of naphthalene, acenaphthene and phenanthrene revealed many oxidized products and included hydroxy, quinone and many ring opening

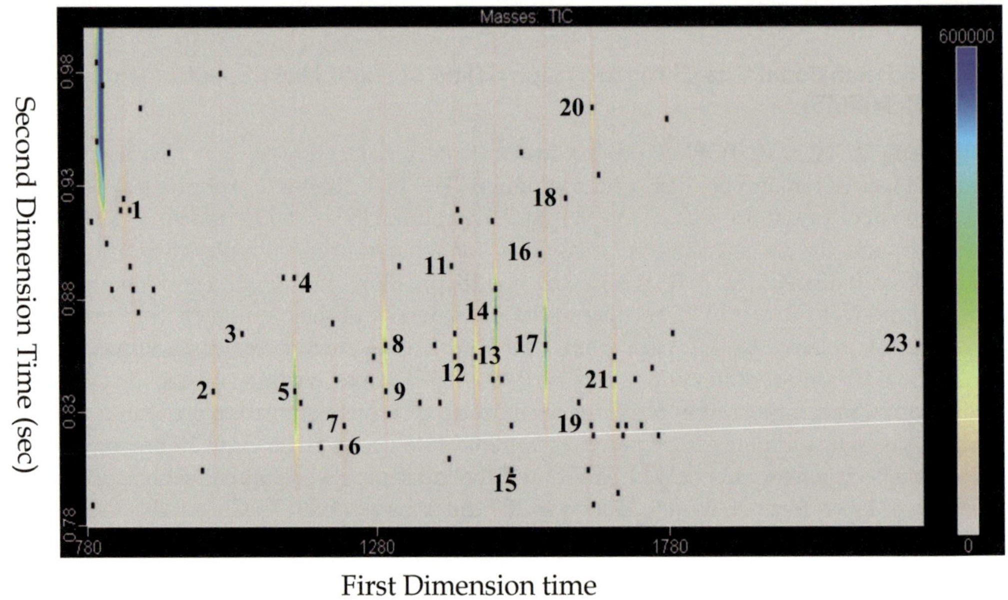

Fig. 1. Showing the 2 dimensional plot of the compounds found in the gas phase and collected by the annular diffusion denuder.

products (Sauret-Szczepanski & Lane, 2004; Lee & Lane, 2009; 2010). The results for the decomposition of naphthalene with the OH radical are shown in the contour chromatographic plots in Figures 1 and 2. In these figures, each black dot represents a distinctly separated and identified product. For some compounds in high concentration the colour surrounding the dot is an indication of intensity. Figure 1 shows the compounds that were trapped on the denuder and, therefore, were in the gas phase. Figure 2 shows the compounds that were extracted from the filter and SIFs. These were particle phase compounds. The numbers on the figures refer to the numbered compounds shown in Table 1 where the compound structure and the retention times on both columns are shown.

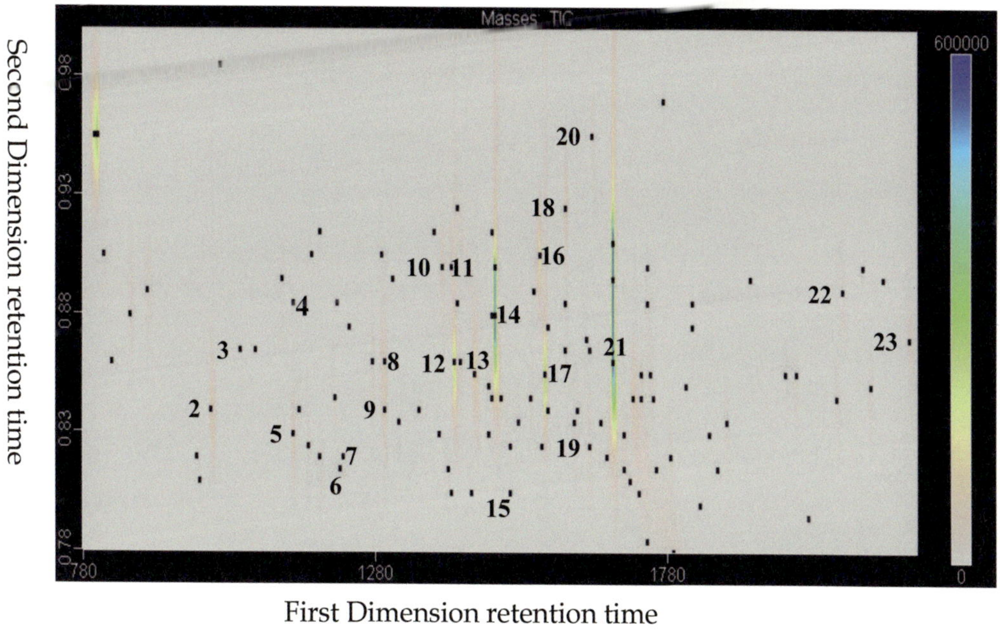

Fig. 2. Showing the contour plot for the particle phase compounds found on the filter and the SIFs.

# in Fig. 1	Products	Structure	Formula	M.W.	1st R.T.*	2nd R.T*
1	1,4-dihydro-1,4-epoxy-Naphthalene	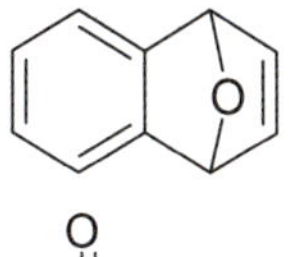				
2	Phthaladehyde	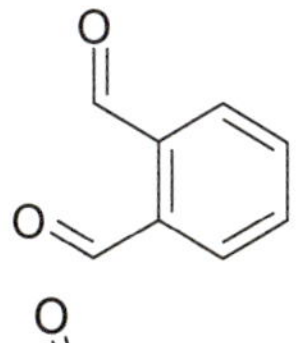	$C_8H_6O_2$	134	1000	0.85
3	Inden-1-one	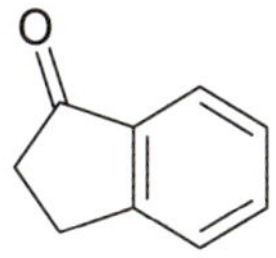	C_9H_8O	132	1050	0.86

# in Fig. 1	Products	Structure	Formula	M.W.	1st R.T.*	2nd R.T*
4	1,2-benzopyrone	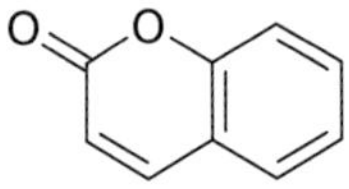	$C_9H_6O_2$	146	1140	0.895
5	Phthalic anhydride	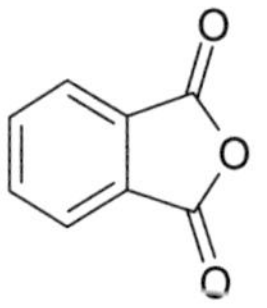	$C_8H_4O_3$	148	1140	0.845
6	Phthalide	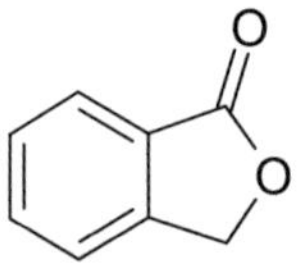	$C_8H_6O_2$	134	1220	0.815
7	1,3-indene-dione	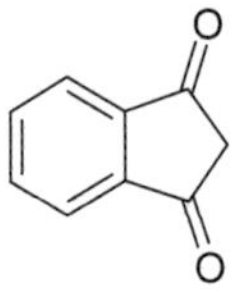	$C_9H_6O_2$	146	1225	0.82
8	1,4-naphthoquinone	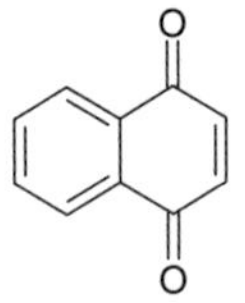	$C_{10}H_6O_2$	158	1295	0.865
9	1-hydroxy-naphthalen-2-one	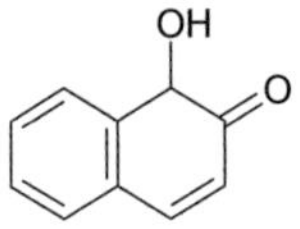	$C_{10}H_8O_2$	160	1295	0.845
10	1-naphthalenol	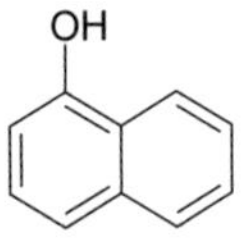	$C_{10}H_8O$	144	1205	0.87
11	2-naphthalenol	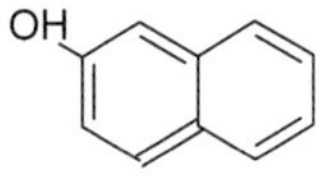	$C_{10}H_8O$	144	1410	0.9
12	1-hydroxy-naphthalen-4-one	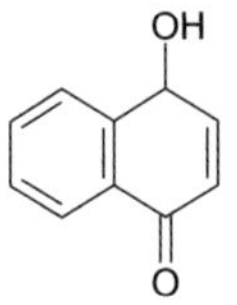	$C_{10}H_8O_2$	160	1415	0.86
13	2,3-epoxy-naphthoquinone	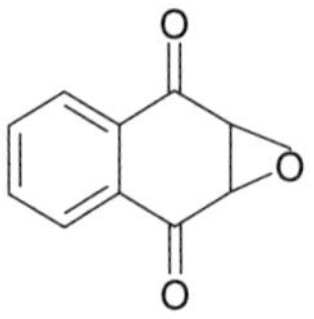	$C_{10}H_6O_3$	174	1450	0.855

# in Fig. 1	Products	Structure	Formula	M.W.	1st R.T.*	2nd R.T*
14	(E)-2-formylcinnamaldehyde	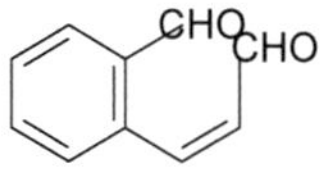	$C_{10}H_8O_2$	160	1485	0.89
15	2,2-dihydroxy-indene-1,3-dione.	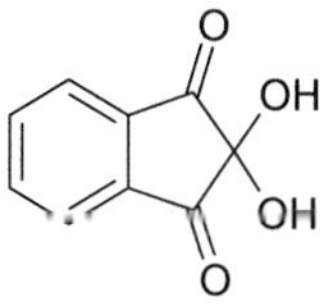	$C_9H_6O_4$	178	1510	0.805
16	1-nitronaphthalene	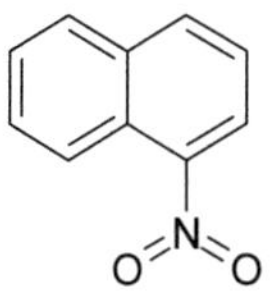	$C_{10}H_7NO_2$	173	1560	0.9
17	(Z)-2-formylcinnamaldehyde	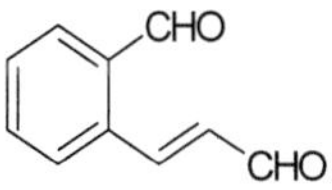	$C_{10}H_8O_2$	160	1570	0.87
18	2-nitronaphthalene	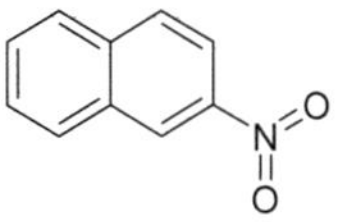	$C_{10}H_7NO_2$	173	1605	0.92
19	1,2-naphthoquinone	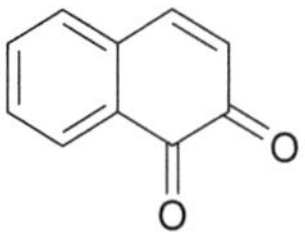	$C_{10}H_6O_2$	158	1645	0.82
20	2-nitro-1-naphthol	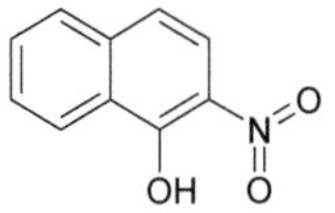	$C_{10}H_7NO_3$	189	1650	0.965
21	2,3-epoxy-1-hydroxy-naphthalen-4-one	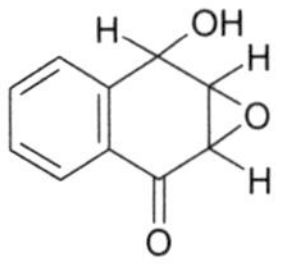	$C_{11}H_{12}O_2$	176	1685	0.84
22	4-nitro-1-naphthalenol	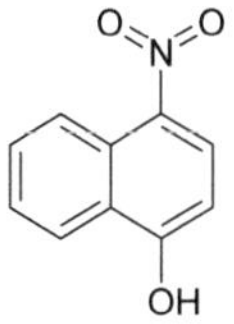	$C_{10}H_7NO_3$	189	2080	0.89
23	1-nitro-2-naphthalenol	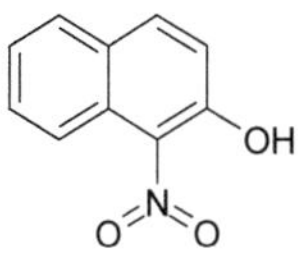	$C_{10}H_7NO_3$	189	2195	0.865

Table 1. The products of the OH radical reaction of Naphthalene with the OH radical are shown together with the retention times on each column.

The products determined for Acenaphthene and Phenanthrene may be found in other publications (Sauret-Szczepanski & Lane, 2004; Lee & Lane 2010).

3.2 Challenges in relating TOFMS data to the NIST and Wiley databases

We soon discovered that TOFMS mass spectral data differ significantly from those in the NIST or WILEY databases that were largely derived from quadrupole mass spectrometric data. This is because of the manner in which the mass scans are obtained. With quadrupole mass spectrometers, maximum practical scanning rates are about 300 daltons.sec^{-1}, or about 1 to 1.5 full scans per second. This means that only about 3 scans can be taken across a single chromatographic peak. Because the mass of material being detected is constantly varying and because the scan takes about a second to be completed, the distribution of the peaks in a mass spectrum are biased high at the upper end of the spectrum as the peak is growing and are biased low at the upper end of the mass spectrum on the descending side of the peak. The reported mass spectra in the commercial databases are, of practical necessity, an average of the 3 or so peaks collected over one chromatographic peak. However, with TOFMS, full mass scans are taken at a rate of 200 per second. At such a rate, relative ion ratios are virtually constant at each point on a chromatographic peak. Approximately 600 spectra are taken over the width of a single chromatographic peak. It was for these reasons that chemical standards were obtained whenever possible and those standards used to generate an in-house library of TOFMS mass spectra. The agreement between samples and the in-house data were well above matches of 990 whereas the best matches with the NIST and WILEY libraries were on the order of 920-940. We had much greater confidence in the determinations of real world samples using our in-house library.

3.3 Retention times and mass spectral identification of products

Many of the products could be identified, although with lower match certainty than desired, through the use of the NIST and Wiley mass spectral databases, however, many more were not found in the databases and had to be determined by other means. For example, if standards or surrogate standards could be obtained mass spectra were obtained. It was also found useful to compare our spectra with mass spectral patterns published in the literature by other investigators. Their suggested identifications were of assistance in our own assignment of identities. Finally, when all else failed we identified the compounds through fundamental analysis of the mass spectra. To improve the match of environmental samples with the known products, we prepared an in-house database of the mass spectra of all reactants and for all products for which standards could be acquired. Thus matches between products in smog samples and database reference standards rose from about 650 to over 990 giving much greater confidence in the identity of the products.

3.4 Analysis of smog samples

A 3D image of the chromatograph of one of the Korean smog samples is shown below in Figure 3. From this sample, we successfully resolved almost 18,000 individual compounds. Many of the peaks could be identified by computer database searching. However, many were unidentified. This was partly due to the incompatibility of NIST and Wiley spectra with TOFMS data as outlined above and, more likely, because the compound in the air did not have a mass spectral signature in the databases.

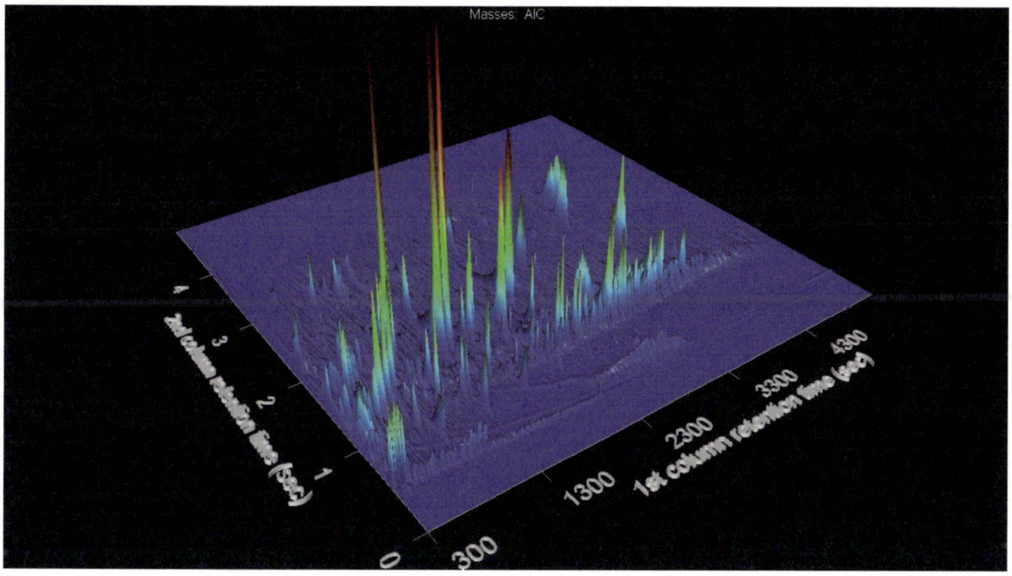

Fig. 3. The GCxGC trace of a Korean smog sample. The first column dimension is the vapor pressure dimension and is governed by chromatography on the DB-5MS column. The second dimension is the polar dimension and is governed by the polar DB-17MS column. The height of the peaks is proportional to the mass of the compound present.

However, for other unknowns or possibly improperly identified peaks, it was clearly a case of the compound not existing in the databases. We succeeded in finding 13 of the photochemical decomposition products from the chamber experiments in the Korean smog samples. The compounds from the decomposition of naphthalene, acenaphthene and phenanthrene that were found in the smog sample are shown in Table 2.

The above compounds were found in the particulate matter collected on the filter of a filter pack from Seoul, Republic of Korea. Since these compounds were found on the filter it is important to note that a) they were in the particle phase, 2) they may have suffered some volatilization or blow-off from the filter, but the degree to which that affected an estimation of the total compound in the atmosphere cannot be determined, and 3) many of the compounds detected such as the quinones, nitro derivatives and the hydroxynitro derivatives are known to be hazardous chemicals (Arey et al., 1989; Atkinson and Arey, 1994; Reisen and Arey, 2005). As many of these compounds, for example 1,2-naphthalenedicarboxaldehyde and (E)-2-formylcinnamaldehyde (Lane & Lee, 2010), are not known to originate in emissions, they must have been the result of atmospheric oxidation and this clearly indicates the formation of secondary aerosol material. It is important to note that there are no known anthropogenic sources of these two compounds. Recently Kroll and Seinfeld (Kroll and Seinfeld, 2008) have demonstrated the importance of atmospheric reaction products in the formation of secondary organic aerosol (SOA) and in the adverse effect of SOA to climate change and visibility in the atmosphere. A recent study (Robinson et al., 2007) has reported that research to estimate the organic aerosol budget, underestimates the production of SOA in the atmosphere when compared to actual field measurements. They suggested that the underestimate was due to the non-inclusion of the SOA produced from atmospheric semivolatile organic compounds.

Parent	Product	Structure
Naphthalene	1-naphthol	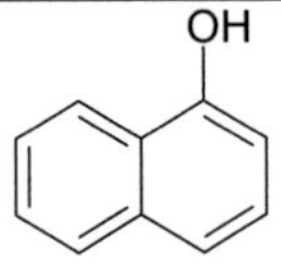
	Phthalide	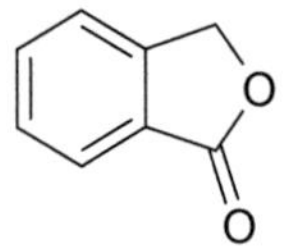
	1,4-naphthalenedione	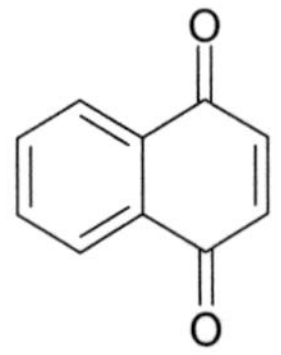
	(E)-2-formylcinnamaldehyde	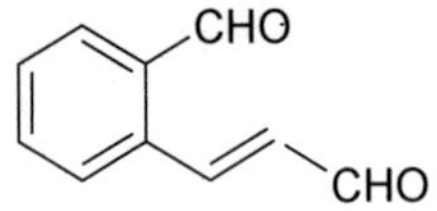
	1,3-indandione	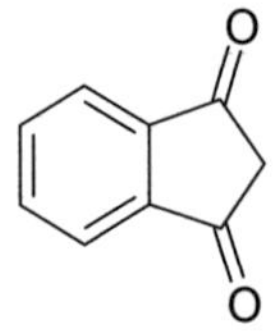
	Phthalic anhydride	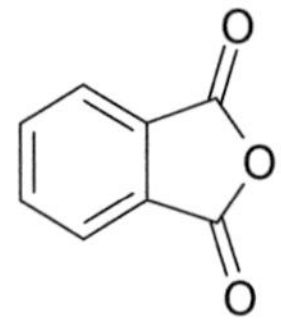
	Indan-1-one	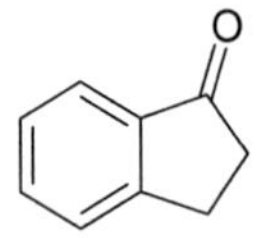
Acenaphthene	1,8-naphthalicanhydride	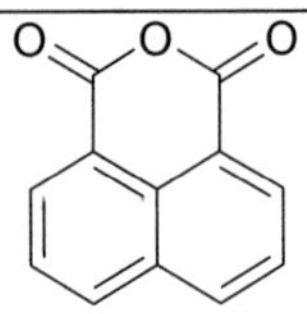
Phenanthrene	9-fluorenone	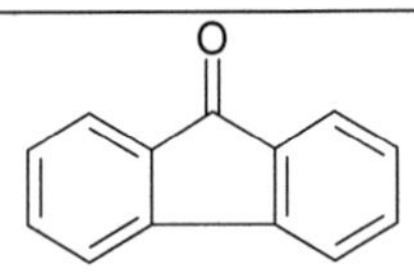
	1,2-naphthalenedicarboxaldahyde	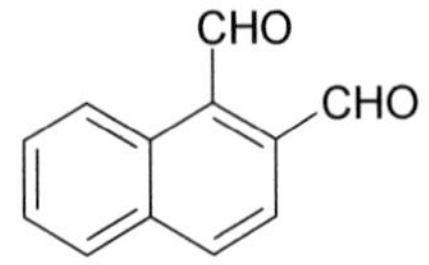

Parent	Product	Structure
	1-phenanthrol	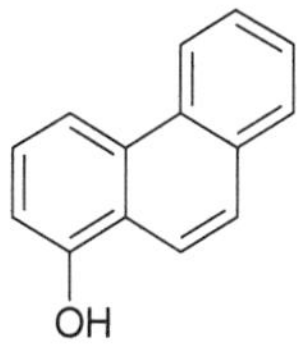
	9-phenanthrol	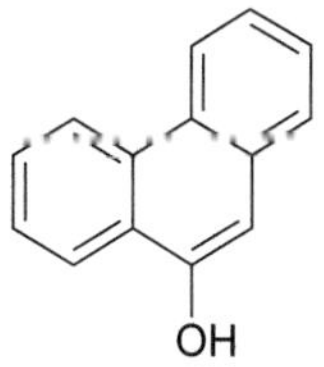
	1,2-naphthalic anhydride	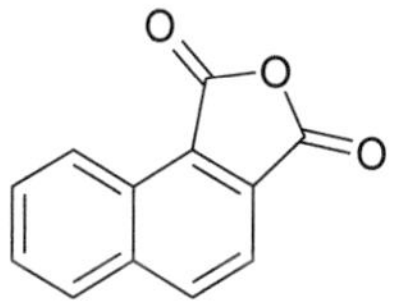
	Dibenzopyranone	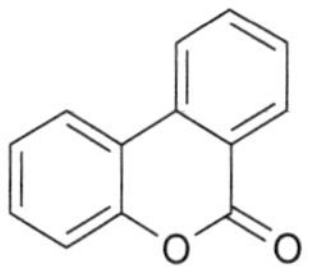

Table 2. Above are presented the products found in the smog chamber during the studies of the reactions of naphthalene, acenaphthene and phenanthrene that were also found in the smog samples.

4. Conclusions

In this chapter, we have demonstrated that GC×GC-TOFMS is an excellent technique to identify the oxidized products of the decomposition of PAH in reactions with the OH radical. When combined with thermal desorption, TD-GC×GC-TOFMS becomes a very powerful tool to examine air extracts for a very wide range of pollutant chemicals. This method provides greatly enhanced sensitivity to chemical components and permits the detection of many, otherwise impossible to detect, compounds. We have demonstrated unequivocally that PAH are oxidized in the atmosphere and form a plethora of oxidized, nitrated and ring-opened products. This lends strong support to the statements of Robinson et al. (Robinson et al., 2007) that the production of SOA is underrepresented in budgets of atmospheric aerosols. We have found known oxidation products, some known only as atmospheric oxidation products with no anthropogenic source, in smog samples. This alone has many implications for the effect of SOA on human health.

5. References

Arey, J., Atkinson, R., Zlelinska, B., and McElroy, P.A. (1989). Diurnal concentrations of volatile polycyclic aromatic hydrocarbons and nitroarenes during a photochemical air pollution episode in Glendora, California. *Environmental Science & Technology*. 23, 321-327.

Atkinson, R., Aschmann, S.M., and Pitts, J.N. Jr. (1984). Kinetics of the reactions of Naphthalene and Biphenyl with OH radicals and with O_3 at 294 ± 1 K. *Environmental Science & Technology.* 18, 110-113.

Atkinson, R. (1988). Estimation of gas-phase hydroxyl radical rate constants for organic chemicals. *Environmental Toxicology & Chemistry.* 7, 435-442.

Atkinson, R., and Arey, J. (1994). Atmospheric chemistry of gas-phase polycyclic aromatic hydrocarbons: formation of atmospheric mutagens. *Environmental Health Perspectives.* 102, 117–126.

Banceu, C.E., Mihele, C.M., Lane, D.A., & Bunce, N.J. (2001). Reactions of Methylated Naphthalenes with Hydroxyl Radicals Under Simulated Atmospheric Conditions. *Polycyclic Aromatic Compounds.* 18, 415-425.

Bunce, N.J., Liu, L., Zhu, J., Lane, D.A., (1997). Reaction of naphthalene and its derivatives with hydroxyl radicals in the gas phase. *Environmental Science and Technology.* 31, 2252–2259.

Esteve, W., Budzinski, H., Villenave, E. (2003). Heterogeneous reactivity of OH radicals with phenanthrene. *Polycyclic Aromatic Compounds.* 23, 441-456.

Falk, H.L., Markul, I., and Kotin, P. (1956). Aromatic hydrocarbons. IV. Their fate following emission into the atmosphere and experimental exposure to washed air and synthetic smog. *A.M.A. Archives of Industrial Health* 13, 13-17.

Falk, H.L., Kotin, P., and Miller, A. (1960). Aromatic polycyclic hydrocarbons in polluted air as indicators of carcinogenic hazards. *International Journal of Air Pollution* 2, 201-209.

Galarneau, E., Harner, T., Shoeib, M., Kozma, M., and Lane, D.A. (2006). A preliminary investigation of sorbent-impregnated filters (SIFs) as an alternative to polyurethane foam (PUF) for sampling gas-phase semivolatile organic compounds in air. *Atmospheric Environment.* 40, 5734-5740.

Gundel, L.A., Lee, V.C., Mahanama, K.R.R., Stevens, R.K., and Daisey, J.M. (1995). Direct determination of the phase distributions of semi-volatile Polycyclic Aromatic Hydrocarbons using annular denuders. Atmospheric Environment. 29, 1719-1733.

Gundel, L.A., and Herring, S.V. (1998). Absorbing filter media for denuder-filter sampling of total organic carbon in airborne particles. Record of invention WIB 1457, Lawrence Berkeley National Laboratory (USA).

Gundel, L.A., and Lane, D.A. (1999). Sorbent-coated diffusion denuders for direct measurement of gas/particle partitioning by semi-volatile organic compounds.. In: Lane, D.A. (Ed.), Gas and Particle Phase Measurements of Atmospheric Organic Compounds. Gordon and Breach Science Publishers, Philadelphia PA, chapter 11.

Helmig, D., and Harger, W.P. (1994). OH radical-initiated gas-phase reaction products of phenanthrene. *Science of the Total Environment.* 148, 11-21.

Johnson, N.D., Barton, S.C., Thomas, G.H.S., Lane, D.A., and W.H. Schroeder. 1985. Development of a gas/particle fractionating sampler for chlorinated organics. In proceedings of the annual meeting of the Air Pollyution Control Association. Paper 85-81-1. June 12-21, Detroit, MI.

Kroll, J.H. and Seinfeld, J.H. (2008). Chemistry of secondary organic aerosol: formation and evolution of low-volatility organics in the atmosphere. *Atmospheric Environment.* 42, 3593-3624.

Lane, D.A. (1975) Gas Chromatographic analysis and photodecomposition studies of atmospheric Polycyclic Aromatic Hydrocarbons. Ph. D. Thesis, York University, Toronto, Canada.

Lane, D.A., and Katz, M. (1977). The photodecomposition of benzo[a]pyrene, benzo[b]fluoranthene and benzo[k]fluoranthene under simulated atmospheric conditions. Advances in Environmental Science & Technology. 8, 137-154.

Lane, D.A., Johnson, N.D., Barton, S.C., Thomas, G.H.S., and W.H. Schroeder. (1988). Development and evaluation of a novel gas and particle sampler for semivolatile chlorinated organic compounds in ambient air. *Environmental Science & Technology.* 22, 941-947.

Lane, D.A., Johnson, N.D., Hanley, M.- J.J., W.H. Schroeder, and Ord, D.T. (1991). Gas- and particle-phase concentrations of α-hexachlorocyclohexane, γ- hexachlorocyclohexane, and hexachlorobenzene in Ontario air. *Environmental Science & Technology.* 26, 126-133.

Lane, D.A., Schroeder, W.H., and Johnson, N.D. (1992). On the spatial and temporal variations in atmospheric concentrations of hexachlorobenzene and hexachlorocyclohexane isomers at several locations in the province of Ontario, Canada. *Atmospheric Environment.* 26A, 31-42.

Lane, D.A. and Tang, H. (1994). Photochemical degradation of Polycyclic Aromatic Mihele, Compounds. I Naphthalene. *Polycyclic Aromatic Compounds.* 5, 131-138.

Lane, D.A., and Gundel, L.A. (1996). Gas and particle sampling of airborne polycyclic aromatic compounds. *Polycyclic Aromatic Compounds.* 9, 67-73.

Lane, D.A., Peters, A.J., Gundel, L.A., Jones, K.C., and Northcott, G.L. (2000). Gas/Particle partition measurements of PAH at Hazelrigg, UK. *Polycyclic Aromatic Compounds.* 20, 225-234.

Lane, D.A. and Lee, J.Y. (2010). Detection of known decomposition products of PAH in Particulate matter from Pollution Episodes in Seoul, Korea. *Polycyclic Aromatic Compounds.* 30, 309-320.

Lee, J.Y., and Lane, D.A. (2009). Unique products from the reaction of naphthalene with the hydroxyl radical. *Atmospheric Environment.* 43, 4886–4893.

Lee, J.Y., and Lane, D.A. (2010). Formation of oxidized products from the reaction of gaseous phenanthrene with the OH radical in a reaction chamber. *Atmospheric Environment.* 44, 2469-2477.

Mihele, C.M., Wiebe, H.A., Lane, D.A., (2002). Particle formation and gas/particle partition measurements of the products of the naphthalene-OH radical reaction in a smog chamber. *Polycyclic Aromatic Compounds.* 22, 729–736.

Nishino, N., Arey, J., and Atkinson, R. (2009). Formation and reactions of 2-formylcinnamaldehyde in the OH radical-initiated reaction of naphthalene. *Environmental Science & Technology.* 43, 1349-1353.

Reisen, F. and Arey, J. (2005). Atmospheric reactions influence seasonal PAH and nitro-PAH concentrations in the Los Angeles Basin. *Environmental Science & Technology* 39, 64-73.

Robinson, A.L., Donahue, N.M., Shrivastava, M.K., Weitkamp, E.A., Sage, A.M., Grieshop, A.P., Lane, T.E., Pierce, J.R. and Pandis, S.N. 2007. Rethinking organic aerosols: semivolatile emissions and photochemical aging. *Science.* 315, 1259-1262.

Sasaki, J., Aschmann, S.M., Kwok, E.S.C., Atkinson, R., & Arey, J. (1997). Products of the gas-phase OH and NO3 radical-initiated reactions of naphthalene. *Environmental Science and Technology*. 31, 3173–3179.

Sauret-Szczepanski, N., & Lane, D.A. (2004). Smog chamber study of acenaphthene: gas/particle partition measurements of the products formed by reaction with the OH radical. *Polycyclic Aromatic Compounds*. 24, 161–172.

Wang, L., Atkinson, R., and Arey, J. (2007). Dicarbonyl products of the OH radical-initiated reactions of naphthalene and the C1- and C2-alkylnaphthalenes. *Environmental Science & Technology*. 41, 2803-2810.

20

Inverse Gas Chromatography in Characterization of Composites Interaction

Kasylda Milczewska and Adam Voelkel
Poznan University of Technology,
Institute of Chemical Technology and Engineering, Poznan,
Poland

1. Introduction

Inverse gas chromatography is a useful and quite versatile technique for materials' characterization, because it can provide information on thermodynamic properties over a wide temperature range. The term "inverse" indicates that the stationary phase of the chromatographic column is of interest, in contrast to conventional gas chromatography. The chromatographic column contains the material under study. The method is simple, fast and efficient. It has been used for the characterization of hyperbranched polymers [Dritsas et al., 2008], block copolymers [Zou et al., 2006], polymer blends [Al-Ghamdi & Al-Saigh, 2000], nanocomposites [Boukerma et al., 2006], fillers [Milczewska & Voelkel, 2002], cement pastes [Oliva et al., 2002], fibers [van Asten et al., 2000] and crude oils [Mutelet et al., 2002].

Mixtures of different types of materials i.e. polymers, blends, modified fillers or compositions are utilized extensively to produce commercially useful materials having combinations of properties not revealed by a single component. Many of the properties and processing characteristics of those mixtures depend on whether they are miscible or not. Theory operates with parameters relating to the pure components [Voelkel et al., 2009 (a)]. The knowledge of the interaction parameters between polymers and solvents is very important in the study of their miscibility and thermodynamic properties of solutions [Huang, 2009].

The interactions between one probe and the polymer are usually characterized by the values of Flory-Huggins interaction parameter [Dritsas et al., 2009]. Only a few techniques can provide quantitative information about the change of free energy when mixing two components. The data from P-V-T experiments might be successfully used in the prediction of the miscibility of polyolefine blends [Han et al., 1999]. Interaction parameter for the components of polymer blends was also determined with the use of small angle x-ray scattering (SAXS) [de Gennes, 1979; Meoer & Strobl, 1987; Ying et al. 1993], thermal induced phase separation (TIPS) [Sun et al., 1999] and small angle neutron scattering (SANS) [Fernandez et al., 1995; Hindawi et al., 1990; Horst & Wolf, 1992; Mani et al., 1992; Schwann et al., 1996]. In last two decades, the Flory-Huggins interaction parameter was also determined using the melting point depression method for crystal-containing polymers by differential scanning calorimetry (DSC) [Lee et al., 1997]. It is worth to note the increasing role of inverse gas chromatography (IGC) [Voelkel et al., 2009 (a)], because of its simplicity, rapidity, and the general availability of GC equipment.

2. Theory of interaction

Inverse Gas Chromatography (IGC) is a gas-phase technique for characterizing surface and bulk properties of solid materials. The principles of IGC are very simple, being the reverse of a conventional gas chromatographic (GC) experiment.

While it is a dynamic method, it was shown many years ago that measurements recorded under the correct conditions could give accurate equilibrium thermodynamic information [Shillcock & Price, 2003]. The retention of a solvent or 'probe' molecule on the material is recorded and the measurement made effectively at infinite dilution of the probe. A range of thermodynamic parameters can then be calculated. One advantage of the method is that it is readily applied to mixtures of two or more polymers.

A cylindrical column is uniformly packed with the solid material of interest, typically a powder, fiber or film. A pulse or constant concentration of gas is then injected down the column at a fixed carrier gas flow rate, and the time taken for the pulse or concentration front to elute down the column is measured by a detector. A series of IGC measurements with different gas phase probe molecules then allows access to a wide range of physico-chemical properties of the solid sample [SMS-*i*GC brochure 2002].

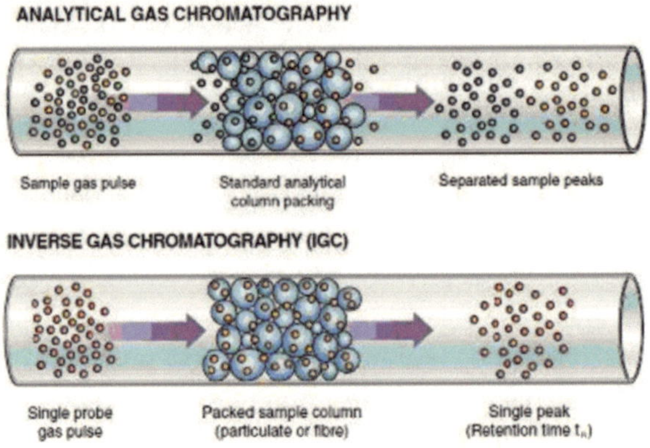

Fig. 1. Analytical vs. Inverse gas chromatography

When a liquid probe is injected into the column, the probe vaporizes and flows with the carrier gas, and a characteristic specific retention volume (V_g) can be measured:

$$V_g = \frac{3}{2} \cdot \frac{t_R' \cdot j \cdot F \cdot 273.15}{m_w \cdot T} \tag{1}$$

where: $t_R' = t_R - t_M$, t_M - gas hold-up time, calculated by Grobler-Balizs procedure [Grobler & Balizs, 1974], j – James-Martin's coefficient [James & Martin, 1952].

3. Flory-Huggins parameters

The properties of polymer blends are determined mainly by the miscibility of the components and structure. Usually thermodynamic miscibility and homogeneity can be attained when the free energy of mixing is negative. The classical thermodynamics of binary polymer–solvent systems was developed independently by P.J. Flory [Flory, 1942] and M.L. Huggins [Huggins, 1942]. It is based on the well-known lattice model qualitatively formulated by K.H. Meyer [Meyer, 1939], who pointed out the effect of the differences in molecular size of polymer and solvent molecules on the entropy of mixing. The quantitative calculation of the entropy of mixing led to the introduction of a dimensionless value, the so-called Flory-Huggins interaction parameter, for the thermodynamic description of polymer solutions [Gundert & Wolf, 1989]. Flory–Huggins interaction parameter (χ) is an important factor of miscibility of polymer blends and solutions.

Using Flory–Huggins theory, the Flory–Huggins interaction parameter between a polymer and probe, χ, can be related to the specific retention volume of probes, V_g, by the following equation [Milczewska & Voelkel, 2002 as cited in Barrales-Rienda, 1988; Voelkel et al., 2009 (b) as cited in Voelkel & Fall, 1995]:

$$\chi_{12}^{\infty} = \ln\left(\frac{273.15 \cdot R}{p_1^o \cdot V_g \cdot M_1}\right) - \frac{p_1^o}{R \cdot T} \cdot \left(B_{11} - V_1^o\right) + \ln\left(\frac{\rho_1}{\rho_2}\right) - \left(1 - \frac{V_1^o}{V_2^o}\right) \tag{2}$$

1 denotes the solute and 2 denotes examined material, M_1 is the molecular weight of the solute, p_1^o is the saturated vapor pressure of the solute, B_{11} is the second virial coefficient of the solute, V_i^o is the molar volume, ρ_i is the density, R is the gas constant.

This equation may be rearranged into form including weight fraction activity coefficient:

$$\ln \Omega_1^{\infty} = \ln\left(\frac{a_1}{w_1}\right) = \ln\left(\frac{273.15 \cdot R}{p_1^o \cdot V_g \cdot M_1}\right) - \frac{p_1^o}{R \cdot T} \cdot \left(B_{11} - V_1^o\right) \tag{3}$$

$$\chi_{12}^{\infty} = \ln \Omega_1^{\infty} + \ln\left(\frac{\rho_1}{\rho 2}\right) - \left(1 - \frac{V_1^o}{V_2^o}\right) \tag{4}$$

When the data of the density and the molecular mass of both the solute and the stationary phase (polymer) are inaccessible it is possible to determine the Flory-Huggins interaction parameter by simplifying Eq. (4):

$$\chi_{12}^{\infty} = \ln \Omega_1^{\infty} - 1 \tag{5}$$

i.e., under the assumption that $\ln\left(\frac{\rho_1}{\rho 2}\right) = 0$ which means that the densities of the solute and the stationary phase are of similar order and $\frac{V_1^o}{V_2^o} \to 0$ (the molar volume of the stationary phase is much higher than that of the test solute) [Voelkel & Fall, 1997].

Etxabarren et al. [Etxabarren et al., 2002] described molecular mass, temperature and concentration dependences of the polymer-solvent interaction parameter. The concentration dependence has been reasonably explained after the consideration of the different compressibilities (or free volumes) of the components. A parabolic dependence of χ with temperature is necessary in order to explain the lower or upper critical solution temperatures characteristic of most of the polymer solutions. In fact, there are experimental evidences of such type of dependence although because of limitations imposed by the degradation of the polymer and the freezing point of the solvent, a limited temperature range can be studied and only a part of this parabolic curve is usually evidenced.

Molecular mass dependence of the interaction parameter has been a recurrent subject in the polymer literature, and Petri et al. [Petri et al., 1995] have reported new experimental results which seem to indicate that there is a real molecular mass dependence of χ, especially in the range of moderate concentrations [Schuld & Wolf, 2001].

Ovejero et al. [Ovejero et al., 2009] determined Flory-Huggins parameter χ_{12}^{∞} for SEBS triblock copolymer. They noticed that Flory-Huggins parameter was defined as independent of concentration, but the effect of concentration is not negligible. Authors tried to develop a thermodynamic tool to simulate a polymer - solvent separation. In their work they also paid attention to temperature dependence. A decrease of Flory-Huggins parameter while increasing temperature was suggested. However they have shown that this dependence is not clear. For investigated rubber values of χ_{12}^{∞} increased slightly with temperature.

When mixture of components is used as a stationary phase in a chromatographic column, subscripts 2 and 3 are used to represent first and second mixtures' component, respectively [Voelkel et al., 2009 (b)]:

$$\chi_{1m}^{\infty} = \ln\left(\frac{273.15 \cdot R}{p_1^o \cdot V_g \cdot M_1}\right) - \frac{p_1^o}{R * T} \cdot \left(B_{11} - V_1^o\right) + \ln\left(\frac{\rho_1}{\rho m}\right) - \left(1 - \frac{V_1^o}{V_2^o}\right) \cdot \varphi_2 - \left(1 - \frac{V_1^o}{V_3^o}\right) \cdot \varphi_3 \tag{6}$$

where φ_2 and φ_3 are the volume fractions of components.

When χ<0.5, the probe liquid is generally characterized as a good solvent for the polymer, whereas χ>0.5 indicates a poor solvent which use may lead to phase separation. In the case of a polymer blend, the parameter χ can still be defined and the miscibility generally occurs when χ<0, because the high molar volume of both components diminishes the combinatorial entropy [Huang, 2009].

When a polymer blend is used the interaction between the two polymers is expressed in terms of $\chi_{23}^{'}$ as an indicator of the miscibility of the components of the polymer blend. If the parameters χ_{12}^{∞} and χ_{13}^{∞} are known (from IGC experiment with appropriate component "2" or "3") the interaction parameter $\chi_{23}^{'}$ may be calculated from equation [El-Hibri et al., 1989; Olabisi, 1975]:

$$\chi_{23}^{'} = \frac{1}{\varphi_2 \cdot \varphi_3} \cdot (\chi_{12}^{\infty} \cdot \varphi_2 + \chi_{13}^{\infty} \cdot \varphi_3 - \chi_{1m}^{\infty}) \tag{7}$$

Here, the second subscript of χ identifies the nature of the column.

The interaction between the two components of composition is expressed in terms of $\chi_{23}^{'}$ may be also calculated from [Milczewska et al., 2001 as cited in Li (Pun Choi), 1996, Milczewska et al., 2003 as cited in Voelkel & Fall, 1997]:

$$\chi_{23}^{'} = \frac{\chi_{23}^{\infty} \cdot V_1}{V_2} = \frac{1}{\varphi_2 \cdot \varphi_3} \cdot \left(\ln \frac{V_{g,m}}{W_2 \cdot v_2 + W_3 \cdot v_3} - \varphi_2 \cdot \ln \frac{V_{g,2}}{v_2} - \varphi_3 \cdot \ln \frac{V_{g,3}}{v_3} \right) \quad (8)$$

Here, the second subscript of V_g identifies the nature of the column. From Eq. (8), $\chi_{23}^{'}$ may be calculated even for probes for which the parameters p_1^o, B_{11} and V_i^o are not known or are known with insufficient accuracy [Al-Saigh & Munk, 1984].

To obtain $\chi_{23}^{'}$ for a polymer blend or composition utilizing IGC, χ_{12}^{∞} values for all components have to be known. Therefore, three columns are usually prepared: two for single components and the third one for a composition of the two components used. A further three columns containing different compositions of components can also be prepared if the effect of the weight fraction of the mixture on the examined property needs to be explored. These columns should be studied under identical conditions of column temperature, carrier gas flow rate, inlet pressure of the carrier gas, and with the same test solutes [Al-Saigh, 1997].

Large positive values of $\chi_{23}^{'}$ indicates the absence or negligible interactions between components, a low value indicates favorable interactions, while negative value indicates strong interactions (the pair of polymers is miscible).

Equations (7) or (8) were frequently used to study the interaction parameter between two stationary phases using the IGC method. In literature data, it was found that, in many miscible systems, $\chi_{23}^{'}$ values were probe dependent. The values of $\chi_{23}^{'}$ were positive when χ_{12}^{∞} and χ_{13}^{∞} were positive, and decreased when χ_{12}^{∞} and χ_{13}^{∞} decreased to negative. Some negative $\chi_{23}^{'}$ values were generally observed for probes with low χ_{12}^{∞} and χ_{13}^{∞} [Huang, 2009].

Nesterov and Lipatov [Nesterov & Lipatov, 1999] studied thermodynamics of interactions in the ternary system: polymer A + polymer B + filler S. In their studies it was shown that the introduction of a third component into the binary immiscible mixture of two polymers, where the third component is miscible with each component of binary mixture, may lead full miscibility of the ternary system. For the immiscible mixtures of polyolefins with polyacrylates and polymethacrylates it was discovered that a mineral filler (e.g. silica) also may serve as compatibilizer.

The compatibilization effect of two immiscible polymers by adding the third polymer (or filler) my be described in the framework of the Flory-Huggins theory extended for describing ternary mixtures. For that mixtures Flory-Huggins parameter can be expressed as:

$$\chi_{A+B+C} \cong \chi_{AB} \cdot \varphi_A \cdot \varphi_B + \chi_{AC} \cdot \varphi_A \cdot \varphi_C + \chi_{BC} \cdot \varphi_B \cdot \varphi_C \qquad (9)$$

A positive value of the parameter χ_{A+B+C} corresponds to an immiscible systems whereas a negative is an indicator of miscibility [Nesterov & Lipatov, 2001].

Values of Flory-Huggins χ'_{23} parameter depend on chemical structure of the solute and it is a common phenomenon, although not allowed by the theory [Fernandez-Sanchez et al., 1988]. It has been interpreted as a result of preferential interactions of the test solute with one of two components. This phenomenon for polymer blends was described independently by Fernandez-Sanchez et al. and Olabisi [Olabisi, 1975]. They attributed this to the non-random distribution of the solute in the stationary phase owing to its preferential affinity for one of the components. Selective solutes do not "sense" the three varieties of intramolecular contacts in the polymer mixture (A-A, A-B, B-B) in proportion to concentration. This modifies the retention volume (and χ'_{23}) values relative to those which would be obtained by truly random mixing of the solute with the polymer. Less selective solvents, on the other hand, exhibit a more random 'sampling' of the molecular environment of the stationary phase owing to the equal affinities they have for both. It is therefore expected that a better measure of the polymer-polymer interaction will be likely with less selective solvents.

Olabisi [Olabisi, 1975] described a polyblend as micro heterogeneous, where the size of the different phases and their interpenetration being limited by a host of factors among which are the extents of mixing, compatibility, molecular weight, clustering behaviour of each polymer, rheological and surface and interfacial properties. He attributed Flory-Huggins parameter dependence on test solute to unequal distribution of the solute in the stationary phase, and to wide range of interactions (polar, nonpolar, hydrogen-bonding and also electronic and electrostatic interactions) [Li, 1996]. Olabisi proposed to use a set of solvents based on their type of interactions with probe: (i) proton accepting strength, probed with chloroform and ethanol; (ii) proton donor strength with methyl-ethyl ketone and pyridine; (iii) polar strength with acetonitrile and fluorobenzene; (iv) nonpolar strength with hexane and carbon tetrachloride.

Prolongo et al. proposed to calculate the polymer-polymer interaction parameter χ from measurements performed on ternary systems composed of the polymer pair plus a solvent or probe. They given the expressions needed to calculate the true polymer-polymer χ based on the equation-of-state theory and they compared that method for PS+PVME data obtained from vapor sorption (VP). The results show that the values obtained from IGC correlation and VP are nearly the same [Prolongo et al., 1989].

Many authors suggest that the χ'_{23} values are solvent (solute) independent for probes giving $\chi_{12}^{\infty} = \chi_{13}^{\infty}$ [Su & Patterson, 1977; Lezcano et al., 1995]. If the difference between the interaction of the components (for blend ≡ polymers 2 and 3) with the solvent is negligible $\left|\chi_{12}^{\infty} - \chi_{13}^{\infty}\right| = \Delta\chi \approx 0$ interaction parameter χ'_{23} should be solvent independent. The equation above is often called "$\Delta\chi$ effect".

Horta's group [Prolongo et al., 1989] have proposed a method based on the equation-of-state theory, which gives a polymer-polymer parameter χ_{23} named 'true', because the assumption that the Gibbs mixing function for the ternary polymer-polymer-solvent system is additive with respect to the binary contributions is avoided. They suggested that it is necessary to substitute the volume fraction φ_i in the Flory-Huggins theory by segment fractions ϕ_i according to:

$$\phi_i = \frac{w_i \cdot v_i^*}{\sum w_i \cdot v_i^*} \tag{10}$$

where v_i^* and w_i represent characteristic specific volume and the weight fraction of the ith component, respectively.

Shi and Shreiber [Shi & Shreiber, 1991] stated that the probe dependence of $\chi_{23}^{'}$ is due to two major contributing factors. Firstly, the surface composition of a mixed stationary phase will rarely, if ever, correspond to the composition of the bulk. Thermodynamic requirements to minimize the surface free energy of the stationary phase will favor the preferential concentration, at the surface, of the component with the lower (lowest) surface free energy. Thus, the values of φ_2 and φ_3, as defined by the bulk composition of mixtures, are inapplicable to Eq. (7). Instead, a graphical method was proposed by Shi and Schreiber to evaluate the effective volume fraction and to correct the problem. Secondly, since χ_{12}^{∞} and χ_{13}^{∞} will not usually be equal, it follows that the volatile phase will partition preferentially to the component that has the lower pertinent χ_{1m}^{∞} value. Thus, the partitioning must vary with each probe, inevitably affecting the $\chi_{23}^{'}$ datum.

Deshpande and Farooque were the first to suggest the use of IGC for studying polymer blends [Deshpandee et al., 1974]. Starting form the Flory-Huggins expression for the change of the free enthalpy in mixing, which was extended to three-component systems, they proposed a method of analysis of IGC measurements on polymer blends which yielded the polymer-polymer interaction parameter $\chi_{23}^{'}$. They also observed probe dependency and tried to develop a method to evaluate probe-independent interaction [Farooque et al., 1992].

Milczewska and Voelkel [Milczewska & Voelkel, 2006] mentioned some of that methods of evaluating probe-independent interaction parameter. One of the solution may be procedure proposed by Zhao and Choi [Zhao & Choi, 2001; Zhao & Choi, 2002]. Authors proposed to use 'common reference volume' which vanishes the problem. As the reference volume they used molar volume of the smallest repeated unit of polymer.

Flory-Huggins parameter for blends can be calculated from equations:

$$\chi_{1m} = \frac{V_o}{V_1} \cdot \left(\ln \frac{273.15 \cdot R}{M_1 \cdot V_g \cdot p_1^o} - 1 + \left(1 - \frac{V_1}{V_2}\right) \cdot \varphi_2 + \left(1 - \frac{V_1}{V_3}\right) \cdot \varphi_3 - \left(\frac{B_{11} - V_1}{R \cdot T}\right) \cdot p_1^o \right) \tag{11}$$

and

$$\chi_{1m} = \varphi_2 \cdot \chi_{12} + \varphi_3 \cdot \chi_{13} - \varphi_2 \cdot \varphi_3 \cdot \chi_{23}^{'} \tag{12}$$

Equation (12) predicts that a plot of χ_{1m} versus ($\varphi_2 \cdot \chi_{12} + \varphi_3 \cdot \chi_{13}$) will give a straight line with a slope 1 and an intercept of $-\varphi_2 \cdot \varphi_3 \cdot \chi_{23}^{'}$.

Jan-Chan Huang [Huang, 2003] and with R. Deanin [Huang & Deanin, 2004] rearranged equation (12) into the following form:

$$\frac{\chi_{1m}}{V_1} = \frac{\varphi_2 \cdot \chi_{12} + \varphi_3 \cdot \chi_{13}}{V_1} - \frac{\varphi_2 \cdot \varphi_3 \cdot \chi_{23}^{'}}{V_2} \tag{13}$$

The polymer-polymer interaction term can be determined from the intercept at $\left(\frac{\varphi_2 \cdot \chi_{12} + \varphi_3 \cdot \chi_{13}}{V_1}\right) = 0$. This modification provided smaller standard deviations for the slope and the polymer-polymer interaction parameter.

Jan-Chan Huang [Huang, 2006] used also solubility parameter model to the study of the miscibility and thermodynamic properties of solutions by means of IGC. Because polymer–polymer mixtures have little entropy of mixing, the miscibility is largely decided by the sign of the heat of mixing. The determination of the heat of mixing becomes the key factor. The heat of vaporization is related to the solubility parameter, δ, of the liquid by the relation:

$$\delta = \left(\frac{\Delta E_{vap}}{V}\right)^{1/2} \tag{14}$$

where ΔE_{vap} is the energy of vaporization and V is the molar volume of the solvent.

The Flory–Huggins interaction parameter can be related to the solubility parameters of the two components by:

$$\chi = \left(\frac{V_1}{RT}\right) \cdot (\delta_1 - \delta_2)^2 \tag{15}$$

where δ_1 and δ_2 are the solubility parameters of the solvent and polymer, respectively, and V_1 is the volume of the solvent.

Guillet and co-workers [DiPaola-Baranyi & Guillet, 1978; Ito & Guillet 1979] have proposed IGC method for estimating of Flory–Huggins interaction parameter and solubility parameter for polymers by the modification of Eq. (15):

$$\left(\frac{\delta_1^2}{RT} - \frac{\chi}{V_1}\right) = \left(\frac{2\delta_2}{RT}\right)\delta_1 - \left(\frac{\delta_2^2}{RT}\right) \tag{16}$$

It is a straight line equation. The left-hand side contains the values of Flory–Huggins interaction parameter of test solute (see Eq. (2)), solubility parameter of test solute (δ_1)) and its molar volume. Plotting the left-hand side of such equation vs. solubility parameter of test solute (δ_1) one obtains the slope ($a = 2\delta_2/RT$) enabling the calculation of the solubility parameter of the examined material. This value should be equal to that found from the intercept and positive [Voelkel et al., 2009 (b)].

When a mixture is used as the stationary phase the solubility parameter of the mixture, δ_m, can be compared with the prediction of the regular solution method, which gives δ_m to be the volume average of the two components [Huang, 2006]:

$$\delta_m = \varphi_A \cdot \delta_A + \varphi_B \cdot \delta_B \quad (17)$$

From this equation the formula of specific heat of mixing in the regular solution theory could be derived. A measurement of the solubility parameter of the polymer mixtures would then be a good indicator to predict their miscibility.

Huang [Huang, 2006] proposed a mechanism of probe dependency. When two polymers with specific interactions are brought together some functional groups interact with each other and are no longer available to the probes. Relative to the volume average of the pure components the probes will feel the mixture becomes lower in polar or hydrogen bonding interaction and more in nonpolar dispersive force. In other words, the mixture becomes more "alkane-like". The polar probes will be squeezed from the stationary phase and the specific retention volume decreased, which increases χ_{1m} through Eq. (5) then decreases χ'_{23} through Eq. (13). Therefore, polar probes have lower retention volume and χ'_{23}, and for n-alkane probes the change is less. This difference between probes is exhibited as the probe dependency.

4. Applications

Authors examined many polymeric materials filled with modified silica or other inorganic fillers. Our measurements were carried out with the use of Chrom5 (Kovo, Prague, Czech.Rep.) gas chromatograph equipped with a flame ionisation detector. Some of the results were presented here.

4.1 Flory-Huggins parameters for polylactic acid compositions

For composition of polylactic acid (P, M=55000), containing different amount (5, 10, 15% wt) of modified silica (B2 and B5) [Jesionowski, 1999] or modified carbonate-silicate fillers (N1 and N2) [Grodzka 2004] we calculated Flory-Huggins parameters χ_{12}^{∞} and χ'_{23}. The influence of the temperature and the amount and type of filler was examined. To eliminate the solvent dependence of χ'_{23} values (from basic Eq. 8) experimental data were recalculated according to Zhao-Choi procedure.

Small volumes (0.5μL) of vapour of the probes were injected manually to achieve the infinite dilution conditions. These were: n-pentane (C5), n-hexane (C6), n-heptane (C7), n-octane (C8), n-nonane (C9), dichloromethane (CH2Cl2), chloroform (CHCl3), carbon tetrachloride (CCl4), 1,2-dichloroethane (Ethyl. Chl.) (all from POCH, Gliwice, Poland).

Values of χ_{12}^{∞} parameter for P-15N1 (it denotes the composition of polylactic acid with 15% of N1 filler) composition we presented in Figure 2. We obtained almost the same values of χ_{12}^{∞} parameter for other investigated compositions.

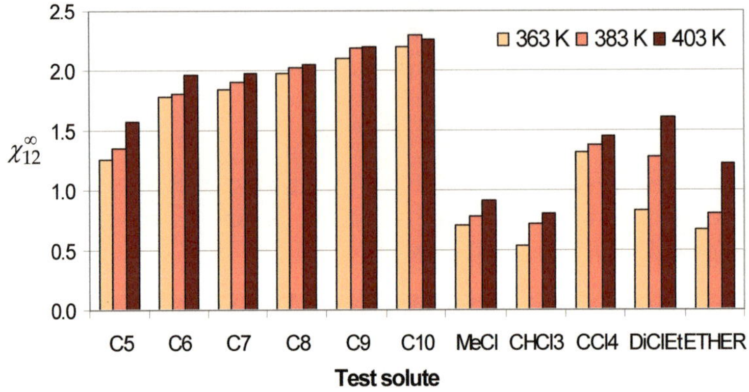

Fig. 2. Values of Flory-Huggins χ_{12}^{∞} parameter for P-15N1 composition

The lowest values of χ_{12}^{∞} parameter were obtained for dichloromethane (MeCl) and chloroform (CHCl3) as the test solute. Values calculated for compositions were almost always lower than those found for pure components, i.e. polymer and/or filler separately. The increase of temperature decreased values of χ_{12}^{∞} only for P-5B5 system, indicating the increase of interactions between composition and test solute. For the other compositions the increase of temperature increased values of χ_{12}^{∞} parameter.

The influence of the amount and type of filler was also examined. The change of these two factors also lead to the changes in the solute-composition interactions (Fig. 3). The influence of the amount of the filler is different for various compositions. For composition with N1 filler (carbonate-silicate filler modified with N-2-aminoethyl-3-aminopropyl-trimethoxysilane) the strongest interaction with solvent was found for the composition containing 5% of the filler. However, for P-N2 the most active is the composition with 15% addition of N2 (carbonate-silicate filler modified with n-octyltriethoxysilane).

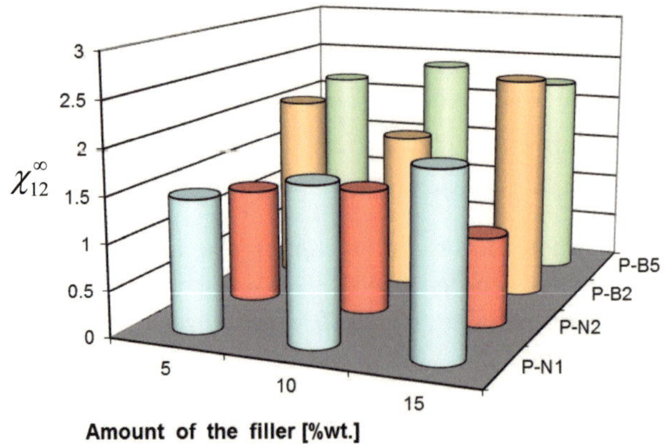

Fig. 3. The influence of the type and the amount of the filler [% wt.] on χ_{12}^{∞} parameters at 403K

Determined values of χ'_{23} depend on the type of the test solute used in IGC experiment (Fig. 4). Influence of the amount of the filler on the Flory-Huggins parameter χ'_{23} was examined and some results are presented in Figure 4. It is depended on test solute used in our study. For C5-C7 and CHCl3 and CCl4 we obtained the lowest values for 5% of N1 filler. For the other solutes - the strongest interaction are observed between polymer and 15% of the filler. Generally, the strongest interaction between components were observed for compositions with 5% of the filler.

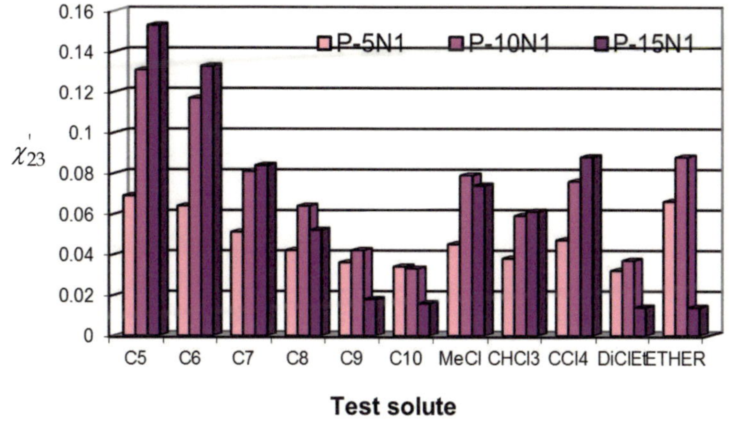

Fig. 4. Values of Flory-Huggins χ'_{23} parameter for compositions with 5%, 10% or 15% of filler

To eliminate the solvent dependence of χ'_{23} values (from basic equation) experimental data were recalculated according to Zhao-Choi procedure (Eq. 12). Values of $ZC\chi'_{23}$ parameter are presented in Figure 5. In all cases only one value for each composition was obtained. All $ZC\chi'_{23}$ values indicated the presence of strong or medium interaction between the modified filler and polymer matrix. This observation is consistent with that formulated after analysis of χ'_{23} data found for most of test solutes in the classic procedure.

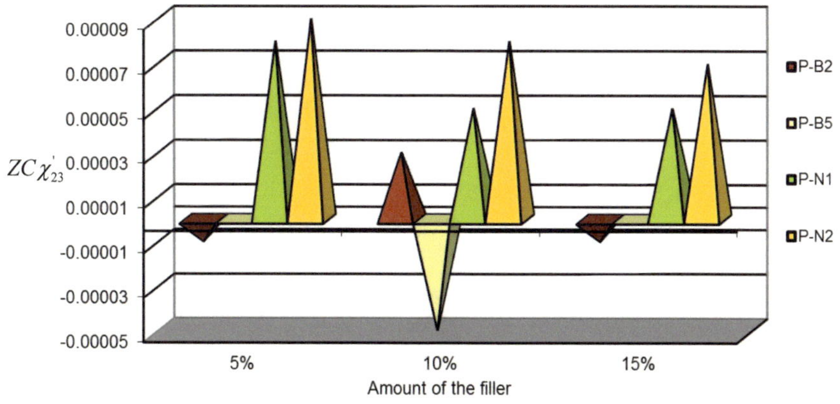

Fig. 5. Values of Flory-Huggins $ZC\chi'_{23}$ parameter for all compositions calculated according to Zhao-Choi procedure

It is worth to note that the increase of the filler content does not enhance the magnitude of interactions. Most often limited (rather negligible) decrease of polymer-filler interactions was observed.

4.2 Chemometric evaluation of IGC data

Principal Component Analysis (PCA) became a popular technique in data analysis for classification for pattern recognition and dimension reduction. It can reveal several underlying components, which explain the vast majority of variance in the data [Héberger, 1999; Malinowski, 1991; Héberger et al., 2001]. The principle is to characterize each object (rows in the input matrix) not by analyzing every variable (columns of the input matrix) but projecting the data in a much smaller subset of new variables (or principal component scores). PCA should facilitate the overcoming of the problem connected with the solute dependence of χ_{23}' parameter.

Values of Flory-Huggins χ_{23}' parameter expressing the magnitude of interactions between the polymer matrix and filler strongly depend on the type of test solute being used in IGC experiment (see Fig. 4). It causes the difficulties in the analysis of the influence of the type and amount of the filler onto the magnitude of these interactions. Such analysis is possible with the help of PCA technique [Voelkel et al., 2006] as presented in Fig. 6 for systems of polyurethane (PU) with modified silica fillers (B2). Materials used in experiments were described elsewhere [Milczewska & Voelkel 2002; Milczewska, 2001]. The magnitude of interactions is similar (the corresponding points belong to one - large cluster) for most of samples. Outside this large cluster the points correspond mainly to the compositions with 5 or 20% of the filler.

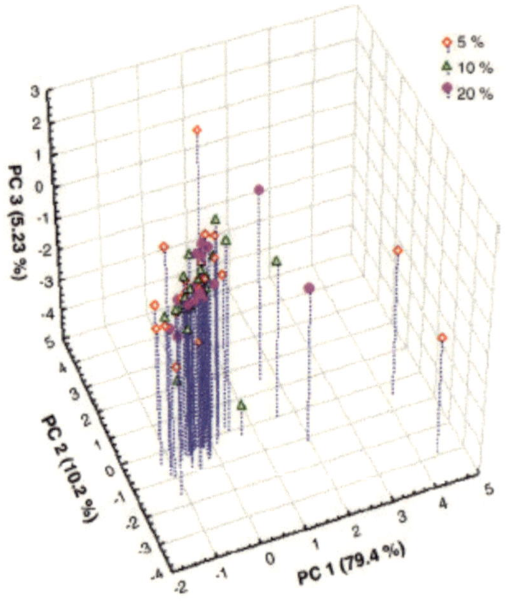

Fig. 6. Scatterplot for Polyurethane (PU) systems [Reprinted from Voelkel et al., 2006 with permission from Elsevier]

IGC procedures discussed earlier allow eliminating the test solutes dependence of χ'_{23} values. However, very often the relatively significant error of the determination was reported.

Values of χ'_{23} parameter calculated according to Zhao-Choi procedure for the examined polymeric composition are presented in Table 1. All values are negative and close to zero. It indicates the existence of polymer–filler interaction although their strength is limited.

Filler	**PU-B2 compositions**		
	PU+5%	PU+10%	PU+20%
IB2	-2*10^{-5}	-2*10^{-5}	-2*10^{-5}
IIB2	-4*10^{-5}	-6*10^{-5}	-3*10^{-5}
IIIB2	-6*10^{-5}	-3*10^{-5}	-7*10^{-5}
IVB2	-3*10^{-5}	-1*10^{-5}	-7*10^{-5}
VB2	-7*10^{-5}	-3*10^{-5}	-1*10^{-5}

Table 1. Values of χ'_{23} parameter calculated by Zhao-Choi method for B2-PU compositions [Reprinted from Voelkel et al., 2006 with permission from Elsevier]

The differences of the magnitude of polymer–filler interactions are significant as the error of determination is equal to approximately 2.5*10^{-7}, i.e. it is at least two orders lower than the determined χ'_{23} values. However, collection of retention data for all test solutes is somewhat time-consuming. It would be useful to select the test solutes carrying the statistically valid information, applied these species in IGC experiments and further use their retention data in calculations of χ'_{23} from Zhao–Choi procedure. The problem was: how the reduction of the number of test solutes will influence the χ'_{23} values as well as error of their determination.

For all PU compositions PCA made possible using three - four test solutes (C6, C8, MeCl and CCl4) for determination of interaction parameters. Recalculating of χ'_{23} from Zhao–Choi procedure for selected test solutes gave values presented in Table 2.

Comparison of values of χ'_{23} calculated by Zhao-Choi method before and after PCA selection of solutes is presented on Figure 7. Corrected values are lower or higher than these found for all test solutes, but they indicate the presence or absence of interaction.

Filler	**PU-B2 compositions**		
	PU+5%	PU+10%	PU+20%
IB2	-3*10^{-5}	-2*10^{-5}	-2*10^{-5}
IIB2	-3*10^{-5}	-5*10^{-5}	-3*10^{-5}
IIIB2	-5*10^{-5}	-3*10^{-5}	-5*10^{-5}
IVB2	-2*10^{-5}	0*10^{-5}	-7*10^{-5}
VB2	-4*10^{-5}	-1*10^{-5}	-1*10^{-5}
Error ~ 2.5*10^{-7}			

Table 2. Values of χ'_{23} parameter for B2-PU compositions calculated by Zhao-Choi method after PCA selection of solutes [Reprinted from Voelkel et al., 2006 with permission from Elsevier]

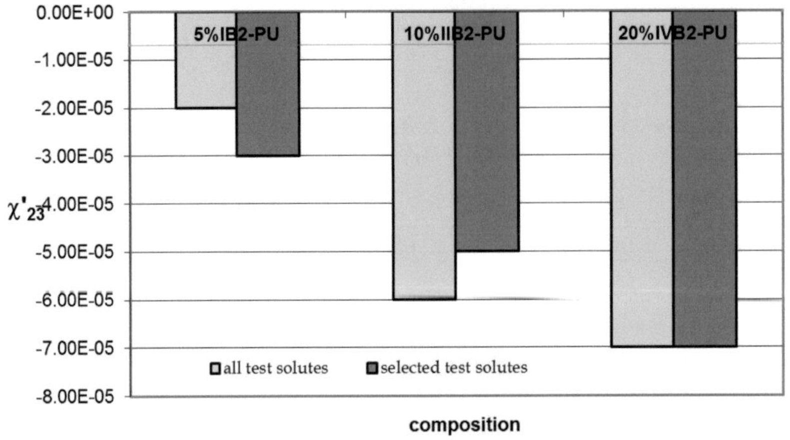

Fig. 7. Comparison of $\chi_{23}^{'}$ calculated by Zhao–Choi procedure before and after PCA selection of test solutes for PU [Reprinted from Voelkel et al., 2006 with permission from Elsevier]

PCA enabled the significant reduction of the number of test solutes required for the proper determination of Flory–Huggins parameter and further the reduction of time required for proper characterization of examined material.

5. Summary

Inverse gas chromatography method has been found to be an effective tool for measurement of the magnitude of interaction (χ_{12}^{∞} interaction material - test solute) in polymers at different temperatures. From technological point of view $\chi_{23}^{'}$ is more interesting. This parameter can be used as indicator of the miscibility of the polymer blend.

Drawback of $\chi_{23}^{'}$ estimated by classical procedure is the test solute dependence. The procedure of its elimination proposed by Zhao-Choi seems to be most appropriate. It leads to realistic values of $\chi_{23}^{'}$ with low error of determination. The full procedure of $\chi_{23}^{'}$ .with the use of large series of the test solutes might be time-consuming. The use of chemometric analysis enabled the reduction of the number of the test solutes without loss of information. One may expect further application of IGC in examination of polymer materials.

6. References

Al-Ghamdi A., Al-Saigh Z.Y. (2000) *J. Polym. Sci., Part B: Polym. Phys.* Vol. 38 pp. 1155-1166.

Al-Saigh Z., Munk P. (1984) *Macromolecules* Vol. 17 pp. 803-809.

Al-Saigh Z.Y. (1997) *TRIP* Vol. 5 pp. 97-101.

Boukerma K., Piquemal J.Y., Chehimi M.M., Mravcakova M., Omastova M., Beaunier P. (2006) *Polymer* Vol. 47 pp. 569-576.

de Gennes P.G. (1979) *Scaling Concepts in Polymer Physics*, Cornell University Press, Ithaca, NY.

Deshpande D.D., Patterson D., Schreiber H.P., Su C.S. (1974) *Macromolecules* Vol. 7 pp. 530-535.

DiPaola-Baranyi G., Guillet JE. (1978) *Macromolecules* Vol. 11 pp. 228–235.

Dritsas G.S., Karatasos K., Panayiotou C. (2008) *J. Polym. Sci., Part B: Polym. Phys.* Vol. 46 pp. 2166-2172.

Dritsas G.S., Karatasos K., Panayiotou C. (2009) *J. Chromatogr. A* Vol. 1216 pp. 8979-8985

El-Hibri M.J., Cheng W., Hattam P., Munk P. (1989) Inverse Gas Chromatography of Polymer Blends. Theory and Practice, *in:* D.R. Lloyd, T.C. Ward, H.P. Schreiber (ed), *Inverse Gas Chromatography. Characterization of Polymers and Other Materials,* ACS Symposium Series, 391, Washington, 121 134.

Farooque A.M., Deshpande D.D. (1992) *Polymer* Vol. 33 pp. 5005-5018.

Fernandez M.L., Higgins J.S., Horst R., Wolf B.A. (1995) *Polymer* Vol. 36 pp. 149-154

Fernandez-Sanchez E., Fernandez-Torres A., Garcia-Dominguez J.A., Santiuste J.M., Pertierra-Rimada E. (1988) *J. Chromatogr.* Vol. 457 pp. 55-71.

Grobler A., Balizs, G. (1974) *J. Chromatogr. Sci.* Vol. 12 p. 57

Grodzka J. (2004) Ph.D. thesis, Poznan University of Technology, Poznan (in polish).

Gundert F., Wolf B. A. (1989) "Polymer-Solvent Interaction Parameters" in: *Polymer Handbook,* Ed. J. Brandrup and E. H. Immergut Wiley Interscience, New York, 3rd edn., pp. VII/173-182

Han S.J., Lohse D.J., Condo P.D., Sperling L.H. (1999) *J. Polym. Sci.: Part B: Polym. Phys.* Vol. 37 pp. 2835-2844.

Héberger K. (1999) *Chemometrics Intell. Lab. Syst.* Vol. 47 pp. 41-49.

Héberger K., Milczewska K., Voelkel A. (2001) *J. Chromatogr. Sci.* Vol. 39 pp. 375-384.

Hindawi J., Higgins J.S., Galambos A.F., Weiss R.A. (1990) *Macromolecules* Vol. 23 pp. 670-674.

Horst R., Wolf B.A. (1992) *Macromolecules* Vol. 25 pp. 5291-5296.

Huang J.-C. (2003) *J Appl. Pol. Sci.* Vol. 90 pp. 671-680.

Huang J.-C. (2009) *J. Appl. Polym. Sci.* Vol. 113 pp. 4085-4091.

Huang J.-C., Deanin R.D. (2004) *J Appl. Pol. Sci.* Vol. 91 pp. 146-156.

Ito K., Guillet J.E. (1979) *Macromolecules* Vol. 12 pp. 1163–1167.

James A.T., Martin A.J.P. (1952) *Biochem. J.* Vol. 50 pp. 679-690.

Jesionowski T. (1999) Ph.D. thesis, Poznan University of Technology, Poznan (in polish).

Lee H.S., Kim W.N., Burns C.M. (1997) *J. Appl. Polym. Sci.* Vol. 64 pp. 1301-1308.

Lezcano E.G., Prolongo M.G., Coll C.S. (1995) *Polymer* Vol. 36 pp. 565-573.

Li Bincai (Pun Choi) (1996) *Rubber Chem. Techn.* Vol. 69 pp. 347-376.

Malinowski E. R. (1991) *Factor Analysis in Chemistry,* 2nd ed., John Wiley and Sons, New York.

Mani S., Malone M.F., Winter H.H. (1992) *Macromolecules* Vol. 25 pp. 5671-5676.

Meoer H., Strobl G.R. (1987) *Macromolecules* Vol. 20 pp. 649-654.

Milczewska K. (2001) Ph.D. thesis, Poznan University of Technology, Poznan (in polish).

Milczewska K., Voelkel A. (2002) *J. Chromatogr. A* Vol. 969 pp. 255-259.

Milczewska K., Voelkel A. (2006) *J Polym Sci Part B: Polym Phys* Vol. 44 pp. 1853–1862

Milczewska K., Voelkel A., Jęczalik J. (2001) *Macromol. Symp.* Vol. 169 pp. 45-55.

Milczewska K., Voelkel A., Jęczalik J. (2003) *Macromol. Symp.*Vol. 194 pp. 305-311.

Mutelet F., Ekulu G., Rogalski M. (2002) *J. Chromatogr. A* Vol. 969 pp. 207-213.

Nesterov A.E. and Lipatov Y.S. (1999) *Polymer* Vol. 40 pp1347-1349

Nesterov A.E., Lipatov Y.S., Ignatova T.D. (2001) *Eur. Polym. J.* Vol. 37 pp. 281-285.
Olabisi O. (1975) *Macromolecules* Vol. 8 pp. 316-322.
Oliva V., Mrabet B., Neves M.I.B., Chehimi M.M., Benzarti K. (2002) *J. Chromatogr. A* Vol. 969 pp. 261-272.
Ovejero G., Pérez P., Romero M.D., Díaz I., Díez E. (2009) *Eur. Polym. J.* Vol.45 pp. 590-594.
Petri H.M., Schuld N., Wolf B.A. (1995) *Macromolecules* Vol. 28 pp. 4975-4980.
Prolongo M.G., Masegosa R.M., Horta A. (1989) *Macromolecules* Vol. 22 pp. 4346-4351.
Schuld N., Wolf B.A. (2001) *J. Polym. Sci.: Part B: Polym. Phys.* Vol. 39 pp. 651-662.
Schwann D., Frielinghaus H., Mortensen K., Almdal K. (1996) *Phys. Rev. Lett.* Vol. 77 pp. 3153-3156.
Shi Z.H., Schreiber H.P. (1991) *Macromolecules* Vol. 24 pp. 3522-3527.
Shillcock I.M., Price G.J. (2003) *Polymer* Vol. 44 pp. 1027-1034.
SMS-*i*GC brochure 2002
Su C.S., Patterson D. (1977) *Macromolecules* Vol. 10 pp. 708-710.
Sun H., Rhee K.B., Kitano T., Mah S.I. (1999) *J. Appl. Polym. Sci.* Vol. 73 pp. 2135-2142.
van Asten A., van Veenendaal N., Koster S. (2000) *J. Chromatogr. A* Vol. 888 pp. 175-196.
Voelkel A., Fall J. (1997) *Chromatographia* Vol. 44 pp. 197-204.
Voelkel A., Milczewska K., Héberger K. (2006) *Analytica Chimica Acta* Vol. 559 pp. 221-226.
Voelkel A., Strzemiecka B., Adamska K., Milczewska K. (2009) *J. Chromatogr. A* Vol. 1216 pp. 1551-1566 (b)
Voelkel A., Strzemiecka B., Adamska K., Milczewska K., Batko K. (2009) Surface and bulk characteristics of polymers by means of inverse gas chromatography *in* A. Nastasovic, S. Jovanovic (eds.) *Polymeric materials*, Research Signpost, pp. 71-102. (a)
Ying Q., Chu B., Wu G., Linliu K., Gao T., Nose T., Okada M. (1993). *Macromolecules* Vol. 26 pp. 5890-5896.
Zhao L., Choi P. (2001) *Polymer* Vol. 42 pp. 1075-1081.
Zhao L., Choi P. (2002) *Polymer* Vol. 43 pp. 6677-6681.
Zou Q.C., Zhang S.L., Wang S.M., Wu L.M. (2006) *J. Chromatogr. A* Vol. 1129 pp. 255-261.

21

Recent Applications of Comprehensive Two-Dimensional Gas Chromatography to Environmental Matrices

Cardinaël Pascal[1], Bruchet Auguste[2] and Peulon-Agasse Valérie[1]
[1]*SMS, Université de Rouen,*
[2]*CIRSEE (Centre International de Recherche Sur l'Eau et l'Environnement),*
France

1. Introduction

Anthropological pressure on environment combined with continuous progress of analytical techniques allows the detection of more micro-pollutants in environmental matrices. Analysis of Persistent Organic Pollutants (POPs) remains a real challenge due to the large number of compounds and the complexity of environmental matrices. Conventional Gas Chromatography (GC) coupled with mass spectrometry (MS) is the reference technique for the analysis and the quantification of volatile and semi-volatile pollutants. Comprehensive two-dimensional gas chromatography (GC×GC) is a relatively new technique, developed in 1991 by Liu and Phillips (Liu & Phillips, 1991). This technique provides high separation power and sensitivity. The principles of multidimensional chromatography were described by Giddings (Giddings, 1984). When a fraction or few fractions of the effluent from a first column, is subsequently injected into a second column with a different selectivity, the multidimensional chromatographic separation techniques are called 'heart cutting'. These methods have proved to be very effective only in target compounds analysis. A two-dimensional separation can be called comprehensive if the three following conditions are established (Schoenmakers et al., 2003). First, every part of the sample is subjected to two different separations. Secondly, equal percentages (either 100% or lower) of all sample components pass through both columns and eventually reach the detector. Finally, the separation (resolution) obtained in the first dimension is essentially maintained. This latter point could be reached if the transfer of the effluent from the first column to the second one was successfully performed by a modulator or column interface. So, the modulator could be considered as the 'heart' of the system and is currently in development. A maximum of retention space could be used especially if compounds are subjected to two independent separations. Orthogonal separation occurs when the two columns use different separation mechanisms, operating independently in the two dimensions. In practice, columns containing chemically different stationary phases are chosen. In normal orthogonality, the first apolar column is coupled to a column containing a stationary phase of equivalent or higher polarity. For reversed orthogonality, the more polar stationary phase is used in first dimension and a less polar one in second dimension. Due to the low peak width, some constraints are imposed for the choice of detector. An ideal data acquisition rate for GC×GC detector is equal or more than 100 Hz to maintain its large separation power. Numerous detectors, conventionally used in GC like Flame Ionization Detector (FID), Electron Capture Detector (ECD) and microECD (μECD), have been widely employed. Concerning the MS, the high speed time-of-flight (TOFMS) with a unit-mass resolution has proved to be the best candidate for GC×GC. Moderate acquisition rate instruments, such as quadrupole mass spectrometer qMS (e.g. 20 Hz) were also used with a limited mass range. Several GC×GC instruments with various modulators have been developed (Semard et al., 2009) and are now commercially available. GC×GC has now demonstrated its capacity of resolution in the field of complex matrices like petroleum products, fragrance (Dallüge et al., 2003). GC×GC is currently one of the most effective techniques for the separation and analysis of environmental samples, offering significantly greater peak capacities than conventional chromatographic methods. GC×GC provides three major benefits, namely, enhanced chromatographic separation, improved sensitivity by effect of cryofocusing with the thermal modulator, and chemical class ordering in the contour plot (Ballesteros-Gomez & Rubio, 2011). Research in GC×GC has recently shifted from instrumental development to application to real samples over the last four years especially with the coupling to MS. Nearly 83% of the over 110 research articles published (specify the type or classification of articles) in 2010 were devoted to applications of GC×GC (Edwards et al., 2011). Moreover, some software improvements have also facilitated GC×GC quantitative analyses.

A few reviews were already published about applications of comprehensive GC in relation to environmental analyses. The most recent review was proposed by Wang et al. (Wang et al., 2010) that covered the works published between 2007 and the beginning of 2009. This chapter focuses on the most important developments in environmental applications of GC×GC, salient advances in GC×GC instrumentation and theoretical aspects reported over the period 2009 to July 2011. Recent applications using GC×GC methods for analysis of environmental toxicants such as PolyChlorinatedDibenzo-p-Dioxins (PCDDs), PolyChlorinatedDibenzoFurans(PCDFs), PolyChlorinatedBiphenyls (PCBs), Polycyclic Aromatic Hydrocarbons (PAHs), pesticides, alkylphenols *etc*... are reviewed. Moreover, this technique appeared especially suitable for the development of multiresidue analytical methods which was the most important trend in GC×GC-MS environmental analysis over the last period. The recent works demonstrated that this technique provides interesting alternative methods in terms of sensitivity and mapping of pollutants and minimizes sample preparation steps. A part of this chapter will be dedicated to the recent screening of emerging contaminants such as pharmaceuticals, plasticizers, personal care products, ... Various matrices (water, soils and sediments) will be considered including river and wastewater. Analysis of air sample were not reported according to the publication of recent reviews dealing with the analysis of Volatile and semivolatile Organic Compounds (VOCs) found in the atmosphere (Arsene et al., 2011; Hamilton, 2010). All aspects of GC×GC will be presented including instrumentation, theoretical considerations and applications. Moreover, novel tools used for optimization of retention space or orthogonality estimation will be discussed.

2. GC×GC and environmental analysis reviews

A non exhaustive review (Ballesteros-Gomez & Rubio, 2011) was focused on main developments and advances in environmental analysis reported over the period 2009-2010.

Numerous aspects of environmental analysis, based on more than 200 articles, were reported including sampling, sample preparation, separation and detection ... Emerging contaminants and atomic spectrometry for the determination of trace metals and metalloids topics were excluded due to the publication of other reviews. Few applications of GC×GC were shortly reported (Eganhouse et al., 2009; Matamoros et al., 2010a; Hilton et al., 2010) and will be discussed in the present chapter. Recently, GC×GC applications devoted to measurement of volatile and semivolatile organic compounds in air and aerosol were reviewed and discussed (Arsene et al., 2011, Hamilton, 2010).

Wang *et al.* (Wang et al., 2010) have reviewed technological advances and applications of GC×GC between 2007 and July 2009. For example, separations of eight persistent organohalogenated classes of pollutants including OrganoChlorinatedPesticides (OCPs), PCBs, PolyBrominatedDiphenylEthers (PBDEs), PolyChlorinatedNaphthalenes (PCNs), PCDDs, PCDFs, PolyChlorinatedTerphenyls (PCTs), and toxaphene (CTT) in environmental samples were reported by Bordajandi *et al.* (Bordajandi et al., 2008). Nine column combinations in normal and reversed orthogonality (ZB-5, HT-8, DB-17 and BP-10, as first dimension column and HT-8, BPX-50 and Carbowax as second dimension one) were tested. The feasibility of the proposed approach for the fast screening of the target classes of pollutants was illustrated by the analysis of food and marine fat samples. A method of quantification of PAHs in air particulates based on a GC×GC isotope dilution mass spectrometry method was also reported by Amador-Munoz *et al.* (Amador-Munoz et al., 2009) with favorable resolution and sensitivity over conventional one dimensional GC. GC×GC-TOFMS method was successfully developed by Skoczynska *et al.* (Skoczynska et al., 2008) to identify 400 compounds in highly polluted sediment sample from the River Elbe (Czech Republic). Several older reviews dealing with the development of GC×GC and its applications including environmental matrices could be mentioned (Ramos et al., 2009; Cortes et al., 2009; Adahchour et al., 2006; Adahchour et al., 2008; Pani & Gorecki., 2006).

3. PCBs

PCBs are composed of 209 distinct congeners and are found in complex mixtures (Aroclors).They were commercially used in a variety of applications, including heat transfer and hydraulic fluids, dielectric fluids for capacitors, and as additives in pesticides, sealants, and plastics (Osemwengie & Sovocool, 2011). The World Health Organization (WHO) has designated twelve PCBs as "dioxin-like", coplanar PCB congeners that exhibited high toxicity.

Recently, a routine accredited method (Muscalu et al., 2011) was presented for analysis of PCBs, chlorobenzenes and other halogenated compounds in soil, sediment and sludge by GC×GC-μECD. A column combination DB1×Rtx-PCB was used to minimize coelution of analytes. The method was developed to analyze these pollutants in a single analytical run and no fractionation of sample extracts prior to instrument analysis, with enhanced selectivity and sensitivity over one dimensional GC method. The method can also be used to perform analytical triage to screen for additional compounds, for additional extract processing and testing or for identification and monitoring of new and emerging halogenated compounds present in sample extracts and to screen other halogenated organics. The optimized method provided quantification of Aroclors and Aroclors mixtures to within 15% of targets values and sub-nanograms per gram detection limits. The authors claimed that GC×GC requires minimal additional training to be used as a routine analytical method for the analysis of halogenated compounds.

Separation of 209 PCB congeners, using a sequence of 1D and 2D chromatographic modes was evaluated (Osemwengie et al., 2011). The authors used a RTX-PCB column as the first column and a DB-17 as the second one. In two consecutive chromatographic runs, 196 PCB congeners were distinguished, including 43 of the 46 pentachlorobiphenyl isomers. PCBs congeners that were not resolved chromatographically were resolved with the deconvolution program (ChromaTOFSoftware). Nevertheless, the 209 congeners have not been successfully separated.

New capillary columns coated with Ionic Liquids (ILs) were used as second columns for the separation of 209 PCBs congeners (Zapadlo et al., 2010; Zapadlo et al., 2011). In the first paper (Zapadlo et al., 2010), the orthogonality of three columns coupled in two series was studied. A non-polar capillary column coated with poly(5%-phenyl–95%-methyl)siloxane was used as the first column in both series. A polar capillary column coated with 70% cyanopropyl-polysilphenylene-siloxane or a capillary column coated with the ionic liquid 1,12-di(tripropylphosphonium)-dodecanebis(trifluoromethanesulfonyl)imide (IL36) was used as the second columns. The authors concluded that column coated with IL was more polar and more selective for the separation of PCBs than BPX-70 column (Figure 1).

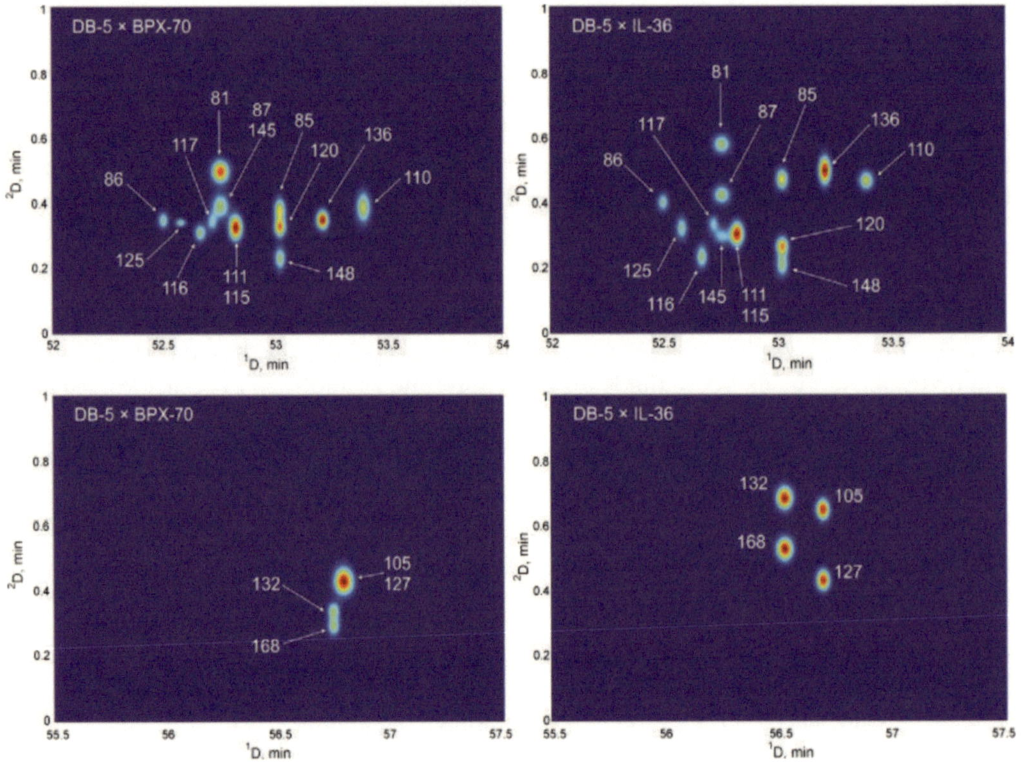

Fig. 1. 2D images for the separation of toxic, dioxin-like PCBs 81 and 105 on DB-5×BPX-70 and DB-5×IL-36 column series. Reprinted from Journal of Chromatography, A, (Zapadlo et al., 2010). Copyright (2010), with permission from Elsevier.

All "dioxin-like" PCBs, with the exception of PCB 118 and PCB 106, were resolved by this set of columns. In the second study (Zapadlo et al., 2011), the separation of 209 PCBs congeners was investigated using GC×GC–TOF-MS with a non-polar/IL column series consisting of poly(50%-n-octyl-50%-methyl)siloxane and (1,12-di(tripropylphosphonium)-dodecanebis(trifluoromethansulfonyl)amide) (SLB-IL59) in the first and second dimensions, respectively. A total of 196 out of 209 PCBs congeners were resolved by separation and/or mass spectral deconvolution using the ChromaTOF software. All "dioxin-like" congeners were separated with no interferences from any PCB congener. The 109 PCBs present in Aroclor 1242 and the 82 PCBs present in Aroclor 1260 were resolved on this column set.

A Quantitative Structure–Retention Relationship (QSRR) method (D'Archivio et al., 2011) was applied to predict the retention times of 209 PCBs in GC×GC. Predicted data were compared to GC×GC retention data taken from the literature. Authors demonstrated that the experimental GC×GC chromatogram of PCBs can be accurately predicted using a QSRR model calibrated with retention data of about 1/3 of the congeners collected under the same separation conditions. The effect of structure on retention time in both dimensions can be successfully encoded by theoretical molecular descriptors quickly available by means of various computational methods.

4. PCDDs and PCDFs

PCDDs and PCDFs constitute two classes of structurally related chlorinated aromatic hydrocarbons that are both highly toxic and produced as by-products during a variety of chemical and combustion processes. Due to their hydrophobic character and resistance to metabolic degradation, these substances exist as complex congener mixtures in the environment and are considered as POPs.

De Vos *et al.* (de Vos et al., 2011a) developed an alternative method of GC coupled with High Resolution Mass Spectrometry (HRMS) for analysis of PCDDs and PCDFs using GC×GC-TOFMS in different matrices. Three GC column combinations (Rtx-Dioxin 2×Rtx-PCB, Rxi-5 SilMS×Rtx-200 and Rxi-XLB×Rtx-200) were evaluated to quantify PCDDs and PCDFs in numerous soil and sediment samples taken from various strategic sites in South Africa with a highest result obtained of 76 ng Toxic Equivalent Quantity/kg. Results were also compared with those obtained using GC–HRMS and a good agreement was observed. The limit of detection (LOD) for the method (300 fg on column for spiked soil samples) was determined using the combination Rxi-XLB×Rtx-200 which provided excellent separation of the compounds mandated for analysis by United States Environmental Protection Agency (US EPA) Method (Figure 2).

Using a multi-step temperature program, all seventeen PCDDs and PCDFs components mandated by EPA Method 1613 were separated. GC×GC–TOFMS appeared to be a viable tool for dioxin screening and quantitation, especially in cases where PCDDs/PCDFs levels are greater than 1 ng.kg^{-1}. The technique proved to be ideal for application in developing countries where GC–HRMS is not available, and can be used to minimize costs by selecting only positive samples for further analysis by GC–HRMS. GC×GC–TOFMS additionally provides full range mass spectra for all sample components, thus allowing for identification of non-target analytes e.g. the brominated dioxins.

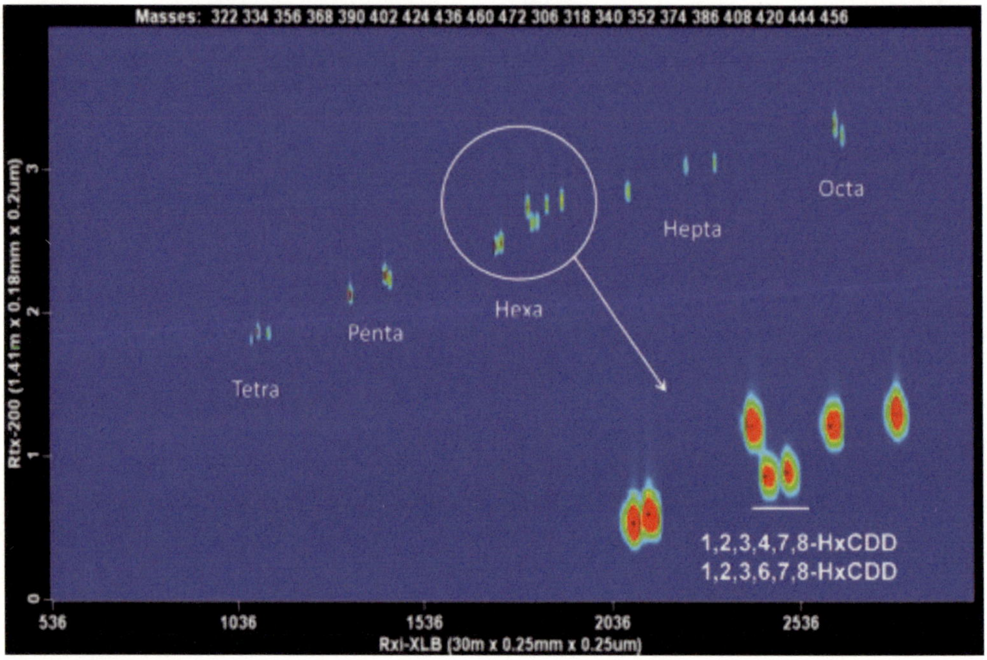

Fig. 2. ²D selected ion contour plot for the 17 priority PCDD/Fs using the Rxi-XLB/Rtx-200 column combination. The PCDD/Fs are again well resolved, especially the 1,2,3,4,7,8-HxCDD and 1,2,3,6,7,8-HxCDD isomers. Reprinted from Journal of Chromatography, A, (de Vos, 2011a). Copyright (2011), with permission from Elsevier.

GC×GC–TOFMS was also applied to investigate (de Vos et al., 2011b) toxic waste. This technique has allowed both comprehensive screening of samples obtained from a hazardous waste treatment facility for numerous classes of POPs and also quantitative analysis for the individual compounds. Various column combinations have been investigated for handling very complex waste samples. The close correlation between values obtained using the GC×GC-TOFMS approach and the GC-HRMS method has confirmed the validity of this technique to quantify PCDDs, PCDFs and four dioxin-like non-ortho substituted PCBs at levels required by regulatory bodies. The authors have obtained consistently higher values with the GC×GC–TOFMS method than those obtained with GC-HRMS. Nevertheless, they considered that the differences were certainly within permissible levels considering that the analyses have been performed in different laboratories.

An original application of the high sensitivity obtained using cryogenic zone compression (CZC) has been described by Patterson *et al.* (Patterson et al., 2011). The use of a GC×GC cryogenic loop modulator to perform CZC-GC coupled with Isotopic Dilution (ID) HRMS has been shown to be the most sensitive method available for the measurement of 2,3,7,8-tetrachlorodibenzo-p-dioxin (2,3,7,8-TCDD) (less than approximately 586,000 2,3,7,8-TCDD molecules) in human samples.

5. PAHs and hydrocarbons

PAHs are organic pollutants generated during the incomplete combustion of different natural and anthropogenic sources. They could enter the environment via municipal/industrial effluents. Exposure to PAHs represents a risk for human health due to their genotoxic and carcinogenic effects. The International Agency for Research on Cancer has classified them as possible and probable carcinogens for humans. The US EPA has included sixteen of them in the list of priority pollutants and establishes a maximum contaminant level for benzo[a]pyrene in drinking water at 0.2 μg.L^{-1}. In the European Union (EU), eight PAHs have been identified as priority hazardous substances in the field of water policy.

Chlorinated or brominated PAHs (Cl-PAHs and Br-PAHs) have been already detected in environmental samples such as fly ash (Horii et al., 2008) and sediment (Ishaq et al., 2003; Horii et al. 2009). Moreover, toxicities of Cl-PAHs have been investigated and reported (Horii et al., 2009). A method (Ieda et al. 2011) using GC×GC coupled with HRTOFMS was developed for the analysis of Cl-PAHs and Br-PAHs congeners in environmental samples. The GC×GC-HRTOFMS method allowed highly selective group type analysis with a very narrow mass window (e.g. 0.02 Da), accurate mass measurements for the full mass range (m/z 35–600) in GC×GC mode, and the calculation of the elemental composition for the detected congeners in the real-world sample. The authors reported, for the first time, the detection of highly chlorinated PAHs, such as $C_{14}H_3Cl_7$ and $C_{16}H_3Cl_7$, and ClBr-PAHs, such as $C_{14}H_7Cl_2Br$ and $C_{16}H_8ClBr$ in the environmental samples (Figure 3).

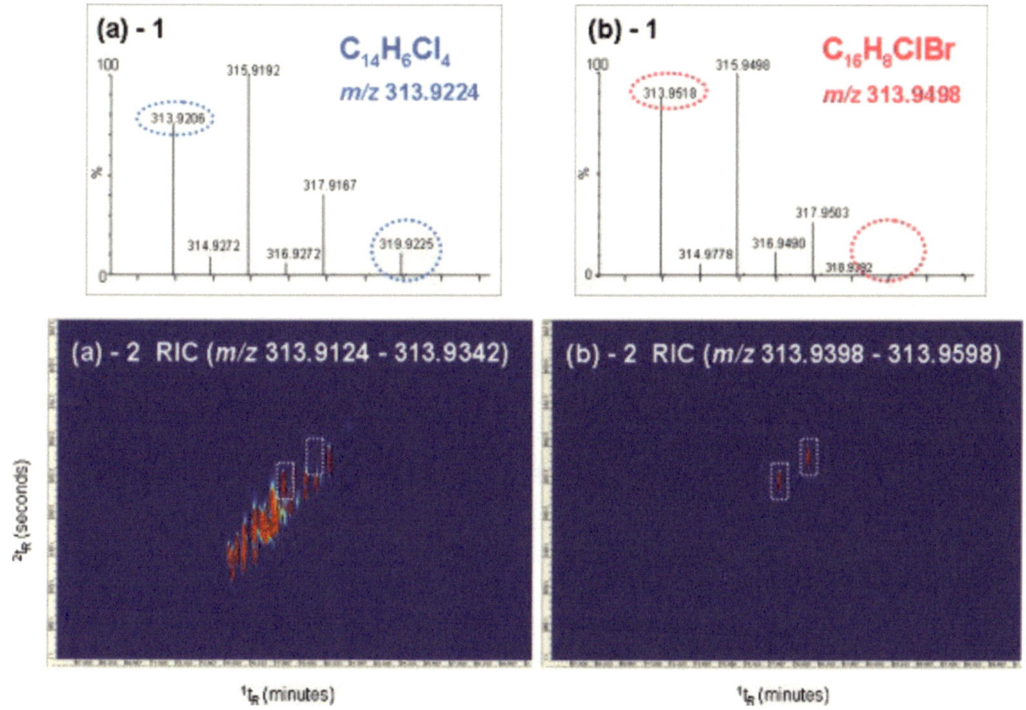

Fig. 3. The difference of isotope patterns between two peaks in the soil extract; (a)-1 $C_{14}H_6Cl_4$ and (b)-1 $C_{16}H_8ClBr$ and GC×GC–HR TOFMS 2D exact mass chromatogram of a 0.02 Da wide windows (a)-2 $C_{14}H_6Cl_4$; m/z 337.9224 and (b)-2 $C_{16}H_8ClBr$; m/z 313.9498. Reprinted from Publication, Journal of Chromatography, A, (Ieda et al., 2011). Copyright (2011), with permission from Elsevier.

Other organohalogen compounds; e.g. PCBs, PCNs, and PCDFs were also detected. This technique provided exhaustive analysis and powerful identification for the unknown and unconfirmed Cl-/Br-PAH congeners in environmental samples.

GC×GC-FID and GC×GC-TOFMS methods (Wardlaw et al., 2011) were used to study the biodegradation of alkylated naphthalenes and benzothiophenes isomers in marine sediment contaminated with crude oil. Their power resolution enabled separation and quantification of multiple structural isomers to determine their first order rate constants for aerobic biodegradation. Rate constants were used as proxies for microbial preference. A strong isomeric biodegradation preference was noted within each of these compound classes, with rate constants varying as much as a factor of 2 for structural isomers of the same compound class.

HPLC-GC×GC-FID and GC×GC-TOFMS were used to study the biodegradation of petroleum hydrocarbons in soil microcosms during 20 weeks (Mao et al., 2009). Aromatic hydrocarbons and n-alkanes were better biodegradable (>60% degraded) than iso-alkanes and cycloalkanes (<40%). GC×GC chromatograms showed that more polar and heavier compounds were formed as biodegradation proceeded.

6. Alkyl phenol isomers

AlkylPhenol EthOxylates (APEOs) are surfactants that have been widely used as detergents, emulsifier and dispersing agents in industrial or household cleaning products including laundry detergents. These compounds are degraded in wastewater treatment plants (WWTPs) in more toxic compounds, such as nonylphenols (NPs) and octylphenols (OPs). NPs and OPs, used for industrial production of APEOs surfactants, are complex mixtures of $C_{3\text{-}10}$-phenols where the main isomers are para-substituted. The interest in NPs and OPs analysis has increased during the last decades due to their capacity to disrupt the endocrine system which varies according to the structure of the branched alkyl group. They have been included in the water framework directive (WFD) as priority hazardous substances.

GC×GC was applied by Eganhouse *et al.* (Eganhouse et al., 2009) to enhance the chromatographic resolution of highly similar compounds such as 4-nonylphenol isomers and facilitate identification of a number of previously unrecognized components. Among the 153-204 peaks attributed to alkylphenol, 59-664-NPs were identified (Figure 4). Seven technical NPs products were analyzed using eight synthetic 4-NP isomers, with significant differences among the products and between two samples from a single supplier. This technique was also applied to environmental samples (wastewater, contaminated groundwater and municipal wastewater). The authors demonstrated that alteration of NPs composition through degradation results in enrichment of the more persistent isomers and removal or reduction of less persistent isomers. So, the estrogenicity may be increased or decreased depending on which 4-NP isomers are removed most rapidly.

The optimization of the separation of complex NPs technical mixtures (Vallejo et al., 2011) has been performed by means of experimental designs using GC×GC–FID and GC×GC–qMS equipped with valve-based modulator. Up to 79 OPs and NPs isomers have been separated using the FID detector and 39 have been undoubtedly identified using the mass spectra obtained from the qMS detector. The 22 OP, 33 OP, 363 NP and 22 NP isomers have been synthesized and quantified in two different technical mixtures from Fluka and Aldrich.

The values obtained for NP isomers were in good agreement with the literature and the values calculated for OP were for the first time reported.

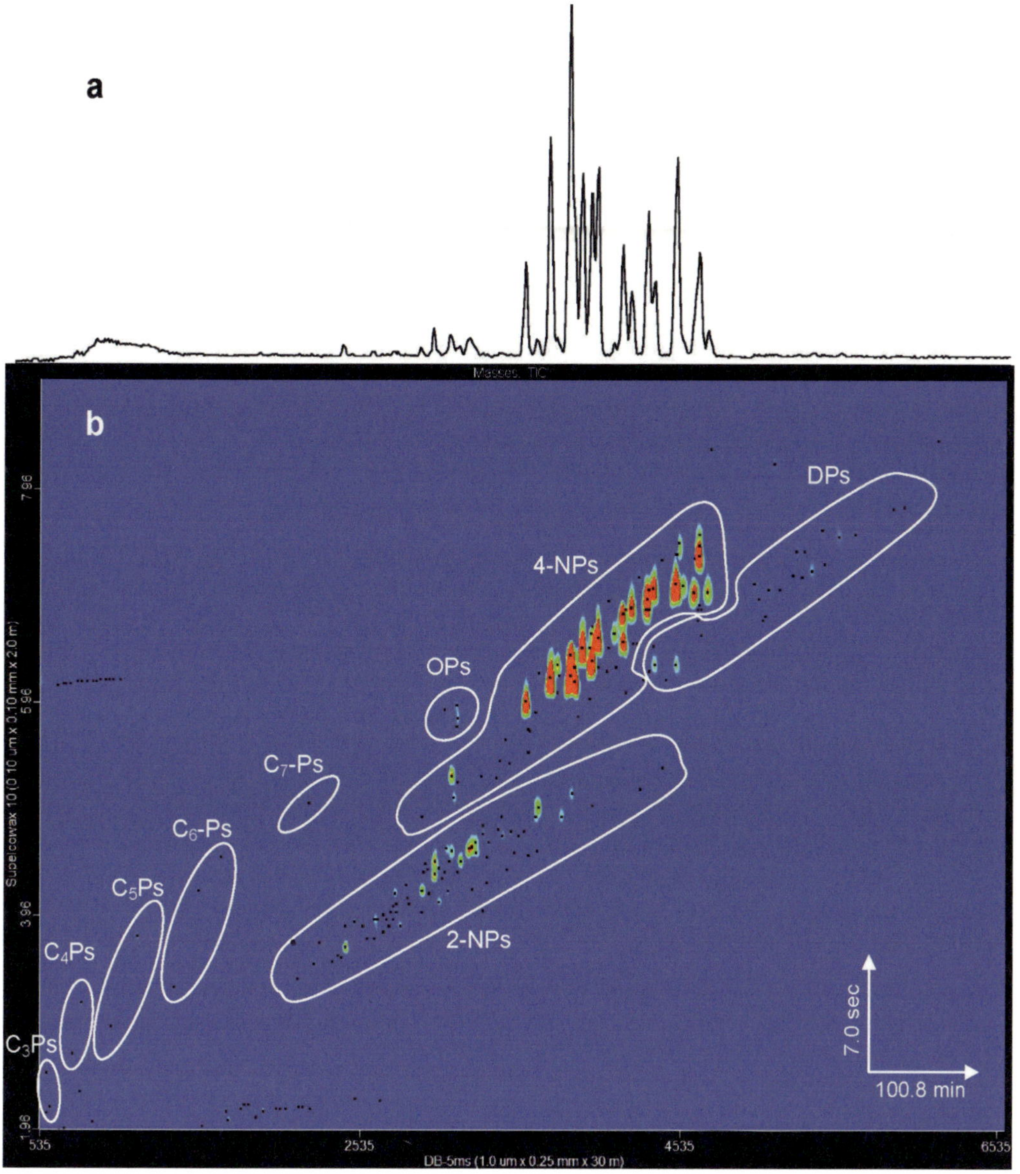

Fig. 4. Total ion chromatograms of technical NP (Fluka) showing (a) reconstructed 1-dimensional plot, and (b) 2-dimensional plot with alkylphenol regions indicated. Cx) C_xH_{2x+1}, OPs) octylphenols, DPs) decylphenols. x-axis represents the separation in the column with the nonpolar stationary phase (DB-5 ms), whereas the y-axis represents the separation in the column with the polar stationary phase (Supelcowax 10); retention time is given in seconds. Reprinted with permission from Environmental Science & Technology (Eganhouse et al., 2009). Copyright 2011 American Chemical Society.

7. Pesticides

Recently, a study (Macedo da Silva et al., 2011) demonstrated the potential of the application of GC×GC-μECD to the analysis of seven pesticide residues (propanil, fipronil, propiconazole, trifloxystrobin, permethrin, difenoconazole and azoxystrobin) in sediments. GC×GC-ECD method improved the separation between analytes and matrix interferences, minimizing the possibility of co-elutions. Its resolution capacity allowed the use of a selective detector instead of the use of a more expensive mass spectrometry detector. Best results were obtained with the set of columns DB-5×DB-17ms. The LODs for GC×GC method were about 36% lower than those obtained for the one dimensional GC method (in the range from 0.08 to 1.07 g.L^{-1}). Accuracy also indicated better results for GC×GC, possibly due to its higher sensitivity and lower contribution of co-eluting matrix components, which was minimized by increased peak capacity.

A GC×GC-qMS method (Purcaro et al., 2011) was developed for the multiresidue analysis of 28 pesticides contained in water. Pesticides extraction was performed by using direct Solid-Phase MicroExtraction (SPME). The rapid-scanning (20 000 amu/s) qMS system was operated using a rather wide *m/z* 50–450 mass range and a 33 Hz spectral production rate. The qMS performances were evaluated in terms of number of data points per peak, mass spectral quality, extent of peak skewing, and consistency of retention times.

A method for the determination of ultra-trace amounts of OCPs in river water was developed by Ochiai *et al.* using GC×GC-HR TOFMS (Ochiai & Sasamoto, 2011). Stir Bar Sorptive Extraction (SBSE) followed by thermal desorption (TD) was used for sample preparation. SBSE conditions including extraction time profiles, phase ratio (sample volume/PolyDiMethylSiloxane (PDMS) volume), and modifier addition were studied. The SBSE-TD-GC×GC-HR TOFMS method was solvent-free and highly selective and sensitive (LOD: 10–44 pg.L^{-1}). The method was successfully applied to the determination of 16 OCPs in river water sample. Authors showed that the results for 8 OCPs were in good accordance with the values obtained by a conventional Liquid–Liquid Extraction (LLE)–GC–HRMS (Selected Ion Monitoring) SIM method. The method also allowed the identification of 20 non-target compounds, e.g. pesticides and their degradation products, PAHs, PCBs and pharmaceuticals and personal care products and metabolites in the same river water sample, by using full spectrum acquisition.

A review dedicated to determination of pyrethroid insecticides in environmental samples was recently published by Feo *et al.* (Feo et al., 2010). The authors discussed the advantages and the disadvantages of the different instrumental techniques including GC×GC.

8. VOCs and other compounds

Benzothiazoles, benzotriazoles and benzosulfonamides are high-production-volume chemicals that are used in industrial and household applications. These compounds were detected in various environmental aqueous samples and were usually quantified by LC-MS/MS. Jover *et al.* (Jover et al., 2009) developed a Solid Phase Extraction (SPE)-GC×GC-TOFMS method for the characterization of benzothiazoles, benzotriazoles and benzosulfonamides in aqueous matrices. Columns combination was optimized to ensure a good separation between target analytes and interfering compounds of the matrix. 12 target analytes were characterized in river water and in wastewater from both the influent and the effluent of a WWTP. Similar method (Matamoros et al., 2010b) was used to study the benzothiazoles and benzotriazoles removal efficiencies of four WWTPs.

The methods for the determination of polycyclic and nitro-aromatic musk compounds as well as those for the respective metabolites are reviewed by Bester (Bester, 2009).The power of GC×GC approaches was demonstrated considering the various production impurities (isomers) of the two polycyclic musks with the highest usage rates.

A methodology to characterize VOCs and semi-volatile compounds from marine salt using HeadSpace (HS)-SPME and GC×GC-TOFMS was developed by Silva *et al.* (Silva et al., 2010). 157 VOCs distributed over the chemical groups of hydrocarbons, aldehydes, esters, furans, haloalkanes, ketones, ethers, alcohols, terpenoids, C_{13} norisoprenoids, and lactones were detected. Furans, haloalkanes and ethers were identified for the first time in marine salt. Contour plot analysis revealed the complexity of marine salt volatile composition and confirmed the importance of a high resolution, sensitive analytical procedure (GC×GC-TOFMS) for this type of analysis. The structured 2D chromatographic profile arising from ^{1}D volatility and ^{2}D polarity was demonstrated, allowing more reliable identifications. Results obtained for analysis of salt from two diverse locations and harvests over three years have suggested loss of volatile compounds according to storage duration of the salt, with environmental factors surrounding the saltpans influencing the volatile composition of the salt. At present the relative contributions of these factors have not been quantified. Origins of newly identified compounds in marine salt were in accordance with previous propositions, with algae, surrounding bacterial community, and environmental pollution being obvious sources.

9. Screening

In environmental monitoring, pollutants lists are periodically updated by regulatory agencies. Both the European Union (EU) and US EPA issued dangerous and hazardous contaminant lists, the so-called priority substances, whose concentration and occurrence in waters were strictly regulated (Directive 2000/60/EC; Decision No.2455/2001/EC and Clean Water Act) (Matamoros et al., 2010a). As the number of environmental regulated pollutants increases, it is necessary to develop global detection methods which can be used to screen a large number of substances simultaneously. This kind of method could reduce cost and time necessary for their detection and quantification. Moreover, there is a diverse group of unregulated pollutants called "emerging" contaminants, including pharmaceuticals and personal care products which were interesting to identify and monitor due to their high mass discharge into the environment. Some emerging contaminants have been recently included in candidate contaminant lists either from US EPA and the EU commission.

Semard *et al.* (Semard et al., 2008a; Semard et al., 2008b) reported a GC×GC-TOFMS method to search 58 target compounds and screen hazardous contaminants including PBDEs, PAHs and pesticides in urban wastewater. A variety of drugs (antidepressants, antibiotics, anticoagulants, *etc*...), personal care products (sunscreens, antiseptics, cosmetics etc.) and carcinogenic compounds, pesticides and compounds toxic for reproduction were identified in the raw wastewater. Most of these compounds were removed or decreased by the WWTP. Four priority substances (1,2,3-trichlorobenzene, 4-tert-butylphenol, benzothiazole and naphthalene) were present with concentrations in the range of 0.05 to 1.5 $mg.L^{-1}$ in the raw wastewater and 0.01 to 0.1 $mg.L^{-1}$ in the treated wastewater.

Household dusts were investigated using GC×GC-TOFMS as efficient screening method (Hilton et al., 2010). PAHs, phthalates, and compounds containing chlorine, bromine, or nitro groups were located on the chromatogram. Household dust (SRM-2585) was extracted with hexane using accelerated solvent extraction (ASE). Large molecules, such as triglycerides and fatty acids were removed with gel permeation chromatography. The resulting peak table was automatically filtered to identify compound classes such as phthalates, PAHs (Figure 5) and their heterocyclic analogs, PCBs, PBDEs, chloroalkyl phosphates, pesticides, and pesticides degradation products... By comparison with concentrations determined by National Institute of Standards and Technology, the technique was able to identify analytes at concentrations as low as 10–20 $ng.g^{-1}$ dust for compounds quantified by NIST (National Institute of Standards and Technology).

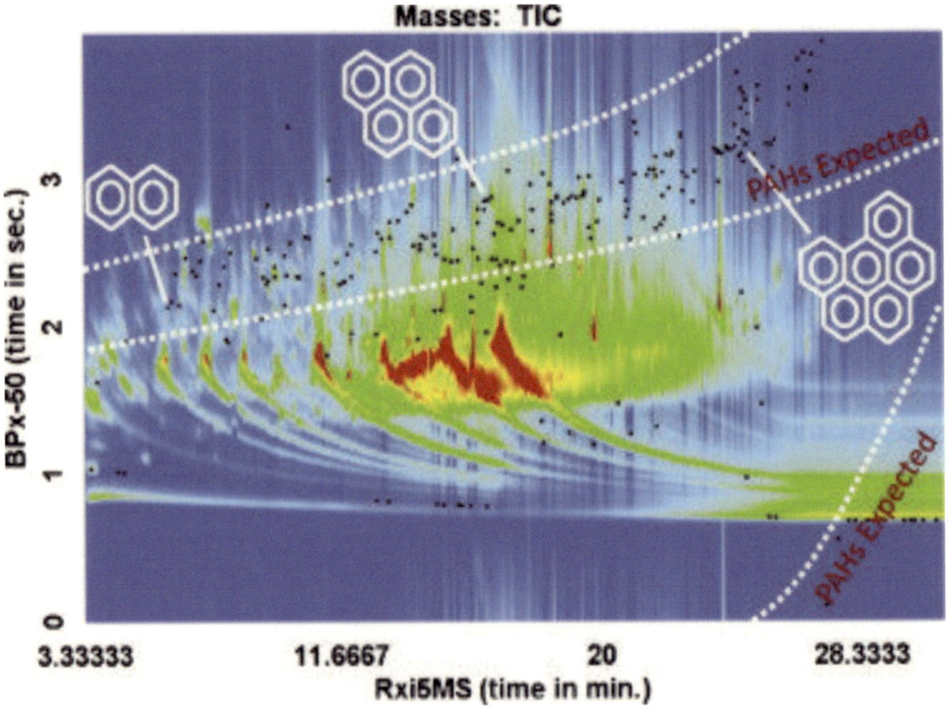

Fig. 5. Location of peaks matching the PAH spectral pattern with those identified by library searching. PAHs can be expected to fall into a band, shown starting at about 2 seconds in the second dimension and, as the chromatogram proceeds, falling later in the second dimension. Reprinted from Journal of Chromatography, A, (Hilton et al., 2010). Copyright (2010), with permission from Elsevier.

A SPE-GC×GC-TOFMS screening method for 97 priority and emerging contaminants in river was developed by Matamoros *et al.* (Matamoros et al., 2010a). The SPE was followed by in GC-port methylation using trimethylsulfonium hydroxide. The target analytes included 13 pharmaceuticals, 18 plasticizers, 8 personal care products, 9 acid herbicides, 8 triazines, 10 organophosphorous compounds, 5 phenylureas, 12 organochlorine biocides, 9

PAHs, 5 benzothiazoles and benzotriazoles. Best resolution between matrix constituents and target analytes was observed with TRB-5MS×TRB-50HT (apolar - polar) columns combination. Moreover, using polar-nonpolar columns combination, a strong correlation between the second dimension retention time and log K_{ow} for the target compounds was observed and was proposed as an additional identification criterion. The method was successfully applied to the analysis of four river water samples with LOD ranging from 0.5 to 100 ng.L^{-1} (Figure 6). Plasticizers (e.g., phthalates and bisphenol A), pharmaceuticals (e.g., naproxen, ibuprofen), and personal care products (e.g., tonalide and methyl dihydrojasmonate) were the most abundant in concentration and detection frequency.

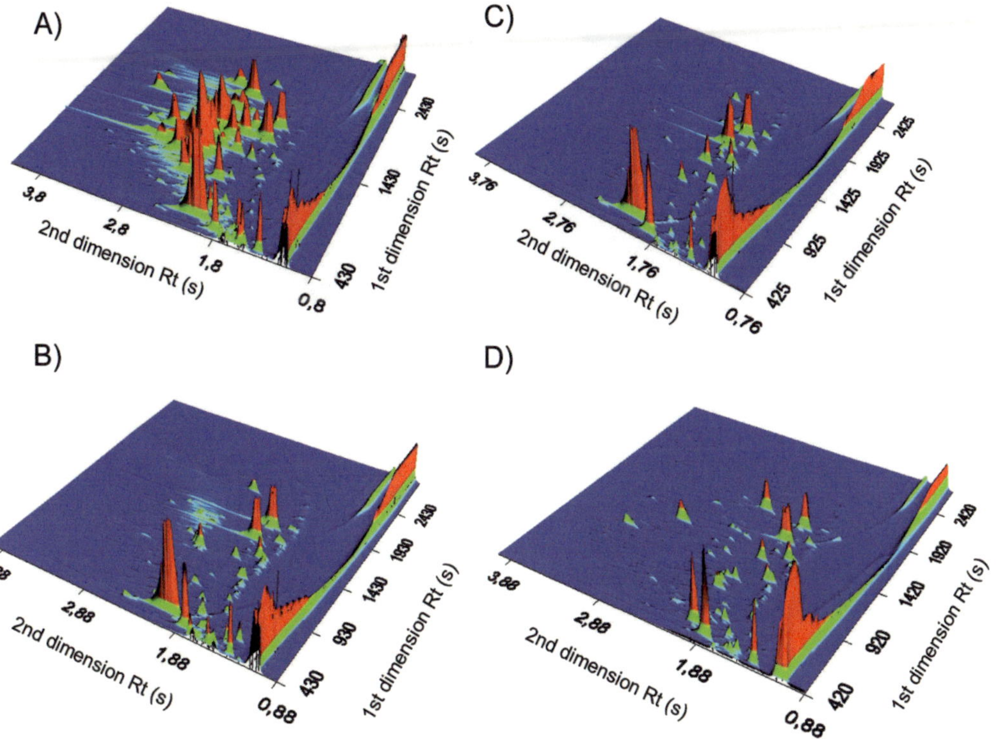

Fig. 6. 3D contour plots of four rivers sampled in this study in which the total ion chromatogram is shown: Ebro (A), Llobregat (B), Ter (C), and Beso`s (D). Reprinted with permission from Analytical Chemistry, (Matamoros et al., 2010a). Copyright 2011 American Chemical Society.

Gomez *et al.* (Gomez et al., 2011) developed a GC×GC-TOFMS method for the automatic searching and evaluation of nonpolar or semipolar contaminants (13 personal care products, 15 PAHs and 27 pesticides) in wastewater and river water. SBSE was selected for sample preparation step. Good results have been obtained in terms of separation efficiency and detection limits at or below 1 ng.L^{-1} for most of the compounds in the MS full scan mode, using only 100 mL of river water sample and 25 mL of wastewater effluent sample. The authors mentioned the possibility to screen for non-target compounds or unknowns. New contaminants have been identified in the wastewater effluents and river water samples, such as cholesterol and its degradation products, pharmaceuticals, illegal drugs, industrial products as well as other pesticides and personal care products. Moreover, GC×GC features were proposed to compare the fingerprinting of different water samples giving valuable information about the contamination status of rivers and wastewaters (Figure 7).

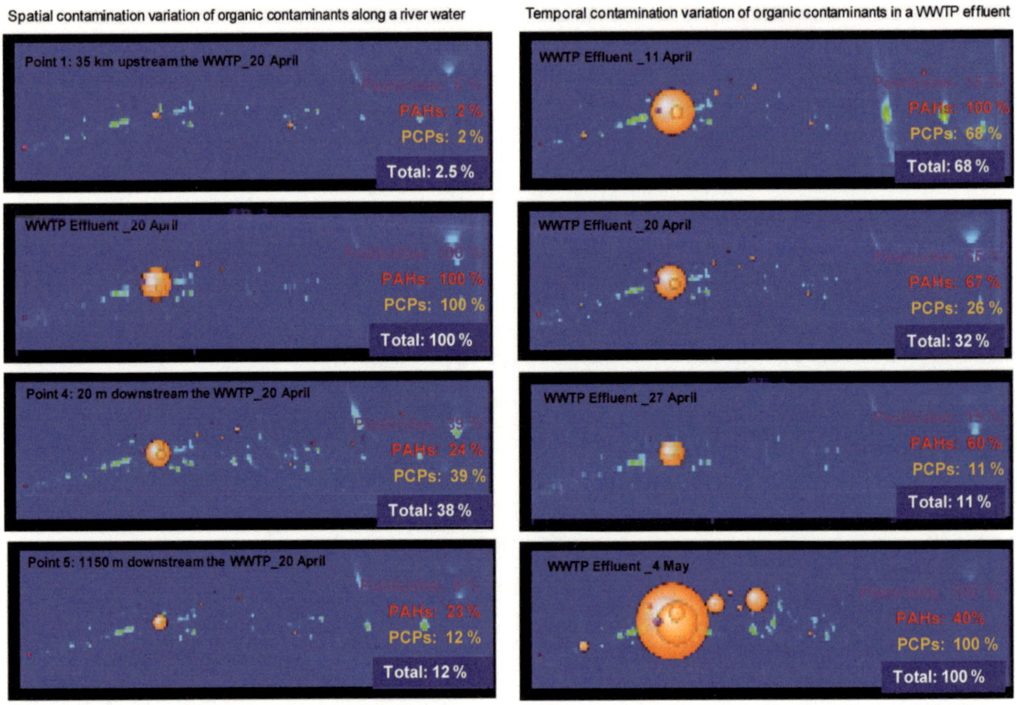

Fig. 7. Contamination status. Automatic searching of temporal and spatial contamination variation of organic contaminants. Reprinted with permission from Analytical Chemistry, (Gómez et al., 2011). Copyright 2011 American Chemical Society.

The most frequently detected contaminants and the contaminants detected at higher concentrations were the personal care products (musk fragrances galaxolide and tonalide). The pesticides and PAHs were detected at much lower concentration.

Halogenated compounds were successfully detected (Hashimoto et al., 2011) from several kinds of environmental samples by using a GC×GC chromatograph coupled with a tandem mass spectrometer (GC×GC-MS/MS). The global and selective detection of halogenated compounds was achieved by neutral loss scans of chlorine, bromine and/or fluorine using an MS/MS which was especially effective for compounds with more than two halogen substituents (Figure 8).

Screening and identification of pollutants was performed with GC×GC-HRTOFMS under the same conditions as those used for GC×GC-MS/MS. A lot of dioxins and PCBs congeners and many other compounds were identified in fly ash extract without any cleanup process and in sediment samples. In the future, the authors expect to achieve the complete global detection of any compound in one measurement of a crude sample simply with a GC×GC-HRTOFMS if it becomes possible to extract the desired information from the GC×GC-HR TOFMS data.

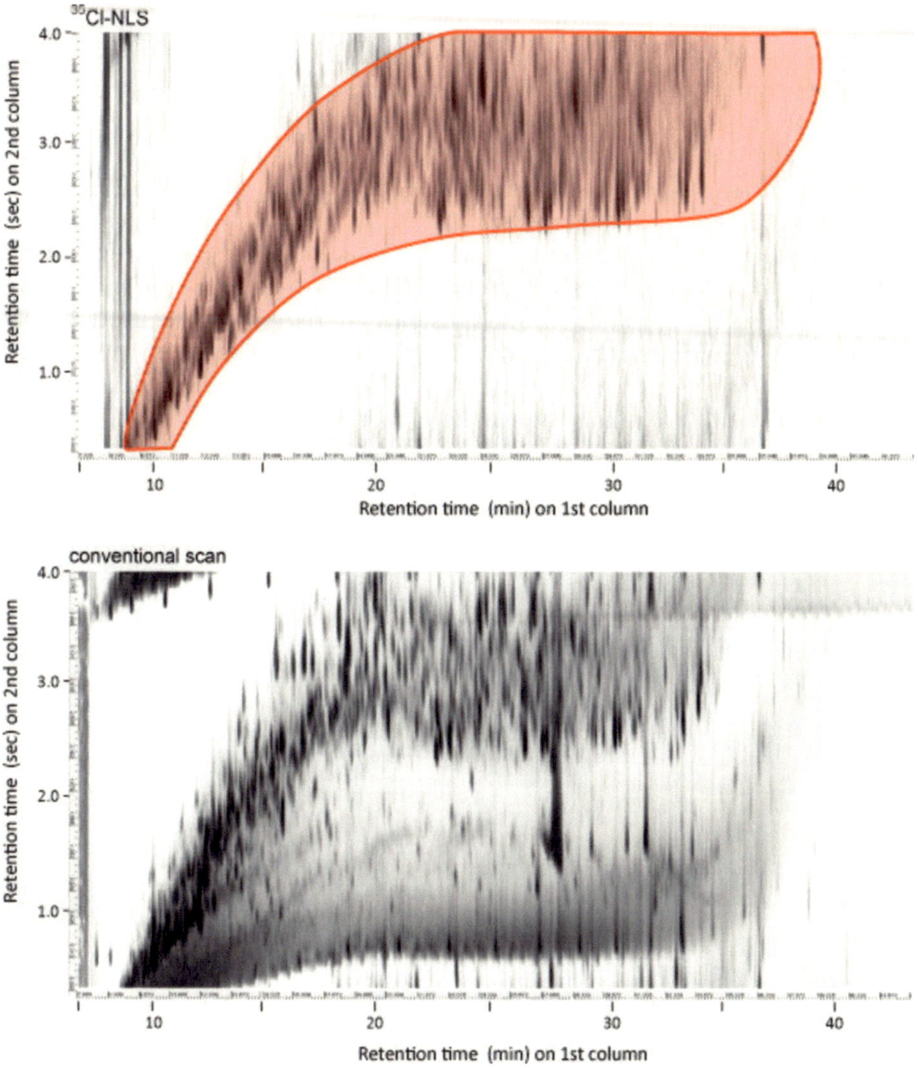

Fig. 8. Two-dimensional TICs of fly ash extract (NIES CRM17) measured with a ^{35}Cl-NLS (upper) and a conventional scan (lower) obtained with the GC×GC-MS/MS. The red translucent shape in the upper chromatogram shows the area where organohalogens are expected to appear. (For interpretation of the references to color in this figure legend, the reader is referred to the web version of the article.). Reprinted from Journal of Chromatography, A, (Hashimoto et al., 2011). Copyright (2011), with permission from Elsevier.

10. GC×GC instrumentation and optimization of operating conditions

Publications and reviews were dedicated to specific aspect of GC×GC. Major innovations in GC×GC modulator development were recently reviewed (Edwards et al., 2011). Cryogenic modulators remain very popular because of their ability to produce very small peak widths at half height and minimize breakthrough. Their commercial availability from several suppliers, has also contributed to their popularity. The use of valve-based modulators is increasing because of their less operating cost and easier maintenance than cryogenic modulator. Nevertheless, their coupling to MS remains problematic due to the large carrier gas flows in the second column and these modulators are not able to produce peaks of the same quality. Thermal modulation with the use of thermoelectric cooling could be a promising alternative if temperatures can be lowered enough to trap VOCs. Tranchida *et al.* (Tranchida et al., 2011a) have also published a review focused on the history (1991–2010) and present trends and future prospects for GC×GC modulation. Authors provided detailed descriptions and discussed the advantages and the disadvantages of the most significant thermal and pneumatic modulators. The authors have concluded that if at the moment, dual stage liquid N_2 systems can still be considered as the most effective modulators, in the next 10 years, the popularity of pneumatic modulators will gradually increase. The authors have included the description of their simple flow modulator, a seven port metallic disc published in 2011 (Tranchida et al., 2011b). A rotary and diaphragm 6-port 2-position valves have been also evaluated as modulators for GC×GC (Lidster et al., 2011).

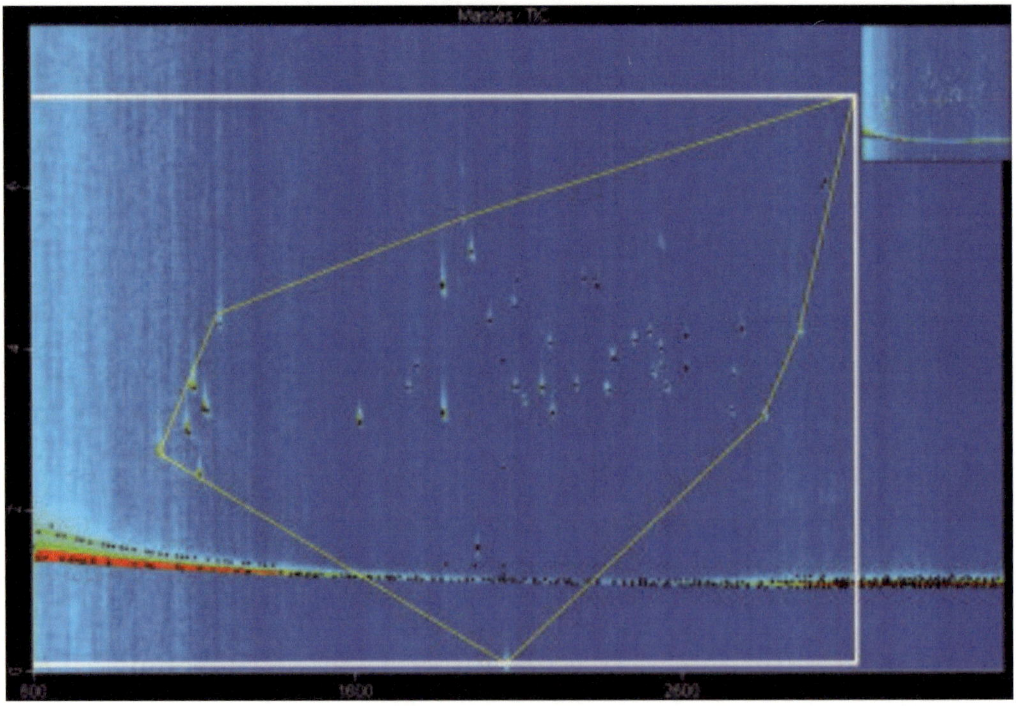

Fig. 9. Chromatogram illustrating the retention space used (white) and the space used calculated with Delaunay's triangulation algorithms (yellow) obtained on HP5-Mega225. Reprinted from Journal of Chromatography, A, (Semard et al., 2010). Copyright (2010), with permission from Elsevier.

In 2011, Panic *et al.* (Panic et al., 2011) developed a new consumable-free thermal modulator for GC×GC. The modulator was constructed from a trapping capillary, installed outside the GC oven, and coated inside with PDMS stationary phase. Dual-stage modulation was accomplished by resistively heating alternate segments of the trap with a custom-designed capacitive discharge power supply. The two unique inventions presented, flattening of the trap and selective removal of the stationary phase, have successfully eliminated the traditional drawbacks of resistively heated modulators.

The identification of compounds by using GC is based on peak retention times and mass spectra which generates uncertainty for the analyst for complex samples containing isomeric species. Retention index procedures were introduced to minimize misidentification of compounds in conventional chromatography. Various approaches to use of the retention index in GC×GC were reviewed and discussed (von Muhlen & Marriott, 2011).

A new method for the calculation of the percentage of separation space used was developed by Semard *et al.* (Semard et al., 2010) using Delaunay's triangulation algorithms (convex hull).

This approach was compared with an existing method and showed better precision and accuracy. It was successfully applied to the selection of the most convenient column set (HP5-Mega225) for the analysis of 49 target compounds including pesticides, HAPs, PCBs, PBDE *etc*... in wastewater. The diameter and length of the second column were optimized to improve the percentage of separation space used up to 40%.

Recently, Omais *et al.* (Omais et al., 2011) have shown that the general notion of orthogonality combining retention mechanisms independence and two dimensional space occupation must be decoupled. They have demonstrated that a non-orthogonal system can offer a good separation and a great space occupation. Moreover, orthogonality is intimately linked to the sample properties and cannot be considered as a sine qua none condition to achieve a good separation.

Table 1 summarizes the acronyms used in this review.

APEOs	AlkylPhenol EthOxylates
CTT	Toxaphene
CZC	Cryogenic Zone Compression
ECD, μECD	Electron Capture Detector, Micro Electron Capture Detector
EU	European Union
EPA	Environmental Protection Agency
FID	Flame Ionization Detector
GC	Gas Chromatography
GC×GC	Comprehensive two-dimensional gas chromatography
HS	HeadSpace
HRMS	High Resolution Mass Spectrometry
HRTOF	High resolution time-of-flight
ID	Isotope Dilution
ILs	Ionic Liquids
LLE	Liquid Liquid Extraction
LOD	Limit of detection
MS	Mass Spectrometer
MS/MS	Tandem mass spectrometer
NIST	National Institute of Standards and Technology
NPs	NonylPhenols

OCPs	OrganoChlorinated Pesticides
OPs	OctylPhenols
PAHs	Polycyclic Aromatic Hydrocarbons
Br-PAHs	Brominated Polycyclic aromatic hydrocarbon
Cl-PAHs	Chlorinated polycyclic aromatic hydrocarbon
PBDEs	PolyBrominated DiphenylEthers
PCBs	PolyChlorinated Biphenyls
PCDDs	PolyChlorinated Dibenzo-p-Dioxins
PCDFs	PolyChlorinated DibenzoFurans
PCNs	PolyChlorinated Naphthalenes
PCTs	PolyChlorinated Terphenyls
PDMS	PolyDiMethylSiloxane
POPs	Persistent Organic Pollutants
qMS	Quadrupole Mass Spectrometer
QSRR	Quantitative Structure–Retention Relationship
SBSE	Stir Bar Sorptive Extraction
SIM	Selected Ion Monitoring
SPE	Solid Phase Extraction
SPME	Solid Phase MicroExtraction
2,3,7,8-TCDD	2,3,7,8-tetrachlorodibenzo-p-dioxin
TD	Thermal Desorption
TEQ/kg	Toxic Equivalent Quantity/kg
TOF	Time-Of-Flight
US EPA	United States Environmental Protection Agency
VOCs	Volatile Organic Compounds
WFD	Water Framework Directive
WHO	World Health Organization
WWTPs	WasteWater Treatment Plants

Table 1. List of acronyms

11. Conclusion

This work reviews about 40 publications over the period 2009-2011 dealing with GC×GC and more especially on environmental applications. This technique coupled to mass spectrometry is an excellent choice for analyzing complex environmental samples and is suitable for multiresidue and non target analyses. As the number of environmental regulated pollutants increases, GC×GC appears ideal for the development of global detection method which can be used to screen a large number of compounds simultaneously. This kind of method could reduce cost and time necessary for their detection and quantification. GC×GC provides the analytical chemist with a new tool for a better separations of organohalogen congeners and, potentially, for more accurate human and environmental exposure data for risk assessments. Moreover, the authors highlighted the improved resolution and sensitivity offered by GC×GC over conventional one dimensional GC. The high selectivity of GC×GC-TOFMS has also facilitated the development of a wide range of analytical methods with minimal sample preparation and allowed the screening of emerging contaminants. Table 2 summarizes the main technical characteristics of methods applied to the various classes of pollutants reviewed in this chapter.

Compounds	Samples	Extraction	Material	Set of columns	references
PCBs and chlorobenzenes	Soil, sediment and sludge	ASE	GC×GC-μECD	DB1×Rtx-PCB	**Muscalu et al., 2011**
PCDDs and PCDFs	soil and sediment samples from South Africa	ASE	GC×GC-TOFMS	Rtx-Dioxin 2xRtx-PCB Rxi-5 Sil MSxRtx-200 Rxi-XLBxRtx-200	**de Vos et al., 2O11a**
PCDDs, PCDFs and four dioxin-like non-ortho substituted PCBs	samples from a hazardous waste treatment facility	Soxhlet	GC×GC-TOFMS	Rtx-5SilmsxRtx-PCB Rtx5xRtx-200 Rtx-XLBxRtx-200 Rtx-Dioxin 2xRtx-PCB.	**de Vos et al., 2O11b**
Chlorinated and brominated polycyclic aromatic hydrocarbon (Cl-/Br-PAHs)	Soil samples from Japan	Soxhlet	GC×GC-HRTOFMS	BPX5×BPX50 BPX5×LC-50HT	**Ieda et al., 2011**
Naphthalene, benzothiophene and their alkylated congeners	Marine sediment	Sonication	GC×GC-FID and GC×GC-TOFMS	not specified	**Wardlaw et al., 2011**
Petroleum hydrocarbons	Soil	ASE	GC×GC-FID and GC×GC-TOFMS	RTX-1×BPX50	**Mao et al., 2001**
Naphthalene, benzothiophene and their alkylated congeners	Marine sediment	Sonication	GC×GC-FID and GC×GC-TOFMS	not specified	**Wardlaw et al., 2011**
4-Nonylphenols	Wastewater	LLE	GC×GC-FID and GC×GC-TOFMS	DB-5 ms×Supelcowax 10	**Eganhouse et al., 2009**
Pesticides (propanil, fipronil, propiconazole, trifloxystrobin, permethrin, difenoconazole and azoxystrobin)	Sediments (Brazil)	Sonication	GC×GC-μECD	DB-5×DB-17ms HP-50×DB-1ms	**Macedo da Silva et al., 2011**
Pesticides	Tap water	SPME	GC×GC-qMS	SLB-5msxSLB-IL59	**Purcaro et al., 2011**
OCPS, PAHs, PCBs and pharmaceuticals and personal care products	River water	SBSE	GC×GC-HRTOFMS	DB-5xBPX-50	**Ochiai & Sasamoto**
benzothiazoles, benzotriazoles and benzosulfonamides	River water and in wastewater from both the influent and the effluent of a WWTP	SPE	GC×GC-TOFMS	TRB-5MSxTRB-50HT ZB-WAXxTRB-5MS TRB-1701MSxTRB-5MS	**Jover et al., 2009**
VOCs and semi-volatile compounds	Marine salts	HS-SPME	GC×GC-TOFMS	BPX5×BP20	**Silva et al., 2010**
PAHs, PBDEs, Pesticides, drugs, personal care products, anti-microbials, carcinogen and reprotoxic compounds	Urban wastewater	LLE	GC×GC-TOFMS	HP5xMegaWaxHT	**Semard et al., 2008**
Phthalates, PAHs and their heterocyclic analogs, PCBs, PBDEs, chloroalkyl phosphates, pesticides, and pesticide degradation products	Household dusts	ASE	GC×GC-TOFMS	Rxi5-MS×BPX50	**Hilton et al., 2010**
Pharmaceuticals, plasticizers, personal care products, acid herbicides, triazines, organophosphorous compounds, phenylureas, organochlorine biocides, PAHs, and benzothiazoles	River	SPE	GCxGC-TOFMS	TRB-5MS×TRB-50HT	**Matamoros et al., 2010a**
Personal care products, PAHs, pesticides, cholesterol and its degradation products ...	River	SBSE	GCxGC-TOFMS	Rtx-5×Rxi-17 Rtx-5×Rt-LC50	**Gomez et al., 2011**
Halogenated compounds including PCBs, PCDDs, PCDFs...	Fly ash extract, Sediment, soil	Soxhlet	GC×GC-MS/MS and GC×GC-HRTOFMS	InertCap 5MS×BPX50	**Hashimoto et al., 2011**

Table 2. characteristics of methods applied to the various classes of pollutants based on compounds, sample preparation techniques, GC×GC method andset of columns used.

12. References

Adahchour, M.; Beens, J.; Vreuls, R. J. J. & Brinkman, U. A. T. (2006). Recent developments in comprehensive two-dimensional gas chromatography (GC×GC). III. Applications for petrochemicals and organohalogens. *Trends in Analytical Chemistry*, Vol.25, No.7, (July-August 2006), pp. 726-741, ISSN 0165-9936

Adahchour, M.; Beens, J. & Brinkman, U. A .T. (2008). Recent developments in the application of comprehensive two-dimensional gas chromatography. *Journal of Chromatography, A*, Vol.1186, No.1-2, (April 2008), pp. 67-108, ISSN 0021-9673

Amador-Munoz, O.; Villalobos-Pietrini, R. ;Aragon-Pina, A.; Tran, T.C.; Morrison, P. & Marriott, P. J. (2008). Quantification of polycyclic aromatic hydrocarbons based on comprehensive two-dimensional gas chromatography-isotope dilution mass spectrometry. *Journal of Chromatography, A,* Vol.1201, No.2, (August 2008), pp. 161-168, ISSN 0165-9936

Arsene, C.; Vione, D.; Grinberg, N. & Olariu, R.I. (2011). GC×GC-MS hyphenated techniques for the analysis of volatile organic compounds in air. *Journal of Liquid Chromatography & Related Technologies*, Vol.34, No.13, (August 2011), pp. 1077-1111, ISSN 1082-6076

Ballesteros-Gomez, A. & Rubio, S. (2011). Recent Advances in Environmental Analysis. *Analytical Chemistry*, Vol.83, No.12, (April 2011), pp. 4579-4613, ISSN 0003-2700

Bester, K. (2009). Analysis of musk fragrances in environmental samples. *Journal of Chromatography, A*, Vol.1216, No.3, (January 2009), pp. 470-480, ISSN 0021-9673

Bordajandi, L. R.; Ramos, J. J.; Sanz, J.; Gonzalez, M. J. & Ramos, L. (2008). Comprehensive two-dimensional gas chromatography in the screening of persistent organohalogenated pollutants in environmental samples. *Journal of Chromatography, A*, Vol.1186, No.1-2, (april 2008), pp. 312-324, ISSN 0021-9673

Cortes, H. J.; Winniford, B.; Luong, J. & Pursch, M. Comprehensive two dimensional gas chromatography review.(2009). *Journal of Separation Science*, Vol.32, No.(5-6), (March 2009), pp 883-904, ISSN 1615-9306

D'Archivio,A. A.; Incani, A. & Ruggieri, F. (2011). Retention modelling of polychlorinated biphenyls in comprehensive two - dimensional gas chromatography. *Analytical and Bioanalytical Chemistry*, Vol.399, No. 2, (January 2011), pp. 903-913. ISSN 1618-2642

Dallüge, J.; Beens, J. & Brinkman U. A. T. (2003). Comprehensive two-dimensional gas chromatography: a powerful and versatile analytical tool. *Journal of Chromatography, A*, Vol.1000, No.1-2, (June 2003), pp. 69-108

de Vos, J.; Gorst-Allman, P. & Rohwer, E. (2011a). Establishing an alternative method for the quantitative analysis of polychlorinated dibenzo-p-dioxins and polychlorinated dibenzofurans by comprehensive two dimensional gas chromatography-time-of-flight mass spectrometry for developing countries. *Journal of Chromatography, A*, Vol.1218, No.21, (May 2011), pp. 3282-3290, ISSN 0021-9673

de Vos, J.; Dixon, R.; Vermeulen, G.; Gorst-Allman, P.; Cochran, J.; Rohwer, E. & Focant, J.-F. (2011b). Comprehensive two-dimensional gas chromatography time of flight mass spectrometry (GC × GC-TOFMS) for environmental forensic investigations in developing countries. *Chemosphere*, Vol.82, No.9, (February 2011), pp. 1230-1239, ISSN 0045-6535

Edwards, M.; Mostafa, A. & Gorecki, T. (2011). Modulation in comprehensive two – dimensional gas chromatography : 20 years of innovation. *Analytical and*

Bioanalytical Chemistry, DOI 10.1007/s00216-011-5100-6, no pp. yet given, ISSN 1618-2650.

Eganhouse, R. P.; Pontolillo, J.; Gaines, R. B.; Frysinger, Glenn S.; Gabriel, F. L. P.; Kohler, H.-P. E.; Giger, W. & Barber, L. B. (2009). Isomer-Specific Determination of 4-Nonylphenols Using Comprehensive Two-Dimensional Gas Chromatography/Time-of-Flight Mass Spectrometry. *Environmental Science & Technology,* Vol.43, No.24, (December 2009), pp. 9306-9313, ISSN 0013-936X

Feo, M. L.; Eljarrat, E.; Barcelo, D. & Barcelo, D. (2010). Determination of pyrethroid insecticides in environmental samples. T*rends in Analytical Chemistry,* Vol.29, No.7, (July-August 2010), pp. 692-705, ISSN 0165-9936

Giddings, J. C. (1984). Two-dimensional separations: concept and promise. *Analytical chemistry,* Vol.56, No.12, (October 1984), pp. 1258A-1260A, 1262A, 1264A, ISSN 0003-2700

Gomez, M. J.; Herrera, S.; Sole, D.; Garcia-Calvo, E.; Fernandez-Alba, A. R. (2011). Automatic Searching and Evaluation of Priority and Emerging Contaminants in Wastewater and River Water by Stir Bar Sorptive Extraction followed by Comprehensive Two-Dimensional Gas Chromatography-Time-of-Flight Mass Spectrometry. *Analytical Chemistry,* Vol.83, No.7, (April 2011), pp. 2638-2647

Hamilton, J. F. (2010). Using comprehensive two-dimensional gas chromatography to study the atmosphere. *Journal of Chromatographic Science,* Vol.48, No.4, (April 2010), pp. 274-282, ISSN 0021-9665

Hashimoto, S.; Takazawa, Y.; Fushimi, A.; Tanabe, K.; Shibata, Y.; Ieda, T.; Ochiai, N.; Kanda, H.; Ohura, T.; Tao, Q. & Reichenbach, S. E. (2011). Global and selective detection of organohalogens in environmental samples by comprehensive two-dimensional gas chromatography-tandem mass spectrometry and high-resolution time-of-flight mass spectrometry. *Journal of Chromatography, A,* Vol.1218, No.24, (June 2011), pp. 3799-3810, ISSN 0021-9673

Hilton, D. C.; Jones, R. S. & Sjoedin, A. (2010). A method for rapid, non-targeted screening for environmental contaminants in household dust. *Journal of Chromatography, A,* Vol.1217, No. 4, (October 2010), pp. 6851-6856, ISSN 0021-9673

Horii, Y.; Ok, G.; Ohura, T. & Kannan, K. (2008). Occurrence and Profiles of Chlorinated and Brominated Polycyclic Aromatic Hydrocarbons in Waste Incinerators. *Environmental Science & Technology,* Vol.42, No.6, (March 2008), pp. 1904-1909, ISSN 0013936X

Horii, Y.; Ohura, T.; Yamashita, N. & Kannan, K. (2009). Chlorinated Polycyclic Aromatic Hydrocarbons in Sediments from Industrial Areas in Japan and the United States. *Archives of Environmental Contamination and Toxicology,* Vol.57, No.4, (November 2009), pp. 651-660, ISSN 0090-4341

Ishaq, R.; Naf, C.; Zebuhr, Y.; Broman, D. & Jarnberg, U. (2003). PCBs, PCNs, PCDD/Fs, PAHs and Cl-PAHs in air and water particulate samples--patterns and variations. *Chemosphere,* Vol.50, No.9, (March 2003), pp. 1131-1150, ISSN 0045-6535

Ieda, T.; Ochiai, N.; Miyawaki, T.; Ohura, T. & Horii, Y. (2011). Environmental analysis of chlorinated and brominated polycyclic aromatic hydrocarbons by comprehensive two-dimensional gas chromatography coupled to high-resolution time-of-flight mass spectrometry. *Journal of Chromatography, A,* Vol.1218, No.21, (May 2011), pp. 3224-3232, ISSN 0021-9673

Jover, E.; Matamoros, V. & Bayona, J. M. (2009). Characterization of benzothiazoles, benzotriazoles and benzosulfonamides in aqueous matrixes by solid-phase extraction followed by comprehensive two-dimensional gas chromatography coupled to time-of-flight mass spectrometry. *Journal of Chromatography, A*, Vol.1216, No.18, pp. 4013-4019, ISSN 0021-9673

Lidster, R. T.; Hamilton, J. F. & Lewis, A. C. (2011). The application of two total transfer valve modulators for comprehensive two-dimensional gas chromatography of volatile organic compounds. *Journal of Separation Science*, Vol.34, No.7, (April 2011), pp. 812-821, ISSN 1615-9306

Liu, Z. & Phillips, J. B. (1991). Comprehensive two-dimensional gas chromatography using an on-column thermal modulator interface. *Journal of Chromatographic Science*, Vol.29, No.6, (June 1991), pp. 227-231, ISSN 0021-9665

Macedo da Silva, J.; Zini, C. A. & Caramao, E. B.(2011). Evaluation of comprehensive two-dimensional gas chromatography with micro-electron capture detection for the analysis of seven pesticides in sediment samples. *Journal of Chromatography, A*, Vol.1218, No.21, (May 2011), pp. 3166-3172, ISSN 0021-9673

Mao, D.; Lookman, R.; Van De Weghe, H.; Weltens, R.; Vanermen, G.; De Brucker, N. & Diels, L. (2009). Estimation of ecotoxicity of petroleum hydrocarbon mixtures in soil based on HPLC-GCXGC analysis. *Chemosphere*, Vol.77, No.11, (December 2009), pp. 1508-1513, ISSN0045-6535

Matamoros, V.; Jover, E. & Bayona, J.M. (2010a). Part-per-Trillion Determination of Pharmaceuticals, Pesticides, and Related Organic Contaminants in River Water by Solid-Phase Extraction Followed by Comprehensive Two-Dimensional Gas Chromatography Time-of-Flight Mass Spectrometry. *Analytical Chemistry*, Vol.82, No.2, (January 2010), pp. 699-706, ISSN0003-2700

Matamoros, V.; Jover, E. & Bayona, J. M. (2010b). Occurrence and fate of benzothiazoles and benzotriazoles in constructed wetlands. *Water Science and Technology*, Vol. 61, No. 1, (January 2010), pp. 191-198, ISSN 0273-1223

Muscalu, A. M.; Reiner, E. J.; Liss, S. N.; Chen, T.; Ladwig, G. & Morse, D. (2011). A routine accredited method for the analysis of polychlorinated biphenyls, organochlorine pesticides, chlorobenzenes and screening of other halogenated organics in soil, sediment and sludge by GCxGC-μECD. *Analytical and Bioanalytical Chemistry*, DOI 10.1007/s00216-011-5114-0, No pp. yet given., ISSN 1618-2642

Ochiai, N. & Sasamoto, K. (2011).Selectable one-dimensional or two-dimensional gas chromatography-olfactometry/mass spectrometry with preparative fraction collection for analysis of ultra-trace amounts of odor compounds.*Journal of Chromatography A*, Vol. 1218, No. 21, pp. 3180-3185, ISSN0021-9673

Omais, B.; Courtiade, M.; Charon, N.; Ponthus,J. & Thiebaut D. (2011). Considerations on Orthogonality Duality in Comprehensive Two-Dimensional Gas Chromatography, Analytical Chemistry, DOI 10.1021/ac201103e, No pp. yet given

Osemwengie, L. I. & Sovocool, G. W. (2011). Evaluation of comprehensive 2D gas chromatography -time-of-flight mass spectrometry for 209 chlorinated biphenyl congeners in two chromatographic runs. Chromatography Research International, doi:10.4061/2011/675920, 14 pp. ISSN 2090-3510

Pani, O. & Gorecki, T. (2006). Comprehensive two-dimensional gas chromatography (GC×GC) in environmental analysis and monitoring. Analytical and Bioanalytical Chemistry, Vol.386, No.4, (October 2006), pp. 1013-1023, ISSN 1618-2642

Panic, O.; Gorecki, T.; McNeish, C.; Goldstein, A. H.; Williams, B. J.; Worton, D. R.; Hering, S. V. & Kreisberg, N. M. (2011). Development of a new consumable-free thermal modulator for comprehensive two-dimensional gas chromatography. *Journal of Chromatography, A*, Vol.1218, No.20, (May 2011), pp. 3070-3079, ISSN 0021-9673.

Patterson, D. G. Jr.; Welch, S. M.; Turner, W. E.; Sjodin, A. & Focant, J.-F. (2011). Cryogenic zone compression for the measurement of dioxins in human serum by isotope dilution at the attogram level using modulated gas chromatographycoupled to high resolution magnetic sector mass spectrometry. *Journal of Chromatography, A*, Vol.1218, No.21, (May 2011), pp. 3274-3281, ISSN 0021-9673

Purcaro, G.; Tranchida, P. Q.; Conte, L.; Obiedzinska, A.; Dugo, P.; Dugo, G. & Mondello, L. (2011). Performance evaluation of a rapid-scanning quadrupole mass spectrometer in the comprehensive two - dimensional gas chromatography analysis of pesticides in water. *Journal of Separation Science*, Vol.34, No.18, (September 2011), pp. 2411-2417, ISSN 1615-9306

Ramos, J. J.; Pena-Abaurrea, M. & Ramos L. (2009). Environmental Analysis, In: *Comprehensive Analytical Chemistry*, Ramos Lourdes (Ed.), pp. 243-280, Elsevier Science, ISBN 0444532374, Amsterdam, The Netherlands

Schoenmakers, P.; Marriott, P. & Beens, J. (2003). Nomenclature and conventions in comprehensive multidimensional chromatography. *LC-GC Europe*, Vol.16, No.6, (June 2003), pp. 335-336 and 338-339, ISSN 1471-6577

Semard, G.; Bruchet, A.; Cardinael P. & Bouillon, J.-P. (2008a). Use of comprehensive two-dimensional gas chromatography for the broad screening of hazardous contaminants in urban wastewaters. *Water Science and Technology*, Vol.57, No.12, (June 2008), pp. 1983-1989, ISSN 0273-1223

Semard, G.; Bruchet, A.; Cardinael, P. & Peulon-Agasse, V. (2008b). Study of the significance of comprehensive two-dimensional gas chromatography for analysis of complex environmental matrices. *Spectra Analyse*, Vol.37, No.261, (April-May 2008), pp. 29-33, ISSN 1635-947X

Semard, G.; Adahchour, M. & Focant, J.-F. (2009). Basic Instrumentation for GC×GC, In: *Comprehensive Analytical Chemistry*, Ramos Lourdes (Ed.), pp. 243-280, Elsevier Science, ISBN 0444532374, Amsterdam, The Netherlands

Semard, G.; Peulon-Agasse, V.; Bruchet, A. ; Bouillon, J.-P. & Cardinael, P. (2010). Convex hull: A new method to determine the separation space used and to optimize operating conditions for comprehensive two-dimensional gas chromatography. *Journal of Chromatography, A*, Vol.1217, No.33, (August 2010), pp. 5449-5454, ISSN 0021-9673

Silva, I.; Rocha, S. M.; Coimbra, M. A. & Marriott, P. J. (2010). Headspace solid-phase microextraction combined with comprehensive two-dimensional gas chromatography time-of-flight mass spectrometry for the determination of volatile compounds from marine salt. *Journal of Chromatography, A*, Vol.1217, No.34, (August 2010), pp. 5511-5521, ISSN 0021-9673

Skoczynska, E.; Korytar, P & de Boer, J. (2008). Maximizing chromatographic information from environmental extracts by GC×GC-ToF-MS. *Environmental Science & Technology*, Vol.42, No.17, (September 2008), pp. 6611–6618, ISSN 0013-936X

Tranchida, P. Q.; Purcaro, G.; Dugo, P. & Mondello, L. (2011a). Modulators for comprehensive two-dimensional gas chromatography. *Trends in Analytical Chemistry*, Vol.30, No.9, (October 2011), pp.1437-1461, ISSN 0165-9936

Tranchida, P. Q.; Purcaro, G.; Visco, A.; Conte, L.; Dugo, P.; Dawes, P. & Mondello, L. (2011). A flexible loop-type flow modulator for comprehensive two-dimensional gas chromatography. *Journal of Chromatography, A*, Vol.1218, No.21, pp. 3140-3145, ISSN 0021-9673

Vallejo, A.; Olivares, M.; Fernandez, L. A.; Etxebarria, N.; Arrasate, S.; Anakabe, E.; Usobiaga, & A.Zuloaga, O. (2011). Optimization of comprehensive two dimensional gas chromatography-flame ionization detection-quadrupole masss pectrometry for the separation of octyl- and nonylphenol isomers. *Journal of Chromatography, A*, Vol. 1218, No.20, pp. 3064-3069, ISSN0021-9673

Wang, Y.; Chen, Q.; Norwood, D. L. & McCaffrey, J. (2010). Recent development in the application of comprehensive two-dimensional gas chromatograph. *Journal of Liquid Chromatography & Related Technologies*, Vol.33, No.9-12, (July 2010), pp. 1082-1115, ISSN 1082-6076

Wardlaw, G. D.; Nelson, R. K.; Reddy, C. M. & Valentine, D. L. (2011). Biodegradation preference for isomers of alkylated naphthalenes and benzothiophenes in marine sediment contaminated with crude oil. *Organic Geochemistry*, Vol.42, No.6, (July 2011), pp. 630-639, ISSN0146-6380

von Muhlen, C. & Marriott, P. J. (2011). Retention indices in comprehensive two - dimensional gas chromatography. *Analytical and Bioanalytical Chemistry*, DOI: 10.1007/s00216-011-5247-1, No pp. yet given. ISSN 1618-2642

Zapadlo, M.; Krupcik, J.; Majek, P.; Armstrong, D. W. & Sandra, P. (2010). Use of a polar ionic liquid as second column for the comprehensive two - dimensional GC separation of PCBs. *Journal of Chromatography, A*, Vol.1217, No.37, (September 2010), pp. 5859-5867, ISSN0021-9673

Zapadlo, M.; Krupcik, J.; Kovalczuk, T.; Majek, P.; Spanik, I.; Armstrong, D. W. & Sandra, P. (2011). Enhanced comprehensive two – dimensional gas chromatographic resolutionof polychlorinatedbiphenyls on a non-polar polysiloxane and an ionic liquid column series. *Journal of Chromatography, A*, Vol.1218, No.5, (February 2011), pp. 746-751, ISSN 0021-9673